de Gruyter Series in Nonlinear Analysis and Applications 8

Jorge Ize
Alfonso Vignoli

Equivariant Degree Theory

Walter de Gruyter · Berlin · New York 2003

Authors

Jorge Ize
Instituto de Investigaciones en Matematicas
Aplicadas y en Sistemas
Universidad Nacional Autonoma de Mexico
01000 MEXICO D. F.
MEXICO

Alfonso Vignoli
Department of Mathematics
University of Rome "Tor Vergata"
Via della Ricerca Scientifica
00133 ROMA
ITALY

Mathematics Subject Classification 2000: 58-02; 34C25, 37G40, 47H11, 47J15, 54F45, 55Q91, 55E09

Keywords: equivariant degree, homotopy groups, symmetries, period doubling, symmetry breaking, twisted orbits, gradients, orthogonal maps, Hopf bifurcation, Hamiltonian systems, bifurcation

⊗ Printed on acid-free paper which falls within the guidelines of the ANSI to ensure permanence and durability.

Library of Congress Cataloging-in-Publication Data

Ize, Jorge, 1946–
Equivariant degree theory / Jorge Ize, Alfonso Vignoli.
p. cm. – (De Gruyter series in nonlinear analysis and applications, ISSN 0941-813X ; 8)
Includes bibliographical references and index.
ISBN 3-11-017550-9 (cloth : alk. paper)
1. Topological degree. 2. Homotopy groups. I. Vignoli, Alfonso, 1940– II. Title. III. Series.
QA612.I94 2003
514′.2–dc21 2003043999

ISBN 3-11-017550-9

Bibliographic information published by Die Deutsche Bibliothek

Die Deutsche Bibliothek lists this publication in the Deutsche Nationalbibliografie; detailed bibliographic data is available in the Internet at <http://dnb.ddb.de>.

Printed in Germany.
Cover design: Thomas Bonnie, Hamburg
Typeset using the authors' TEX files: I. Zimmermann, Freiburg
Printing and binding: Hubert & Co. GmbH & Co. Kg, Göttingen

J. I. expresses his love to his wife, Teresa, and his sons, Pablo, Felipe and Andres.

A. V. wishes to dedicate this book to his beloved wife Lucilla, to his son Gabriel and to Angela, who kept cheering him up in times of dismay and frustration becoming more frequent at sunset.

Preface

The present book grew out as an attempt to make more accessible to non-specialists a subject – Equivariant Analysis – that may be easily obscured by technicalities and (often) scarcely known facts from Equivariant Topology. Quite frequently, the authors of research papers on Equivariant Analysis tend to assume that the reader is well acquainted with a hoard of subtle and refined results from Group Representation Theory, Group Actions, Equivariant Homotopy and Homology Theory (and co-counter parts, i.e., Cohomotopy and Cohomology) and the like. As an outcome, beautiful theories and elegant results are poorly understood by those researchers that would need them mostly: applied mathematicians. This is also a self-criticism.

We felt that an overturn was badly needed. This is what we try to do here. If you keep in mind these few strokes you most probably will understand our strenuous efforts in keeping the mathematical background to a minimum. Surprisingly enough, this is at the same time an easy and very difficult task. Once we took the decision of expressing a given mathematical fact in as elementary as possible terms, then the easy part of the game consists in letting ourselves to go down to ever simpler terms. This way one swiftly enters the realm of stop and go procedures, the difficult part being when and where to stop. In our case, we felt relatively at ease only when we arrived at the safe harbor of matrices. Of course, you have to buy a ticket to enter. The fair price is to become a jingler with them. After all, nothing is given for free.

We have enjoyed (and suffered) with the fact that so many beautiful results can be obtained with so little mathematics. Our hope is that you will enjoy (and not suffer) reading this book.

Acknowledgments. We would like to thank our families for their patience and support during the, longer than expected, process of writing the book. Very special thanks to Alma Rosa Rodríguez for her competent translation of ugly hieroglyphics to beautiful LaTeX. Thanks to our colleagues, Clara Garza, for reading the manuscript, to Arturo Olvera for devising and running some of the numerical schemes which have given evidence to some of our results and to Ana Cecilia Pérez for her computational support. We are grateful to L. Vespucci, Director of the Library at La Sapienza, for her help in our bibliographical search. Last but not least, let us mention the contributions of our friend and collaborator Ivar Massabó with whom we started, in 1985, the long journey through equivariant degree.

During the last two years, the authors had the partial support of the CNR, of the University of Rome, Tor Vergata, given through the scientific agreement between IIMAS-UNAM and Tor Vergata, and of several agencies on the Italian side, including

CANE, and from CONACyT (grant G25427-E, Matemáticas Nolineales de la Física y la Ingeniería, and the agreement KBN-CONACyT) on the Mexican side.

México City and Rome, February 2003

Jorge Ize
Alfonso Vignoli

Contents

Introduction

Nonlinearity is everywhere. But few nonlinear problems can be solved analytically. Nevertheless much qualitative information can be obtained using adequate tools. Degree theory is one of the main tools in the study of nonlinear problems. It has been extensively used to prove existence of solutions to a wide range of equations.

What started as a topological (or combinatorial) curiosity has evolved into a variety of flavors and represents, nowadays, one of the pillars, together with variational methods, of the qualitative treatment of nonlinear equations.

In the simplest situation, the "classical" degree of a continuous map $f(x)$ from $\mathbb{R}^n$ into itself with respect to a bounded open set Ω such that $f(x)$ is non-zero on $\partial\Omega$ is an integer, $\deg(f; \Omega)$, with the following properties:

(a) **Existence.** If $\deg(f; \Omega) \neq 0$, then $f(x) = 0$ has a solution in Ω.

(b) **Homotopy invariance.** If one deforms continuously $f(x)$, without zeros on the boundary, then the degree remains constant.

(c) **Additivity.** If Ω is the union of two disjoint open sets, then $\deg(f; \Omega)$ is the sum of the degrees of $f(x)$ with respect to each of the pieces.

If one has in mind studying a set of equations, those properties have a striking conceptual importance: a single integer gives existence results by loosening the rigidity of the equations and allowing deformations (and not only small ones). In other words, one does not need to solve explicitly the equations in order to get this information and one may obtain it by deforming the equations to a simpler set for which one may easily compute this integer. Furthermore, one has a certain localization of the solutions or one may obtain multiplicity results for these solutions.

Thus, in dimension one, the degree is another way to view the Intermediate Value Theorem of Calculus and, in dimension two, it is nothing else than the winding number of a vector field, familiar from Complex Analysis.

If, furthermore, one requires the property

(d) **Normalization.** The degree of the identity with respect to a ball containing the origin is 1,

then, one may show that the degree is *unique*.

Now there are many ways to construct the degree. As a consequence of the uniqueness, they are all equivalent and depend more on the possible application or on the particular taste of the user. For instance, one may take a combinatorial approach, or analytical (through perturbations or integrals), or topological (homotopical, cohomological) or an approach from fixed point theory.

Classical degree theory, or Brouwer degree, would have remained a simple curiosity if it were not for the extension to infinite dimensional problems, in particular to non-linear differential equations. This extension has required some compactness, starting from the Leray–Schauder degree with compact (or completely continuous) perturbations of the identity, continuing with k-set contractions, A-proper maps, 0-epi maps (these terms will be defined in Chapter 1) and so on. In most of these extensions the compactness is used to construct a good approximation by finite dimensional maps. One of the by-products of the construction presented here is to pinpoint a new way to see where the compactness is used.

Now the subject of this book is also that of symmetry. This is a basic concept in mathematics and words like symmetry breaking, period doubling or orbits are familiar even outside our discipline. In fact, many problems have symmetries: in the domains and in the equations. Very often these symmetries are used in order to reduce the set of functions to a special subclass: for instance look for odd (or even) solutions, or radial, or independent of certain variables. They are also used to avoid certain terms in series expansions or, in connection with degree theory, in order to get some information on this integer, the so-called Borsuk–Ulam results. However, since any continuous (i.e., not necessarily respecting the symmetry) perturbation is allowed, the ordinary degree will not give a complete topological information. This very important point will be clearer once the equivariant degree is introduced and computed in many examples.

In this book we shall integrate both concepts, that of a degree and that of symmetry, by defining a topological invariant for maps which commute with the action of a group of symmetries and for open sets which are invariant under these symmetries, i.e., for equivariant maps and invariant sets.

More precisely, a map $f(x)$, from $\mathbb{R}^n$ to $\mathbb{R}^m$ for instance or between two Banach spaces, is said to be *equivariant* under the action of Γ (a compact Lie group, for technical reasons) if

$$f(\gamma x) = \tilde{\gamma} f(x)$$

for all γ in Γ, where γ and $\tilde{\gamma}$ represent the action of the element γ in $\mathbb{R}^n$ and $\mathbb{R}^m$ respectively. Think of odd maps ($\gamma = \tilde{\gamma} = -\operatorname{Id}$) or even maps ($\gamma = -\operatorname{Id}$, $\tilde{\gamma} = \operatorname{Id}$), or any matrix γ expressed in two bases. The set Ω will be called *invariant* if, whenever x is in Ω, then the whole *orbit* Γx is also in Ω. By looking only at maps with these properties, including deformations of such maps, one gets an invariant, $\deg_\Gamma(f; \Omega)$, which is not an integer anymore (unless $m = n$, $\Gamma = \{e\}$, in which case one recovers the Brouwer degree) but with properties (a)–(c) valid and (d) replaced by a universality property.

Since the construction of this equivariant degree is quite simple, we shall not resist the temptation to present it now. Let $f(x)$ be an equivariant map, with respect to the actions of a group Γ, defined in an open bounded invariant set Ω and non-zero on $\partial\Omega$. Since Ω is bounded, one may choose a very large ball B containing it. Then one constructs an equivariant extension $\tilde{f}$ of f to B. The new map $\tilde{f}(x)$ may have new zeros outside Ω. One takes an invariant partition of unity $\varphi(x)$ with value 0 in $\bar{\Omega}$ and 1 outside a small neighborhood N of Ω, so small that on $N\backslash\Omega$ the map $\tilde{f}(x)$ is non-zero (it is non-zero on $\partial\Omega$). Take now a new variable t in $I = [0, 1]$ and define

$$\hat{f}(t, x) = (2t + 2\varphi(x) - 1, \tilde{f}(x)).$$

It is then easy to see that $\hat{f}(t, x) = 0$ only if x is in Ω with $\tilde{f}(x) = f(x) = 0$ and, since $\varphi(x) = 0$, one has $t = 1/2$. In particular, the map $\hat{f}(t, x)$ is non-zero on $\partial(I \times B)$ and defines an element of the abelian group (this group will be studied in Chapter 1)

$$\Pi^{\Gamma}_{S^n}(S^m)$$

of all Γ-equivariant deformation (or homotopy) classes of maps from $\partial(I \times B)$ into $\mathbb{R}^{m+1}\backslash\{0\}$. We define the Γ-*equivariant degree* of $f(x)$ with respect to Ω as the class of $\hat{f}(t, x)$ in $\Pi^{\Gamma}_{S^n}(S^m)$:

$$\deg_{\Gamma}(f; \Omega) = [\hat{f}]_{\Gamma}.$$

This degree turns out to have properties (a)–(c), where having non-zero degree here means that the class $[\hat{f}]_{\Gamma}$ is not the trivial element of $\Pi^{\Gamma}_{S^n}(S^m)$. Furthermore, by construction, this degree has the *Hopf property*, which is that if Ω is a ball and $[\hat{f}]_{\Gamma}$ is trivial, then $f|_{\partial\Omega}$ has a non-zero Γ-equivariant extension to Ω. In other words, $\deg_{\Gamma}(f; \Omega)$ gives a complete classification of Γ-homotopy types of maps on spheres. This property implies also that $\deg_{\Gamma}(f; \Omega)$ is *universal* in the sense that, if one has another theory which satisfies (a)–(c) such that, for a map f and a set Ω, one has a non-trivial element, then $\deg_{\Gamma}(f; \Omega)$ will be non-zero.

The simplest example is that of a non-equivariant map from $\mathbb{R}^n$ into itself. Then we shall see that $[\hat{f}]_{\Gamma}$ is the Brouwer degree of $\hat{f}$ with respect to $I \times B$. Since $\hat{f}$ is not zero on $I \times (B\backslash\Omega)$, this degree is that of $\hat{f}$ with respect to $I \times \Omega$, where $\hat{f}$ is a product map. A simple application of the product theorem implies that $[\hat{f}]_{\Gamma} = \deg(f; \Omega)$, a result which is, of course, not surprising but which indicates that our approach has the advantage of a very quick definition, with an immediate extension to the case of different dimensions, including infinite ones.

A second simple example is that of a $\mathbb{Z}_2$-action on $\mathbb{R}^n = \mathbb{R}^k \times \mathbb{R}^{n-k} = \mathbb{R}^m$, where $x = (y, z)$ and $f(y, z) = (f_0(y, z), f_1(y, z))$ with f_0 even in z and f_1 odd in z. It turns out that in this case $\Pi^{\mathbb{Z}_2}_{S^n}(S^n) \cong \mathbb{Z} \times \mathbb{Z}$, and that $\deg_{\mathbb{Z}_2}(f; \Omega)$ is given by two integers: $\deg(f_0(y, 0); \Omega \cap \mathbb{R}^k)$ and $\deg(f; \Omega)$. As a consequence of the oddness of f_1, with respect to z, one has $f_1(x, 0) = 0$ and it is clear that these two integers are well defined. The set $\{x, 0\}$ is the *fixed point* subspace of the action of $\mathbb{Z}_2$ and it is not surprising that these two integers are important. What is less intuitive is that if Ω is a ball then these two integers characterize completely all $\mathbb{Z}_2$-maps defined on Ω.

A third example is that of an S^1-action on $\mathbb{R}^k \times \mathbb{C}^{m_1} \times \cdots \times \mathbb{C}^{m_p}$, where S^1 leaves $\mathbb{R}^k$ fixed and acts as $\exp(in_j\varphi)$, for $j = 1, \dots, p$, on each complex coordinate of $\mathbb{C}^{m_j}$. This is an important example because if one writes down the autonomous equation

$$\frac{dX}{dt} - f(X) = 0, \quad X \text{ in } \mathbb{R}^k,$$

for $X(t) = \sum X_n e^{int}$, that is for 2π-periodic functions, then the fact that $f(X)$ does not depend on t implies that its component $f_n(X)$ on the n-th mode has the property that

$$f_n(X(t+\varphi)) = e^{in\varphi} f_n(X(t)),$$

i.e., the equation is equivalent to an S^1-equivariant problem (infinite dimensional). It turns out that, in this case, $\deg_{S^1}(f;\Omega)$ is a single integer given by $\deg(f|_{\mathbb{R}^k};\Omega|_{\mathbb{R}^k})$, i.e., by the invariant part of f. This is a slightly disappointing result but it can also be viewed as indicating that points with large orbits, in the sense of positive dimension, corresponding to the complex coordinates do not count when classifying the Γ-equivariant classes. This is a general fact which will be true for any group. Thus, in this particular example, one will have new invariants if the domain has (at least) one more dimension than the range, i.e., f is a function of a parameter ν and of X. In the case of differential equations, the extra parameter ν may come from a rescaling of time and represent the frequency. This occurs when one looks for periodic solutions of unknown period. In that case, it turns out that

$$\Pi^{S^1}_{S^{n+1}}(S^n) \cong \mathbb{Z}_2 \times \mathbb{Z} \times \mathbb{Z} \times \cdots$$

with one $\mathbb{Z}$, giving an integer for each type of one-dimensional orbits, and $\mathbb{Z}_2$, an orientation, corresponding to the invariant part. It is clear that we now have a much richer structure, which will lead to a host of applications, ranging from Hopf bifurcation to period doubling and so on. For instance, one may perturb an autonomous differential equation by a small time-periodic function. Then one may see what happens to the invariants in $\Pi^{S^1}_{S^{n+1}}(S^n)$, where one forgets about the S^1-action, i.e., in $\Pi_{S^{n+1}}(S^n) \cong \mathbb{Z}_2$. Of course, one could also break the symmetry by adding a $(2\pi/p)$-periodic perturbation, giving rise to other types of invariants.

A last example would be that of the action of a torus T^n, or of the largest torus in a general group. If this torus is generated by the phases $\varphi_1, \dots, \varphi_n$, each in $[0, 2\pi]$, one may look at Γ-equivariant maps $f(x)$ which have the additional property of being *orthogonal*. This means that

$$f(x) \cdot A_j x = 0, \quad j = 1, \dots, n,$$

where A_j is the infinitesimal generator corresponding to φ_j. This situation occurs when one considers gradients of invariant functionals: if $f(x) = \nabla\varphi(x)$, where $\varphi(\gamma x) = \varphi(x)$, then, by differentiating with respect to φ_j, one obtains this orthogonality. For instance, this is the situation for Hamiltonian systems, where one of the orthogonality relations is the conservation of energy. For such Γ-orthogonal maps one may repeat the construction of the degree and obtain a new invariant

$$\deg_\perp(f;\Omega) \quad \text{in } \Pi^{\Gamma}_{\perp S^n}(S^n),$$

a group which is much larger than $\Pi^{\Gamma}_{S^n}(S^n)$. In fact, it is a product of $\mathbb{Z}$'s, one for each orbit type, independent of the dimension of the orbit, as we shall describe below, by relating this new degree to "Lagrange multipliers". One may look at zeros of the map

$$f(x) + \sum \lambda_j A_j x = 0,$$

where if one takes the scalar product with $f(x)$ one obtains a zero of $f(x)$ and the relation $\sum \lambda_j A_j x = 0$. In particular, if, for some x, the $A_j x$'s are linearly independent, this implies that $\lambda_j = 0$. Of course, this linear independence depends on x, but the introduction of these multipliers will enable us to compute completely the group $\Pi^{\Gamma}_{\perp S^n}(S^n)$.

It is now time to have a closer look at the content of the book. We shall do so by pointing out the parts which may be of special interest to a given group of readers. As explained in the Preface, we have tried to write a book as self-contained as possible. This implies that the first chapter is devoted to a collection of some simple facts from different fields which are needed in the book. Thus we introduce group actions, equivariant maps, averaging and irreducible representations, in particular, Schur's Lemma and its consequences. This is all which will be needed from Representation Theory.

From the point of view of Topology, one of our main tools will be that of extensions of equivariant maps. There is a special extension for orthogonal maps. A full proof is given in Theorem 7.1, using the Gram–Schmidt orthogonalization process. We give also the definition and some basic properties of equivariant homotopy groups of spheres, the groups where our degrees live. The last section in the chapter is a review of some of the results from Analysis, in particular, Ordinary Differential Equations, which will be needed in the last chapter. Thus we integrate a quick survey of Bifurcation Theory, Floquet Theory (also expanded in Appendix B), Hamiltonian systems and the special form of orbits arising in these problems (*twisted orbits*).

Hence, an expert in any of these fields should only glance at some of these results in order to get acquainted with our notation, and look at some of the examples. For a reader who is not familiar with these subjects, we hope that (s)he will find all the necessary tools and acquire a working knowledge and a good intuition from this chapter.

In this brief description of the first chapter, we left out the second section on the fundamental cell. This construction, explained here for abelian groups, is the key to most of the work on equivariant homotopy groups. It says that one may find a region in $\mathbb{R}^n$, made of sectorial pieces, such that, if one has any continuous function defined on the cell with some symmetry properties on its boundary, then one may extend the map to the whole space, using the action of the group. Think of a map defined on a half-space and extended as an odd map or of a map defined on a sector in $\mathbb{C}$ of angle $(2\pi/n)$.

The second chapter is devoted to the definition and study of the basic properties of the equivariant degree. Furthermore, we show how this degree may be extended to infinite dimensions by approximations by finite dimensional maps, *à la Leray–Schauder*, and how one may define the orthogonal degree. Next, we present abstract applications to continuation and bifurcation problems and, finally, we study the usual operators on our degree: symmetry breaking, products and composition, operations which will be studied more deeply in the next chapter and applied in the last chapter. Of course, this chapter is the abstract core of the book.

Chapter 3 has a more topological flavor. In it we compute the equivariant homotopy groups of spheres, in the particular case of abelian groups. The reason for this choice is that we are able to give explicit constructions of the generators for the groups with elementary arguments (although sometimes lengthy). Thus, anyone should be able to follow the proofs. The basic idea is that of *obstruction theory*, that is, of extension of maps. The program is to start from an equivariant function which is non-zero on a sphere ∂B and see under which conditions one may construct an extension inside the sphere, first to the fundamental cell where one has either an extension, if the dimension is low enough, or a first obstruction given by some Brouwer degree, or secondary obstructions which are not unique but may be completely determined. Then, one uses the group action to extend the map to the whole ball B. Finally, the homotopy group structure enables one to subtract a certain number of generators and write down any map as a sum of multiples of explicit generators. These multiples will be the essence of the degree.

In order to make this program a reality, we work stage by stage. (Here, we ask the reader to allow us to use some technical arguments so that we may illustrate the range of ideas developed in the book.) The first step is to consider a map which is Γ-equivariant and non-zero on ∂B^H and on the union of all B^K, such that H is a subgroup of K, and where B^H stands for the ball in the subspace fixed by H. In particular, all points in $B^H \setminus \bigcup B^K$ have the same orbit type H, and extensions are completely determined by the behavior on the boundary of the fundamental cell. Hence, if the map is between the spaces V^H and W^H, the fundamental cell has dimension equal to $\dim V^H - \dim \Gamma/H$, and, if this difference is less than $\dim W^H$, one always has a non-zero extension, while if one has equality one obtains a first obstruction: the degree of the map on the boundary of the fundamental cell. This is the content of Theorem 1.1. The next step is to give conditions under which this obstruction is independent of the previous extensions. One obtains a well-defined *extension degree.*

The next step is to continue this extension process to non-zero Γ-maps defined from $\bigcup \partial B^H$ with $\dim \Gamma/H = k$, which are also non-zero on $\bigcup B^K$ for K with $\dim \Gamma/K < k$. For this purpose the concept of *complementing maps* is quite important. We show that essentially this set of maps behaves as a direct sum of maps characterized by the extension degrees. The final step is to go on for all k's which meet the hypothesis. For instance, if $V = \mathbb{R}^k \times W$, then one proves that

$$\Pi^{\Gamma}_{S^V}(S^W) = \Pi_{k-1} \times \mathbb{Z} \times \mathbb{Z} \times \cdots,$$

with one $\mathbb{Z}$ for each orbit type H with $\dim \Gamma/H = k$ and Π_{k-1} concerns only orbits of dimension lower than k.

The next question is the following. Given a map, how does one compute its decomposition into the direct sum? This is done in two different ways: either by approximations by *normal maps* (a topological substitute to Sard's lemma) or by looking at *global Poincaré sections*. One may relate the $\mathbb{Z}$-components in the above decomposition to ordinary degrees (see Corollary 3.1 in Chapter 3).

The fourth section is devoted to Borsuk–Ulam results, that is to the computation of the ordinary degree of an equivariant map. The purpose of this section is to show how the extension ideas can be used in this sort of computations.

The next section treats the case of maps from $\mathbb{R} \times W$ into W, which is particularly important when one breaks the S^1-symmetry, for instance for an autonomous differential equation with unknown period by perturbing it by a $(2\pi/p)$-periodic field. We compute then Π_0, in the above formula, and prove that now there are obstructions for extensions to the faces of the fundamental cell and to the body of that cell. For each H with Γ/H finite one has a classification of the secondary obstructions in a group isomorphic to $\mathbb{Z}_2 \times \Gamma/H$, with explicit generators according to the different presentations of Γ/H.

The sixth section deals with the computation of the homotopy group of spheres for Γ-orthogonal maps, proving that

$$\Pi^{\Gamma}_{\perp S^V}(S^V) = \mathbb{Z} \times \mathbb{Z} \times \cdots ,$$

with one $\mathbb{Z}$ for each orbit type, independent of its dimension. This is done via the Lagrange multipliers already mentioned, and the reader will guess why the case of Γ-equivariant maps with parameters, from $\mathbb{R}^k \times W$ onto W, is important here.

The last section of Chapter 3 deals with operations: suspension, products, composition and symmetry breaking. That is, what happens to the explicit generators under one of these operations.

As we have already said this third chapter is more topologically inclined. A reader more interested in applications should only look at the statements of the results, which will be used in the last chapter, and see some of the examples.

However, we would like to make a few points. Our entire construction relies on a single basic fact: a map from a sphere into a higher dimensional sphere has a non-zero extension to the ball, while, if the dimensions are equal, one has a unique "obstruction", an integer, for extension (and other invariants if the dimension of the range is lower). From this, with "elementary" but explicit arguments, and with no algebraic machinery, we obtain surprising new results which may be understood by any non-specialist. Of course, there is a price to be paid: our actions are linear and the groups are abelian (the non-abelian case may be dealt with in a similar, but less explicit way). On the other hand, our pedestrian approach stresses some new concepts, like those of complementing maps, normal maps and global Poincaré sections, which may be useful in a more abstract context. In short, independently of the reader's background, we believe that this chapter may be useful and interesting to anyone.

The last chapter is essentially devoted to applications, although the first section states that any element in $\Pi^{\Gamma}_{S^V}(S^W)$ is the Γ-degree of a map defined on a reasonable Ω. Now, in order to be useful, a degree should be computable in some simple generic cases, for instance for an isolated orbit or an isolated loop of orbits. For the case of an isolated orbit, the natural hypothesis is to assume that 0 is a regular value. (We recall here our introductory remarks: one does not have to consider the nonlinear equation under study, but a, hopefully, simpler equation where one may look at these generic situations.) This leads to approximation by the linearization of the map at the orbit. The simplest case is when one has a stationary solution, or, even better, a family of such solutions, leading to bifurcation. In this case, the Γ-index is given by the sign of determinants of the linearization on the fixed point subspace of Γ and on the subspaces where Γ acts as $\mathbb{Z}_2$, giving conditions for period doubling. The next case is when the isolated orbit has an orbit type which is not the full group. For this sort of solution, we obtain an abstract result (Theorem 2.4) and the Γ-index is given in terms of the spectrum of the linearization, *à la Leray–Schauder*, but with many of these indices. This abstract result is applied to autonomous differential equations of unknown period or of fixed period but with an extra parameter, or with a first integral. One may then perturb this autonomous differential equation with a time-periodic function and obtain subharmonics or phase locking phenomena. If the autonomous differential equation has also a geometrical symmetry, then one obtains twisted orbits.

We are phrasing this part of the introduction in a way which will be easily recognizable by a reader familiar with low dimensional dynamical systems. However, each specific behavior will be explained in that chapter.

A similar situation occurs for orthogonal maps. In that case the orthogonal index has components which are of the previous type (i.e., leading to period doubling) and a new type given by a full Morse index, i.e., the number of negative eigenvalues of a piece of the linearization. This is applied to Hamiltonian systems of different types, where variational methods give also invariants depending on Morse numbers. In the present case it is the orthogonality which brings in this invariant.

In order to show how to apply our degree, we give the complete study of two spring-pendulum systems. We hope that this example makes the point of the usefulness of the equivariant degree approach and we challenge the reader to guess (a priori) the type of solution we obtain.

The final section deals with the index of a loop of stationary solutions, with applications to Hopf bifurcation, systems with first integrals and so on. It is important to point out that all our examples (except a very simple retarded differential equation) come from Ordinary Differential Equations. The main reason for this choice is to avoid technicalities. It should be clear to anyone interested in Partial Differential Equations, for instance, how to adapt these result to many situations. For example, replace Fourier series by eigenfunctions expansions or other Galerkin-type approximations. Another reason for this choice is that the reader may easily see how the degree arguments are used to obtain information on the solutions of a nonlinear equation in an integrated way, that is, with the same tool in different situations (and not with ad hoc degrees), and see what happens if one modifies the conditions of the problem, as in symmetry breaking. Here, we would like to stress the Hopf property, i.e., that, if the degree is zero, then it is likely that one may perturb the problem (in the sense of extensions of maps) so that the new problem has no solutions. This property and the global picture which enables one to relate two different solutions or two different problems, is one of the main conceptual contributions of degree theory. Of course, we are not computing the actual solutions (nothing is for free), although it would be interesting to adapt the homotopy numerical continuation methods to equivariant problems.

Each chapter has a final section on bibliographical remarks. We have tried to indicate some other approaches to the subject matter of this book. However, it is clear that most of this book is based on the authors' research in the last 15 years. It is also clear that there is still much to do. For instance, perform similar computations for actions of non-abelian groups with its endless list of applications. Similarly, there are more or less straightforward extensions (we have mentioned several times the word k-set contraction) or applications to P.D.E.'s (essentially some technical problems) and many more. We hope that this book will serve as an incentive for the reader to follow up in that direction.

A last technical point: theorems, lemmas, remarks and examples are listed independently. For instance Theorem 5.2 refers to the second theorem in Section 5 of the chapter. When referring to a result from another chapter, this is done explicitly: for instance, Theorem 5.2 of Chapter 1. On the other hand, our notations are standard, but we would like to emphasize a particular one (maybe not too familiar): $H < K$ means that H is a subgroup of K (and could be K itself).

Chapter 1
Preliminaries

As mentioned in the Introduction, the main purpose of this chapter is to collect some of the most useful definitions and properties of actions of compact Lie groups on Banach spaces, as well as the elements of homotopy theory and some facts about operators which will be most frequently used in this book. Thus, the reader will find here almost all the results needed in this text. The expert will have only to glance at the definitions in order to get acquainted with our notation.

1.1 Group actions

In the whole book Γ will stand for a compact Lie group (the reader will see below which properties of a Lie group are used here).

Definition 1.1. A Banach space E is a Γ-*space* or a *representation* of the group Γ, if there is a homeomorphism ρ of Γ into $\mathrm{GL}(E)$, the general linear group of (linear) isomorphisms over E. In this case, we say that Γ *acts linearly* on E, via the *action* $\rho(\gamma)x$, such that

$$\rho(\gamma\gamma') = \rho(\gamma)\rho(\gamma'),$$
$$\rho(e) = \mathrm{Id}\,.$$

When no confusion is possible, we shall denote the action simply by γ.

Example 1.1. Let $E = \mathbb{R}^n \times \mathbb{R}^m$ and $\Gamma = \mathbb{Z}_2 = \{-I, I\}$ with

$$\rho(-I)(X, Y) = (-X, Y). \tag{1.1}$$

Example 1.2. If $E = \mathbb{C}$ and $\Gamma = \mathbb{Z}_m = \{0, 1, \ldots, m-1\}$ is the additive group of the integers modulo m, let

$$\rho(k)z = e^{2\pi ikp/m}z, \quad \text{where } p \text{ is a fixed integer.} \tag{1.2}$$

Example 1.3. If $E = \mathbb{C}$ and $\Gamma = S^1 = \mathbb{R}/2\pi = \{\varphi \in [0, 2\pi)\}$, then one may have

$$\rho(\varphi)z = e^{in\varphi}z \tag{1.3}$$

for some integer n.

Example 1.4. If $E = \mathbb{C}$ and $\Gamma = T^n \times \mathbb{Z}_{m_1} \times \cdots \times \mathbb{Z}_{m_s} = \{(\varphi_1, \ldots, \varphi_n, k_1, \ldots, k_s)$ with $\varphi_j \in [0, 2\pi), 0 \le k_j < m_j\}$, then one may have

$$\rho(\gamma)z = \exp i\Big(\sum_1^n n_j\varphi_j + 2\pi \sum_1^s k_j l_j/m_j\Big)z, \tag{1.4}$$

where n_j and l_j are given integers.

Remark 1.1. We shall see below that this is the general case of an irreducible representation of any compact abelian Lie group. It is easy to see that if $\mathbb{Z}_m$ acts on $\mathbb{C}$, then $\rho(m) = 1 = \rho(1)^m$ and $\rho(1)$ must have the form given in (1.2). Since the same argument applies to S^1 acting on $\mathbb{C}$, then any Γ given by an abelian product as in (1.4), must act on $\mathbb{C}$ as in that formula.

On the other hand if $\mathbb{Z}_m$ acts non-trivially on $\mathbb{R}$, then m is even and $\rho(1) = -1$, while S^1 may act only trivially on $\mathbb{R}$, i.e., $\rho(\varphi) = \rho(\varphi/N)^N$, take N so large that the continuity of ρ implies that $\rho(\varphi/N)$, being close to 1, must be positive. Hence, $\rho(\varphi)$ is always a positive number. Since $\rho(2\pi) = 1 = \rho(2\pi/N)^N$ one gets $\rho(2\pi/N) = 1$ and $\rho(2\pi p/q) = \rho(2\pi/q)^p = 1$ and by denseness of $\mathbb{Q}$ in $\mathbb{R}$, one obtains $\rho(\varphi) \equiv 1$. For convenience in the notation, we shall very often use (1.4) to denote also the action of Γ on $\mathbb{R}$, with the convention that, in that case, $n_j = 0$, l_j is a multiple of $m_j/2$ if m_j is even, or $l_j = 0$ if m_j is odd.

Example 1.5. Let $E = C^0_{2\pi}(\mathbb{R}^N)$ be the space of continuous, 2π-periodic functions on $\mathbb{R}^N$ with the uniform convergence norm. The group $\Gamma = S^1$ may act on E as

$$\rho(\varphi)X(t) = X(t + \varphi)$$

i.e., as the time shift.

One may also set this action in terms of Fourier series by writing

$$X(t) = \sum_{-\infty}^{\infty} X_n e^{int},$$

with $X_n \in \mathbb{C}^N$, $X_{-n} = \bar{X}_n$ (since $X(t) \in \mathbb{R}^N$). For the Fourier coefficients X_n one has the equivalent action:

$$\rho(\varphi)X_n = e^{in\varphi}X_n. \tag{1.5}$$

Definition 1.2. Let E be a Γ-space and $x \in E$ be given. The *isotropy subgroup of* Γ *at* x is the set $\Gamma_x = \{\gamma \in \Gamma : \gamma x = x\}$, which is a closed subgroup of Γ.

Definition 1.3. The action of Γ on E is said to be *free* if $\Gamma_x = \{e\}$ for any $x \in E\backslash\{0\}$. The action is *semi-free* if $\Gamma_x = \{e\}$ or Γ for any $x \in E$.

For instance, in Example 1.1, $\Gamma_{(X,Y)} = \mathbb{Z}_2$ if and only if $X = 0$ and the action is semi-free. In Example 1.2, the action is free only if p and m are relatively prime (denoted as $(p : m) = 1$), while if $p/m = q/n$ with $(q : n) = 1$, then $\Gamma_z = \mathbb{Z}_{m/n} = \{k = sn, s = 0, \dots, m/n - 1\}$. In Example 1.3, one has $\Gamma_z = \mathbb{Z}_N = \{\varphi = k/N, k = 0, \dots, N - 1\}$. The case of Example 1.4 will be given below in Lemma 1.1.

Definition 1.4. The element $x \in E$ is called a *fixed point* of Γ if $\Gamma_x = \Gamma$. The *subspace of fixed points of* Γ *in* E is denoted by E^Γ. If H is a subgroup of Γ then $E^H = \{x \in E : \gamma x = x \text{ for any } \gamma \in H\}$ is a closed linear subspace of E.

Notation 1.1. If H is a subgroup of K, we shall write $H < K$. Note that if $H < K$, then $E^K \subset E^H$.

Definition 1.5. If $H < \Gamma$, the *normalizer* $N(H)$ of H is

$$N(H) = \{\gamma \in \Gamma : \gamma^{-1} H \gamma \subset H\}$$

and the *Weyl group* $W(H)$ of H is

$$W(H) = N(H)/H.$$

Note that if Γ is abelian, then $N(H) = \Gamma$.

Also, if $x \in E^H$, then $\gamma x \in E^H$ for any $\gamma \in N(H)$, since $\gamma_1 \gamma x = \gamma \gamma_2 x = \gamma x$ for some γ_1 and γ_2 in H. Hence γx is fixed by the action of H. Furthermore, if $H = \Gamma_x$ for some x and $\gamma x \in E^H$ for some γ, then it is easy to see that γ belongs to $N(H)$, i.e., $N(H)$ is the largest group which leaves E^H invariant. Moreover, if Γ is abelian, then $N(H) = \Gamma$ and E^H is Γ-invariant.

Let us now consider the case of Example 1.4.

Lemma 1.1. *Let* $\Gamma = T^n \times \mathbb{Z}_{m_1} \times \cdots \times \mathbb{Z}_{m_s}$ *act on* $\mathbb{C}$ *via*

$$\exp i(\langle N, \Phi\rangle + 2\pi \langle K, L/M\rangle),$$

where $\langle N, \Phi\rangle = \sum_1^n n_j \varphi_j$ *and* $\langle K, L/M\rangle = \sum_1^s k_j l_j / m_j$. *If* $l_j/m_j = \tilde{l}_j/\tilde{m}_j$, *with* $\tilde{l}_j$ *and* $\tilde{m}_j$ *relatively prime, let* $\tilde{m}$ *be the least common multiple of the* $\tilde{m}_j$*'s (l.c.m) and set* $|N| = \sum_1^n |n_j|$. *Then:*

(a) *If* $L \neq 0$, *there is* K_0 *such that* $\langle K_0, L/M\rangle \equiv 1/\tilde{m}, [2\pi]$, *and any other* K *gives an action of the form* $q/\tilde{m}$ *for some* $q \in \{0, \dots, \tilde{m} - 1\}$. *In particular, if* $N = 0$ *and* H *is the isotropy subgroup, then* $W(H) \cong \mathbb{Z}_{\tilde{m}}$.

(b) *If* $N \neq 0$, *the congruence* $\langle N, \Phi\rangle \equiv 0, [2\pi]$, *gives* $|N|$ *hyperplanes in* T^n. *In particular, if* $L = 0$, *then* $W(H) \cong S^1 = T/\mathbb{Z}_{|N|}$.

(c) *If* $L \neq 0$ *and* $N \neq 0$, *then* $W(H) \cong S^1 = T/\mathbb{Z}_{\tilde{m}|N|}$.

Proof. (a) If $s = 1$, then $k\tilde{l}/\tilde{m}$ is an integer if and only if k is a multiple of $\tilde{m}$ and $e^{2\pi ik\tilde{l}/\tilde{m}}$ gives $\tilde{m}$ distinct roots of unity, hence the result is clear.

If $s = 2$, from the preceding case, one has $k_j\tilde{l}_j/\tilde{m}_j \equiv \tilde{k}_j/\tilde{m}_j$, with $0 \le \tilde{k}_j < \tilde{m}_j$ and one has to consider $\tilde{k}_1/\tilde{m}_1 + \tilde{k}_2/\tilde{m}_2$. Now, $\tilde{m} = p_1\tilde{m}_1 = p_2\tilde{m}_2$, with p_1 and p_2 relatively prime by the definition of a l.c.m. Thus, there are integers α_1, α_2 such that $\alpha_1 p_1 + \alpha_2 p_2 = 1$, where α_1 and α_2 have opposite signs. Assume that $\alpha_1 > 0$. Divide α_1 by $\tilde{m}_1$ and get $\alpha_1 = a_1\tilde{m}_1 + k_1^0$, with $a_1 \ge 0$ and $0 \le k_1^0 < \tilde{m}_1$. Likewise, $-\alpha_2 = (a_2+1)\tilde{m}_2 - k_2^0$, with $a_2 \ge 0$ and $0 \le k_2^0 < \tilde{m}_2$. Then, $p_1k_1^0 + p_2k_2^0 = \alpha_1 p_1 + \alpha_2 p_2 + (a_2 + 1 - a_1)\tilde{m}$, defining K_0 in this case. For any other pair $(\tilde{k}_1, \tilde{k}_2)$, we have $\tilde{k}_1/\tilde{m}_1 + \tilde{k}_2/\tilde{m}_2 = (p_1\tilde{k}_1 + p_2\tilde{k}_2)/\tilde{m} \equiv (p_1\tilde{k}_1 + p_2\tilde{k}_2)(k_1^0/\tilde{m}_1 + k_2^0/\tilde{m}_2)$, proving the result for $s = 2$.

For the general case, assume the result true for $s - 1$. Let $\hat{m}$ be the l.c.m. of $(\tilde{m}_1, \dots \tilde{m}_{s-1})$ and $\tilde{m}$ be the l.c.m. of $\hat{m}$ and m_s. We have

$$\sum_1^{s-1} k_j\tilde{l}_j/\tilde{m}_j + k\tilde{l}/\tilde{m}_s \equiv q_0/\hat{m} + k\tilde{l}/\tilde{m}_s,$$

where q_0 is given by the induction hypothesis in such a way that

$$\sum_1^{s-1} k_j^0\tilde{l}_j/\tilde{m}_j \equiv 1/\hat{m} \quad \text{and} \quad k_j = q_0k_j^0.$$

One is then reduced to the two "modes" case.

(b) For the action of T^n, one has that $\langle N, \Phi\rangle$ spans an interval of length $2\pi|N|$. The congruence $\langle N, \Phi\rangle \equiv 0$, $[2\pi]$, gives $|N|$ parallel hyperplanes in T^n. One may change φ_j to $2\pi - \varphi_j$ whenever N_j is negative, defining an isomorphism of T^n for which all N_j's are positive. Then, $\langle N, \Phi\rangle = |N|\varphi$, with $0 \le \varphi < 2\pi/|N|$, will give that, if $L = 0$, then $H \cong T^{n-1} \times \mathbb{Z}_{|N|} \times \mathbb{Z}_{m_1} \times \cdots \times \mathbb{Z}_{m_s}$ with $W(H) \cong S^1 = T/\mathbb{Z}_{|N|}$.

(c) In general, one may write $\langle N, \Phi\rangle + 2\pi\langle K, L/M\rangle$ as $|N|\varphi + 2\pi q/\tilde{m}$, with $0 \le q < \tilde{m}$, φ in $[0, 2\pi/|N|)$. The relation $|N|\varphi + 2\pi q/\tilde{m} = 2k\pi$ will give $\varphi = \langle N, \Phi\rangle/|N| = 2k\pi/|N| - 2\pi q/\tilde{m}|N|$ which represents $\tilde{m}|N|$ different parallel hyperplanes in T^n. Thus, $H \cong T^{n-1} \times \mathbb{Z}_{\tilde{m}|N|}$ and $W(H) \cong S^1 = T/\mathbb{Z}_{\tilde{m}|N|}$. □

Definition 1.6. An isotropy subgroup H is *maximal* if H is not contained in a proper isotropy subgroup of Γ.

Lemma 1.2 (Golubitsky). *If H is a maximal isotropy subgroup of Γ and $E^\Gamma = \{0\}$, then $W(H)$ acts freely on $E^H\backslash\{0\}$.*

Proof. In fact, if $\gamma x = x$ for some $x \ne 0$ in E^H and some $\gamma \in N(H)/H$, then $\Gamma_x \supset H \cup \{\gamma\}$. Hence, from the maximality of H, one has $\Gamma_x = \Gamma$, but then $x \in E^\Gamma = \{0\}$. □

Remark 1.2. The groups which act freely on Euclidean spaces have been completely classified: a reduced number of finite groups, S^1 and $N(S^1)$ in S^3 and S^3 (see [Br] p. 153). For an abelian group with an action given by (1.4), one has $H = \{e\}$, i.e., $W(H) = \Gamma$, only if either $n = 0, s = 1$ and the action of $\mathbb{Z}_m$ is given by $e^{2\pi ikp/m}$, with p and m relatively prime (hence $\tilde{m} = m$), or $n = 1, s = 0, |N| = 1$, with an action of S^1 given by $e^{i\varphi}$ (see Lemma 1.1).

Definition 1.7. The *orbit* of x under Γ is the set $\Gamma(x) = \{\gamma x \in E : \gamma \in \Gamma\}$.

It is easy to see that $\Gamma(x)$ is homeomorphic to Γ/Γ_x, that $\Gamma_{\gamma x} = \gamma\Gamma_x\gamma^{-1}$ (in particular $\Gamma_{\gamma x} = \Gamma_x$ if Γ is abelian) and that the orbits form a partition of E. The set E/Γ is the *orbit space* of E with respect to Γ.

Definition 1.8. Two points x and y have the same *orbit type* H if there are γ_0 and γ_1 such that $H = \gamma_0^{-1}\Gamma_x\gamma_0 = \gamma_1^{-1}\Gamma_y\gamma_1$.

If E is finite dimensional, then it is clear that there are only a finite number of orbit types.

Definition 1.9. The set of isotropy subgroups for the action of Γ on E will be denoted by $\mathrm{Iso}(E)$.

1.2 The fundamental cell lemma

In this section we shall assume that one has a finite dimensional representation V of the abelian group $\Gamma = T^n \times \mathbb{Z}_{m_1} \times \cdots \times \mathbb{Z}_{m_s}$ in such a way that any X in V is written as $X = \sum x_j e_j$, where $x_j \in \mathbb{C}$ if $W(\Gamma_{e_j}) \cong \mathbb{Z}_p$ or S^1, $p > 2$, or $x_j \in \mathbb{R}$ if $W(\Gamma_{e_j}) = \{e\}$ or $\mathbb{Z}_2$. The action of Γ on the elements of the basis is given by

$$\gamma e_j = \exp i(\langle N^j, \Phi\rangle + 2\pi\langle K, L^j/M\rangle)e_j,$$

as in (1.4) and Remark 1.1, with

$$N^j = (n_1^j, \dots, n_n^j)^T \quad \text{and} \quad L^j/M = (l_1^j/m_1, \dots, l_s^j/m_s)^T.$$

Then $\gamma X = \sum x_j\gamma e_j$ and $\gamma X = X$ gives $\gamma e_j = e_j$ if $x_j \neq 0$. Hence, $\Gamma_X = \bigcap \Gamma_{e_j}$, where the intersection is over those j's for which $x_j \neq 0$. Thus, $W(\Gamma_{e_j}) < W(\Gamma_X)$.

Lemma 2.1. $V^{T^n} = \{X \in V : W(\Gamma_X) < \infty\}$.

Proof. If $W(\Gamma_X)$ is finite, then $W(\Gamma_{e_j})$ is a finite group and Γ_{e_j} contains T^n. In this case, Γ_X contains also T^n, that is, X belongs to V^{T^n}. Conversely, if X is fixed by T^n, then $W(\Gamma_X)$ is a factor of $\mathbb{Z}_{m_1} \times \cdots \times \mathbb{Z}_{m_s}$ and hence is finite. □

Denote by $H_j = \Gamma_{e_j}$ and define $\tilde{H}_{j-1} = H_1 \cap \cdots \cap H_{j-1}$, $H_0 = \Gamma$. Then H_{j-1} acts on the space V_j generated by e_j ($V_j \cong \mathbb{R}$ or $\mathbb{C}$), with isotropy $\tilde{H}_{j-1} \cap H_j = \tilde{H}_j$, if

$x_j \neq 0$, and $\tilde{H}_{j-1}/\tilde{H}_j$ acts freely on $V_j\backslash\{0\}$. Then, from Lemma 1.1, this Weyl group is isomorphic either to S^1, to $\{e\}$, or to $\mathbb{Z}_p$, $p \geq 2$. Let k_j be the cardinality of this group: $k_j = |\tilde{H}_{j-1}/\tilde{H}_j|$. If the group is S^1, then $k_j = \infty$, while $k_j = 1$ means that $\tilde{H}_{j-1} = \tilde{H}_j$. If $k_j = 2$ and V_j is complex, then V_j splits into two real representations of $\tilde{H}_{j-1}/\tilde{H}_j \cong \mathbb{Z}_2$, while if V_j is real, then $k_j = 1$ or 2.

Consider $\mathcal{C} = \{X \in V : |x_j| = 1 \text{ for any } j\}$, a torus in V. Let $H = H_1 \cap H_2 \cap \cdots \cap H_{m+r}$ be the isotropy type of $\mathcal{C}$, where there are m of the V_j' s which are complex and r which are real (hence $\dim V = 2m + r$). Let k be the number of j's with $k_j = \infty$. Let

$$\Delta = \{X \in \mathcal{C} : 0 \leq \operatorname{Arg} x_j < 2\pi/k_j \text{ for all } j = 1, \ldots, m + r\}.$$

That is, if $k_j = 1$ there is no restriction on x_j (in $\mathbb{C}$ or $\mathbb{R}$), while, if $k_j = \infty$, then $x_j \in \mathbb{R}^+$ and, if $x_j \in \mathbb{R}$ and $k_j = 2$, then x_j is positive. Let

$$\Delta_V = \{X \in V : 0 \leq \operatorname{Arg} x_j < 2\pi/k_j\}.$$

Then Δ_V is a cone of dimension equal to $\dim V - k$. The set Δ_V will be called the *fundamental cell.* It will enable us to compute all the equivariant homotopy extensions and to classify their classes in Chapter 3.

Lemma 2.2 (Fundamental cell lemma). *The images of Δ under Γ/H cover properly $\mathcal{C}$ (i.e., in a 1-1 fashion).*

Proof. The proof will be by induction on $m + r$. If there is only one coordinate, then Γ/H_1 acts freely on $V_1\backslash\{0\}$. If this group is S^1, then the image of e_1 under it will generate $\mathcal{C}$, while if this group is $\mathbb{Z}_{k_1}$, $k_1 \geq 1$, then one has to cut $\mathcal{C}$ into k_1 equal pieces in order to generate $\mathcal{C}$.

If the result is true for $n - 1$, let $\mathcal{C} = \mathcal{C}_{n-1} \times \{|x_n| = 1\}$, $\Delta = \Delta_{n-1} \times \{0 \leq \operatorname{Arg} x_n < 2\pi/k_n\}$ and write $\Gamma/H = (\Gamma/\tilde{H}_{n-1})(\tilde{H}_{n-1}/H)$, recalling that these groups are abelian. By the induction hypothesis, the images of Δ_{n-1} under $\Gamma/\tilde{H}_{n-1}$ cover properly $\mathcal{C}_{n-1}$. Furthermore, from the case $n = 1$, the set $\{x_n : |x_n| = 1\}$ is covered properly by the images of $\{x_n : 0 \leq \operatorname{Arg} x_n < 2\pi/k_n\}$ under $\tilde{H}_{n-1}/H$, a group which fixes all points of $\mathcal{C}_{n-1}$. Hence, if (X_{n-1}, x_n) is in $\mathcal{C}$, there are γ_{n-1} in $\Gamma/\tilde{H}_{n-1}$ and γ_n in $\tilde{H}_{n-1}/H$ such that $X_{n-1} = \gamma_{n-1}X^0_{n-1}$, with X^0_{n-1} in $\mathcal{C}_{n-1}$, $\gamma^{-1}_{n-1}x_n = \gamma_n x^0_n$, with $0 \leq \operatorname{Arg} x^0_n < 2\pi/k_n$ and $\gamma_n X_{n-1} = X_{n-1}$.

Then $(X_{n-1}, x_n) = (\gamma_{n-1}X^0_{n-1}, \gamma_{n-1}\gamma^{-1}_{n-1}x_n) = \gamma_{n-1}\gamma_n(X^0_{n-1}, x^0_n)$, i.e., $\mathcal{C}$ is covered by the images of Δ under Γ/H.

If $(X_{n-1}, x_n) = \gamma_1(X^1, x^1) = \gamma_2(X^2, x^2)$, with (X^j, x^j) in Δ and γ_j in Γ/H, then $(X^1, x^1) = \gamma_1^{-1}\gamma_2(X^2, x^2)$. Thus, $X^1 = \gamma X^2$, $x^1 = \gamma x^2$. By the induction hypothesis, $X^1 = X^2$ and γ belongs to $\tilde{H}_{n-1}$, but then $x^1 = x^2$ and γ belongs to H. □

This fundamental cell lemma will be the key tool in computing the homotopy groups of Chapter 3.

Example 2.1. Let S^1 act on e_j via $e^{n_j\varphi}$, with $n_j > 0$. Then, $H_j = \{\varphi = 2\pi k/n_j,\ k = 0, \dots, n_j - 1\} \cong \mathbb{Z}_{n_j}$. Let $\tilde{n}_j = (n_1 : \cdots : n_j)$ be the largest common divisor (l.c.d.) of $n_1, \dots, n_j$, then $\tilde{H}_j = \{\varphi = 2\pi k/\tilde{n}_j,\ k = 0, \dots, \tilde{n}_j - 1\} \cong \mathbb{Z}_{\tilde{n}_j}$. Thus, $k_1 = \infty$, $k_j = \tilde{n}_{j-1}/\tilde{n}_j$.

Note that, since $\Gamma/H = (\Gamma/\tilde{H}_1) \times (\tilde{H}_1/\tilde{H}_2) \times \cdots \times (\tilde{H}_{m+r-1}/H)$ if $\dim \Gamma/H = k$, then there are exactly k coordinates (which have to be complex) with $k_j = \infty$. In fact, since $\tilde{H}_j$ is the isotropy subgroup for the action of $\tilde{H}_{j-1}$ on x_j, each factor, by Lemma 1.1, is at most one-dimensional.

Lemma 2.3. *Under the above circumstances, one may reorder the coordinates in such a way that $k_j = \infty$ for $j = 1, \dots, k$ and $k_j < \infty$ for $j > k$.*

Proof. Assuming $k > 0$, there is at least one coordinate with $\dim \Gamma/H_j = 1$: if not, $H_j > T^n$ for all j's and hence $H > T^n$ with $|\Gamma/H| < \infty$. Denote by z_1 this coordinate, then $\Gamma/H = (\Gamma/H_1)(H_1/H)$, with $\dim H_1/H = k - 1$. If H_1/H is a finite group, i.e., $k = 1$, then one has a decomposition into finite groups with $\tilde{k}_j < \infty$ for $j > 1$. On the other hand, if $k > 1$, then, by repeating the above argument, one has a coordinate z_2 with $H_1/\tilde{H}_2$ of dimension 1. □

The following result will be used very often in the book.

Lemma 2.4. *Let T^n act on $V = \mathbb{C}^m$ via $\exp i\langle N^j, \Phi\rangle$, $j = 1, \dots, m$. Let A be the $m \times n$ matrix with N^j as its j-th row. Then:*

(a) *$\dim \Gamma/H = k$ if and only if A has rank k.*

(b) *Assuming $k_j = \infty$ for $j = 1, \dots, k$ and that the $k \times k$ matrix B with $B_{ij} = n^i_j$, $1 \le i, j \le k$, is invertible, then one may write $A\Phi = \binom{B}{D}\tilde{\Psi}$, with $\tilde{\Psi} = \tilde{\Phi} + \Lambda\hat{\Phi}$, where $\Phi^T = (\tilde{\Phi}^T, \hat{\Phi}^T)$ and $\tilde{\Phi}^T = (\varphi_1, \dots, \varphi_k)$.*

(c) *With the same hypothesis, there is an action of T^k on $\mathbb{C}^m$, generated by $\Psi^T = (\Psi_1, \dots, \Psi_k)$ such that $\langle N^j, \Phi\rangle = \langle M^j, \Psi\rangle$, with $M^j = \langle m^j_1, \dots, m^j_k\rangle$ such that $m^j_l = \delta_{lj} M_j$ for $j = 1, \dots, k$, i.e., the action of T^k on the first k coordinates reduces to $e^{iM_j\Psi_j}$.*

Proof. (a) The relation $\langle N^j, \Phi\rangle \equiv 0, [2\pi]$ gives parallel hyperplanes in $\mathbb{R}^n$ with normal N^j. Thus, $\dim H = n - k$ is equivalent to $\dim \ker A = n - k$.

(b) Write $A = \begin{pmatrix} B & C \\ D & E \end{pmatrix}$ and let $\Lambda = B^{-1}C$. Then, $A\Phi = 0$ means $\tilde{\Phi} = -\Lambda\hat{\Phi}$ and $(E - D\Lambda)\hat{\Phi} = 0$. Since $\dim \ker A = n - k$, one has $E = D\Lambda$, $\ker A = \langle -\Lambda\hat{\Phi}, \hat{\Phi}\rangle$ and $A\Phi$ has the form given in the lemma.

(c) Let M be a $k \times k$ diagonal matrix such that $B^{-1}M$ has integer entries. Define $\Psi = M^{-1}B\tilde{\Psi}$. Then, $A\Phi = \binom{B}{D}B^{-1}M\Psi = \binom{M\Psi}{DB^{-1}M\Psi}$ gives the action of T^k, once one has noticed that the entries of $DB^{-1}M$ are integers. □

Another simple but useful observation is the following

Lemma 2.5. *Let T^n act on V as before. Then there is a morphism $S^1 \to T^n$ given by $\varphi_j = M_j\varphi$, M_j integers, such that $\langle N^j, M\rangle \not\equiv 0, [2\pi]$, unless $N^j = 0$ and $V^{S^1} = V^{T^n}$. The vector M is $(M_1, \ldots, M_n)^T$.*

Proof. As before, the congruences $\langle N^j, \Phi\rangle \equiv 0, [2\pi]$ give families of hyperplanes with normal N^j, if this vector is nonzero. From the denseness of $\mathbb{Q}$ in $\mathbb{R}$ it is clear that one may find integers $(M_1, \ldots, M_n)$ such that the direction $\{\varphi_j = M_j\varphi\}$ is not on any of the hyperplanes $\langle N^j, \Phi\rangle = 0$, for $j = 1, \ldots, m$. Thus, $\sum n_l^j M_l \neq 0$ and, being an integer, this number cannot be another multiple of 2π, unless $N^j = 0$ and the corresponding coordinate is in V^{T^n}. □

Definition 2.1. Let K be a subgroup of Γ (not necessarily an isotropy subgroup) and let $H = \bigcap \Gamma e_j \supset K$, where $\{e_j\}$ span V^K. We shall call H the *isotropy subgroup* of V^K. Note that $K < \Gamma_{e_j}$ and that $V^H = V^K$.

A final technical result is the following:

Lemma 2.6. *Let H be an isotropy subgroup with* $\dim W(H) = k$. *Then there are two isotropy subgroups $\underline{H}$ and $\bar{H}$, both with Weyl group of dimension k, such that $\underline{H} < H < \bar{H}$. The group $\bar{H}$ is maximal among such subgroups and $\underline{H}$ is the unique minimal such subgroup. $\underline{H}$ will be called the* **torus part** *of H.*

Proof. Let $\bar{H}$ be such a maximal element, for example given by $H_1 \cap \cdots \cap H_k$ as in Lemma 2.3. Then, $\Gamma/H = (\Gamma/\bar{H})(\bar{H}/H)$ and $\bar{H}/H$ is a finite group. If $H = T^{n-k} \times \mathbb{Z}_{n_1} \times \cdots \times \mathbb{Z}_{n_l}$, then, from Lemma 2.1 applied to $\bar{H}$, one has that $V^{T^{n-k}}$ is the linear space of all points with $W(H_x)$ finite. If $\underline{H}$ is the isotropy subgroup of $V^{T^{n-k}}$, then, since V^H is contained in $V^{T^{n-k}}$, one has that $\underline{H}$ is a subgroup of H and contains T^{n-k} (from Definition 2.1) and is clearly unique. □

Remark 2.1. If A is the matrix generated by the action of T^n on V and A^H its restriction on V^H (as in Lemma 2.4), then A^H and $A^{\underline{H}}$ have rank k. Furthermore, from Lemma 2.4 (b), $A^{\underline{H}}\Phi = \binom{B}{D^{\underline{H}}}\tilde{\Psi}$ with $\tilde{\Psi} = \tilde{\Phi} + \Lambda\hat{\Phi}$ and the torus part corresponds to $\tilde{\Psi} \equiv 0$. It is easy to see that on $V^{\underline{H}}$ one has exactly $n_i^j = \sum_{l=1}^k n_l^j \lambda_i^l$ for $i > k$ and $j = 1, \ldots, \dim V^{\underline{H}}$, where $\lambda_i^l, l = 1, \ldots, k, i = k+1, \ldots, n$ are the elements of the $k \times (n-k)$ matrix Λ.

1.3 Equivariant maps

A look at the heading of this book tells us that perhaps it is time to get started with some formal definitions.

Definition 3.1. Let E be a Γ-space. A subset Ω of E is said to be Γ-*invariant* if for any x in Ω, the orbit $\Gamma(x)$ is contained in Ω.

Definition 3.2. If B and E are Γ-spaces, with actions denoted by γ and $\tilde{\gamma}$ respectively, then a map $f : B \to E$ is said to be Γ-*equivariant* if

$$f(\gamma x) = \tilde{\gamma} f(x)$$

for all x in B.

Definition 3.3. Let Γ act trivially on E. A map $f : B \to E$ is said to be Γ-*invariant* if $f(\gamma x) = f(x)$, for all $x \in B$.

Example 3.1. Let $\mathbb{Z}_2$ act on $B = E$ as $-I$, then an odd map, $f(-x) = -f(x)$, is $\mathbb{Z}_2$-equivariant. On the other hand, an even map, $f(-x) = f(x)$, with a trivial action on E is Γ-invariant. In general, if $B = B^{\mathbb{Z}_2} \oplus B_1$, $E = E^{\mathbb{Z}_2} \oplus E_1$, with an action of $\mathbb{Z}_2$ as $-I$ on B_1 and E_1, then an equivariant map $f(x_0, x_1) = (f_0, f_1)(x_0, x_1)$, will have the property that $f_0(x_0, -x_1) = f_0(x_0, x_1)$ and $f_1(x_0, -x_1) = -f_1(x_0, x_1)$. In particular, $f_1(x_0, 0) = 0$, that is, f maps $B^{\mathbb{Z}_2}$ into $E^{\mathbb{Z}_2}$. We shall see below that this is a general property of equivariant maps.

Example 3.2. Let $C^0_{2\pi}(\mathbb{R}^N)$, respectively $C^1_{2\pi}(\mathbb{R}^N)$, be the space of continuous, respectively differentiable, 2π-periodic functions $X(t)$ in $\mathbb{R}^N$, with the action $\rho(\varphi)X(t) = X(t + \varphi)$. Let $f(X)$ be a continuous vector field on $\mathbb{R}^N$, independent of t. Then

$$F(X) = \frac{dX}{dt} - f(X)$$

is S^1-equivariant.

In terms of Fourier series, $X(t) = \sum X_n e^{int}$ with $X_{-n} = \bar{X}_n$, one has the equivalent formulation

$$inX_n - f_n(X_0, X_1, \dots), \quad n = 0, 1, 2, \dots,$$

with $f_n(X_0, X_1, \dots) = \frac{1}{2\pi} \int_0^{2\pi} f(X(t))e^{-int} dt$. In this case the action of S^1 on X_n is given by $e^{in\varphi} X_n$, and it is an easy exercise of change of variables to see that $f_n(X_0, e^{i\varphi} X_1, e^{i2\varphi} X_2, \dots) = e^{in\varphi} f_n(X_0, X_1, X_2, \dots)$, i.e., that the map F is equivariant.

Note that the isotropy group of X_n is the set $H = \{\varphi = 2k\pi/n,\ k = 0, \dots, n-1\} \cong \mathbb{Z}_n$ and that $V^H = \{X_m,\ m = 0 \text{ or a multiple of } n\}$.

Example 3.3. Let Γ_0 be a group acting on $\mathbb{R}^N$ and let $f(\gamma_0 X) = \gamma_0 f(X)$ be a Γ_0-equivariant vector field.

If $\Gamma = S^1 \times \Gamma_0$ one may consider the Γ-equivariant map

$$F(X) = \frac{dX}{dt} - f(X)$$

on the space of 2π-periodic functions in $\mathbb{R}^N$. If H is the isotropy subgroup of a Fourier component X_n, then the space V^H of "twisted orbits" has an interesting description given in the last section of this chapter.

We are going now to describe some of the simplest consequences of the equivariance.

Property 3.1 (Orbits of zeros). *If $f(\gamma x) = \tilde{\gamma} f(x)$ and $f(x_0) = 0$, then $f(\gamma x_0) = 0$, for all γ in Γ.*

Property 3.2 (Stratification of the space). *If $f : B \to E$ is Γ-equivariant, then if $H < \Gamma$, f maps B^H into E^H. The map $f^H \equiv f|_{B^H}$ is $N(H)$-equivariant.*

Proof. For x in B^H and γ in H, one has $f(\gamma x) = f(x) = \tilde{\gamma} f(x)$. Hence, $f(x)$ is fixed by H, i.e., it belongs to E^H. Now, since $N(H)$ is the largest group which keeps B^H invariant, this implies that γx is in B^H for γ in $N(H)$ and x in B^H, and the remaining part of the statement follows. □

Note that, in particular, if Γ is abelian, then f^H is Γ-equivariant. This simple property implies that one may try to study f by looking for zeros with a given symmetry (for example, radial solutions). It is then convenient to reduce the study to the smallest possible B^H, i.e., the largest H, in particular to maximal isotropy subgroups, where one knows that $W(H)$ acts freely on B^H and which are completely classified. If, furthermore, one decomposes B^H into irreducible representations of $W(H)$ (see Section 5), one may determine, not only the linear terms, but also higher order terms in the Taylor series expansion, if the number of representations is small. These ideas have been used extensively, in particular in the physics literature, in order to give normal form expansions. The information obtained this way is very precise but, from the requirements of genericity and low dimension, it does not allow for a complete study of stability, symmetry breaking or period doubling, when one has to consider perturbations with a symmetry different from the one for the given solutions. Hence, in these cases, it is convenient not to fix a priori the symmetry of the solution and to treat the complete equivariant problem. *Then one will have a more general vision, but probably less precise. This is the point of view adopted in this book.*

Property 3.3 (Linearization). *If $f(\gamma x) = \tilde{\gamma} f(x)$ and f is C^1 at x_0, with $\Gamma_{x_0} = H$, then*

$$Df(\gamma x_0)\gamma = \tilde{\gamma} Df(x_0),$$

for all γ in Γ. In particular, $Df(x_0)$ is H-equivariant.

Proof. Since $f(\gamma x_0 + \gamma x) - f(\gamma x_0) = \tilde{\gamma}(f(x_0 + x) - f(x_0)) = \tilde{\gamma} Df(x_0)x + \cdots$, one has that f is linearizable at γx_0 and the above formula holds. □

This implies, if $B = E = \mathbb{R}^N$, that $Df(\gamma x_0)$ is conjugate to $Df(x_0)$ with the same determinant.

On the other hand, if the dimension of the orbit of x_0 is positive, i.e., if $\dim \Gamma/H = k$ with $H = \Gamma_{x_0}$, then one may choose a differentiable path $\gamma(t)$, with $\gamma(0) = I, \gamma'(0) \neq 0$, such that $f(\gamma(t)x_0) = f(x_0)$. Differentiating with respect to t and evaluating at $t = 0$, one has

$$Df(x_0)\gamma'(0)x_0 = 0.$$

Hence, $\gamma'(0)x_0$ is in the kernel of $Df(x_0)$, for each direction $\gamma'(0)$ such that $\gamma'(0)x_0 \neq 0$. Since the orbit is a differentiable manifold, this will be true for any direction tangent to the orbit. Hence one has at least a k-dimensional kernel. For example, if Γ is abelian and T^n acts, as in Example 1.4, by $\exp i\langle N^j, \Phi\rangle$, then one may take $\gamma(t) = (0, \dots, t, 0, \dots)$ i.e., $\varphi_j = 0$ except $\varphi_l = t$. In this case, $\gamma'(0)x_0$ is $i(n_l^1 x_1, \dots, n_l^m x_m)^T$.

A property which will be used frequently in this book is the following:

Property 3.4 (Diagonal structure). *If $B = B^H \oplus B_\perp$, $E = E^H \oplus E_\perp$ with $B_\perp$ and $E_\perp$ being $N(H)$-topological complements and $f = f^H \oplus f_\perp$, then at any x_H in B^H one has*

$$Df(x_H) = \begin{pmatrix} D_H f^H & 0 \\ 0 & D_\perp f_\perp \end{pmatrix},$$

where $x = x_H \oplus x_\perp$ and D_H, $D_\perp$ stand for differentials with respect to these variables.

Proof. One has that

$$Df(x_H) = \begin{pmatrix} D_H f^H & D_\perp f^H \\ D_H f_\perp & D_\perp f_\perp \end{pmatrix}.$$

From the fact that $f_\perp(x_H) = 0$, one has $D_H f_\perp(x_H) = 0$. Since the decomposition of B and E is $N(H)$-invariant (hence H-invariant), the action of H on these spaces is diagonal. The H-equivariance of $Df(x_H)$ implies that $D_\perp f_\perp \gamma = \tilde{\gamma} D_\perp f_\perp$, and $D_\perp f^H = D_\perp f^H \gamma$ for any γ in H. Let A denote $D_\perp f^H$, then, since $A\gamma = A$, one has that $\ker A$ is a closed H-invariant subspace of $B_\perp$. Assume there is $x_\perp$ with $Ax_\perp \neq 0$. Let V be the subspace of $B_\perp$ generated by $x_\perp$ and $\ker A$. Defining z by the relation $\gamma x_\perp = x_\perp + z(\gamma)$, one has that z is in $\ker A$ and for any $x = ax_\perp \oplus y$ in V (i.e., y belongs to $\ker A$) one gets $\gamma x = ax_\perp + az(\gamma) + y$, proving that V is also an H-invariant subspace, with $\ker A$ as a one-codimensional subspace. This implies (see any book on Functional Analysis) that there is a continuous projection P from V onto $\ker A$. As a matter of fact, we shall prove below (in Lemma 4.4.) that one may take P to be equivariant. Then, if $\tilde{x}_\perp = (I - P)x_\perp$, one has $A\tilde{x}_\perp = Ax_\perp$ (since $Px_\perp$ belongs to $\ker A$) and $\gamma\tilde{x}_\perp = (I - P)\gamma x_\perp$, from the equivariance of P, and $\gamma\tilde{x}_\perp = k(\gamma)\tilde{x}_\perp$ since $(I - P)V$ is one-dimensional. Applying A to this relation, one obtains $k(\gamma) = 1$ and $\tilde{x}_\perp$ is fixed by H, i.e., $\tilde{x}_\perp$ belongs to $B^H \cap B_\perp = \{0\}$, a contradiction. Hence $A = 0$. □

For the last property of this section, we shall assume that E is a Γ-Hilbert space and the action of Γ is via orthogonal operators, i.e., $\gamma^T\gamma = I$ (in finite dimensional spaces one may always redefine the scalar product in such a way that the representation turns out to be orthogonal: see below, Lemma 5.1).

Property 3.5 (Gradients). *If $J : E \to \mathbb{R}$ is a C^1, Γ-invariant functional, then $f(x) = \nabla J(x)$ is equivariant.*

Proof. Since $J(\gamma x) = J(x)$, one has, from Property 3.3, that $DJ(\gamma x)\gamma = DJ(x)$, since the action $\tilde{\gamma}$ on $\mathbb{R}$ is trivial. But, $DJ(x) = \nabla J(x)^T$, hence $\nabla J(x) = \gamma DJ^T(\gamma x)$, giving the result. □

Remark 3.1. If Γ has positive dimension and one takes a path $\gamma(t)$ with $\gamma(0) = I$, then, differentiating the identity $J(\gamma(t)x) = J(x)$, one obtains

$$\nabla J(x) \cdot \dot{\gamma}(0)x = 0,$$

that is $\dot{\gamma}(0)x$ is orthogonal to the field $\nabla J(x) = f(x)$. If one looks for critical points of J, i.e., such that $\nabla J(x) = 0$, this orthogonality may be regarded as a reduction in the number of "free" equations. From the analytical point of view, one may use some analogue of the Implicit Function Theorem and reduce the number of variables. Or, one may use, as in conditioned variational problems, a "Lagrange multiplier", i.e., one may add a new variable μ and look for zeros of the equation

$$f(x) + \mu\dot{\gamma}(0)x = 0.$$

In fact, if $f(x) = 0$, then $\mu = 0$ gives a solution of the above equation. Conversely, if (μ, x) is a solution, then by taking the scalar product with $\dot{\gamma}(0)x$, one has $\mu\|\dot{\gamma}(0)x\|^2 = 0$, hence $f(x) = 0$ and $\mu\dot{\gamma}(0)x = 0$, in particular $\mu = 0$ if $\dot{\gamma}(0)x \neq 0$.

This argument can be repeated for each subgroup $\gamma(t)$ and one obtains $\dot{\gamma}_j(0)$ for $j = 1, \ldots, \dim \Gamma$. Considering the equation

$$f(x) + \sum \mu_j \dot{\gamma}_j(0)x = 0,$$

one obtains a problem with several parameters. A solution of this problem will give that

$$\text{(a)}\ \ f(x) = 0 \quad \text{and} \quad \text{(b)}\ \ \sum \mu_j \dot{\gamma}_j(0)x = 0.$$

One will conclude that $\mu_j = 0$ if $\dot{\gamma}_j(0)x$ are linearly independent. This will depend on the isotropy subgroup of x. This point of view will be taken when studying orthogonal maps (see § 7).

1.4 Averaging

At this stage the reader may be puzzled why we insist on working with compact Lie groups. As a matter of fact, up to now, the compactness of the Lie group Γ was not used in our considerations and seems to bear only a decorative aspect in the whole business. Almost the same can be said about linear actions. Now, the consistency of these two features namely, compactness of Γ and linearity of the actions, becomes evident when you realize that, under these two conditions, a powerful instrument is at hand. Precisely, the existence of an integration on Γ, the *Haar integral*, such that $\int_\Gamma d\gamma = 1$, which is Γ-invariant on the class of continuous real-valued functions g on Γ, under both left and right actions, i.e.,

$$\int_\Gamma g(\gamma'^{-1}\gamma)\,d\gamma = \int_\Gamma g(\gamma)\,d\gamma = \int_\Gamma g(\gamma\gamma')\,d\gamma.$$

The first important consequence of this fact is that, provided E is a Banach Γ-space, one may define a new norm, say

$$|||x||| = \int_\Gamma \|\gamma x\|\,d\gamma,$$

satisfying, $|||\gamma' x||| = |||x|||$, i.e., the action of Γ is an isometry.

This allows us to assume *in the rest of the book* that the *action is an isometry*. In particular, the ball

$$B_R = \{x : \|x\| < R\} \text{ is } \Gamma\text{-invariant.}$$

Using *Pettis integrals* and standard averaging, one has the following remarkable result.

Lemma 4.1 (Gleason's Lemma). *If B and E are Γ-spaces and $f(x)$ is a continuous map from B into E, then*

$$\tilde{f}(x) \equiv \int_\Gamma f(\gamma x)\,d\tilde{\gamma} \text{ is } \Gamma\text{-invariant}$$

and

$$\hat{f}(x) \equiv \int_\Gamma \tilde{\gamma}^{-1} f(\gamma x)\,d\tilde{\gamma} \text{ is } \Gamma\text{-equivariant.}$$

Furthermore, if f is compact, then so are $\tilde{f}$ and $\hat{f}$.

Proof. From the change of variables $\gamma\gamma'$, one has

$$\tilde{f}(\gamma' x) = \int_\Gamma f(\gamma\gamma' x)\,d\tilde{\gamma} = \int_\Gamma f(\gamma'' x)\,d\gamma'' = f(x).$$

Also, $\hat{f}(\gamma' x) = \int_\Gamma \tilde{\gamma}^{-1} f(\gamma\gamma' x)\,d\tilde{\gamma} = \tilde{\gamma}' \int_\Gamma (\tilde{\gamma}\tilde{\gamma}')^{-1} f(\gamma\gamma' x)\,d\tilde{\gamma} = \tilde{\gamma}' \hat{f}(x)$, under the same change of variables. See [Br. p. 36].

The continuity of $\tilde{f}$ and $\hat{f}$ follows from the compactness of Γ. In fact, the orbit Γx_0 is compact and hence f is uniformly continuous on it. Moreover, if x is close to x_0 (therefore, the orbit Γx is close to Γx_0, taking into account that the action is an isometry), one gets

$$\hat{f}(x) - \hat{f}(x_0) = \int_\Gamma \tilde{\gamma}^{-1}(f(\gamma x) - f(\gamma x_0))\, d\tilde{\gamma}.$$

Also,

$$\|\hat{f}(x) - \hat{f}(x_0)\| \leq \max_\Gamma \|f(\gamma x) - f(\gamma x_0)\|.$$

As far as compactness is concerned, recall that f is said to be *compact* if it is continuous and if $\overline{f(K)}$ is compact, for any bounded set K in B.

Therefore, the sets $\tilde{A} \equiv \bigcup_{\Gamma\times K} f(\gamma x)$ and $\hat{A} \equiv \bigcup_{\Gamma\times K} \tilde{\gamma}^{-1} f(\gamma x)$ are precompact. In fact, if you have a sequence $\{\tilde{\gamma}_n^{-1} f(\gamma_n x_n)\}$ in $\hat{A}$ then, by the compactness of Γ, you get a subsequence $\{\gamma_{n_j}\}$ converging to some γ and $\{f(\gamma_{n_j} x_{n_j})\}$, converging to some y. Thus,

$$\tilde{\gamma}_{n_j}^{-1} f(\gamma_{n_j} x_{n_j}) - \tilde{\gamma}^{-1} y = (\tilde{\gamma}_{n_j}^{-1} - \tilde{\gamma}^{-1}) f(\gamma_{n_j} x_{n_j}) + \tilde{\gamma}^{-1}(f(\gamma_{n_j} x_{n_j}) - y)$$

yields the convergence, since $\|\tilde{\gamma}_{n_j}^{-1} - \tilde{\gamma}^{-1}\|$ tends to 0, as operators, and, since ΓK is bounded, $\tilde{A}$ is compact and $\hat{A}$ is bounded.

Now, cover $\tilde{A}$ and $\hat{A}$ with balls of radius $1/2^{N+1}$ and extract a *finite* subcover based at $f(\gamma_j x_j)$, $j = 1, \ldots, k$, and $\tilde{\gamma}_l^{-1} f(\gamma_l x_l)$, $l = 1, \ldots, r$, respectively. Let $\{\varphi_j\}$ be a partition of unity associated to the covering, i.e., $\varphi_j : E \to [0, 1]$, with support in a ball centered at $y_j \equiv f(\gamma_j x_j)$, respectively $\tilde{\gamma}_j^{-1} f(\gamma_j x_j)$, of radius $1/2^N$ and such that $\sum \varphi_j(y) = 1$.

Define,

$$\tilde{f}_N(x) = \int_\Gamma \sum \varphi_j(f(\gamma x)) f(\gamma_j x_j)\, d\tilde{\gamma},$$

$$\hat{f}_N(x) = \int_\Gamma \sum \varphi_l(\tilde{\gamma}^{-1} f(\gamma x)) \tilde{\gamma}_l^{-1} f(\gamma_l x_l)\, d\tilde{\gamma}.$$

Then, $\tilde{f}_N(x)$ belongs to the space generated by $\{f(\gamma_j x_j)\}$, while $\hat{f}_N(x)$ belongs to the finite dimensional space generated by $\{\tilde{\gamma}_l^{-1} f(\gamma_l x_l)\}$. Hence, $\tilde{f}_N(K)$ and $\hat{f}_N(K)$ are precompact. Furthermore,

$$\tilde{f}(x) - \tilde{f}_N(x) = \int_\Gamma \sum \varphi_j(f(\gamma x))(f(\gamma x) - f(\gamma_j x_j))\, d\tilde{\gamma},$$

$$\hat{f}(x) - \hat{f}_N(x) = \int_\Gamma \sum \varphi_l(\tilde{\gamma}^{-1} f(\gamma x))(\tilde{\gamma}^{-1} f(\gamma x) - \tilde{\gamma}_l^{-1} f(\gamma_l x_l))\, d\tilde{\gamma}.$$

Now, since $\varphi_j(y)$ is non-zero only if $\|y - y_j\| < 1/2^N$ and $\sum \varphi_j(y) = 1$, one gets

$$\|\tilde{f}(x) - \tilde{f}_N(x)\| \leq 1/2^N \quad \text{and} \quad \|\hat{f}(x) - \hat{f}_N(x)\| \leq 1/2^N.$$

But then, for any bounded sequence $\{x_n\}$, one has a subsequence $\{x_{n(N)}\}$ such that $\tilde{f}_N(x_{n(N)})$, respectively $\hat{f}_N(x_{n(N)})$, is convergent. Using a Cantor diagonal process, one obtains, due to the uniform approximation of $\tilde{f}(x)$ by $\tilde{f}_N(x)$, respectively of $\hat{f}(x)$ by $\hat{f}_N(x)$, a convergent subsequence for $\tilde{f}(x_{N(N)})$, respectively $\hat{f}(x_{N(N)})$. □

Remark 4.1. If $f(\gamma x) = f(x)$, then $\tilde{f}(x) = f(x)$, while, if $f(\gamma x) = \tilde{\gamma} f(x)$, then $\hat{f}(x) = f(x)$.

Example 4.1. If $\Gamma = S^1$ acts on $C^0_{2\pi}(\mathbb{R})$ via time translation as in Example 1.5, and $f(t, x)$ is continuous and 2π-periodic in t, then f induces a mapping from $C^0_{2\pi}(\mathbb{R})$ into itself, via $f(t, x(t))$. Then

$$\tilde{f}(x(t)) = (1/2\pi)\int_0^{2\pi} f(t, x(t+\varphi))\,d\varphi = (1/2\pi)\int_0^{2\pi} f(t, x(\varphi))\,d\varphi,$$

$$\hat{f}(x(t)) = (1/2\pi)\int_0^{2\pi} f(t-\varphi, x(t))\,d\varphi.$$

Example 4.2. If $\Gamma = \mathbb{Z}_m$ is generated by γ_0, then

$$\int_\Gamma g(\gamma)\,d\gamma = (1/m)\sum_0^{m-1} g(\gamma^j).$$

Remark 4.2. In the proof of the compactness of $\hat{f}(x)$ and $\tilde{f}(x)$, we have seen that a map f is compact if and only if it can be uniformly approximated on bounded sets by finite dimensional maps. The reader may recover this important result by forgetting the action of Γ. Now, for the case of a non-trivial action of Γ on E, a word of caution is necessary: The map $\tilde{f}_N(x)$ is invariant and belongs to a finite dimensional subspace. However, $\hat{f}_N(x)$ is *not* equivariant. One could have tried to use the set $\tilde{A}$ also for this case and define

$$\hat{f}'_N(x) = \int_\Gamma \sum \varphi_j(f(\gamma x))\tilde{\gamma}^{-1} f(\gamma_j x_j)\,d\tilde{\gamma}$$

which is Γ-equivariant and approximates, within $1/2^N$ on K, the map $\hat{f}(x)$, but which is *not* necessarily finite dimensional, as the following example shows, since the orbit of $f(\gamma_j x_j)$ may *not* span a finite dimensional space.

Example 4.3. On $l_2 = \{(x_0, x_1, x_2, \dots), x_0 \in \mathbb{R}, x_j \in \mathbb{C}$ for $j \geq 1$ with $\sum |x_j|^2 < \infty\}$, consider the action of S^1 given by

$$e^{i\varphi}(x_0, x_1, x_2, \dots) \equiv (x_0, e^{i\varphi}x_1, e^{2i\varphi}x_2, \dots).$$

Consider the point $x_0 = (1, 1/2, 1/2^2, \dots, 1/2^n, \dots) = (a^0, a, a^2, a^3, \dots)$. Then, for any n, $e^{i\varphi_1}x_0, \dots, e^{i\varphi_n}x_0$, for $\varphi_1, \dots, \varphi_n$ different, are linearly independent. In fact, taking the first n components, one obtains a Van der Monde matrix, with

j-th row equal to $(1, a_j, a_j^2, \dots a_j^{n-1})$, where $a_j = e^{i\varphi_j}/2$ and determinant equal to $\prod_{i>j}(a_i - a_j)$. Hence, the closure of the linear space generated by the orbit of x_0 is l_2.

However, we will show in the next section that the set of points in E whose orbit is contained in a finite dimensional Γ-invariant subspace is dense in E. Thus, in the definition of $\hat{f}'_N$ take y_j such that $\Gamma y_j \subset M_j$, a finite dimensional Γ-invariant subspace, with $\|y_j - f(\gamma_j x_j)\| \le 1/2^N$, and define

$$\hat{f}''_N(x) = \int_\Gamma \sum \varphi_j(\gamma x)\tilde{\gamma}^{-1} y_j \, d\tilde{\gamma}.$$

Thus, since $\tilde{\gamma}^{-1} y_j \subset M_j$, the Γ-map $\hat{f}''_N$ has range in the finite dimensional Γ-invariant subspace generated by the M_j's and $\|\hat{f}(x) - \hat{f}''_N(x)\| \le 1/2^{N-1}$.

We have thus proved the following result, which will be crucial for the extension of the Γ-degree to the infinite dimensional setting.

Theorem 4.1. *A continuous Γ-equivariant map f from B into E is compact if and only if, for each bounded subset K of B, there is a sequence of Γ-equivariant maps f_N, with range in a finite dimensional Γ-invariant subspace M_N of E, such that, for all x in K, one has*

$$\|f(x) - f_N(x)\| \le 1/2^N.$$

In our construction of the Γ-degree, we shall also need the following consequences of averaging:

Lemma 4.2 (Invariant Uryson functions). *If A and B are closed Γ-invariant subsets of E, with $A \cap B = \phi$, then there is a continuous Γ-invariant function $\tilde{\varphi} : E \to [0, 1]$, with $\tilde{\varphi}(x) = 0$ if $x \in A$ and $\tilde{\varphi}(x) = 1$ if $x \in B$.*

Proof. Indeed, let φ be any Uryson function relative to A and B (for instance $\operatorname{dist}(x, A)/(\operatorname{dist}(x, A) + \operatorname{dist}(x, B))$), then

$$\tilde{\varphi}(x) = \int_\Gamma \varphi(\gamma x)\, d\gamma$$

has the required properties. Note that, if one has renormed E in such a way that the action is an isometry, then $\operatorname{dist}(x, A) = \operatorname{dist}(\gamma x, A)$ and $\tilde{\varphi}(x)$ can be chosen to be the above map. □

Lemma 4.3 (Invariant neighborhood). *If $A \subset E$ is a Γ-invariant closed set and U, containing A, is an open, Γ-invariant set, then there is a Γ-invariant open subset V such that $A \subset V \subset \bar{V} \subset U$.*

Proof. In fact, let $\tilde{\varphi} : E \to [0, 1]$ be a Γ-invariant Uryson function with $\tilde{\varphi}|_A = 0$ and $\tilde{\varphi}|_{U^C} = 1$. Then, $V = \varphi^{-1}([0, 1/2))$ has the required properties. □

Lemma 4.4 (Equivariant projections). *If E_0 is a closed Γ-invariant subspace of E and P is a continuous projection from E onto E_0, then*

$$\tilde{P}x \equiv \int_\Gamma \gamma^{-1} P \gamma x \, d\gamma$$

is a Γ-equivariant projection onto E_0. If $E_0 = E^\Gamma$, then

$$\bar{P}x \equiv \int_\Gamma \gamma x \, d\gamma$$

is a Γ-invariant projection onto E^Γ. Moreover, $E_1 \equiv (I - \tilde{P})E$ and $(I - \bar{P})E$ are closed Γ-invariant complements of E_0 and E^Γ.

Proof. The first part is clear since $\int_\Gamma d\gamma = 1$ and E_0 is Γ-invariant. As far as the second part is concerned, notice that $\bar{P}x$ is in E^Γ and $\bar{P}x = x$ for x in E^Γ. □

1.5 Irreducible representations

A good deal of this book is based on the decomposition of finite dimensional representations into irreducible subrepresentations and the corresponding form of linear equivariant maps.

Definition 5.1. Two representations of B and E are *equivalent* if there is a continuous linear invertible operator T from B onto E such that $\tilde{\gamma}T = T\gamma$.

Lemma 5.1. *Every finite dimensional representation is equivalent to an orthogonal representation, i.e., with $\tilde{\gamma}$ in $O(n)$.*

Proof. In fact, the bilinear form

$$B(x, y) = \int_\Gamma (\gamma x, \gamma y) \, d\gamma$$

is positive definite, symmetric and invariant. Hence, there is a positive definite matrix A such that $B(x, y) = (Ax, y)$. One may define a positive symmetric matrix T such that $T^2 = A$, by diagonalizing A. Hence $B(x, y) = (Tx, Ty)$. Since $B(\gamma x, \gamma y) = B(x, y)$, one has that $(T\gamma T^{-1}x, T\gamma T^{-1}y) = B(\gamma T^{-1}x, \gamma T^{-1}y) = (x, y)$, which implies that $T\gamma T^{-1}$ is in $O(n)$. □

Remark 5.1. The same result is true in any Hilbert space. The existence of the self-adjoint bounded positive operator A follows from Riesz Lemma and that of T from the spectral decomposition of A.

Definition 5.2. A representation E of Γ is said to be *irreducible* if E has no proper invariant subspace (not necessarily closed).

This implies that $E^\Gamma = \{0\}$ unless Γ acts trivially on E and dim $E = 1$.

Definition 5.3. A *subrepresentation* E_0 of Γ in E is a closed proper invariant subspace E_0 of E.

Lemma 5.2. *If E is a finite dimensional representation of Γ, then there are irreducible subrepresentations $E_1, \ldots, E_k$, such that $E = E_1 \oplus \cdots \oplus E_k$.*

Proof. From Lemma 5.1 it is enough to consider the case where the representation is orthogonal. Then, if E_1 is Γ-invariant, the orthogonal complement $E_1^\perp$ is also Γ-invariant, since $(\gamma x, y) = (x, \gamma^T y) = (x, \gamma^{-1} y)$. Hence, if $x \in E_1^\perp$ and y is in E_1 (hence also $\gamma^{-1} y \in E_1$), this scalar product is 0 and γx is in $E_1^\perp$. Applying this argument a finite number of times one obtains a complete reduction of E. □

The above arguments can be extended to the infinite dimensional setting in the following form.

Lemma 5.3. (a) *If E_0 is an invariant subspace of the representation E, then $\bar{E}_0$ is a subrepresentation. If furthermore E is a Hilbert space, then $E = \bar{E}_0 \oplus E_1$, where E_1 is also a subrepresentation.*

(b) *If E is an orthogonal representation (hence E is Hilbert) and E_0 is an invariant subspace, then $E_0^\perp$ is a subrepresentation.*

Proof. (a) If $\{x_n\}$ in E_0 converges to x, then $\{\gamma x_n\}$, which is in E_0, converges to γx and $\bar{E}_0$ is invariant. The second part follows from Lemma 4.4, since there is always a projection on $\bar{E}_0$.

(b) follows from the argument used in Lemma 5.2 and the fact that $E_0^\perp$ is closed. □

Lemma 5.4 (Schur's Lemma). *If B and E are irreducible representations of Γ and there is a linear equivariant map A from B into E, such that $A\gamma = \tilde{\gamma} A$ for all γ in Γ, then either $A = 0$, or A is invertible.*

Proof. Note first that the statement is purely algebraic and no topology is used. Since the domain of A is linear and Γ-invariant (so that the equivariance makes sense), one has that the domain of A is all of B. Furthermore, since ker A is Γ-invariant, then either it is B (and $A = 0$) or it reduces to $\{0\}$ and A is one-to-one. But then Range A which is also Γ-invariant and non-trivial (since $A \neq 0$) must be E. Hence A is also onto and invertible. □

Remark 5.2. If E is not irreducible, then either $A = 0$, or A is one-to-one and onto Range A. This last subspace is (algebraically) irreducible since A^{-1} is clearly equivariant.

Corollary 5.1. (a) *If E is an irreducible representation of Γ and A is a Γ-equivariant linear map from E into E, i.e., $A\gamma = \gamma A$ with a real eigenvalue λ, then $A = \lambda I$.*

(b) *If E has no proper subrepresentations and A is a bounded Γ-equivariant linear map with eigenvalue λ, then $A = \lambda I$. Any bounded Γ-equivariant linear map B is either 0 or one-to-one.*

(c) *If furthermore E is a Hilbert space with no proper subrepresentations and equivalent to an orthogonal representation of Γ (i.e., there is a continuous isomorphism T on E such that, if $\tilde{\gamma} \equiv T^{-1}\gamma T$, then $\tilde{\gamma}^T\tilde{\gamma} = I$), and A is a bounded Γ-equivariant linear map from E into E, then*

$$T^{-1}AT = \mu I + \nu B$$

with $B^2 = -I$, $B + B^T = 0$. Moreover, $T = I$ if the representation is orthogonal.

Proof. (a) In fact, $A - \lambda I$ is Γ-equivariant, with a non-trivial kernel, hence, from Schur's Lemma, it must be 0.

(b) Since $\ker(A - \lambda I)$ is closed, the previous argument gives the result. Similarly, if $\ker B \neq \{0\}$, then $B = 0$.

(c) One has $T^{-1}AT\tilde{\gamma} = T^{-1}A\gamma T = \tilde{\gamma}T^{-1}AT$, hence $T^{-1}AT$ is Γ-equivariant with respect to the orthogonal representation. Let $\tilde{A} = T^{-1}AT$, then $\tilde{A} + \tilde{A}^T$ and $\tilde{A}^T\tilde{A}$ are self-adjoint and equivariant. Hence, $2\mu = \pm\|\tilde{A} + \tilde{A}^T\|$ is an eigenvalue for $\tilde{A} + \tilde{A}^T$. From (b), one has $\tilde{A} + \tilde{A}^T = 2\mu I$ or, else $(\tilde{A} - \mu I) + (\tilde{A} - \mu I)^T = 0$. Furthermore, $(\tilde{A} - \mu I)^T(\tilde{A} - \mu I) = \nu^2 I$, since this operator is either positive, or identically 0 if it has a kernel (again from (b)). If $\nu = 0$, then $(\tilde{A} - \mu I)^2 = 0$ and $\tilde{A} - \mu I$ must have a non-trivial kernel, i.e., from (b), $\tilde{A} = \mu I$. On the other hand, if $\nu \neq 0$, let $B = (\tilde{A} - \mu I)/\nu$. Then, $B^T + B = 0$ and $B^T B = I$, i.e., $B^2 = -I$. □

Corollary 5.2. *If E is a finite dimensional irreducible representation of an abelian group Γ, then either $E \cong \mathbb{R}$ and Γ acts trivially or as $\mathbb{Z}_2$, or $E \cong \mathbb{C}$ and Γ acts as in* (1.4).

Proof. Since Γ is abelian, one has that $\tilde{\gamma}\tilde{\gamma}_1 = \tilde{\gamma}_1\tilde{\gamma}$, where $\tilde{\gamma}$ is the equivalent orthogonal representation given in the preceding corollary. Furthermore $\tilde{\gamma}$, a matrix, is Γ-equivariant, hence

$$\tilde{\gamma} = \mu I + \nu B,$$

where μ, ν, B depend on $\tilde{\gamma}$. Since $\tilde{\gamma}^T\tilde{\gamma} = I$ one has $\mu^2 + \nu^2 = 1$. If $\tilde{\gamma}_1 = \mu_1 I + \nu_1 B_1$ and $\tilde{\gamma}_2 = \mu_2 I + \nu_2 B_2$, from $\tilde{\gamma}_1\tilde{\gamma}_2 = \tilde{\gamma}_2\tilde{\gamma}_1$, one obtains, if $\nu_1\nu_2 \neq 0$, that $B_1B_2 = B_2B_1 \equiv B$. But then, $B^T = B$ and $B^2 = I$. From Schur's Lemma, the self-adjoint matrix B must be of the form λI, with $\lambda^2 = 1$. If $\lambda = 1$, then $B_1B_2 = I$ implies (by multiplying with B_1) that $B_2 = -B_1$ and then one may change ν_2 to $-\nu_2$. While, if $\lambda = -1$, then one obtains $B_2 = B_1$. That is, one has a unique B such that any $\tilde{\gamma}$ is written as $\mu I + \nu B$.

Now, if for all γ's the corresponding ν is 0, then $\tilde{\gamma} = \pm I$ (since $\mu^2 = 1$) and any one-dimensional subspace is invariant. Then $E = \mathbb{R}$ and Γ acts trivially if $\mu = 1$ for all γ, or Γ acts as $\mathbb{Z}_2$ if, for some γ, μ is -1.

On the other hand, if there is a non-zero ν, then from $B^2 = -I$, one has $(\det B)^2 = (-1)^{\dim E}$ and hence E is even-dimensional. Furthermore, if $e \neq 0$, then the subspace generated by e and Be is Γ-invariant and of dimension 2, since Be is orthogonal to $e : (e, Be) = (B^T e, e) = -(Be, e)$. Thus, from the irreducibility of E, one has that E is equal to this subspace. Take e of length 1 and define a complex structure by defining $Be = i$. Then, $\tilde{\gamma} = \mu + \nu i$, with $\mu^2 + \nu^2 = 1$, is a unit complex number. Remark 1.1 and the fact that any compact abelian group can be represented as a product, ends the proof. □

Remark 5.3. Another way of seeing the above argument is the following: $\tilde{\gamma}$, as an orthogonal real matrix, has two-dimensional invariant eigenspaces, where $\tilde{\gamma}$ acts as a rotation. Since $\tilde{\gamma}_1$ commutes with $\tilde{\gamma}$, these invariant subspaces are also invariant for $\tilde{\gamma}_1$. Hence, the action of Γ on this subspace can be written as $R_\varphi\binom{x}{y}$, where R_φ is a rotation by an angle φ. Writing $z = x + iy$, this vector can be identified with $e^{i\varphi} z$.

Clearly, we could have taken $\bar{z} = x - iy$. Then this action would have been $e^{-i\varphi}\bar{z}$. These two representations are equivalent as real representations, since the map $T\binom{x}{y} = \binom{x}{-y}$, corresponding to conjugation, is equivariant. Of course, they are not equivalent as complex representations.

The next set of results in this section will concern the fact that any irreducible representation (in the sense of our definition) of a compact Lie group is finite dimensional. We shall begin with the Hilbert space case.

Theorem 5.1. *If E is an orthogonal irreducible representation of Γ, with no proper subrepresentations, then E is finite dimensional. Furthermore, one has the equality*

$$\int_\Gamma ((\gamma x_1, y_1)(\gamma x_2, y_2) + (\gamma x_1, y_2)(\gamma x_2, y_1))\, d\gamma = 2(x_1, x_2)(y_1, y_2)/\dim E,$$

for all x_1, x_2, y_1, y_2.

Proof. The left hand side of the above equality is a continuous linear functional on E, as a function of x_1 alone. Hence, from Riesz Lemma, it has the form (x_1, z) for some z which depends upon y_1, x_2, y_2. For fixed y_1, y_2, the vector z depends linearly and continuously on x_2. Therefore one may write $z = Ax_2$, where the operator A depends on y_1 and y_2. From the invariance of the Haar integral, one has that

$$(\tilde{\gamma} x_1, A\tilde{\gamma} x_2) = (x_1, Ax_2),$$

hence $\tilde{\gamma}^T A \tilde{\gamma} = A$ and A is equivariant. Furthermore, by interchanging x_1 with x_2, one has that $A = A^T$. Thus, from Corollary 5.1 (b), one has that $A = \lambda I$, where, of course, λ depends on y_1 and y_2 but the left hand side is $\lambda(x_1, x_2)$.

By using the same argument with y_1 and y_2, one has that the left hand side is $\mu(y_1, y_2)$, hence it is of the form $c(x_1, x_2)(y_1, y_2)$, where c is independent of x_1, x_2, y_1, y_2. Taking $x_1 = x_2$, $y_1 = y_2$, the left hand side is $\int_\Gamma 2(\gamma x_1, y_1)^2 d\gamma$ and c is positive.

Take now, $e_1, e_2, \ldots, e_N$ an arbitrary collection of orthonormal vectors in E. Then, from Parseval's inequality, one has

$$\sum_1^N (\gamma x, e_j)^2 \leq \|\gamma x\|^2 \leq \|x\|^2.$$

Taking $x_1 = x_2 = x$ and $y_1 = y_2 = e_j$, and integrating the above equality, one obtains

$$2\sum_1^N \int_\Gamma (\gamma x, e_i)^2 \, d\gamma = Nc\|x\|^2 \leq 2\|x\|^2.$$

Hence, $c \leq 2/N$. From this it follows that E is finite dimensional. Furthermore, if $\dim E = N$, one gets an equality, and one obtains $c = 2/N$. □

Corollary 5.3. *If E is a Γ-Banach space with no proper subrepresentations, then E is finite dimensional.*

Proof. For a general Banach space E, take X a non-zero element of E^*, i.e., a continuous linear functional on E. Consider

$$(x, y)_X = \int_\Gamma X(\gamma x) X(\gamma y) \, d\gamma.$$

Then, $(x, y)_X$ is bilinear, continuous in x and y and $(x, x)_X \geq 0$. Hence, E is given the structure of a pre-Hilbert space: define the equivalence relation $x \underset{X}{\sim} y$ if and only if $(x - y, x - y)_X = 0$, i.e., iff $X(\gamma(x - y)) \equiv 0$ for all γ in Γ. Taking the set of equivalence classes and completing with respect to the $\| \ \|_X$-norm, one obtains a Hilbert space H_X and a natural mapping φ_X from E into H_X. Define an action $\tilde{\gamma}$ of Γ on H_X by factorization and extension by continuity of the action of Γ on E. Since, $(\tilde{\gamma} x, \tilde{\gamma} y)_X = (x, y)_X$, one has that H_X is an orthogonal representation of Γ. Furthermore, $\varphi_X \gamma = \tilde{\gamma} \varphi_X$, by construction, and φ_X is a linear mapping, with $\|\varphi_X(x)\|_X^2 = \int_\Gamma X(\gamma x)^2 d\gamma \leq \|X\|^2 \|x\|^2$, i.e., φ_X is continuous ($\|X\|$ is the norm of X in E^*).

Now, since E has no proper subrepresentations, one has, from Schur's Lemma, that φ_X is one-to-one (since $X \neq 0$, at least for some x one has $\varphi_X(x) \neq 0$). Now, if H_X contains a proper subrepresentation M, we may assume that M is finite dimensional (the precise argument will be given in the next corollary). Let P be an equivariant orthogonal projection from H_X onto M (see Lemma 4.4.). Then, $P\varphi_X$ is a continuous linear map from E into M. From Corollary 5.1 (b), $P\varphi_X$ is either one-to-one, or identically 0. In the first case, this implies that E is finite dimensional. In the second case, $\varphi_X(E) \subset M^\perp$, which contradicts the fact that $\varphi_X(E)$ is dense in H_X. □

Note that, if $E = C^0_{2\pi}(\mathbb{R})$ and $X(x(t)) \equiv x(0)$, then, under the time shift, one has $\|x\|_X^2 = \frac{1}{2\pi}\int_0^{2\pi} x^2(\varphi)d\varphi$ and H_X is $L^2[0, 2\pi]$.

Corollary 5.4. (a) *Any infinite dimensional Banach Γ-space E contains finite dimensional irreducible representations.*

(b) *The set of points whose orbits are contained in a finite dimensional invariant subspace is dense in E.*

Proof. (a) If E has all its subrepresentations of infinite dimension, take a sequence $M_1 \supset M_2 \supset \cdots$ of subrepresentations and let $M_\infty = \bigcap M_n$. Then, M_∞ is a closed linear invariant subspace of E. By ordering such sequences by inclusion, one should have, by Zorn's Lemma, a maximal element. For this element, the corresponding M_∞ is an infinite dimensional subrepresentation. If E is a Hilbert space (with orthogonal action), the above conclusion contradicts the maximality, since either M_∞ has a proper subrepresentation M' and then $\{M_n \cap M'\}$ is strictly "larger" than $\{M_n\}$, or, M_∞ is finite dimensional. This implies that the argument in Corollary 5.3 is complete and one may repeat it for a general Banach space.

(b) Take a finite dimensional subrepresentation M_1 of E and N_1 an invariant closed complement (which exists, by Lemma 4.4). Since N_1 is an infinite dimensional representation, it contains a finite dimensional representation M_2 (of course, if E is finite dimensional, there is nothing to prove). Let N_2 be an invariant closed complement of M_2 in N_1. Continuing this process, one obtains a sequence M_n of finite dimensional invariant subspaces and complements N_n such that $M_{n+1} \oplus N_{n+1} = N_n$. Moreover, there are equivariant projections P_n from E onto $\bigoplus_1^n M_j$ such that $I - P_n$ projects onto N_n. Let $N \equiv \bigcap N_n$. Then, it is easy to see that N is a closed, linear and invariant subspace of E. Ordering sequences of such $\{N_n\}$ by inclusion, construct the corresponding N for a maximal sequence. Then, if $N \neq \{0\}$, N contains a finite dimensional subrepresentation M and its corresponding complement $\tilde{N}$ (take $M = N$ in case N is finite dimensional). But then $\{N_n \cap \tilde{N}\}$ is strictly "larger" than $\{N_n\}$, contradicting the maximality. Hence, $N = \{0\}$ and, for any x in E, one has that $(I - P_n)x$ goes to 0, i.e., $P_n x$, which belongs to $\bigoplus_1^n M_j$, approximates x. Note that, for a Hilbert space, one may take the space E_0 of all points whose orbits lie in a finite dimensional invariant subspace. Clearly, E_0 is an invariant linear subspace and $\bar{E}_0$ is a closed invariant subrepresentation. If $\bar{E}_0$ is a proper subrepresentation, then $E_0^\perp$ contains a finite dimensional subrepresentation $\tilde{N}$, which is a contradiction, since $\tilde{N}$ should be in E_0. Hence, $\bar{E}_0$ is E. Here the maximal N is $E_0^\perp$, the intersection of all the orthogonal complements of finite dimensional invariant subspaces. □

Remark 5.4. In a finite dimensional irreducible representation, the set of finite linear combinations of points on a given orbit is dense: if not, the closure of the linear space generated by such combinations would be a proper subrepresentation.

Our last set of results of this section concerns the form of a linear equivariant map between two finite dimensional representations V and W.

Let $V = V_1 \oplus \cdots \oplus V_q$ and $W = W_1 \oplus \cdots \oplus W_l$ be a decomposition of V and W into irreducible subspaces. Let $P_i : V \to V_i$ and $Q_j : W \to W_j$ be equivariant projections, i.e., $\gamma P_i = P_i \gamma$ and $\tilde{\gamma} Q_j = Q_j \tilde{\gamma}$. Assume that there is a linear map $A : V \to W$, such that $A\gamma = \tilde{\gamma} A$. Let $A_{ij} = Q_j A P_i : V_i \to W_j$. Then, $A_{ij}\gamma = \tilde{\gamma} A_{ij}$ and, from Schur's Lemma, either $A_{ij} = 0$ or A_{ij} is an isomorphism, in which case $\dim V_i = \dim W_j$ and V_i and W_j are equivalent representations. Hence, if one considers all possible A's, it follows that one has to look only at the subrepresentations of V which are equivalent to those of W. Furthermore, since an equivalent representation amounts to a choice of bases (in V and W) and since $\ker A$ as well as Range A are also representations, with complements which are representations, the problem can be reduced to the study of A from V into itself, with $\gamma A = A\gamma$ and $A_{ij} = 0$ if V_i and W_j are not equivalent.

As in Corollary 5.1, one may assume that γ is in $O(V)$ (again a choice of basis). Then $A_{ij} = \mu_{ij} I + \nu_{ij} B_{ij}$, with $B_{ij}^2 = -I$ and $B_{ij} + B_{ij}^T = 0$.

Theorem 5.2. *Let V be a finite dimensional irreducible orthogonal representation. Then exactly one of the following situations occurs.*

(a) *Any equivariant linear map A is of the form $A = \mu I$, i.e., V is an absolutely irreducible representation.*

(b) *There is only one equivariant map B, such that $B^2 = -I$, $B^T + B = 0$. Then, any equivariant linear map A has the form $A = \mu I + \nu B$. In this case, V has a complex structure for which $A = (\mu + i\nu)I$.*

(c) *There are precisely three equivariant maps B_1, B_2, B_3 with the above properties. Then, $B_i B_j = -B_j B_i$ and $B_3 = B_1 B_2$. In this case, V has a quaternionic structure and any equivariant linear map A can be written as $A = \mu I + \nu_1 B_1 + \nu_2 B_2 + \nu_3 B_3 = qI$, where $q = \mu + \nu_1 i_1 + \nu_2 i_2 + \nu_3 i_3$ is in $\mathbb{H}$.*

Proof. If Γ is abelian, this result was proved in Corollary 5.2, where only (a) and (b) occur. Since the abelian case is the main topic of our book, we shall not give the proof of Theorem 5.2 here. However, an elementary proof is not easy to find. Thus, we give a proof in Appendix A. □

In the same vein, one has the following result (with an easy proof in the abelian case) which will be proved in Appendix A.

Theorem 5.3. *Let V be decomposed as*

$$\bigoplus_{i=1}^{i=I} (V_i^{\mathbb{R}})^{n_i} \bigoplus_{j=1}^{j=J} (V_j^{\mathbb{C}})^{n_j} \bigoplus_{l=1}^{l=L} (V_l^{\mathbb{H}})^{n_l},$$

where $V_i^{\mathbb{R}}$ are the absolutely irreducible representations of real dimension m_i repeated n_i times, $V_j^{\mathbb{C}}$ are complex irreducible representations of complex dimension m_j repeated n_j times, while $V_l^{\mathbb{H}}$ are quaternionic representations of dimension (over $\mathbb{H}$) m_l and repeated n_l times. Then, there are bases of V such that any equivariant matrix has a block diagonal form

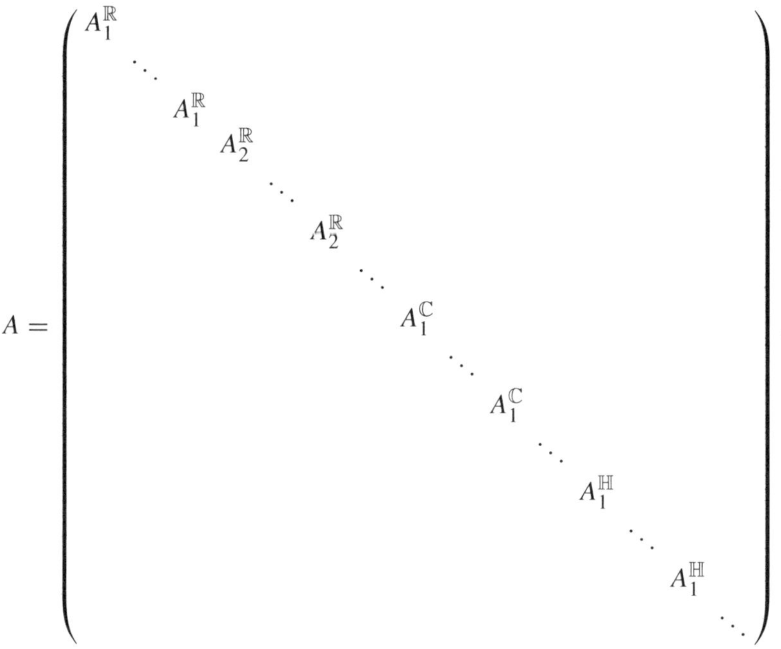

$$A = \begin{pmatrix} A_1^{\mathbb{R}} & & & & & & & & & & & \\ & \ddots & & & & & & & & & & \\ & & A_1^{\mathbb{R}} & & & & & & & & & \\ & & & A_2^{\mathbb{R}} & & & & & & & & \\ & & & & \ddots & & & & & & & \\ & & & & & A_2^{\mathbb{R}} & & & & & & \\ & & & & & & \ddots & & & & & \\ & & & & & & & A_1^{\mathbb{C}} & & & & \\ & & & & & & & & \ddots & & & \\ & & & & & & & & & A_1^{\mathbb{C}} & & \\ & & & & & & & & & & \ddots & \\ & & & & & & & & & & & A_1^{\mathbb{H}} \\ & & & & & & & & & & & & \ddots \\ & & & & & & & & & & & & & A_1^{\mathbb{H}} \\ & & & & & & & & & & & & & & \ddots \end{pmatrix},$$

where $A_i^{\mathbb{R}}$ are real $n_i \times n_i$ matrices repeated m_i times, $A_j^{\mathbb{C}}$ are complex $n_j \times n_j$ matrices, repeated m_j times and $A_l^{\mathbb{H}}$ are $n_l \times n_l$ quaternionic matrices repeated m_l times.

On the new basis, the equivariance of A and the action have the following form: γ is block diagonal on each subspace corresponding to the repetition of the same matrix, i.e., if $B_{n\times n}$ is repeated m times, on W corresponding to the same representation, then $\gamma = (\gamma_{ij} I)_{1\le i,j\le m}$, with γ_{ij} in $\mathbb{K} = \mathbb{R}$, $\mathbb{C}$ or $\mathbb{H}$, and I the identity on $\mathbb{K}^n$, where the product, for the quaternionic case, is on the right.

Remark 5.5. If Γ is abelian, the irreducible representations of Γ are either one-dimensional and Γ acts trivially or as $\mathbb{Z}_2$, or two-dimensional and Γ acts as $\mathbb{Z}_n$, $n \ge 3$ or S^1. Of course, in this case there are no quaternionic components.

Note also that the equivariance of A and the action of Γ on the new basis will be important when considering Γ-equivariant deformations of A; any deformation of $A_i^{\mathbb{R}}$, $A_j^{\mathbb{C}}$ or $A_l^{\mathbb{H}}$, in the corresponding field, will give rise, by repeating the deformation on the m replicae, to a Γ-deformation of A. This will be the situation when computing the Γ-index of 0, when A is invertible, or when studying the Γ-bifurcation with several parameters, as in [*I*].

1.6 Extensions of Γ-maps

Many of our constructions are based upon extensions of equivariant maps, in particular when possible, by non-zero maps. As a matter of fact, the equivariant degree will consist of obstructions to such non-zero equivariant extensions. Thus, the key to our computations of homotopy groups will be a step by step extension of Γ-maps, subtracting "topologically" multiples of generators along the way, in order to get a formula for the class of each map.

Our first result is a simple extension of Dugundji's theorem.

Theorem 6.1 (Dugundji–Gleason extensions). *Let $A_1 \subset A_2$ be Γ-invariant closed subsets of B. If $f : A_1 \to E$ is a Γ-equivariant continuous map, then there is a Γ-equivariant continuous extension $\tilde{f} : A_2 \to E$. Furthermore, $\tilde{f}$ is compact if so is f.*

Proof. From Dugundji's theorem, f has a continuous extension $\hat{f}$ from A_2 into E which is compact if f is compact. From Lemma 4.1, the map

$$\tilde{f}(x) \equiv \int_\Gamma \tilde{\gamma}^{-1} \hat{f}(\gamma x)\, d\tilde{\gamma}$$

is Γ-equivariant (and compact if $\hat{f}$ is compact). Furthermore, if x is in A_1, then $\hat{f}(\gamma x) = f(\gamma x) = \tilde{\gamma} f(x)$ and $\tilde{f}(x) = f(x)$. □

In case B and E are infinite dimensional, we shall look at maps with the following compactness property.

Definition 6.1. If $B = U \times W$ and $E = V \times W$, where U, V are finite dimensional representations of Γ and W is an infinite dimensional representation, an equivariant map f, from a closed Γ-invariant subset A of B into E is called a Γ-*compact perturbation of the identity* if f has the form

$$f(u, w) = (g(u, w), w - h(u, w)),$$

where g is Γ-equivariant from B into V and h in W is compact and Γ-equivariant.

Definition 6.2. If f_0 and f_1 are Γ-maps from a closed invariant subset A of B into $E\backslash\{0\}$ (Γ-compact perturbations of the identity if B and E are infinite dimensional), then f_0 is said to be Γ-*homotopic* to f_1, if there is $f(t, x)$, Γ-equivariant, from $I \times A$ into $E\backslash\{0\}$ (and a Γ-compact perturbation of the identity), where $I = [0, 1]$, with $f(0, x) = f_0(x)$ and $f(1, x) = f_1(x)$.

One then has the following crucial result:

Theorem 6.2 (Equivariant Borsuk homotopy extension theorem). *Let $A_1 \subset A_2$ be Γ-invariant closed subsets of B. Assume that f_0 and f_1, from A_1 into $E\backslash\{0\}$, are*

Γ-equivariant maps which are Γ-homotopic. Then f_0 extends Γ-equivariantly to A_2 without zeros if and only if f_1 does. If this is the case, then the extensions are Γ-homotopic. Similarly, if f_0, f_1 and the Γ-homotopy are Γ-compact perturbations of the identity, then the extensions and the homotopy must be taken Γ-compact perturbations of the identity.

Proof. Let $\hat{f}_0 : A_2 \to E\backslash\{0\}$ be the Γ-extension of f_0 and $f(t, x) : I \times A_1 \to E\backslash\{0\}$ be the Γ-homotopy from f_0 to f_1. Let, by Dugundji–Gleason Theorem 6.1, $g(t, x)$ be any Γ-equivariant extension to $I \times A_2$ of the map defined as $f(t, x)$ on $I \times A_1$ and $\hat{f}_0(x)$ on $\{0\} \times A_2$.

It is easy to see that, in the infinite dimensional case, one preserves the compactness of the perturbations.

Let A be the subset of A_2 consisting of all x for which there is a t with $g(t, x) = 0$.

Then, by construction, $A \cap A_1 = \phi$. Furthermore, from the compactness of $[0, 1]$, if $\{x_n\}$ is in A, converging to x in A_2, then $g(t_n, x_n) = 0$, $\{t_n\}$ has a subsequence converging to some t and $g(t, x) = 0$. Thus, A is closed. Furthermore, the equivariance of g, with respect to x, implies that A is invariant.

From Lemma 4.2, there is an invariant Uryson function $\varphi : A_2 \to [0, 1]$ such that $\varphi(A) = 0$ and $\varphi(A_1) = 1$.

Define $\hat{f}(t, x) = g(\varphi(x)t, x)$. Then the Γ-equivariance of $\hat{f}$ follows from that of g (and of the invariance of φ), as well as the compactness property. Furthermore, $\hat{f}(0, x) = g(0, x) = \hat{f}_0(x)$. Finally, if $\hat{f}(t, x) = 0$ for some t, then x belongs to $A, \varphi(x) = 0$, but $g(0, x) = \hat{f}_0(x) \neq 0$. The map $\hat{f}(t, x)$ gives a Γ-homotopy on A_2, from $\hat{f}_0(x)$ to $\hat{f}_1(x) = g(\varphi(x), x)$, which provides an extension of f_1, since, on $A_1, \varphi(x) = 1$. □

Another useful fact is the following observation:

Lemma 6.1. *Let S^n be the unit sphere in the Γ-space $V \cong \mathbb{R}^{n+1}$ and $f : S^n \to W\backslash\{0\}$ (another finite dimensional representation) a Γ-map. Then any Γ-equivariant extension $\hat{f}$ of f to the unit ball has a zero if and only if f is* ***not*** *Γ-deformable to a non-zero constant map.*

Proof. Note first that a non-zero constant equivariant map may exist only if $W^\Gamma \neq \{0\}$. In other words, if W^Γ is reduced to 0, any equivariant extension $\hat{f}$ must have $\hat{f}(0) = 0$ (see also Property 3.2).

Now, if $\hat{f}$ is such an extension, define the Γ-homotopy $f : I \times S^n \to W\backslash\{0\}$, by $f(t, x) = \hat{f}((1-t)x)$, deforming radially and equivariantly $f(0, x) = f(x)$ to $\hat{f}(0)$. On the other hand, if $f(t, x)$ Γ-deforms $f(x)$, for $t = 1$, to the constant $f(0, x)$, define $\hat{f}(x) = f(\|x\|, x/\|x\|)$ which will provide the appropriate Γ-extension of f. □

One of the key tools which will be used in our computations of equivariant homotopy groups of spheres is the existence of *complementing maps*, which will play the role of a *suspension* (defined in Section 8). In order to be more specific, let us assume that U and W are finite dimensional orthogonal representations of an abelian compact Lie group Γ, with action given as in Example 1.4. Suppose that an equivariant map is given from U^H into W^H, for some subgroup H of Γ. The problem is then the following: is it possible to give a "complementing" Γ-equivariant map from $(U^H)^\perp$ into $(W^H)^\perp$ which is zero only at zero? Recall that, since Γ is abelian and the action is orthogonal, all the above subspaces are representations of Γ. The answer to the question is in general negative, as the following example shows.

Example 6.1. On $\mathbb{C}^2$, consider the following action of $\mathbb{Z}_{p^2q}$, where p and q are relatively prime: On (z_1, z_2) in U, Γ acts via $(e^{2\pi ik/p^2}, e^{2\pi ik/(pq)})$ for $k = 0, \dots, p^2q-1$. On (ξ_1, ξ_2) in W, Γ acts as $(e^{2\pi ik/p}, e^{2\pi ik/(p^2q)})$. The isotropy subgroups for the action of Γ on U are as follows:

$$\begin{aligned} H &\cong \mathbb{Z}_q, && \text{for } k \text{ a multiple of } p^2 \text{ and } U^H = \{(z_1, 0)\}, \\ K &\cong \mathbb{Z}_p, && \text{for } k \text{ a multiple of } pq \text{ and } U^K = \{(0, z_2)\}, \\ L &\cong \{e\}, && \text{for } k = 0 \text{ and } U^{\{e\}} = U. \end{aligned}$$

One has $W^H = W^K = \{(\xi_1, 0)\}$, but there is no non-zero equivariant map between $(U^H)^\perp$ and $(W^H)^\perp$, since $(U^H)^\perp \cap U^K = U^K$ and $(W^H)^\perp \cap W^K = \{0\}$. On the other hand, if $\alpha q + \beta p = 1$, the map

$$F(z_1, z_2) = (z_1^p + z_2^q, z_1^\alpha z_2^\beta)$$

(where a negative power is interpreted as a conjugate: $z^{-1} \equiv \bar{z}$), is an equivariant map from U into W with only one zero at the origin.

One of our *main hypotheses* in Chapter 3 will be the following:

For any pair of isotropy subgroups H and K for U, one has

$$\text{(H)} \quad \dim U^H \cap U^K = \dim W^H \cap W^K.$$

Note that in Example 6.1, hypothesis (H) fails, although there $\dim U^H = \dim W^H$, for all isotropy subgroups of Γ on U.

Lemma 6.2. *Hypothesis* (H) *holds if and only if both* (a) *and* (b) *hold:*

(a) $\dim U^H = \dim W^H$, *for all isotropy subgroups H on U.*

(b) *There are integers $l_1, \dots, l_s$ such that the map $F : (x_1, \dots, x_s) \to (x_1^{l_1}, \dots, x_s^{l_s})$ is Γ-equivariant. Here x_j is a (real or complex) coordinate of U on which Γ acts as in Example* 1.4, *and a negative power means a conjugate. Furthermore, for all γ in Γ one has* $\det \gamma \det \tilde{\gamma} > 0$.

Proof. Let H_j be the isotropy subgroup of x_j and $H_0 = \bigcap H_j$. Then $U^{H_0} = U$ and any isotropy subgroup $H = \Gamma_x = \bigcap H_j$, where the intersection is on the j's for which the coordinate x_j of x is non-zero (see § 2), is such that $H_0 < H$.

Hence, if (H) holds, one obtains (a), since $W^H \subset W^{H_0}$. Note that any equivariant map from U into W will have its image in W^{H_0}. For notational purposes, define, for $K > H$, $(U^K)^{\perp_H}$ as $U^H \cap (U^K)^{\perp}$. Then, hypothesis (H) implies that $\dim(U^K)^{\perp_H} = \dim(W^K)^{\perp_H}$.

Now, if $\Gamma/H_j \cong \mathbb{Z}_2$ and γ acts as $-I$ on $(U^\Gamma)^{\perp_{H_j}}$, then on $(W^\Gamma)^{\perp_{H_j}}$, $\tilde{\gamma}$ must also act as $-I$, since if not one would violate the equality of the dimensions. Since the action on a complex coordinate is a multiplication by a unit complex number, i.e., corresponding to a rotation with determinant equal to 1, then $\det\gamma$ and $\det\tilde{\gamma}$ (restricted to W^{H_0}) have the same sign.

We may now begin to build up the map F. We shall identify U^Γ and W^Γ and take $l_j = 1$ for these components.

Let H be maximal among the H_j's. Then, from Lemma 1.1, $\Gamma/H \cong \mathbb{Z}_n, n \geq 2$ or S^1 and acts freely on $(U^\Gamma)^{\perp_H}\backslash\{0\}$ and without fixed points on $(W^\Gamma)^{\perp_H}\backslash\{0\}$, as it follows from Lemma 1.2, since no point in the second set, fixed by H, may be fixed by Γ/H without being in W^Γ.

Thus, if γ generates $\mathbb{Z}_n$, one has $\gamma x_j = e^{2\pi i m_j/n}x_j$ with $1 \leq m_j < n, m_j$ and n relatively prime and $\tilde{\gamma}\xi_j = e^{2\pi i n_j/n}\xi_j$, with $1 \leq n_j < n$. Now, there is a unique $p_j, 1 \leq p_j < n$, such that $p_j m_j \equiv 1, [n]$. Let l_j be the residue class, modulo n, of $p_j n_j$. Then, $(\gamma x_j)^{l_j} = \tilde{\gamma}x_j^{l_j}$. Note that, if $n = 2$, then $n_j = m_j = 1$ and $l_j = 1$. That is, on the real representations of Γ, where it acts as $\mathbb{Z}_2$, the map F is the identity.

On the other hand, if $\Gamma/H \cong S^1$, acting as $e^{i\varphi}$ (or $e^{-i\varphi}$) on $(U^\Gamma)^{\perp_H}$ and as $e^{in_j\varphi}$ on $(W^\Gamma)^{\perp_H}$, then $l_j = n_j$ (or $-n_j$) will give the equivariant map (with negative l_j meaning conjugates).

Let now K and L be isotropy subgroups for $(U^H)^\perp$. Let H_1 be the isotropy subgroup for $U^K \cap U^L$, i.e., H_1 is the intersection of the isotropy subgroups for all the coordinates in that subspace. Then, $U^K \cap U^L \subset U^{H_1}$. Since K and L are also intersections of the corresponding subgroups, it is clear that K and L are subgroups of H_1 and then $U^{H_1} \subset U^K \cap U^L$, that is $U^{H_1} = U^K \cap U^L$, while $W^{H_1} \subset W^K \cap W^L$. But, from (H), one has $\dim U^{H_1} = \dim W^{H_1}$ and $\dim U^K \cap U^L = \dim W^K \cap W^L$, then $W^{H_1} = W^K \cap W^L$. Since $\dim(U^H)^\perp \cap U^K \cap U^L = \dim U^K \cap U^L - \dim U^H \cap U^K \cap U^L$, one obtains that the hypothesis (H) is valid on $(U^H)^\perp$ and $(W^H)^\perp$. Then, one may repeat the above argument by choosing a maximal isotropy subgroup among the remaining H_j's, proving the implication in a finite number of steps.

Conversely, if the map F exists, it is clear that $\dim U^H \leq \dim W^H$ (and it is easy to give examples with a strict inequality). While, if (a) and (b) hold, it is easy to see, by direct inspection, that (H) is true. □

In order to construct the generators of the equivariant homotopy groups, in Chapter 3, we shall need some invariant monomials. We shall again assume that the abelian group Γ acts on U, with coordinates $\{x_1, \ldots, x_s\}$, with $H_j = \Gamma_{x_j}$. Let H_0 be a subgroup of Γ and define, as in § 2, $\tilde{H}_j = H_0 \cap H_1 \cap \cdots \cap H_j$. Let $k_j = |\tilde{H}_{j-1}/\tilde{H}_j|$.

Lemma 6.3. *There are integers $\alpha_1, \ldots, \alpha_s$ such that $x_j^{\alpha_j} \ldots x_s^{\alpha_s}$ is $\tilde{H}_{j-1}$-invariant. (If α is negative, x^α means $\bar{x}^{|\alpha|}$). If $k_s < \infty$, then one may take $\alpha_s = k_s$, while if $k_s = \infty$, then $\alpha_s = 0$. Furthermore, if $k_j = 1$ for $j < s$, then one may take $\alpha_j = 0$.*

Proof. The proof will be by induction on j. If $j = s$ and $k_s = \infty$, any constant is Γ-invariant, hence $\alpha_s = 0$ will do. While, if k_s is finite, then $\tilde{H}_{s-1}/\tilde{H}_s$ acts freely on x_s (as in § 2) and any γ in $\tilde{H}_{s-1}$ can be written as $\gamma = \beta_s^\alpha \delta$, for some δ in $\tilde{H}_s$ and a fixed β_s such that $\beta_s x_s = e^{2\pi i/k_s} x_s$. Hence, $(\gamma x_s)^{k_s} = \beta_s^{\alpha k_s} x_s^{k_s} = x_s^{k_s}$ is $\tilde{H}_{s-1}$-invariant.

Assume now that $P(x_{j+1}, \ldots, x_s) \equiv x_{j+1}^{\alpha_{j+1}} \ldots x_s^{\alpha_s}$ is $\tilde{H}_j$-invariant, for some $j \geq 1$. Then, if $\tilde{H}_{j-1}/\tilde{H}_j \cong S^1$, this group acts freely on x_j and as $e^{in_l\varphi}$ on x_l, for $l = j, \ldots, s$, with $n_j = 1$. Since $P(e^{in_{j+1}\varphi} x_{j+1}, \ldots, e^{in_s\varphi} x_s) = e^{\sum n_l \alpha_l} P(x_{j+1}, \ldots, x_s)$, one may choose $\alpha_j = -\sum n_l \alpha_l$ and $x_j^{\alpha_j} \ldots x_s^{\alpha_s}$ will be $\tilde{H}_{j-1}$-invariant. On the other hand, if k_j is finite, then any γ in $\tilde{H}_{j-1}$ is written as $\gamma = \beta_j^\alpha \delta$, with β_j generating $\tilde{H}_{j-1}/\tilde{H}_j$ and acting as $e^{2\pi i/k_j}$ on x_j, $0 \leq \alpha < k_j$ and δ in $\tilde{H}_j$. Then,

$$P(\gamma x_{j+1}, \ldots, \gamma x_s) = \beta_j^{\alpha\alpha_{j+1}} (\delta x_{j+1})^{\alpha_{j+1}} \ldots \beta_j^{\alpha\alpha_s} (\delta x_s)^{\alpha_s}.$$

Now, as before, $\beta_j = \beta_k^{\varepsilon_k} \eta_k$, where β_k generates $\tilde{H}_{j-1}/(H_k \cap \tilde{H}_{j-1})$, $\beta_k x_k = e^{2\pi i/n_k} x_k$, where n_k is the order of this group if finite (or $\beta_k x_k = e^{i\varphi} x_k$ if the group is isomorphic to S^1 and $\beta_k^{\varepsilon_k}$ means $e^{2\pi i \varepsilon_k/n_k}$ for some n_k: since $\beta_j^{k_j}$ is in $\tilde{H}_j$, the $\tilde{H}_j$-invariance of P implies that the corresponding φ is a rational multiple of 2π); one has $0 \leq \varepsilon_k < n_k$ and η_k is in $H_k \cap \tilde{H}_{j-1}$.

Thus, $\beta_j^{\alpha\alpha_k} (\delta x)^{\alpha_k} = e^{2\pi i \alpha\alpha_k \varepsilon_k/n_k} (\delta x_k)^{\alpha_k}$. Hence,

$$P(\gamma x_{j+1}, \ldots, \gamma x_s) = e^{2\pi i \varepsilon\alpha} P(\delta x_{j+1}, \ldots, \delta x_s) = e^{2\pi i \varepsilon\alpha} P(x_{j+1}, \ldots, x_s),$$

with $\varepsilon = \sum_{k=j+1}^s \alpha_k \varepsilon_k / n_k$.

Now, if $\gamma = \beta_j^{k_j}$, i.e., $\alpha = k_j$, then this γ belongs to $\tilde{H}_{j-1} \cap H_j = \tilde{H}_j$ and the corresponding εk_j must be an integer. Let ε_0 be the non-integer part of ε and define $\alpha_j = -k_j \varepsilon_0$ (it is an integer and $\alpha_j = 0$ if $k_j = 1$). Then, if $P(x_j, \ldots, x_s) = x_j^{\alpha_j} \ldots x_s^{\alpha_s}$, one has

$$\begin{aligned} P(\gamma x_j, \ldots, \gamma x_s) &= (\beta_j x_j)^{\alpha_j} e^{2\pi i \varepsilon\alpha} P(x_{j+1}, \ldots, x_s) \\ &= e^{2\pi i \alpha\alpha_j/k_j} e^{2\pi i \varepsilon\alpha} P(x_j, \ldots, x_s) = P(x_j, \ldots, x_s) \end{aligned}$$ □

1.7 Orthogonal maps

In the last chapters of the book, we shall be interested in a particular class of maps, which we shall call *orthogonal maps*. The setting is the following: let Γ be a compact abelian group acting on the finite dimensional orthogonal representation V. Thus, if $\Gamma = T^n \times \mathbb{Z}_{m_1} \times \cdots \times \mathbb{Z}_{m_s}$, with the torus T^n generated by $(\varphi_1, \ldots, \varphi_n)$, φ_j in $[0, 2\pi]$, we shall define by

$$A_j x = \frac{\partial}{\partial \varphi_j} (\gamma x)\Big|_{\gamma = \mathrm{Id}},$$

the infinitesimal generator corresponding to φ_j.

Hence, if the action of φ_j on the coordinate x_l is as $e^{in_j^l \varphi_j}$, then

$$A_j x = (in_j^1 x_1, \ldots, in_j^m x_m)^T,$$

where inx stands for $(-n \operatorname{Im} x, n \operatorname{Re} x)^T$.

Lemma 7.1. *Let $H = \Gamma_{x_0}$. Then:*

(a) *There are exactly k linearly independent $A_j x_0$ if and only if* $\dim \Gamma/H = k$.

(b) *In this case, if $\underline{H}$ is the torus part of H and $\bar{H}$ corresponds to the first k (non-zero) coordinates of x_0, then for any x in $V^{\underline{H}}$ one has $A_j x = \sum_{l=1}^k \lambda_j^l A_l x$ and $A_1 x, \ldots, A_k x$ are linearly independent whenever $x_1, \ldots, x_k$ are non-zero.*

Proof. (a) Since $H = \bigcap H_j$, for the non-zero coordinates of x_0, one has from Lemma 2.4 (a), that $\dim \Gamma/H = k$ if and only if A^H has rank k, where A^H is the matrix formed by n_j^i.

(b) follows from Remark 2.1 and the definition of λ_j^l as given in Lemmas 2.4 (b) and 2.6. Note that one may reparametrize T^n by choosing $\Psi_j = \varphi_j + \sum_{l=k+1}^n \lambda_l^j \varphi_l$, for $j = 1, \ldots, k$ and taking $\Psi_{k+1}, \ldots, \Psi_n$ acting trivially on $V^{\underline{H}}$. In this case, if $\tilde{A}_j$ is the diagonal matrix corresponding to the action of H, that is, to the derivative with respect to Ψ_j, for $j = k+1, \ldots, n$ (since $\underline{H}$ corresponds to $\Psi_j = 0$, $j = 1, \ldots, k$), then $\tilde{A}_j$ is 0 on $V^{\underline{H}}$ and, on any irreducible representation of H in $(V^{\underline{H}})^\perp$, one of the $\tilde{A}_j$, $j = k+1, \ldots, n$, will be invertible. □

Definition 7.1. A Γ-equivariant map f, from V into itself, is said to be Γ-*orthogonal* if $f(x) \cdot A_j x = 0$, for all $j = 1, \ldots, n$ and all x in the domain of definition of f. Here the dot stands for the *real* scalar product. In terms of complex scalar product one has $\operatorname{Re}(f(x) \cdot \overline{A_j x}) = 0$.

Example 7.1 (Gradient maps). If $f(x) = \nabla J(x)$, where $J(\gamma x) = J(x)$ is an invariant function, we have seen in Remark 3.1, that $f(\gamma x) = \gamma f(x)$ and that $f(x) \cdot A_j x = 0$, i.e., that the gradient of an invariant function is an orthogonal map.

Linearizations of orthogonal maps have quite interesting properties. In fact:

Lemma 7.2. *Assume that the Γ-orthogonal map f is C^1 at x_0, with a k-dimensional orbit. Let $H = \Gamma_{x_0}$ and denote by D the matrix $Df(x_0)$. Then:*

(a) *D is H-orthogonal. If $K < H$ is any isotropy subgroup, then $D^K = Df^K(x_0) = \operatorname{diag}(D^H, D_{\perp K})$. For $K < \underline{H}$, the torus part of H, we have that $D^K = \operatorname{diag}(D^H, D_{\perp \underline{H}}, D'_{\perp K})$, where $D'_{\perp K}$ is a complex self-adjoint matrix which is H-orthogonal.*

(b) *If $f(x_0) = 0$, then $A_j x_0$ are in* $\ker D$ *and are orthogonal to* Range D. *In particular, if* $\dim \ker D = k$, *then, for any $K < H$, the matrix $D_{\perp K}$ is invertible and the algebraic multiplicity of D is k.*

Proof. The fact that D is H-equivariant was proved in Property 3.3. The diagonal structure comes from Property 3.4 and Theorem 5.3. In particular, if $K < \underline{H}$, then $D'_{\perp K}$ is a complex matrix and $\dim \underline{H}/K \geq 1$.

Now, since f is Γ-orthogonal it is also H-orthogonal. If $f^K = (f^{\underline{H}}, f_\perp)$, then $f^K(x) \cdot \tilde{A}_j x = f_\perp(x) \cdot \tilde{A}_j x_\perp = 0$ for any $x = x_{\underline{H}} + x_\perp$ in V^K, where $\tilde{A}_j$ are the generators for the action of H, since $\tilde{A}_j$ is 0 on $V^{\underline{H}}$. From $f_\perp(x_H) = 0$, one obtains $(Df_\perp(x_{\underline{H}})x_\perp + R(x_\perp)) \cdot \tilde{A}_j x_\perp = 0$, where $R(x_\perp) = o(\|x_\perp\|)$. Dividing by $\|x_\perp\|^2$ and taking the limit when $x_\perp$ goes to 0, one has that $Df_\perp(x_{\underline{H}})x_\perp \cdot \tilde{A}_j x_\perp = 0$. In particular, $D'_{\perp K}$ is H-orthogonal.

Take K corresponding to an irreducible representation of H on $(V^{\underline{H}})^\perp$ and choose j such that $\tilde{A}_j$ is invertible on it (and hence it is just a multiplication by im, for some integer m). Set $B \equiv D'_{\perp K}$. Since B is H-equivariant, one has $B\tilde{A}_j = \tilde{A}_j B$ and $Bx \cdot \tilde{A}_j x = 0$. Furthermore, from $B(x + x_0) \cdot \tilde{A}_j(x + x_0) = 0$ for any x and x_0 in the representation, one has $\tilde{A}_j^T B + B^T \tilde{A}_j = 0$. But $\tilde{A}_j^T = -\tilde{A}_j$, hence, $B^T = \tilde{A}_j B \tilde{A}_j^{-1} = B\tilde{A}_j \tilde{A}_j^{-1} = B$ on that representation. Now, since the action of H on x in that representation is as S^1, B, in fact, is a real matrix of the form $\begin{pmatrix} \mathcal{A} & -\mathcal{B} \\ \mathcal{B} & \mathcal{A} \end{pmatrix}$. Then, $B = B^T$ implies $\mathcal{A} = \mathcal{A}^T$ and $\mathcal{B} = -\mathcal{B}^T$, that is $(\mathcal{A} + i\mathcal{B})^* = \mathcal{A} + i\mathcal{B}$.

For the second part of the lemma, differentiating the relation $f(\gamma x_0) = 0$ with respect to φ_j, one obtains $DA_j x_0 = 0$. Furthermore, from $f(x) \cdot A_j x = 0$, one has, for all x and x_0

$$Df(x_0)x \cdot A_j x_0 + f(x_0) \cdot A_j x = 0.$$

In particular, if $f(x_0) = 0$, then $A_j x_0$ is orthogonal to Range D. Also, if $\dim \ker D = k$, since $\{A_1 x_0, \ldots, A_k x_0\}$ are linearly independent, then $V = \ker D \oplus \operatorname{Range} D$, the algebraic multiplicity of D is k and $D_{\perp K}$ is invertible, for any $K < H$. □

A crucial property of Γ-orthogonal maps is the following: they can be extended as Γ-orthogonal maps. Namely,

Theorem 7.1. *Let $A_1 \subset A_2$ be Γ-invariant closed subsets of V. If $f : A_1 \to V$ is a continuous Γ-orthogonal map, then there is a continuous Γ-orthogonal extension $\tilde{f}$ to A_2, which is obtained by a Gram–Schmidt orthogonalization process.*

Proof. Let $\tilde{f}_0$ be a Γ-equivariant extension of f, given in Theorem 6.1. Since $\tilde{f}_0$ is not necessarily orthogonal to $A_j x$, we shall use the following orthogonalization:

Let

$$\tilde{A}_1(x) = \begin{cases} A_1 x/\|A_1 x\|, & \text{if } A_1 x \neq 0 \\ 0, & \text{if } A_1 x = 0, \end{cases}$$

$$\hat{A}_j(x) = A_j x - \sum_1^{j-1} (A_j x, \tilde{A}_i(x))\tilde{A}_i(x)$$

and

$$\tilde{A}_j(x) = \begin{cases} \hat{A}_j(x)/\|\hat{A}_j(x)\|, & \text{if } \hat{A}_j(x) \neq 0 \\ 0, & \text{if } \hat{A}_j(x) = 0. \end{cases}$$

Clearly, the $\tilde{A}_j(x)$ are orthogonal and $\tilde{A}_j(x) = 0$ if and only if $A_j x$ is a linear combination of $A_1 x, \dots, A_{j-1}x$. Furthermore, A_j is Γ-equivariant as well as $\tilde{A}_j(x)$ and $\hat{A}_j(\lambda x) = \lambda \hat{A}_j(x)$, for λ in $\mathbb{R}$. All these facts can be easily proved by induction. Let

$$\tilde{f}(x) = \tilde{f}_0(x) - \sum_1^s (\tilde{f}_0(x), \tilde{A}_j(x))\tilde{A}_j(x).$$

By construction, $\tilde{f}(x)$ is orthogonal to $\tilde{A}_j(x)$ for all j's and hence to all $A_j x$, which are linear combinations of them. Furthermore, $\tilde{f}(x)$ is Γ-equivariant and if x is in A_1, then $\tilde{f}_0(x) = f(x)$ which is orthogonal to all $\tilde{A}_j(x)$ and $\tilde{f}(x) = f(x)$.

Thus, the more delicate part is the continuity of $\tilde{f}(x)$, that is the continuity of $(\tilde{f}_0(x), \tilde{A}_j(x))\tilde{A}_j(x)$. Let $\{x_n\}$ be a sequence converging to x_0 such that $\hat{A}_j(x_n)$ is non-zero and converges to 0 (the other cases are trivial). Then, since $\tilde{A}_j(x_n)$ has norm 1, there is a subsequence such that $\tilde{A}_j(x_n)$ converges to some v, with norm 1, and the above expression converges to $(\tilde{f}_0(x_0), v)v$.

Assume now that j is the first index for which $\tilde{A}_j(x_0) = 0$. Then $A_j x_0 = \sum_1^{j-1} \lambda_i^j A_i x_0$, that is x_0 belongs to $\ker(A_j - \sum_1^{j-1} \lambda_i^j A_i) \equiv V_1$. But V_1 is invariant under Γ and in fact $V_1 = V_1^{T_1}$, where T_1 is the torus $(-\lambda_1^j \varphi, \dots, -\lambda_{j-1}^j \varphi, \varphi, 0, \dots, 0)$. Hence, from the equivariance, $\tilde{f}_0(x_0)$ belongs to V_1 and one would have proved the continuity by showing that v is in $V_2 \equiv V_1^{\perp}$.

Write any x in V as $x_1 + x_2$, with x_i in V_i and $A(x)$ as $A(x)_1 + A(x)_2$. Since $\tilde{A}_i$ is equivariant one has that $\tilde{A}_i(x_1)$ is in V_1. Furthermore, since $A_j x_1$ is a linear combination of $A_1 x_1, \dots, A_{j-1}x_1$, it follows that $\hat{A}_j(x_1) = 0$. Note also that, due to the linearity of A_i, one gets that $\hat{A}_k(x_2)$ is in V_2, $\tilde{A}_k(x)_1$ is a linear combination of $A_l x_1$, for $l \leq k$, while $\tilde{A}_k(x)_2$ is a linear combination of $A_l x_2$, for $l \leq k$.

Since j is the first index for which $\tilde{A}_j(x_0) = 0$, then, in a neighborhood of x_0 and $k < j$, it follows that $\tilde{A}_k(x)$ is non-zero and continuous, in particular, $\|\tilde{A}_k(x)_2\| \leq$

$c\|x_2\|$. Now, we claim that

$$\hat{A}_j(x) = \mathcal{A}_j x - \sum_{k<j} (\mathcal{A}_j x, \tilde{A}_k(x))\tilde{A}_k(x),$$

where $\mathcal{A}_j = A_j - \sum_{i<j} \lambda_i^j A_i$. In fact, since $A_i x = \hat{A}_i(x) + \sum_{l<i} (A_i x, \tilde{A}_l(x))\tilde{A}_l(x)$, one deduces, from $(\tilde{A}_l x, \tilde{A}_k x) = \delta_{lk}$, the cancellation of the extra terms. Since $\mathcal{A}_j x = \mathcal{A}_j x_2$, one obtains, from the above bound for $\|A^k(x)_2\|$, the existence of two constants, c and C, such that $c\|\mathcal{A}_j x\| \le \|\hat{A}_j(x)\| \le C\|\mathcal{A}_j x\|$, for x close enough to x_0 (in fact for x_2 small enough). Hence,

$$\tilde{A}_j(x) = \mathcal{A}_j x / \|\mathcal{A}_j x\| + 0(\|x_2\|).$$

In particular, any limit point for $\tilde{A}_j(x)$ will be in $(\ker \mathcal{A}_j)^\perp$.

Assume now, that j is not the first index for which $\hat{A}_j(x) = 0$. In fact, let I be the set of indices for which $\hat{A}_i(x_0) = 0 = (A_i - \sum_{k<i} \lambda_k^i A_k)x_0 \equiv \mathcal{A}_i x_0$. Then, x_0 belongs to $\ker \mathcal{A}_i$, which is the fixed point subspace of a one-dimensional torus. Hence, x_0 is in the fixed point subspace of a m-torus T_m, where m is the cardinality of I. Denote by V_i the intersection of all $\ker \mathcal{A}_k$, for $k \le i$, both in I. Hence, $\{V_i\}$ is a decreasing sequence of subspaces which are fixed point subspaces of tori. Since x_0 is in the smallest one then, by equivariance, this is also the case for $f(x_0)$. One would have proved the continuity if one could show that any limit point of $\tilde{A}_j(x)$, for j in I, is in $V_j^\perp$.

Since the proof, by induction, is rather involved we shall break it up in several lemmata.

Lemma 7.3. *Define $\mathcal{A}_j$ as above if j is in I and as A_j if j is in I^c. Then, one may change, in the formulae for $\hat{A}_j(x)$, A_j by $\mathcal{A}_j$, without changing $\hat{A}_j(x)$.*

Proof. For $j = 1$, then $\mathcal{A}_1 = A_1$ and $\hat{A}_1(x)$ is unchanged. Assume, that the lemma is true up to $j-1$. Then, if j is I^c, there is nothing to prove, while if j is in I, it is enough to repeat the above argument. □

Lemma 7.4. *Define inductively the following linear operators for j in I:*

(a) *For the first element of I: $\mathcal{B}_j = \mathcal{A}_j$.*

(b) *For the subsequent elements of I: $\mathcal{B}_j = \mathcal{A}_j - \sum_{i<j} \left(\mathcal{A}_j x, \frac{\tilde{A}_i(x)}{\|\hat{A}_i(x)\|}\right)\mathcal{B}_i$, where in the sum one has only elements of I.*

Then

$$\hat{A}_j(x) = \mathcal{B}_j x - \sum_{\substack{k\in I^c \\ k<j}} (\mathcal{B}_j x, \tilde{A}_k(x))\tilde{A}_k(x).$$

Proof. For the first element of I, the result has already been proved. Assume, by induction, that it is true for $i < j$, then,

$$\hat{A}_j(x) = \mathcal{A}_j x - \sum_{\substack{k \in I^c \\ k<j}} (\mathcal{A}_j x, \tilde{A}_k(x))\tilde{A}_k(x) - \sum_{\substack{i \in I^c \\ i<j}} (\mathcal{A}_j x, \tilde{A}_i(x))\tilde{A}_i(x).$$

Using the induction hypothesis in the second sum, one has that

$$\tilde{A}_i(x) = \|\hat{A}_i(x)\|^{-1}(\mathcal{B}_i x - \sum_{\substack{k \in I^c \\ k<i}} (\mathcal{B}_i x, \tilde{A}_k(x))\tilde{A}_k(x)).$$

Collecting the terms with $\mathcal{B}_i x$, one recognizes $\mathcal{B}_j x$ and one will get the result provided the double sum of terms $(\mathcal{A}_j x, \tilde{A}_i(x))(\mathcal{B}_i x, \tilde{A}_k(x))\tilde{A}_k(x)$ is the same when k is in I^c, i in I, $i < j$, and either $k < i$ (which is what the substitution gives) or $k < j$ (if the formula is to be verified). The difference between the two sums corresponds to those k's with $i \leq k < j$ (in fact a strict inequality since i and k are in disjoint sets). But there, by the induction hypothesis, $\mathcal{B}_i x$ is a linear combination of $\hat{A}_i(x)$ and $\tilde{A}_l(x)$, for $l < i$, hence orthogonal to $\tilde{A}_k(x)$, proving the result. □

Lemma 7.5. *If x is written as $x = x_1 \oplus x_2$, where x_1 is in V_j and x_2 in $V_j^{\perp}$, then, for x close to x_0, there are constants c, C, C_2 and D such that, for j in I:*

(a) $\|\tilde{A}_k(x)_2\| \leq C_2\|x_2\|$, *for k in I^c, $k < j$.*

(b) $c\|\mathcal{B}_j x\| \leq \|\hat{A}_j(x)\| \leq C\|\mathcal{B}_j x\| \leq D\|\mathcal{A}_j x\|$.

Proof. The proof will be again by induction, where the first step has already been done. If the result is true for $i < j$, then for k in I^c (hence one does not worry about $\|\hat{A}_k(x)\|$), one gets: $\hat{A}_k(x)_2 = A_k x_2 - \sum_{l<k}(A_k x, \tilde{A}_l(x))\tilde{A}_l(x)_2$.

If l is not in I, then the bound is valid by induction, while if l is in I, then

$$\begin{aligned}(A_k x, \tilde{A}_l(x)) &= (A_k x_2, \mathcal{B}_l x_2)/\|\hat{A}_l(x)\| \\ &\quad - \sum_{\substack{n \in I^c \\ n<l}} (\mathcal{B}_l x_2, \tilde{A}_n(x)_2)(\tilde{A}_n(x), A_k x)/\|\hat{A}_l(x)\|,\end{aligned}$$

since $\mathcal{B}_l x_1 = 0$. Hence, again by induction, one has (a).

For (b) all the inequalities, but the first, are now straightforward. For the first one, $\|\hat{A}_j(x)\| \geq \|\mathcal{B}_j x\| - \tilde{c}\|\mathcal{B}_j x\|\|x_2\|$, where one uses (a) and $\mathcal{B}_j x_1 = 0$. □

End of the proof of the theorem. Since $\hat{A}_j(x) = \mathcal{B}_j x_2 + 0(\|x_2\|\|\mathcal{B}_j x_2\|)$, then limit points of $\tilde{A}_j(x)$ will be of the form $\alpha\mathcal{B}_j\eta$, with $\|\eta\| = 1$, hence in $V_j^{\perp}$. □

Corollary 7.1 (Orthogonal Borsuk homotopy extension theorem). *Let $A_1 \subset A_2$ be Γ-invariant closed subsets of V. Assume that f_0 and f_1, from A_1 to $V\backslash\{0\}$, are Γ-orthogonal maps which are Γ-homotopic, with an orthogonal homotopy. Then f_0 extends Γ-orthogonally to A_2 without zeros if and only if f_1 does. In this case the extensions are Γ-orthogonally homotopic.*

Proof. It is enough to check that the proof of Theorem 6.2 is still valid, and one uses Theorem 7.1 instead of Theorem 6.1. □

1.8 Equivariant homotopy groups of spheres

Our equivariant degree, which will be defined in the next chapter, will be an element of the group of equivariant homotopy classes of Γ-maps between two spheres in two Γ-representations. In this section, we shall recall some known results of the ordinary case, i.e., without a Γ-action, and give some preliminary results in the equivariant case.

The setting is the following: let V and W be two finite dimensional Γ-representations (hence, from Theorem 5.1, one may assume that they are orthogonal). Let B_R be the ball $\{x \in V : \|x\| < R\}$ and consider the set $\mathcal{C}$ of all equivariant maps

$$F : [0, 1] \times B_R \to \mathbb{R} \times W$$
$$F : S^V = \partial([0, 1] \times B_R) \to \mathbb{R} \times W\backslash\{0\}.$$

Thus, $F(t, x)$ has the form $(\Psi(t, x), f(t, x))$, where Ψ is invariant and f is equivariant with respect to x. If $W^\Gamma = \{0\}$ we shall restrict $\mathcal{C}$ to the maps which have $\Psi(0, 0)$ and $\Psi(1, 0)$ both positive.

These mappings are divided in Γ-homotopy classes: $F \overset{\Gamma}{\sim} G$ if there is a continuous Γ-homotopy

$$H : [0, 1] \times S^V \to \mathbb{R} \times W\backslash\{0\}$$

such that:

(a) $H(0, t, x) = F(t, x)$, $H(1, t, x) = G(t, x)$, for (t, x) in S^V;

(b) $H(\tau, \cdot, \cdot)$ belongs to $\mathcal{C}$ for any τ in $[0, 1]$.

Definition 8.1. The set of all such Γ-homotopy classes will be denoted by $\Pi^\Gamma_{S^V}(S^W)$. The class of F will be denoted by $[F]_\Gamma$.

Remark 8.1. If V and W are trivial representations of Γ, then $\Pi^\Gamma_{S^V}(S^W)$ is nothing else than the abelian group $\Pi_n(S^m)$, where $n = \dim V$ and $m = \dim W$, for which the following facts are well known (see [Gr]).

(a) $\Pi_n(S^m) = 0$, if $n < m$, in which case any map f from S^n into $\mathbb{R} \times W\backslash\{0\}$ has a non-zero extension to $[0, 1] \times B_R$.

(b) $\Pi_n(S^n) \cong \mathbb{Z}$ and $[F]$ is its Brouwer degree.

(c) $\Pi_{n+1}(S^n) = \begin{cases} 0, & \text{if } n = 1 \\ \mathbb{Z}, & \text{if } n = 2 \\ \mathbb{Z}_2, & \text{if } n > 2, \end{cases}$ where the generator for $n = 2$ is the *Hopf map*: $\mathbb{C} \times \mathbb{C} \to \mathbb{R}^3$,

$$\eta(\lambda_1, \lambda_2) = (2\lambda_1\bar{\lambda}_2, |\lambda_1|^2 - |\lambda_2|^2),$$

and, for $n > 2$, is the *suspension* of the Hopf map: $\mathbb{C}\times\mathbb{C}\times\mathbb{R}^{n-3} \to \mathbb{R}^3\times\mathbb{R}^{n-3}$,

$$\Sigma_{n-3}\eta(\lambda_1, \lambda_2, \lambda) = (\eta(\lambda_1, \lambda_2), \lambda).$$

Now, the set $\Pi^{\Gamma}_{S^V}(S^W)$ has also a group structure. In order to define an addition we shall use the following result.

Lemma 8.1. *For any F in $\mathcal{C}$, there is a G in $\mathcal{C}$, such that $F \overset{\Gamma}{\sim} G$ and $G(t, x) = (1, 0)$ for $t = 0$ or 1.*

Proof. Define the following closed Γ-invariant set

$$A = \{0\} \times B_R \cup \{1\} \times B_R.$$

Clearly, the Γ-homotopy $F(t, \tau x)$ is admissible on A for any τ in $[0, 1]$. Then the restriction of F to A is Γ-homotopic to $H(t, x) = F(t, 0) = (\Psi(t, 0), f(t, 0))$, which is in $\mathbb{R} \times W^{\Gamma}\backslash\{0\}$, for $t = 0$ or 1. If $\dim W^{\Gamma} > 0$, one may choose two non-zero paths from $F(0, 0)$ and $F(1, 0)$ to $(1, 0)$. If $W^{\Gamma} = \{0\}$ (and hence $f(t, 0) = 0$), one may achieve the same goal since $\Psi(0, 0)$ and $\Psi(1, 0)$ are both positive.

The composition of both maps provides a deformation on A from $F(t, x)$ to $G(t, x) = (1, 0)$ on A.

Now, using the Γ-equivariant Borsuk extension theorem, the map F will be Γ-homotopic to a map G in $\mathcal{C}$, extending $(1, 0)$ on A to all of S^V. □

To proceed further, we need a concept of addition in $\Pi^{\Gamma}_{S^V}(S^W)$. To this end let F and G be any two maps belonging to $\mathcal{C}$. By virtue of Lemma 8.1, we may assume that $F|_A = G|_A = \{1, 0\}$. Define their sum $F \oplus G$ as the map

$$(F \oplus G)(t, x) = \begin{cases} F(2t, x), & \text{if } 0 \le t \le \frac{1}{2} \\ G(2t - 1, x), & \text{if } \frac{1}{2} \le t \le 1. \end{cases}$$

Clearly, $F \oplus G$ belongs to $\mathcal{C}$.

Definition 8.2. The addition in $\Pi^{\Gamma}_{S^V}(S^W)$ is given by

$$[F]_{\Gamma} + [G]_{\Gamma} = [F \oplus G]_{\Gamma}.$$

This addition turns out to be associative (see [Gr, p. 7]) and the class 0_Γ of the map $(1, 0)$ is the neutral element of the group. Note that, from Lemma 6.1, 0_Γ is the class of all maps which have a non-vanishing Γ-equivariant extension to the cylinder $I \times B_R$. Furthermore, the inverse element of $[F]_\Gamma$ is the class of $[F(1-t, x)]_\Gamma$. In fact, the Γ-homotopy $H_\tau(t, x)$, for $0 \le \tau \le 1$, defined as

$$H_\tau(t,x) = \begin{cases} F(2t, x), & \text{for } 0 \le 2t \le \tau \\ F(\tau, x), & \text{for } \tau \le 2t \le 2-\tau \\ F(2-2t, x), & \text{for } 2-\tau \le 2t \le 2 \end{cases}$$

is a valid Γ-deformation from $(1, 0)$, for $\tau = 0$, to $[F]_\Gamma - [F]_\Gamma$, for $\tau = 1$ (here we have assumed that $F(0, x) = F(1, x) = (1, 0)$).

Therefore, $\Pi^\Gamma_{S^V}(S^W)$ is a group under the addition defined above.

Lemma 8.2. *If* $\dim V^\Gamma > 0$, *then* $\Pi^\Gamma_{S^V}(S^W)$ *is an abelian group.*

Proof. Let x_0 be a coordinate, in V^Γ, of x. Let $A^+ = A \cup \{(t, x) : 0 \le t \le 1, \|x\| = R, x_0 \ge 0\}$, where A is the set used in the preceding lemma. Then, A^+ is closed and Γ-invariant. If F is in $\mathcal{C}$, with $F(t, x) = (1, 0)$ on A, i.e., if $t = 0$ or 1, consider the Γ-deformation of F restricted to A^+: $H_\tau(t, x_0, y) = F(t, \alpha(\tau)x_0 + \beta(\tau), \alpha(\tau)y)$, where $\alpha(\tau) = (1+R^{-1}x_0 \sin \pi\tau)^{-\frac{1}{2}} \cos \tau\pi/2$ and $\beta(\tau) = (1+R^{-1}x_0 \sin \pi\tau)^{-\frac{1}{2}} \sin \tau\pi/2$, which are chosen in such a way that the arguments in V have norm R if $\|x\| = R$. Thus, H_τ is a valid deformation on A^+, from F for $\tau = 0$, to $F(t, R, 0)$ for $\tau = 1$. One deforms next on A^+ via $F(t(1-\tau), R, 0)$, to $(1, 0)$.

Hence, $F|_{A^+}$ is Γ-homotopic to $(1, 0)$. Then, using the Γ-equivariant Borsuk homotopy extension theorem, the map F is Γ-homotopic to a map having value $(1, 0)$ on A^+.

Note that one could have performed the same procedure on A^-, corresponding to $x_0 \le 0$, by changing R to $-R$ in the deformation H_τ.

We are now in a position to prove the lemma. Indeed, consider two maps F_1 and F_2 such that $F_1(t, x) = (1, 0)$ for (t, x) in A^+ and $F_2(t, x) = (1, 0)$ for (t, x) in A^-. Define the following Γ-equivariant homotopy on S^V:

$$H_\tau(t,x) = \begin{cases} F_1(2t-\tau, x), & \text{for} (t,x) \text{ in } A^- \text{ and } 0 \le 2t-\tau \le 1 \\ F_2(2t-(1-\tau), x), & \text{for } (t,x) \text{ in } A^+ \text{ and } 0 \le 2t-(1-\tau) \le 1 \\ (1,0), & \text{otherwise.} \end{cases}$$

Simple computations give that H_τ is admissible. Moreover, H_0 is in $[F_1]_\Gamma + [F_2]_\Gamma$ and H_1 is in $[F_2]_\Gamma + [F_1]_\Gamma$. Thus $\Pi^\Gamma_{S^V}(S^W)$ is abelian. □

Part of Chapter 3 will be devoted to the computation of $\Pi^\Gamma_{S^V}(S^W)$. See also the Bibliographical remarks at the end of this chapter.

A construction that we shall use very often is that of the *suspension*, more precisely, that of an *equivariant suspension*: Let U, V and W be Γ-representations and $f : V \to W$ be a Γ-equivariant map.

Definition 8.3. The Γ-*suspension* of f is the map $\Sigma^U f = (f(x), u)$, from $V \times U$ into $W \times U$.

It is clear that if F belongs to $\mathcal{C}$, giving an element of $\Pi^{\Gamma}_{S^V}(S^W)$, then $(F(t,x), u) \equiv \Sigma^U F$ will provide an element of $\Pi^{\Gamma}_{S^{V\times U}}(S^{W\times U})$ and Σ^U will be a morphism between these two groups.

Remark 8.2. If Γ acts trivially on V and W and U is $\mathbb{R}$, then the Freudenthal suspension theorem asserts that $\Sigma : \Pi_n(S^m) \to \Pi_{n+1}(S^{m+1})$ is onto if $n = 2m - 1$ and an isomorphism if $n < 2m - 1$.

The situation for the equivariant case is more complicated. In the case of an abelian group, we shall prove, in Chapter 3, the appropriate result. In the general case we state, without proof, the corresponding result. We shall only indicate the references since we shall not use, in this book, the result in its full generality.

The following theorem is due to Namboodiri (cfr. [N]).

Theorem 8.1. *Assume* $V = \mathbb{R}^k \times W$. *Then* Σ^U *is one-to-one if for all isotropy subgroups* H *of* W *one has*

(α) $\dim W^H \geq k + 2$;

(β) $\dim W^{H\cap K} - \dim W^H \geq k+2$, *for any* K *isotropy subgroup for* U *which does not contain* H *or any conjugate of* H.

Moreover, if $k + 2$ *is replaced by* $k + 1$ *in the above inequalities, then* Σ^U *is onto.*

Note that if Γ acts trivially on W and $U = \mathbb{R}$, then the only condition is (α), which amounts to the standard Freudenthal suspension theorem.

In the case of an abelian action, with $V = \mathbb{R}^k \times W$, we shall prove, in Chapter 3, the stronger result:

Theorem 8.2. (a) Σ^U *is one-to-one provided*

$$\dim W^H \geq k + 2 - \dim \Gamma/H$$
$$\dim W^H - \dim W^K \geq k + 2 - \dim \Gamma/H,$$

for all isotropy subgroups H *and* K *of* W *such that* H *is strictly contained in* K *and* $K \cap H_0 = H$, *for some isotropy subgroup* H_0 *of* U.

(b) *If there are no new isotropy subgroups for* U, *then* Σ^U *is onto, replacing* $k+2$ *by* $k + 1$ *in the above inequalities. Otherwise, this will not be the case, in general, unless* $k = 0$ *and the new isotropy subgroups* H_0 *are such that* $\Gamma/H_0 \cong S^1$.

Note that (α) requires that $\dim W^\Gamma \geq k+2$ (unless $W^\Gamma = \{0\}$), while (a) gives a better result if one has $U^\Gamma = \{0\}$. Both conditions coincide if $U^\Gamma \neq \{0\}$. Note also that if one adds enough dummy variables to W (so that one gets to the point of $\dim W^\Gamma \geq k+2$ and $\dim W^H - \dim W^K \geq k+2-\dim \Gamma/H$, for any pair K, H in $\mathrm{Iso}(W)$, with $K > H$), then, in the abelian case, Σ^U will be one-to-one under any suspension. This stabilization process will be important when computing the degree through finite dimensional approximations. On the other hand, if in Theorem 8.1, one takes H to be the isotropy subgroup of W^K, when K is in $\mathrm{Iso}(U)$ but not in $\mathrm{Iso}(W)$, then $K < H$ and $W^K = W^H$: see Definition 2.1. In this case condition (β) is never satisfied.

Another argument which we will use often, and which is fundamental for bifurcation, is the deformation of families of linear maps. More precisely, assume that $B(\lambda)$ is a family of Γ-equivariant matrices, defined for $\|\lambda\| \leq \rho$, λ in $\mathbb{R}^k$, and invertible for $\|\lambda\| = \rho$, an S^{k-1}-sphere. One has an application:

$$S^{k-1} \to \mathrm{GL}_\Gamma(V),$$

the set of invertible Γ-equivariant matrices. If one considers all Γ-deformations of such matrices, one obtains an element of $\Pi_{k-1}(\mathrm{GL}_\Gamma(V))$.

Now, from Theorem 5.3, we know that $B(\lambda)$ has a block diagonal structure on the irreducible subrepresentations of V, that is, any Γ-deformation will have to preserve the structure and should be generated by deformations of families of restrictions $A^{\mathbb{R}}(\lambda)$, $A^{\mathbb{C}}(\lambda)$ or $A^{\mathbb{H}}(\lambda)$, as given in that theorem. The facts which will be used in this book are the following:

Theorem 8.3. (a) $\mathrm{GL}(\mathbb{R}^d)$ *has two components characterized by the sign of the determinant. Thus,* $\Pi_0(\mathrm{GL}(\mathbb{R}^d)) \cong \mathbb{Z}_2$, *where* $A^{\mathbb{R}}(\lambda)$ *is non-trivial if and only if its determinant changes sign.*

(b) *If* $\det A^{\mathbb{R}}(\lambda) > 0$, *then, for* $d = 2$, $\Pi_1(\mathrm{GL}^+(\mathbb{R}^d)) \cong \mathbb{Z}$ *and is generated by* $A(\lambda_1, \lambda_2) \equiv \begin{pmatrix} \lambda_1 & -\lambda_2 \\ \lambda_2 & \lambda_1 \end{pmatrix}$. *For* $d > 2$, $\Pi_1(\mathrm{GL}^+(\mathbb{R}^d)) \cong \mathbb{Z}_2$ *and is generated by* $\mathrm{diag}(A(\lambda_1, \lambda_2), I_{d-2})$. $\Pi_{k-1}(\mathrm{GL}^+(\mathbb{R}^d))$ *is an abelian group with* $[B]+[D] \equiv [BD]$.

(c) $\mathrm{GL}(\mathbb{C}^d)$ *and* $\mathrm{GL}(\mathbb{H}^d)$ *are connected, hence* $\Pi_0 = 0$ *for them. Also,*

$$\Pi_1(\mathrm{GL}(\mathbb{C}^d)) \cong \mathbb{Z},$$

where $A^{\mathbb{C}}(\lambda)$ *is deformable to* $\mathrm{diag}(\det A^{\mathbb{C}}(\lambda), I_{d-1})$ *and two families are homotopic if and only if their complex determinants, as maps from* S^1 *into* $\mathbb{C}\backslash\{0\}$, *are homotopic, i.e., they have the same winding number. Finally,* $\mathrm{GL}(\mathbb{H}^d)$ *is simply-connected.*

The proof of this result can be found in any book on Lie groups. Notice, that for $k > 1$ one has the Bott periodicity results, see [*I*], but in the present book we shall limit ourselves to the case $k \leq 1$. We shall see in particular how the non-connectedness of $\mathrm{GL}(\mathbb{R}^d)$ affects the computations of $\Pi^\Gamma_{S^V}(S^W)$.

A fundamental tool in bifurcation theory is the following extension of the Whitehead homomorphism: consider, as before, a family of Γ-matrices $B(\lambda)$, invertible for $\lambda \neq 0$ and such that $B(0) = 0$ (for instance $\rho^{-1}\|\lambda\| B(\lambda\rho/\|\lambda\|)$). In $\mathbb{R} \times V = \mathbb{R}^k \times W$ consider the ball $\mathcal{B} = \{(\lambda, x) : \|\lambda\| < 2\rho, \|x\| < 2R\}$ and the map

$$J^\Gamma(B(\lambda)x) = (\|x\| - R, B(\lambda)x).$$

Then, $J^\Gamma(B(\lambda)x)$ is non-zero on the boundary of the ball $\mathcal{B}$ and if $B(\lambda)$, defined as in the example from $\|\lambda\| = \rho$, is Γ-homotopic to $C(\lambda)$, then $J^\Gamma(B(\lambda)x)$ is Γ-homotopic to $J^\Gamma(C(\lambda)x)$, i.e., one has an induced map:

$$J^\Gamma : \Pi_{k-1}(\mathrm{GL}_\Gamma(W)) \to \Pi^\Gamma_{S^V}(S^W).$$

(Here, the variable t is given by λ_1).

Furthermore, if all $A^{\mathbb{R}}(\lambda)$ have positive determinant, then J^Γ is a morphism of abelian groups.

In the case $\Gamma = \{e\}$, the above construction is called the *Hopf construction*, J is the *Whitehead homomorphism* and has been thoroughly studied (for $k < d$) by Bott and Adams. Of particular importance is the kernel of J^Γ. In fact, if $J^\Gamma[B(\lambda)] \neq 0$, then, from Lemma 6.1, any Γ-extension of $J^\Gamma[B(\lambda)]$, or of any map Γ-deformable to it on $\partial\mathcal{B}$, from $\partial\mathcal{B}$ to $\mathcal{B}$ must have a zero. Now, if $g(\lambda, x) = 0(\|x\|^2)$, then for R small enough, the couple $(\|x\| - R, B(\lambda)x + g(\lambda, x))$ is Γ-deformable to $J^\Gamma[B(\lambda)]$ on $\partial\mathcal{B}$, provided $B(\lambda)$ is invertible for $\|\lambda\| = 2\rho$. Hence, if $J^\Gamma[B(\lambda)] \neq 0$, the couple will have zeros and the map $B(\lambda)x + g(\lambda, x) \equiv f(\lambda, x)$ will have zeros in $\mathcal{B}$, with $\|x\| = R$, for any small R, besides $(\lambda, 0)$.

Note that, on ∂B, the Γ-homotopy $((1 - \tau)(\|x\| - R) + \tau(\rho - \|\lambda\|), B(\lambda)x)$ is admissible. In fact, on $\partial\mathcal{B}$, if $B(\lambda)x = 0$, then either $\lambda = 0$ and $\|x\| = 2R$ (and the first component is positive), or $x = 0$ and $\|\lambda\| = 2\rho$ (and the first component is negative). Thus,

$$J^\Gamma[B(\lambda)] = [\rho - \|\lambda\|, B(\lambda)x].$$

For a more detailed exposition of J^Γ, see [*I*].

Our last set of preliminaries concerns Γ-orthogonal maps. As in § 7, let Γ be abelian and W be an orthogonal representation of Γ. Let V be $\mathbb{R}^k \times W$, where k may be 0. Then, one may consider the set $\mathcal{C}_\perp$ of all Γ-orthogonal maps F from $[0, 1] \times B_R$ into $R \times W$, which are not zero on the boundary of the cylinder.

Definition 8.4. The set of all Γ-orthogonal homotopy classes in $\mathcal{C}_\perp$ is denoted by $\Pi^\Gamma_{\perp S^V}(S^W)$.

Lemma 8.3. *If* $\dim V^\Gamma > 0$, *then* $\Pi^\Gamma_{\perp S^V}(S^W)$ *is an abelian group, where the addition is that of Definition 8.2.*

Proof. It is enough to check that one may repeat the arguments of Lemma 8.1 (i.e., that $F(t, x)$ can be taken orthogonally as $(1, 0)$ for $t = 0$ and $t = 1$) and that of

Lemma 8.2 (for the deformation on $A^{\pm}$). Now, both of those arguments were based on the equivariant Borsuk extension theorem, which is valid for orthogonal maps (see Corollary 7.1). □

We shall see, in Chapter 3, Section 6, that $\Pi^{\Gamma}_{\perp S^V}(S^W)$ has a much richer structure than $\Pi^{\Gamma}_{S^V}(S^W)$. One has also a J-homomorphism. In fact, let $B(\lambda)$ be a family of Γ-orthogonal matrices, then, from Lemma 7.2,

$$B(\lambda) = \operatorname{diag}(A_1^{\mathbb{R}}(\lambda), \dots, A_m^{\mathbb{R}}(\lambda), A_1^{\mathbb{C}}(\lambda), \dots, A_s^{\mathbb{C}}(\lambda)),$$

where $A_i^{\mathbb{R}}(\lambda)$ correspond to the irreducible real representations in V^{T^n}, while $A_j^{\mathbb{C}}(\lambda)$ are complex self-adjoint matrices in $(V^{T^n})^{\perp}$. As before, one has a map

$$J_{\perp}^{\Gamma} : \Pi_{k-1}(\mathrm{GL}_{\Gamma}^{\perp}(W)) \to \Pi^{\Gamma}_{S^V}(S^W).$$

Now, the connected components of $\mathrm{GL}_{\Gamma}^{\perp}(W)$ are characterized by the *Morse index* of $A_j^{\mathbb{C}}$ (i.e., the dimension of the space where $A_j^{\mathbb{C}}$ is negative definite). Note, that the addition in $\Pi_{k-1}(\mathrm{GL}_{\Gamma}^{\perp}(W))$ is given as in Definition 6.2 and does not correspond to a product (which is of course not self-adjoint). The base point (corresponding to the map $(1, 0)$ for $t = 0$ and $t = 1$) will be a matrix of the form $(-I_1, I_2)$ for each A_j, where I_1 is the identity on a space of dimension equal to the Morse index of A_j.

In this book we will only treat the case $k = 1$. For the general case see [IV3].

Remark 8.3. The reader may wonder where the finite dimensionality of the spaces was used, in particular in the definition of $\Pi^{\Gamma}_{S^V}(S^W)$. The answer is that it was never used and we invite the reader to go over the arguments and check that this group may be defined also in the case of infinite dimensional spaces.

The problem is that it is likely that this group would be trivial as the following example shows. Take $l^2 = \{(x_1, x_2, x_3, \dots) \text{ with } \sum x_i^2 < \infty\}$ and let S be the unit sphere in l^2. Now, the homotopy

$$h(\tau, x) = \tau(x_1, x_2, x_3, \dots) + (1 - \tau)(0, x_1, x_2, \dots)$$

is valid on S, since it is not 0 there. Hence, the identity on S is homotopic to $(0, x_1, x_2, \dots)$, which, via $((1 - \tau^2)^{\frac{1}{2}}, \tau x_1, \tau x_2, \dots)$ with norm 1, is in turn homotopic to $(1, 0, 0, \dots)$. Note that if there is a group action, the first homotopy will not be equivariant, unless there is only one isotropy subgroup. Hence, the identity is homotopic to $(1, 0, 0, \dots)$. Thus, any map $f(x_1, x_2, x_3, \dots)$, which is non-zero on S will be homotopic to $f(1, 0, 0, \dots)$, via $f(h(\tau, x)/\|h(\tau, x)\|)$ followed by the second homotopy. Hence, any map is homotopic to a constant and $\Pi_S(S) = \{0\}$.

This is one of the reasons for introducing, in Definition 6.1, compact perturbations of the identity: that is, if $B = U \times W$ and $E = V \times W$, with U and V finite dimensional representations of Γ, then the class $\mathfrak{C}$, is reduced to maps of the form

$$F(t, x) = (\Psi(t, x), g(t, u, w), w - h(t, u, w)),$$

where h is compact and $x = (u, w)$. The homotopies have to be in this class $\mathcal{C}$. In order to define an addition on the set $\Pi^{\Gamma}_{SB}(S^E)$, the most economical way is to use the approximation by finite dimensional compact Γ-maps of Theorem 4.1. In fact, if $F(t, x) \neq 0$ on S^B, then there is N such that $\|F(t, x)\| \geq 1/2^{N-1}$. If not, there would be (t_N, x_N) in S^B such that $\|F(t_N, x_N)\| < 1/2^{N-1}$, for all N. From the finite dimensionality of U, we may assume that (t_N, u_N) converge to (t, u) and, from the compactness of h, the sequence $h(t_N, u_N, w_N)$ would converge to some w. Then, w_N would converge to w and one would have $F(t, u, w) = 0$ for some point in S^B.

Take then h_N, with $h_N(I \times B_R) \subset M_N$ a finite dimensional subrepresentation of W, such that $\|h(t, x) - h_N(t, x)\| \leq 1/2^N$. Then, $F(t, x)$ is compactly Γ-homotopic on S^B to

$$F_N(t, u, w_N, \tilde{w}_N) = (\Psi(t, x), g(t, x), w - h_N(t, x)),$$

where w_N is in M_N and $\tilde{w}_N$ is in a complement subrepresentation. The map F_N is in turn compactly Γ-homotopic to

$$\tilde{F}_N(t, u, w_N, \tilde{w}_N) = (\Psi(t, u, w_N), g(t, u, w_N), w_N - h_N(t, u, w_N), \tilde{w}_N)$$

by deforming $\tilde{w}_N$ in the arguments of Ψ, g and h_N to 0.

Thus, $\tilde{F}_N$ is a suspension of a finite dimensional map by $\tilde{w}_N$. From Lemma 8.1, one may assume that $\tilde{F}_N = (1, 0, \tilde{w}_N)$ for $t = 0$ and $t = 1$, which is compactly Γ-homotopic to $(1, 0, w)$ for $t = 0$ or $t = 1$. Two such maps may be added, as in Definition 8.2, and Lemma 8.2 goes through for such maps, replacing $(1, 0)$ by $(1, 0, w)$ for the finite dimensional approximation h_N. Hence, $\Pi^{\Gamma}_{SB}(S^E)$, with compact perturbations of the identity, is an abelian group if $\dim B^{\Gamma} > 0$.

We leave to the reader the task of considering other classes of maps, such as k-set-contractions (see e.g. [IMPV]).

1.9 Symmetries and differential equations

The applications in this book will be mainly to ordinary differential equations. Although it is not difficult to see how to apply the equivariant degree to nonlinear PDE's or delay equations, we have chosen, in order to keep the spirit of the preface, to try to minimize the technical aspects which could obscure the interplay between Symmetry and Analysis.

The reader is invited to keep in mind the following example: Find 2π-periodic solutions to the equation

$$\frac{dX}{dt} = f(t, X, \lambda),$$

where X is in $\mathbb{R}^N$, λ is in the space of parameters, f is 2π-periodic in t (for instance f may be autonomous) and equivariant with respect to a group Γ_0, i.e., $f(t, \gamma_0 X, \lambda) = \tilde{\gamma}_0 f(t, X, \lambda)$.

For instance, if one wishes to find periodic solutions (of unspecified period) of the equation

$$\frac{dX}{d\tau} = f(X),$$

then, the time scaling $t = \nu\tau$ gives the equivalent system

$$\nu \frac{dX}{dt} = f(X)$$

and $2\pi/\nu$-periodic solutions of the first system correspond to 2π-periodic solutions of the second and the frequency ν appears as an extra-parameter.

As we have seen in Example 1.5, we may write $X(t)$ as

$$X(t) = \sum_{-\infty}^{\infty} X_n e^{int},$$

with X_n in $\mathbb{C}^N$, $X_{-n} = \bar{X}_n$ and obtain an equivalent formulation

$$inX_n - f_n(X_0, X_1, \ldots, \lambda) = 0, \quad n = 0, 1, 2, \ldots$$

where the Fourier coefficients will be Γ_0-equivariant, $S^1 \times \Gamma_0$-equivariant (as in Example 3.3), if f is autonomous, or $\mathbb{Z}_p \times \Gamma_0$-equivariant if $f(\cdot, X, \lambda)$ is $2\pi/p$-periodic in t.

The expression $\frac{dX}{dt} - f(t, X, \lambda)$ may be regarded as a nonlinear map from $C^1_{2\pi}(\mathbb{R}^N)$ into $C^0_{2\pi}(\mathbb{R}^N)$, or between the Sobolev spaces $H^1(S^1)$ to $L^2(S^1)$, where, for $p > 0$,

$$H^p(S^1) = \left\{X(t) = \textstyle\sum_{-\infty}^{\infty} X_n e^{int} : \sum_0^{\infty} |X_n|^2(1 + n^{2p}) < \infty\right\}.$$

Recall that $H^p(S^1) \subset C^0_{2\pi}(\mathbb{R}^N)$ for $p > \frac{1}{2}$.

Notice that, if f has a linearization $A(t)$ at some X_0, then $\frac{dX}{dt} - A(t)X$ is a *Fredholm operator* of index 0 between any of the above spaces.

Definition 9.1. Let B and E be Banach spaces and L be a linear continuous operator from B into E. Then L is said to be a *Fredholm operator* of index i if and only if:

(a) $\dim\ker L = d < \infty$,

(b) Range L is closed and has finite codimension d^*.

The index i of L is the difference $d - d^*$.

If, in the above example, one assumes that A is a constant matrix (for instance the 0 matrix), then $Lx = g$ is equivalent to

$$inX_n - AX_n = g_n,$$

which is always solvable for any g in L^2 (and then X is in H^1), provided g_n is in the range of $inI - A$ (always true for n large enough) i.e., if g_n is orthogonal to $\ker(-inI - A^T)$. From Linear Algebra one has that the index is 0. For the case of a non-constant A, periodic solutions will correspond to starting points X_0 such that $\Phi(2\pi)X_0 = X_0$, where $\Phi(t)$ is a fundamental matrix or one may use a deformation of $A(t)$ to 0, using the Ljapunov–Schmidt reduction.

In fact, one of the important properties of maps which have linearizations which are Fredholm operators, is the reduction to a finite dimensional local problem.

Assume that B and E are Banach spaces and consider the equation

$$F(\lambda, x) = Ax - T(\lambda)x - g(\lambda, x)$$

from $\mathbb{R}^k \times B$ into E, A a Fredholm operator, $T(\lambda)$ is a family of continuous linear operators with $T(0) = 0$, $\|T(\lambda)\| \to 0$ as $\lambda \to 0$ and $g(\lambda, x) = o(\|x\|)$, uniformly on λ. If B and E are Γ-spaces we shall assume that A, $T(\lambda)$ and g are Γ-equivariant.

Let then P and Q be two projections (which we may assume to be equivariant, since $\ker A$ and Range A are subrepresentations) P from B onto $\ker A$ and Q from E onto Range A. Then,

$$\begin{aligned} B &= \ker A \oplus B_2 \\ E &= E_2 \oplus \text{Range } A \end{aligned}$$

with B_2 a closed subspace (a subrepresentation by Lemma 4.4) and E_2 of dimension d^*. Any x in B is written as $x = x_1 + x_2$, with $x_1 = Px$.

Since A is continuous, one-to-one from B_2 onto Range A, there is a continuous inverse from Range A onto B_2, that is,

$$AKQ = Q, \quad KA(I - P) = I - P.$$

One may write the equation as

$$(A - QT(\lambda))(x_1 + x_2) - Qg(\lambda, x_1 + x_2) \ominus (I - Q)(T(\lambda)(x_1 + x_2) + g(\lambda, x_1 + x_2))$$

and, using the facts that $A - QT(\lambda) = A(I - KQT(\lambda))$, where for λ small, $I - KQT(\lambda)$ is an invertible mapping from B into itself, with an inverse which is given by power series and that

$$T(\lambda)(I - KQT(\lambda))^{-1}KQ = T(\lambda)KQ + (T(\lambda)KQ)^2 + \cdots = (I - T(\lambda)KQ)^{-1} - I,$$

as a mapping from E into E, one has that

$$\begin{aligned} F(\lambda, x) = (A - QT(\lambda))[x_2 - (I - KQT(\lambda))^{-1}KQ(T(\lambda)x_1 + g(\lambda, x))] \\ \ominus (I - Q)[T(\lambda)((I - KQT(\lambda))^{-1}x_1 + x_2 \\ - (I - KQT(\lambda))^{-1}KQ(T(\lambda)x_1 + g(\lambda, x))) \\ + (I - T(\lambda)KQ)^{-1}g(\lambda, x)]. \end{aligned}$$

In order to better appreciate this formula, define

$$\begin{aligned} H(\lambda, x_1, x_2) &= x_2 - (I - KQT(\lambda))^{-1} KQ(T(\lambda)x_1 + g(\lambda, x)) \\ B(\lambda) &= -(I - Q)T(\lambda)(I - KQT(\lambda))^{-1} P \\ G(\lambda, x) &= -(I - Q)(I - T(\lambda)KQ)^{-1} g(\lambda, x). \end{aligned}$$

One then has

$$F(\lambda, x) = (A - QT(\lambda))H(\lambda, x_1, x_2) \oplus B(\lambda)x_1 + G(\lambda, x) - (I - Q)T(\lambda)H(\lambda, x_1, x_2).$$

It is clear that, if $F(\lambda, x) = 0$ and for small λ and x, then $H(\lambda, x_1, x_2) = 0$ has a unique solution $x_2 = x_2(\lambda, x_1)$, with $\|x_2\| \leq C\|x_1\|(\|\lambda\| + 0(\|x_1\|))$, provided $g(\lambda, x)$ is C^1 and $\|T(\lambda)\| \leq C\|\lambda\|$, by using any contraction mapping argument.

Then the zeros of F coincide with those of the *bifurcation equation*

$$B(\lambda)x_1 + G(\lambda, x_1 + x_2(\lambda, x_1)) = 0,$$

where $B(0) = 0$, $B(\lambda)$ is a $d \times d^*$ matrix and $G(\lambda, x_1) = o(\|x_1\|)$.

Taking $g = 0$, one has that $\dim\ker(A - T(\lambda)) = \dim\ker B(\lambda)$, while for g any element of E, one gets $\operatorname{codim}\operatorname{Range}(A - T(\lambda)) = \operatorname{codim}\operatorname{Range} B(\lambda)$, that is, the spectral properties of $A - T(\lambda)$ can be recovered from those of $B(\lambda)$. In particular, $\operatorname{Range}(A - T(\lambda))$ is closed and $A - T(\lambda)$ is a Fredholm operator of index $d - d^*$.

An important particular case, which will be used throughout the book, is when $B = E$, $A = I - T$ and $T(\lambda) = \lambda T$, with T a compact operator. In this case one may build up the projections P and Q in two stages. In fact, since T is compact, one has that $\ker(I - T)^\alpha = \ker(I - T)^{\alpha+\beta}$ for all $\beta > 0$, so, for an α, called the ascent of $I - T$, the dimension m of $\ker(I - T)^\alpha$ is the algebraic multiplicity and one has

$$E = \ker(I - T)^\alpha \oplus \operatorname{Range}(I - T)^\alpha.$$

Both subspaces are invariant under T and $A = I - T$ is nihilpotent on $\ker(I - T)^\alpha$, hence one may choose a basis such that A is in Jordan form, with d blocks of dimension m_j. On a typical block of dimension m, one has

$$A = I - T = \begin{pmatrix} 0 & 1 & & 0 \\ & \ddots & \ddots & \\ & & \ddots & 1 \\ 0 & & & 0 \end{pmatrix} = J, \qquad Q = \begin{pmatrix} 1 & & & 0 \\ & \ddots & & \\ & & 1 & \\ 0 & & & 0 \end{pmatrix},$$

$$K = \begin{pmatrix} 0 & & & 0 \\ 1 & \ddots & & \\ & \ddots & \ddots & \\ 0 & & 1 & 0 \end{pmatrix} = J^T, \qquad I - P = \begin{pmatrix} 0 & & & 0 \\ & 1 & & \\ & & \ddots & \\ 0 & & & 1 \end{pmatrix},$$

$$T(\lambda) = \lambda T = \lambda I - \lambda J.$$

Since $QJ = J$, one obtains $J^T Q = J^T$ and $J^T J = I - P$, then $KQT(\lambda) = \lambda J^T Q(I - J) = \lambda J^T - \lambda(I - P)$ and it is easy to check directly that

$$(I - KQT(\lambda))^{-1} = \begin{pmatrix} 1 & 0 & 0 & 0 & \dots & 0 \\ \frac{\lambda}{1+\lambda} & \frac{1}{1+\lambda} & 0 & 0 & \dots & 0 \\ \frac{\lambda^2}{(1+\lambda)^2} & \frac{\lambda}{(1+\lambda)^2} & \frac{1}{1+\lambda} & 0 & \dots & 0 \\ \frac{\lambda^3}{(1+\lambda)^3} & \frac{\lambda^2}{(1+\lambda)^3} & \frac{\lambda}{(1+\lambda)^2} & \frac{1}{1+\lambda} & \dots & 0 \\ \vdots & \vdots & \vdots & \vdots & \ddots & \\ \frac{\lambda^{m-1}}{(1+\lambda)^{m-1}} & \frac{\lambda^{m-2}}{(1+\lambda)^{m-1}} & \dots & \dots & \dots & \frac{1}{1+\lambda} \end{pmatrix}$$

Hence, the first column of $T(\lambda)(I - KQT(\lambda))^{-1}$ will be

$$\left(\frac{\lambda}{1+\lambda}, \frac{\lambda^2}{(1+\lambda)^2}, \dots, \frac{\lambda^{m-1}}{(1+\lambda)^{m-1}}, \frac{\lambda^m}{(1+\lambda)^{m-1}} \right)^T$$

and $B(\lambda)$ on this block will be $-\lambda^m/(1+\lambda)^{m-1}$.

Hence $B(\lambda)$ will be a diagonal matrix with components $-\lambda^{m_j}/(1+\lambda)^{m_j-1}$, for $j = 1, \dots, d$ and $\sum m_j = m$, the algebraic multiplicity.

Another case, which is used mainly for bifurcation purposes, is when $B \subset E$, A is a Fredholm operator of index 0, with 0 as an isolated eigenvalue and $T(\lambda) = \lambda I$.

In this case, one has a finite ascent α and

$$B = B \cap \text{Range } A^\alpha \oplus \ker A^\alpha, \quad E = \text{Range } A^\alpha \oplus \ker A^\alpha.$$

On Jordan blocks as before, A, Q, K, and $I - P$ have the same form, while

$$(I - KQT(\lambda))^{-1} = \begin{pmatrix} 1 & 0 & \dots & \dots & 0 \\ \lambda & 1 & 0 & \dots & 0 \\ \vdots & \ddots & \ddots & & \\ \vdots & \ddots & \ddots & \ddots & \\ \lambda^{m-1} & \dots & \dots & \lambda & 1 \end{pmatrix},$$

since $I - \lambda KQ = I - \lambda J$. Then, on the block, if x in $\ker A^\alpha$ has coordinates $(x_1, 0, \dots, 0)$ and g has components $(g_1, \dots, g_m)$, one obtains

$$B(\lambda)x + G(\lambda, x)|_{\text{Block}} = \lambda^m x_1 + \lambda^{m-1} g_1 + \dots + g_m.$$

Then $B(\lambda)$ is a diagonal matrix with entries $\lambda^{m_1}, \dots, \lambda^{m_d}$, with $\sum m_j = m = \dim \ker A^\alpha$, the algebraic multiplicity.

Example 9.1 (Equivariant maps). If A, $T(\lambda)$ and g are equivariant, then, by choosing equivariant projections, one sees easily that K is also equivariant, the uniqueness of $x_2(\lambda, x_1)$ will imply that $x_2(\lambda, \gamma x_1) = \gamma x_2(\lambda, x_1)$ and the bifurcation equation is equivariant. In particular, $B(\lambda)$ has a block-diagonal form, in case $B(\lambda)$ is invertible.

Example 9.2 (Gradient maps). If B is continuously imbedded in the Hilbert space E and $F(\lambda, x)$ is the gradient of a C^2 functional $\Phi(\lambda, x)$, i.e., $\Phi_x(\lambda, x)h = (F(\lambda, x), h)$, for all h in B, then A and $T(\lambda)$ are self-adjoint operators. Assume that A is a Fredholm operator (hence of index 0).

One may choose $B_2 = B \cap \text{Range } A$, $E_2 = \ker A$ and $Q = I - P$. It is easy to see that K is symmetric and that $B(\lambda)^T = B(\lambda)$.

Furthermore, it is clear that $x_2(\lambda, x_1)$ is C^1. Let

$$\Psi(\lambda, x_1) = \Phi(\lambda, x_1 + x_2(\lambda, x_1)),$$

then the Frechet derivative of Ψ is such that, for h in $\ker A$

$$\begin{aligned}
&\Psi(\lambda, x_1 + h) - \Psi(\lambda, x_1) \\
&\quad = \Phi_x(\lambda, x_1 + x_2(\lambda, x_1))(h + x_2(\lambda, x_1 + h) - x_2(\lambda, x_1) + o(h)) \\
&\quad = \Phi_x(\lambda, x_1 + x_2(\lambda, x_1))(I - KQT(\lambda))^{-1}h + o(h) \\
&\quad = (F(\lambda, x_1 + x_2(\lambda, x_1)), h) + o(h) \\
&\quad = (B(\lambda)x_1 + G(\lambda, x_1 + x_2(\lambda, x_1)), h) + o(h),
\end{aligned}$$

where, in the last equality, one uses that $F(\lambda, x_1 + x_2(\lambda, x_1))$ belongs to $\ker A$, while $KQT(\lambda)h$ belongs to Range A. Hence,

$$\nabla\Psi(\lambda, x_1) = B(\lambda)x_1 + G(\lambda, x_1 + x_2(\lambda, x_1)).$$

Example 9.3 (Orthogonal maps). Assume that $B \subset E$ are both Γ-Hilbert spaces and let $F(\lambda, x)$ be Γ-orthogonal, with respect to the scalar product in E and Γ is abelian. As above, A is a Fredholm operator of index 0.

Lemma 9.1. *Under the above hypothesis, one may choose P and Q such that the bifurcation equation is Γ-orthogonal.*

Proof. From Lemma 7.2, the orthogonality of $F(\lambda, x)$ implies that A, $T(\lambda)$ and $g(\lambda, x)$ are also Γ-orthogonal. In particular, $A - T(\lambda)$ has a diagonal structure on equivalent irreducible representations of Γ and, on $(E^{T^n})^\perp$, its restriction has a complex self-adjoint form $\tilde{A} - \tilde{T}(\lambda)$ and the above space has the decomposition $\ker \tilde{A} \oplus \text{Range } \tilde{A}$. Choose P and Q equivariant, hence K and $B(\lambda)$ will be equivariant and will commute with A_j. Furthermore, one may choose an orthogonal projection $\tilde{P}$ onto $\ker \tilde{A}$ with $\tilde{Q} = I - \tilde{P}$, hence the part of $B(\lambda)$ on $\ker A \cap (E^{T^n})^\perp$ will be $\tilde{B}(\lambda) = -\tilde{P}\tilde{T}(I - \tilde{K}(I - \tilde{P})\tilde{T})^{-1}\tilde{P}$ which commutes with A_j and is self-adjoint (expand the inverse in power series). Hence, $B(\lambda)$ is Γ-orthogonal. On the other hand,

$$\begin{aligned}
-(G(\lambda, x), A_j x_1) &= ((I - \tilde{T}\tilde{K}\tilde{Q})^{-1}g, A_j x_1) \\
&= (g, (I - \tilde{Q}\tilde{K}\tilde{T})^{-1}A_j x_1) \\
&= (g, A_j x_1) + (\tilde{Q}g, \tilde{K}\tilde{T}(I - \tilde{Q}\tilde{K}\tilde{T})^{-1}A_j x_1),
\end{aligned}$$

by using the fact that A_j is 0 on E^{T^n} and that it has a diagonal structure. Since g is orthogonal, one may replace the first term by $-(g, A_j x_2)$. But $x_2(\lambda, x_1)$ solves $Qg = (A - QT)(x_1 + x_2)$, hence, using the fact that A and T are orthogonal and Q commutes with A_j, one obtains

$$(g, A_j x_1) = (QTx_1, A_j x_2).$$

The same substitution in the second term yields

$$((I - \tilde{T}\tilde{K}\tilde{Q})^{-1}\tilde{T}\tilde{K}(A - QT)x_2, A_j x_1) - (x_1, \tilde{T}\tilde{Q}\tilde{K}\tilde{T}(I - \tilde{Q}\tilde{K}\tilde{T})^{-1}A_j x_1),$$

where the first term reduces to $(\tilde{T}x_2, A_j x_1)$, by writing $\tilde{T}\tilde{K}(\tilde{A} - \tilde{Q}\tilde{T}) = \tilde{T} - \tilde{T}\tilde{K}\tilde{Q}\tilde{T} = (I - \tilde{T}\tilde{K}\tilde{Q})\tilde{T}$, since on E^{T^n} one has $A_j x_1 = 0$. The second term is of the form $(x_1, LA_j x_1)$, with L self-adjoint (expand again the inverse in power series) and hence 0, since we have seen that orthogonality is equivalent to self-adjointness for linear operators. Thus

$$-(G(\lambda, x), A_j x_1) = (Tx_1, A_j x_2) + (Tx_2, A_j x_1) = 0,$$

since T is Γ-orthogonal. □

In the case of autonomous differential equations, we shall assume that the equation

$$F(X, \lambda) \equiv \frac{dX}{dt} - f(X, \lambda) = 0,$$

for λ in $\mathbb{R}^k$ and X in $\mathbb{R}^N$, is such that there are bounded sets Λ in $\mathbb{R}^k$ and Ω in $\mathbb{R}^N$, with the following properties:

1. $f(\gamma_0 X, \lambda) = \gamma_0 f(X, \lambda)$, for γ_0 in Γ_0, a compact abelian Lie group, of dimension n.

2. Ω is invariant under Γ_0 and any 2π-periodic solution in $\bar{\Omega}$ is in fact in Ω, for any λ in Λ.

Let then

$$\tilde{\Omega} \equiv \{X \in H^1(S^1) : \|X\|_1 < R, X(t) \in \Omega\},$$

where R is chosen so large that any periodic solution in Ω has $\|X\|_1 < R/2$: R depends upon bounds on f over $\Lambda \times \Omega$ and Sobolev constants. Then, $F(X, \lambda) \neq 0$ on $\partial\tilde{\Omega}$ and is equivariant with respect to $\Gamma \equiv S^1 \times \Gamma_0$.

As particular cases we shall also consider the problem of finding 2π-periodic solutions to the following *Hamiltonian system*

$$\mathcal{H}(X, \lambda) = JX' + \nabla H(X, \lambda) = 0,$$

where X is in $\mathbb{R}^{2N}$, J is the standard symplectic matrix $\begin{pmatrix} 0 & -I \\ I & 0 \end{pmatrix}$ and H is C^2 and Γ_0-invariant.

In this case we shall assume that Γ_0 acts *symplectically on* $\mathbb{R}^{2N}$, i.e., it commutes with J. Then, again, $\mathcal{H}(X, \lambda)$ is Γ-equivariant, with $\Gamma = S^1 \times \Gamma_0$. In fact, the following result holds.

Proposition 9.1. *The mapping $\mathcal{H}$ is Γ-orthogonal with respect to the $L^2(S^1)$ scalar product.*

Proof. Here the infinitesimal generators for Γ will be $AX \equiv X'$ for the action of S^1 and $A_j X$, $j = 1, \dots, n$, if the rank of Γ_0 is n (i.e., $\dim \Gamma_0 = n$). Then

$$(\mathcal{H}(X, \lambda), AX) = \int_0^{2\pi} (JX' \cdot X' + \nabla H(X, \lambda) \cdot X')\, dt = 0,$$

since $JY \cdot Y = 0$ and the second term integrates to $H(X(t), \lambda)$, giving 0 on periodic functions.

On the other hand $\nabla H(X, \lambda) \cdot A_j X = 0$, since H is Γ_0-invariant, and

$$(JX', A_j X) = \int_0^{2\pi} -(X^T J A_j X)'\, dt/2 = 0,$$

where we have used the relations $J^T = -J$, $A_j^T = -A_j$, $JA_j = A_j J$.

Thus, $\mathcal{H}(X, \lambda)$ is Γ-orthogonal. □

The second particular case is that of a second order Hamiltonian

$$E(X, \lambda) = -X'' + \nabla V(X, \lambda) = 0,$$

under the same assumptions on the potential V. One has the same infinitesimal generator $AX = X'$ and $A_j X$, if V is Γ_0-invariant. Here, of course, B is $H^2(S^1)$ and, as before, one has

Proposition 9.2. *$E(X, \lambda)$ is Γ-orthogonal with respect to the $L^2(S^1)$ scalar product.*

Note that we have taken $-X''$ so that the associated operator is non-negative on $L^2(S^1)$.

Recall that the equation $F(X, \lambda) = 0$ is equivalent to

$$inX_n - f_n(X, \lambda) = 0, \quad n = 0, 1, 2, \dots,$$

where $X(t) = \sum X_n e^{int}$ with $X_{-n} = \bar{X}_n$ in $\mathbb{C}^N$. Recall also that the action of Γ_0 on $\mathbb{R}^N$ decomposes this space in irreducible subrepresentations of Γ_0 and one may write any X in $\mathbb{R}^N$ as $(x^1, \dots, x^s)$, with x^j in $\mathbb{R}$ or $\mathbb{C}$, and the action of Γ_0 on x^j is of the form

$$\gamma_0^j \equiv \exp i(\langle N^j, \Phi \rangle + 2\pi \langle K, L^j/M \rangle),$$

(see Example 1.4 and §2). Then, on the j-th coordinate of X_n, the action of $\Gamma = S^1 \times \Gamma_0$, will be of the form

$$\gamma_n^j = \gamma_0^j e^{in\varphi}.$$

Remark 9.1. In the Hamiltonian case, Γ_0 commutes with J, hence if $X = (Y, Z)$, with Y and Z in $\mathbb{R}^N$, then the actions of Γ_0 on Y and Z are the same. If one of the complex irreducible representations of Γ_0 associates one coordinate of Y with one of Z, then J on this pair takes the form of a multiplication by i.

Remark 9.2. For the general case it is easy to see that $F(X, \lambda) = 0$ may be written as $X_n - f_n/(in)$, i.e., of the form Id-compact on $H^1(S^1)$, a situation where one will be able to use the equivariant degree in infinite dimension. In the Hamiltonian case, one could use the same argument (by multiplying by $J/(in)$) but then one looses the orthogonality. One has then to keep the strongly indefinite operator JX' and use a global Ljapunov–Schmidt reduction in the following form: On a large ball in $H^1(S^1)$, one has that $X(t)$ is bounded as well as $D^2H(X, \lambda)$ (thus, we need that H is C^2). Write $X = X_1 \oplus X_2$, where $X_1 = PX$ corresponds to modes n, with $|n| \leq N_1$ and X_2 to the others. Note that JX' is self-adjoint on $L^2(S^1)$ and a Fredholm operator of index 0 from $H^1(S^1)$ into $L^2(S^1)$. The equation

$$(I - P)JX' + (I - P)\nabla H(X, \lambda) = 0$$

is uniquely solvable for X_2 as a C^1-function of X_1, for N_1 large enough. In fact, the linearization at any X_0 in the ball has the property that

$$\|JX_2' + (I - P)D^2H(X_0, \lambda)X_2\|_{L^2} \geq (1 - M/N_1)\|X_2\|_{H^1},$$

where M is a uniform bound for $\|D^2H(X_0, \lambda)\|$. Hence, the global implicit function theorem may be applied. Furthermore, since $(\nabla H(X, \lambda), AX) = 0$, where AX is either X', or A_jX, one has that the scalar product

$$(P\nabla H(X_1+X_2(X_1, \lambda)), AX_1) = -((I-P)\nabla H, AX_2) = ((I-P)JX_2', AX_2) = 0.$$

Thus, the reduced equation is Γ-orthogonal and one may look at

$$JX_1' + P\nabla H(X_1 + X_2(X_1, \lambda), \lambda) = 0,$$

in the finite dimensional space $PH^1(S^1)$, where the second term inherits the gradient structure, as in Example 9.2.

Let now $X_0(t)$ be a 2π-periodic function such that $F(X_0(t), \lambda) = 0$, with $\Gamma_{X_0} \equiv H$. Then, if $\dim \Gamma/H = k$, one has that $X_0'(t)$, $A_jX_0(t)$ belong to ker $DF(X_0, \lambda)$, by Property 3.3, and exactly k of these vectors are linearly independent (Lemma 2.4). In other words, each of the above vectors is a solution of the equation

$$Y' - Df(X_0(t), \lambda)Y = 0,$$

respectively $JY' + D^2H(X_0(t), \lambda)Y = 0$, or $-Y'' + D^2V(X_0(t), \lambda)Y = 0$.

We would have to identify V^H, $V^{\underline{H}}$ and V^K, where $\underline{H}$ is the torus part of H and $H/K \cong \mathbb{Z}_2$, where the subgroups K with that property will lead to *period doubling* and *"twisted orbits"* as explained below.

We shall consider three possible cases:

(a) *A* time-stationary $X_0(t)$

(b) *A* rotating wave $X_0(t)$

(c) *A* truly time periodic $X_0(t)$.

Case 9.1. *Time-stationary* $X_0(t)$. If $X_0'(t) \equiv 0$, then $H = S^1 \times H_0$, with $H_0 < \Gamma_0$ such that $\dim \Gamma_0/H_0 = k$ and $\underline{H} = S^1 \times T^{n-k}$. Thus, $V^{\underline{H}}$ is contained in $\mathbb{R}^N$, the space of constant functions. Recall that $Df(X_0, \lambda)$ is H-equivariant and has the diagonal structure of Property 3.4 and Theorem 5.3. Since this matrix is constant, one has for each mode n, the linearization

$$(inI - Df(X_0, \lambda))X_n.$$

The spectral properties of $Df(X_0, \lambda)$ will be crucial when discussing the *Hopf bifurcation*, i.e., bifurcation of truly periodic solutions near the constant solution X_0.

Case 9.2. *Rotating wave* $X_0(t)$. Assume that X_0' is a linear combination of the A_jX_0's Writing this relation on Fourier series and taking into account that A_j is diagonal (being equivariant), it is easy to see that for each coordinate z_s of $\mathbb{R}^N$, with a non-trivial action of T^n, there is at most one mode n_s such that X_0' is non-zero on that mode.

Consider then the matrix

$$A(t) = \operatorname{diag}(\dots, e^{-in_st}, \dots),$$

written this way according to the action of Γ_0, i.e., each exponential corresponds to a rotation for a pair of real coordinates of X.

Let then $Y(t) = A(t)X(t)$. If $Y_0(t) = A(t)X_0(t)$, then, since the s'th component of $X_0(t)$ is $e^{in_st}X_{n_s}$, one has that $Y_0'(t) = 0$, i.e., the rotating wave $X_0(t)$ has been frozen.

Furthermore,

$$Y' = A'(t)A^{-1}(t)Y(t) + A(t)f(A^{-1}(t)Y(t), \lambda),$$

for any solution of $F(X(t), \lambda) = 0$. It is easy to see that $A'A^{-1} = A'(0)$. Also, since $n_s = \sum \lambda_j n_j^s$, where the action of T^n on z_s is via $\exp(i\langle N^s, \Phi\rangle)$, with $N^s = (n_1^s, \dots, n_n^s)^T$, then, from the equivariance of f with respect to Γ_0, one has that $f(A^{-1}(t)Y(t), \lambda) = A^{-1}f(Y(t), \lambda)$, by taking $\varphi_j = \lambda_j/t$. Hence,

$$Y'(t) = A'(0)Y(t) + f(Y(t), \lambda),$$

and one has a reduction to the previous case.

Remark 9.3. If the s'th coordinate of X is $X^s = x + iy$, then $e^{-in_s t}X^s$ has to be interpreted as

$$\begin{pmatrix} \cos n_s t & \sin n_s t \\ -\sin n_s t & \cos n_s t \end{pmatrix} \begin{pmatrix} x \\ y \end{pmatrix}$$

and $X^s(t) = e^{in_s t}X_{n_s}$ has the same decomposition. This can also be seen from the fact that

$$x(t) = \sum_{n\geq 0}(x_n e^{int} + \bar{x}_n e^{-int})$$
$$y(t) = \sum_{n\geq 0}(y_n e^{int} + \bar{y}_n e^{-int}).$$

Then one has $x(t) + iy(t) = X^s(t) = \sum_{n\geq 0}(x_n + iy_n)e^{int} + (\bar{x}_n + i\bar{y}_n)e^{-int}$. If $n_s > 0$, then $x_n + iy_n = 0$ for $n \neq n_s$ and $\bar{x}_n + i\bar{y}_n = 0$ for all $n \geq 0$. In particular, $x_{n_s} = iy_{n_s} = X_{n_s}/2$. Hence, one recovers the above expression for $X^s(t)$.

For a Hamiltonian system, the coordinates z_s come in pairs or J acts as i on a complex coordinate. This implies that J commutes with $A(t)$. One arrives at $Y'(t) = A'(0)Y(t) + J\nabla H(Y(t), \lambda)$, i.e., to

$$JY' - JA'(0)Y + \nabla H(Y(t), \lambda) = 0,$$

and a new Hamiltonian $\tilde{H}(Y, \lambda) = H(Y, \lambda) - (JA'(0)Y, Y)/2$. (Note that, since J and $A'(0)$ commute and both are antisymmetric, their product is self-adjoint).

For a second order Hamiltonian, the above transformation gives rise to

$$-Y'' - A'(0)^2Y + 2A'(0)Y' + \nabla V(Y, \lambda) = 0.$$

We leave to the reader to check that this equation is Γ-orthogonal.

Case 9.3. *Truly periodic solutions.* If X_0', $A_1X_0, \ldots, A_{k-1}X_0$ are linearly independent, we may assume, from Case 9.2, that $A_kX_0, \ldots, A_nX_0$ are linear combinations of $A_1X_0, \ldots, A_{k-1}X_0$ only. In particular, if $k = 1$, then $A_jX_0 = 0$ and X_0 belongs to V^{T^n}. In general, from Lemma 2.4, one may reparametrize T^n such that on $V^{\underline{H}}$ one has $A_jX = 0$, for $j \geq k$, where $\underline{H}$ is the torus part of H (see Lemma 2.6).

Assume that $X_0(t)$ is $2\pi/p$-periodic in time, hence $H = \mathbb{Z}_p \times H_0$, with modes which are multiples of p. One has $\dim \Gamma_0/H_0 = k - 1$ and $\underline{H} = \underline{H}_0 = T^{n-k+1}$.

Lemma 9.2. *Let* $V_0 \equiv (\mathbb{R}^N)^{\underline{H}_0}$, *then* $V^{\underline{H}} = \{X(t) \in V_0, \forall t\}$ *and* $(V^{\underline{H}})^{\perp} = \{X(t) \in V_0^{\perp}, \forall t\}$.

Proof. In fact, $H = \{(\varphi, \Phi, K) : n\varphi + \langle N^j, \Phi\rangle + \langle K, L^j/M\rangle \in \mathbb{Z}$, for each non-zero component X_n^j of $X_0\}$, where N^j and L^j have to be interpreted as in §2. (Here $\varphi, \varphi_1, \ldots, \varphi_n$ are in [0,1] and N^j, K, L^j, M are integer-valued vectors). From the reparametrization of T^n, the phases $\varphi_k, \dots, \varphi_n$ do not appear in $\langle N^j, \Phi\rangle$ and the fact that X_0' is linearly independent from $A_j X_0$, restricts $\varphi, \varphi_1, \dots, \varphi_{k-1}$ to a discrete set in the above expression. Hence, the torus part of H and H_0 corresponds to $(\varphi_k, \dots, \varphi_n)$. The lemma is then clear. □

Lemma 9.3. *There is a γ_0 in Γ_0 such that $\gamma_0^{q_0} X_0 = X_0$ and $X_0(t) = \gamma_0 X_0(t+2\pi/q)$, with $q = pq_0$.*

Proof. As noted above, the set of $(\varphi, \varphi_1, \dots, \varphi_{k-1})$ in H is discrete. Since Γ is compact, there is a positive minimum φ_0 such that (φ_0, Ψ_0, K_0) is in H, where Ψ_0 corresponds to the reparametrization of Lemma 2.4. From the congruences, φ_0, as well as each component of Ψ_0, is a rational, of the form r/q, with r and q coprime. If $r > 1$, then there are integers k and a such that $kr + aq = 1$ and, changing φ_0 to $k\varphi_0$, one may take $\varphi_0 = 1/q$.

Then, $X_0(t) = \gamma_0 X_0(t + 2\pi/q)$, where γ_0 corresponds to (Ψ_0, K_0). Now, any other element of H gives $X_0(t) = \gamma X_0(t + 2\pi\varphi)$, with γ corresponding to (Ψ, K). For such an element, let k be such that $0 \le \varphi - k\varphi_0 < \varphi_0$. Then, $X_0(t) = \gamma\bar{\gamma}_0^k X_0(t + 2\pi(\varphi - k\varphi_0))$ and $(\varphi - k\varphi_0, \Psi - k\Psi_0, K - kK_0)$ belongs to H, contradicting the minimality of φ_0, unless $\varphi = k\varphi_0$ and $\gamma = \gamma_0^k$.

Recall that $H_0 < \Gamma_0$ is the isotropy subgroup of the geometrical coordinates of $X_0(t)$. Since $\varphi_0 = 1/q$, one has that γ_0^q is in H_0 and then

$$H = \{k(\varphi_0, \Psi_0, K_0), k = 0, \dots, q-1\} \cup \{(\Psi, K) \in H_0\}.$$

Let q_0 be the smallest integer such that $\gamma_0^{q_0} \in H_0$: from the minimality $q = pq_0$, one has $\gamma_0^{q_0} X_0 = X_0$ and the lemma is proved. □

Lemma 9.4. *The space V^H consists of all $2\pi/p$-periodic functions $X(t)$ with $X(t) \in V_0^{H_0}$ for all t and $X(t) = \gamma_0 X(t + 2\pi/q)$.*

Proof. On the component X_n^j the action of H is as $\gamma_n^j \equiv \exp 2\pi i(kn/q + k\langle N^j, \Psi_0\rangle + k\langle K_0, L^j/M\rangle + \langle N^j, \Psi\rangle + \langle K, L^j/M\rangle)$, with (Ψ, K) in H_0. Taking $k = 0$, one needs that (Ψ, K) is in H_j, the isotropy subgroup of the j'th coordinate, hence $H_0 < H_j$ and $X(t)$ is in $V_0^{H_0}$. In particular, $\gamma_0^{q_0}$ acts trivially on X^j. Hence, taking $k = q_0$, one concludes that n has to be a multiple of p and $X(t)$ is $2\pi/p$-periodic. The inverse inclusion is clear. □

Consider now K such that $H/K \cong \mathbb{Z}_2$. Since $K = \bigcap H_{jn}$, the inclusions $K < H \bigcap H_{jn} < H$, imply that either $H < H_{jn}$, or $K = H \bigcap H_{jn}$, where H_{jn} is the isotropy of X_n^j. In the second case, which must hold at least for one H_{jn}, one has that for any γ in H, then γ^2 must be in H_{jn}. In particular, for $\varphi = 0$ and $\tilde{\gamma}$ in H_0, one needs $\tilde{\gamma}^2 \in H_j$ and $H_0/(H_0 \cap H_j)$ has at most order 2. Let $K_0 = H_0 \cap H_j$, for all such j's, then, either $K_0 = H_0$, or $H_0/K_0 \cong \mathbb{Z}_2$. In the second case, let $V_1 \equiv (\mathbb{R}^N)^{K_0}$, then there is a γ_1 in H_0, with γ_1^2 in K_0, i.e., γ_1 acts as Id on $V_0^{H_0}$ and as $-$ Id on $V_1 \cap V_0^{H_0\perp}$. Since $\gamma_0^{q_0}$ is in H_0, one has that $\gamma_0^{2q_0}$ acts as Id on V_1. Let $V_1^{\pm}$ be the subspaces of V_1 where $\gamma_0^{q_0}$ acts as $\pm$ Id. Hence, $V_0^{H_0} \subset V_1^{+}$.

Lemma 9.5. *V^K consists of all 2π-periodic functions $X(t)$, in V_1 for all t, of the form $X(t) = X_+(t) + X_-(t)$, with $X_\pm(t) = \pm\gamma_0 X_\pm(t + 2\pi/q)$. In particular, if q is odd, then $X_\pm(t)$ are in $V_1^{\pm}$ and both are $2\pi/p$-periodic. If q is even and p is odd, then $X(t)$ is in V_1^{+} and it is $2\pi/p$-periodic. The components of $X_+(t)$ in V_1^{+} are $2\pi/p$-periodic and those in V_1^{-} are $2\pi/p$-antiperiodic. The behavior of the components of $X_-(t)$ differs by a factor $(-1)^{q_0}$.*

Proof. For the coordinate X_j, we know that $2q_0(\langle N^j, \Psi_0\rangle + \langle K_0, L^j/M\rangle) = a_j$ is an integer, which is even if X_j is in V_1^{+} and odd if X_j is in V_1^{-}. Since $(2\varphi_0, 2\Psi_0, 2K_0)$ fixes X_n^j, one has that $2n/q + a_j/q_0 = b$ is an integer. From $n = bq/2 - a_j p/2$, one has that, if q is odd, then b has the parity of a_j, while, if q is even and a_j is odd, then p has to be even. Even b's will give $X_+(t)$ and odd b's give $X_-(t)$. There are minima $n_j^{\pm}$ such that the modes of $X_\pm^j$ are of the form $n^{\pm} = n_j^{\pm} + cq$, for any integer c. The numbers $n_j^{\pm}$ are multiples of p, except if p is even and (for X_+^j) a_j is odd, or (for X_-^j) a_j and q_0 have opposite parities, in which case $n_j^{\pm}$ are odd multiples of $p/2$. These elements prove one inclusion. The reverse inclusion is clear. □

Remark 9.4. (a) If $X_\pm(t) = \pm\gamma_0 X_\pm(t + 2\pi/q)$ then, for $X(t) = X_+(t) + X_-(t)$, one has

$$X(t) = \gamma_0^2 X(t + 4\pi/q)$$

and the relations

$$X_\pm(t) = \frac{1}{2}(X(t) \pm \gamma_0^{-1} X(t - 2\pi/q)).$$

Conversely, if $X(t) = \gamma_0^2 X(t + 4\pi/q)$ then, defining $X_\pm(t)$ by these last relations, one obtains $X_\pm(t) = \pm\gamma_0 X_\pm(t + 2\pi/q)$. Hence, V^K is the set of all 2π-periodic functions, with $X(t)$ in V_1 for all t, such that $X(t) = \gamma_0^2 X(t + 4\pi/q)$.

(b) If $H_0/K_0 \cong \mathbb{Z}_2$, then, for any γ in H_0, there is $\alpha = 0$ or 1 such that $\gamma = \gamma_1^{\alpha}\delta$ with δ in K_0. Thus, since $\gamma_0^{q_0}$ is in H_0, one has either $\alpha = 0$ and $\gamma_0^{q_0}$ is in K_0, in particular $V_1^{-} = \{0\}$, or $\alpha = 1$ and $\gamma_0^{q_0}$ has the same action as γ_1 on V_1, i.e., one may take $\gamma_1 = \gamma_0^{q_0}$, in particular one has, in this case, $V_0^{H_0} = V_1^{+}$. Thus, $V_0^{H_0}$ is strictly contained in V_1^{+} if and only if $V_1^{-} = \{0\}$.

(c) The components of $X_+(t)$ which lie in $V_0^{H_0}$ give an element of V^H. In particular, if $K_0 = H_0$ and $H/K \cong \mathbb{Z}_2$, one cannot have q odd (since then $X_-(t)$ would be in $V_1^{-} = \{0\}$ and $V_0^{H_0} = V_1^{+}$).

A last result in this section will be the identification of the irreducible representations of H in $(V^{\underline{H}})^{\perp}$.

Lemma 9.6. *Assume $X^0, X^1, \dots, X^r$ are the coordinates of a set of equivalent irreducible representations of H_0 in $V_0^{\perp}$. Then, for each $n_0 = 0, \dots, q-1$, there is a different set of equivalent irreducible representations of H, with isotropy K_{n_0}, in $(V^{\underline{H}})^{\perp}$, with $V^{K_{n_0}}$ given by functions $X^j(t)$, $j = 0, \dots, r$ such that*

$$R_{-2\pi(n_0/q+\alpha_0)}\gamma_0 X^j(t + 2\pi/q) = X^j(t),$$

where R_φ is a rotation of an angle φ of the coordinates of X^j, or equivalently $R_\varphi X^j = e^{i\varphi}X^j$, and α_0 is given by $\gamma_0 X^0 = e^{2\pi i\alpha_0}X^0$. Defining $a_j = q_0(\alpha_j - \alpha_0)$ with $\gamma_0 X^j = e^{2\pi i\alpha_j}X^j$, one has more precisely

$$X^j(t) = x^j(t) + iy^j(t),$$

with $x^j(t)$ and $y^j(t)$ real, $x^j(t) = x_1^j(t) + x_2^j(t)$, with $x_2^j(t) = \bar{x}_1^j(t)$. Furthermore the following holds:

1. *If $2(n_0 - a_j p)$ is not a multiple of q, then*

$$x_1^j(t) = \sum_{-\infty}^{\infty} x_m e^{i(n_0 - a_j p)t} e^{imqt} \quad \text{and} \quad y^j(t) = i(x_2^j(t) - x_1^j(t)),$$

that is $X^j(t) = 2x_1^j(t)$.

2. *If $2(n_0 - a_j p) = lq$, then*

$$x_1^j(t) = \sum_{m+l/2\geq 0} x_m e^{i(lq/2)t} e^{imqt} \quad \text{and} \quad y_1^j(t) = \sum_{m+l/2\geq 0} y_m e^{i(lq/2)t} e^{imqt}.$$

If $X^j(t)$ is in $V^{K_{n_0}}$, then $e^{-2ia_j pt}\bar{X}^j(t)$ is in $V^{K_{q-n_0}}$.

Proof. The action of H on X_n^j is of the form $\gamma_n^j \exp 2\pi i\langle N^j, \tilde{\Psi}\rangle$, where γ_n^j has the form given in the proof of Lemma 9.4, while $\langle N^j, \tilde{\Psi}\rangle = \sum_k^n n_l^j \Psi_l$ is non-trivial in $V_0^{\perp}$. One will have the same action for different (n, j)'s if the following happens: taking $\gamma_n^j = \mathrm{Id}$ (i.e., $k = 0$ and $\Psi = 0$, $K = 0$) then n_l^j have to be the same for all j's for $l = k, \dots, n$. Taking $k = 0$ in γ_n^j, then one needs the same action for all (Ψ, K). Hence, the different X^j's are in the same set of equivalent irreducible representations of H_0 in $V_0^{\perp}$. If $\alpha_j = \langle N^j, \Psi_0\rangle + \langle K_0, L^j/M\rangle$ gives the action of γ_0, then, since $\gamma_0^{q_0}$ is in H_0, one needs that $q_0(\alpha_j - \alpha_l)$ is an integer a_{jl}. Then, for $X_{n_j}^j$ and $X_{n_l}^l$, one has that $(n_j - n_l)/q + a_{jl}/q_0$ is an integer. Hence, for X^0 one has the modes $n_0 + mq$, where $0 \leq n_0 < q$ and m is any integer. For a fixed n_0, the modes for X^j will be of the form $n_j = n_0 - a_j p + mq$, where $a_j = a_{jo} = (\alpha_j - \alpha_0)q_0$. The action of H on that mode is of the form

$$e^{2\pi i(kn_j/q+k\alpha_j+\varphi)} = e^{2\pi i(kn_0/q+k\alpha_0+\varphi)}$$

where φ corresponds to the action, as S^1, of H_0 on the coordinate X^j. Hence this mode is fixed by the group

$$K_{n_0} = \{(k, \varphi = -k(n_0/q + \alpha_0), \text{ mod } 1), k = 0, \dots, q-1\} \cong \mathbb{Z}_q.$$

From here, it is easy to see that, taking $k = 1$, one has the relation

$$R_{2\pi\varphi}\gamma_0 X^j\big(t + 2\pi/q\big) = X^j(t).$$

The converse inclusion is clear.

Now, one needs that $X^j(t) = x^j(t) + iy^j(t)$, with $x^j(t)$ and $y^j(t)$ real, that is

$$x^j(t) = \sum_{n\geq 0}(x_n e^{int} + \bar{x}_n e^{-int})$$
$$y^j(t) = \sum_{n\geq 0}(y_n e^{int} + \bar{y}_n e^{-int}).$$

Thus, the pair of modes, e^{int} and e^{-int}, will be fixed by K_{n_0} if and only if either

1. $n = n_0 - a_j p + mq \geq 0$ and $\bar{x}_n + i\bar{y}_n = 0$ (unless $-n$ has the same form, that is if $2(n_0 - a_j p)$ is a multiple of q), or
2. $-n = n_0 - a_j p + mq \leq 0$ and $x_n + iy_n = 0$ (unless $2(n_0 - a_j p)$ is a multiple of q). Thus

$$x^j(t) = \sum_{n=n_0-a_j p+mq\geq 0}(x_n e^{int} + \bar{x}_n e^{-int}) + \sum_{n=n_0-a_j p+mq\leq 0}(\bar{\tilde{x}}_n e^{int} + \tilde{x}_n e^{-int}),$$

while $y^j(t)$ has $y_n = -ix_n$ in the first sum and $\tilde{y}_n = i\tilde{x}_n$ in the second sum. Writing

$$x_1^j(t) = \sum_{n\geq 0} x_n e^{int} + \sum_{n\leq 0} \bar{\tilde{x}}_n e^{int}$$

and $x_2^j(t) = \bar{x}_1^j(t)$, one has $x^j(t) = x_1^j(t) + x_2^j(t)$, while $y^j(t) = i(x_2^j(t) - x_1^j(t))$, since $\bar{\tilde{y}}_n = -i\bar{\tilde{x}}_n$.

If $2(n_0 - a_j p) = lq$, then

$$x^j(t) = \sum_{n=n_0-a_j p+mq\geq 0}(x_n e^{int} + \bar{x}_n e^{-int})$$

and $y^j(t)$ of the same form and independent from $x^j(t)$.

Note that, if one sets

$$\tilde{x}_1^j(t) = \sum_{n=n_0-a_j p+mq\geq 0} (x_n + \tilde{x}_{-n})e^{int}$$

$$\tilde{y}_1^j(t) = i \sum_{n=n_0-a_j p+mq\geq 0} (\tilde{x}_{-n} - x_n)e^{int},$$

then $x^j(t) = \tilde{x}_1^j + \bar{\tilde{x}}_1^j$ and $y^j(t) = \tilde{y}_1^j + \bar{\tilde{y}}_1^j$, i.e., the two formulations of the lemma are equivalent.

Finally, if $X^j(t)$ belongs to V^{Kn_0}, one has

$$R_{2\pi(n_0/q+\alpha_0)}\bar{\gamma}_0\bar{X}^j(t+2\pi/q) = \bar{X}^j(t).$$

For $Y^j(t) = e^{-2ia_j pt}\bar{X}^j(t)$, one computes easily that

$$R_{-2\pi(-n_0/q+\alpha_0)}\gamma_0 Y^j(t+2\pi/q) = Y^j(t).$$

Note that, if $a_j = 0$, then $Y^j(t) = \bar{X}^j(t)$, a fact which can also be seen from the Fourier series expansion.

Note that for $q = 1$, then $n_0 = 0$ and the unique set of equivalent irreducible representations is $\{Y(t), 2\pi\text{-periodic in } V_0^{\perp}\}$. □

1.10 Bibliographical remarks

In this short section, we would like to give some references to the results in this chapter and to some more advanced texts.

1. *Group actions.* There are many books on representations of groups, with a variety of flavors. Closest to the spirit of the present text are the following:

A. A. Kirillov: *Elements of the theory of representations*, Springer-Verlag, 1976.

G. E. Bredon: *Introduction to compact transformation groups*, Academic Press, 1980.

T. Broecker and T. tom Dieck: *Representations of compact Lie groups*, Springer-Verlag, 1985.

T. tom Dieck: *Transformation groups and representation theory*, Springer-Verlag, 1979.

The last three are more inclined towards topology. The results of this section are taken from [IV1].

2. *Fundamental cell lemma.* This construction is taken from [IV1] and, in the particular case of S^1, from [IMV2].

A similar construction is developed, for a general Lie group, in

A. Kushkuley and Z. Balanov: *Geometric methods in degree theory for equivariant maps*, Springer-Verlag, 1996.

Lemmas 2.4–2.6 are extracted from [I.V. 2 and 3].

3. *Equivariant maps.* Some of the results are taken from [I].

4. *Averaging.* This important tool is taken from Bredon's book. The discussion on approximation by finite dimensional equivariant maps is new, although this fact was used in [IMV1]. The presentation of the other facts is close to [I].

5. *Irreducible representations.* The material on Schur's Lemma is standard (see for instance Kirillov's book). Corollaries 5.1 and 5.2 follow [I]. Theorem 5.1 is an adaptation of the standard result. The presentation of Theorems 5.2 and 5.3 follows [I], with a proof adapted from Pontrjagin's book: *Topological groups*, 1939.

6. *Extension of* Γ*-maps.* This is the substance of obstruction theory. Here we have only used the most basic elements, taken from [I]. for further reading, one may look at the books of Bredon, tom Dieck and Kushkuley–Balanov. The lemmas in this section are taken from [I.V. 1 and 2].

7. *Orthogonal maps.* The material presented here is taken, with some modifications, from [IV3]. The notion of orthogonal map has also been used, for $\Gamma = S^1$, by S. Rybicki.

8. *Equivariant homotopy groups of spheres.* The construction and basic properties are adaptations of the non-equivariant case: see the books of Greenberg or G. W. Whitehead: *Elements of homotopy theory*, Springer-Verlag, 1978.

The most useful results can be found in the books by tom Dieck and papers by Namboodiri and Hauschild. The J^Γ-homomorphism is taken from [I], and the results on Γ-orthogonal maps from [IV3].

9. *Symmetries and differential equations.* In order to apply our techniques to elliptic equations the reader may consult, for instance, the book by A. Friedman. For the case of O.D.E's any book with some Floquet theory may be useful. For the case of Hamiltonian systems, the book by I. Ekeland: *Convexity methods in Hamiltonian mechanics*, Springer-Verlag, 1990, will provide a good introduction to Conley index methods.

The Ljapunov–Schmidt reduction is from [I]. The applications to Hamiltonian systems follows the ideas of Amann–Zehnder and is taken from [IV3], as well as the classification of "twisted orbits".

Chapter 2
Equivariant Degree

In this chapter we are entering the main part of equivariant degree: we shall construct this degree, first in finite dimension, give its first properties and examples. Then, we shall extend it to infinite dimension and apply it to bifurcation and continuation problems. We shall also give the construction for orthogonal maps.

2.1 Equivariant degree in finite dimension

As explained in the Introduction to this book, a definition of an equivariant degree through a geometric construction, as in the case of the classical Brouwer degree, meets several serious difficulties: a "good" definition of genericity, a density result similar to Sard's lemma, a consistent definition of the invariants and of their sum. The construction below avoids most of these difficulties and may also be used in the non-equivariant case.

The setting of this section is the following: Let B and E be two finite dimensional Γ-spaces, where Γ is a compact Lie group acting via isometries on B and E as in Chapter 1. We shall indicate by remarks the few places where the finite dimensionality is used and how to put special hypotheses in order to validate the arguments in infinite dimensions.

Let Ω be a bounded, open, Γ-invariant subset of B and consider a continuous map $f(x)$, from $\bar{\Omega}$ into E, such that

(a) $f(x) \neq 0$ for x on $\partial\Omega$.

(b) $f(\gamma x) = \tilde{\gamma} f(x)$, for all γ in Γ and x in $\bar{\Omega}$.

Since Ω is bounded, let B_R be a closed ball of radius R and centered at the origin, containing Ω. Since the action on B is an isometry, B_R is Γ-invariant. Then, there is a Dugundji–Gleason Γ-extension $\tilde{f}(x)$, from B_R into E, of $f(x)$ (see Theorem 6.1 of Chapter 1).

Let then N be a Γ-invariant neighborhood of $\partial\Omega$, such that N is open, contained in B_R and $\tilde{f}(x) \neq 0$ on $\bar{N}$: the existence of N follows from Lemma 4.3 of Chapter 1, using the fact that $\partial\Omega \subset U$, where U is the open Γ-invariant subset of B_R such that $f(x) \neq 0$ (one may also restrict to a small neighborhood of $\partial\Omega$). Let $\varphi(x)$, from B_R into [0,1], be a Γ-invariant Uryson function with value 0 in $\bar{\Omega}$ and 1 outside $\Omega \cup N$.

Let $F(t, x) : [0, 1] \times B_R \to \mathbb{R} \times E$ be the map defined by

$$F(t, x) = (2t + 2\varphi(x) - 1, \tilde{f}(x)).$$

It is clear that F is Γ-equivariant, where the action on t in [0,1] and on the first component of $\mathbb{R} \times E$ is trivial. Furthermore, $F(t, x) \neq 0$ if x is in $\bar{N}$ (since $\tilde{f}(x) \neq 0$ there) and if x is outside $\Omega \cup N$ (there $\varphi(x) = 1$ and the first component of $\bar{F}$ reduces to $2t + 1 \geq 1$). Hence, if $F(t, x) = 0$, then x is in Ω, $\tilde{f}(x) = f(x) = 0$, $\varphi(x) = 0$ and $t = \frac{1}{2}$. In particular,

$$F(t, x) : S^B \equiv \partial([0, 1] \times B_R) \to \mathbb{R} \times E \backslash \{0\}$$

defines an element, $[F]_\Gamma$, of $\Pi^\Gamma_{S^B}(S^E)$, as defined in §8 of Chapter 1. Note that, if $E^\Gamma = \{0\}$, then, since $\tilde{f}(0) = 0$ in this case, our problem will be interesting only if 0 is not in Ω. Then one has $\varphi(0) = 1$, and the first component is always positive, as required in §8 of Chapter 1.

Definition 1.1. The *equivariant degree* of f with respect to Ω, is defined as $[F]_\Gamma$ in $\Pi^\Gamma_{S^B}(S^E)$, which is an abelian group provided $\dim B^\Gamma > 0$.

Remark 1.1. It is clear that up to here we have not used the finite dimensionality of B and E. Thus, one may define the Γ-degree either in general or, as in Remark 8.3 of Chapter 1, for maps which are compact perturbations of the identity (or k-set-contractions).

The next step in our construction is to show that the class of F is independent of R, N and φ.

Proposition 1.1. *The homotopy class* $[F]_\Gamma$ *does not depend on*

(a) *the* Γ*-invariant Uryson function* φ*,*

(b) *the choice of the* Γ*-invariant neighborhood* N *of* $\partial\Omega$*,*

(c) *the* Γ*-equivariant extension* $\tilde{f}$ *of* f*,*

(d) *the choice of the ball* B_R *containing* $\bar{\Omega}$*.*

Proof. (a) Let $\varphi_0, \varphi_1 : B_R \to [0, 1]$ be two Uryson functions with values 0 in $\bar{\Omega}$ and 1 outside $\Omega \cup N$. Let $\varphi_\tau(x) = \tau\varphi_1(x) + (1 - \tau)\varphi_0(x)$, τ in $[0, 1]$, which is also a Uryson function with the same properties. Let

$$F_\tau(t, x) - (2t + \varphi_\iota(x) \quad 1, \tilde{f}(x)).$$

Then F_τ is an admissible Γ-homotopy between F_0 and F_1, therefore $[F_0]_\Gamma = [F_1]_\Gamma$.

(b) Let us first assume that there are two invariant open neighborhoods N_0 and N_1, of $\partial\Omega$, such that $N_0 \subset N_1 \subset B_R$. Let φ_0 and φ_1 be the Γ-invariant Uryson functions associated to N_0 and N_1, respectively. Let $\varphi_\tau(x) = \tau\varphi_1(x) + (1-\tau)\varphi_0(x)$ and $F_\tau(t, x)$ be defined as above. Since $\varphi_\tau(x) = 0$ for x in $\bar{\Omega}$ and $\varphi_\tau(x) = 1$ for x outside $\Omega \cup N_1$, the Γ-homotopy F_τ is admissible and $[F_0]_\Gamma = [F_1]_\Gamma$.

In the case where N_0 and N_1 are arbitrary, one can use the previous argument applied to $N_0 \cap N_1$ and to each N_0 and N_1.

(c) Given two Γ-equivariant extensions $\tilde{f}_0$ and $\tilde{f}_1$ of f, one can choose a Γ-invariant open neighborhood N of $\partial\Omega$ on which the Γ-equivariant extensions $\tilde{f}_\tau(x) \equiv \tau\tilde{f}_1(x) + (1-\tau)\tilde{f}_0(x)$ is not vanishing for τ in $[0, 1] = I$, applying Lemma 4.3 of Chapter 1 to $\partial\Omega$ and to $U = \{x \text{ in } B_R : \tilde{f}_\tau(x) \neq 0 \text{ for all } \tau \text{ in } I\}$: in fact U^C is closed from the compactness of I and the continuity of f_0 and f_1. This map will induce an admissible Γ-homotopy and the assertion follows.

(d) Let $R_0 < R$ with $\Omega \subset B_{R_0}$. Let $\tilde{f}_0$ and $\tilde{f}$ be two Γ-equivariant extensions of f to B_{R_0} and B_R, respectively. By (b) and (c) we may assume that $\tilde{f}_0$ and $\tilde{f}$ do not vanish on a common Γ-invariant open neighborhood $N \subset B_{R_0}$ of $\partial\Omega$ and such that $\tilde{f}|_{B_{R_0}} = \tilde{f}_0$. Let $\varepsilon > 0$, be such that $\|x\| \leq R_0 - \varepsilon$, if x is in $\bar{N}$. For any τ in I consider the Γ-map

$$\tilde{f}_\tau(x) = \tilde{f}(\alpha(\tau, x)x)/\alpha(\tau, x),$$

where

$$\alpha(\tau, x) = \begin{cases} 1, & \text{if } \|x\| \leq R_0 - \varepsilon \\ 1 + \tau(R - R_0)(\|x\| - R_0 + \varepsilon)/(\varepsilon R_0), & \text{if } R_0 - \varepsilon \leq \|x\| \leq R_0. \end{cases}$$

The scaling $\delta_\tau(x) = \alpha(\tau, x)x$ is a Γ-equivariant homeomorphism from B_{R_0} into B_R, leaving fixed $B_{R_0-\varepsilon}$ and $\delta_1(B_{R_0}) = B_R$. Hence, $\tilde{f}_\tau(x)$ is a Γ-equivariant extension of f to B_{R_0}, for any τ in I. Thus, from (c), since $\tilde{f}_0 = \tilde{f}|_{B_{R_0}}$, the Γ-homotopy class $[F_0]_\Gamma$ of F_0 induced by $\tilde{f}_0$ coincides with the class $[F_1]_\Gamma$, where F_1 is induced by $\tilde{f}_1$.

Moreover, if we extend $\tilde{f}_1(x)$ as $\tilde{f}(Rx/\|x\|)(\|x\|/R)$ for $R_0 \leq \|x\| \leq R$, we obtain a Γ-equivariant extension of f to B_R. Thus, once again applying (c), we have that $[F_1]_\Gamma = [F]_\Gamma$, where we have identified, via the scaling, the two groups of Γ-homotopy classes of maps defined on the two cylinders $I \times B_{R_0}$ and $I \times B_R$. □

Remark 1.2. Proposition 1.1 is also valid in the case of infinite dimensions, if there are no restrictions on the maps. Furthermore, in the case of Γ-compact perturbations of the identity, $\tilde{f}(x)$ is untouched in the proofs of (a) and (b) and it is easy to see that the linear homotopy of (c) and the scaling of (d) will preserve the character of perturbations of the identity. Hence, in both cases, the Γ-degree is well defined.

2.2 Properties of the equivariant degree

In this section we shall see that the equivariant degree has all the properties of the Brouwer degree (up to a slight condition for the addition and noticing that in general this degree will not be a single integer). As before, we shall leave to remarks the case of infinite dimensions.

Property 2.1 (Existence). *If* $\deg_\Gamma(f;\Omega)$ *is non-trivial, then there exists* x *in* Ω *such that* $f(x)=0$.

Proof. As noted in §8 and Lemma 6.1 of Chapter 1, the neutral element 0_Γ in $\Pi_{SB}^\Gamma(S^E)$ consists of the class of all maps which have a non-vanishing Γ-extension to $I\times B_R$. Thus, if $f(x)\neq 0$ in Ω, then $F(t,x)\neq 0$ on $I\times B_R$ and $[F]_\Gamma=0$. □

Note that, due to the equivariance, $f(\gamma x)=0$, that is, solutions come in orbits.

Property 2.2 (Γ-homotopy invariance). *Let* $f_\tau:\bar{\Omega}\to E$, $0\le\tau\le 1$, *be a continuous one-parameter family of* Γ*-equivariant maps not vanishing on* $\partial\Omega$ *for all* τ *in* I. *Then the* Γ*-degree* $\deg_\Gamma(f_\tau;\Omega)$ *does not depend on* τ.

Proof. Immediate from the fact that the construction of $F_\tau: I\times B_R\to\mathbb{R}\times E$ can be performed uniformly with respect to τ. □

Remark 2.1. In the infinite dimensional case, one may construct $\tilde{f}_\tau$ an equivariant extension to $I\times B_R$ of f_τ on $I\times\bar{\Omega}$. Then, taking $A=\{x \text{ in } B_R:\tilde{f}_\tau(x)=0 \text{ for some } \tau\}$, which is a closed set, since if $f_{\tau_n}(x_n)=0$ and $\{x_n\}$ converges to x, then one may assume that for some subsequence, also denoted by τ_n, one has $\{\tau_n\}$ converging to τ and $\tilde{f}_\tau(x)=0$. Then, $U=A^C$ is open and contains $\partial\Omega$. Applying Lemma 4.3 of Chapter 1, one obtains a common N for all τ's.

In the particular case of Γ-compact perturbations of the identity, one has that $\|f_\tau(x)\|>\varepsilon>0$ for (τ,x) in $I\times\partial\Omega$ and for some ε: if not one would have a sequence (τ_n,u_n,w_n), with $f_{\tau_n}(u_n+w_n)=(g_{\tau_n}(u_n+w_n), w_n-h_{\tau_n}(u_n+w_n))$ going to 0. The compactness of $h_\tau(x)$ and the finite dimensionality of $I\times P\bar{\Omega}$, P the projection on U will imply the convergence of some subsequence and a zero of $f_\tau(u+w)$ on $\partial\Omega$. The same argument will show that there is an invariant η-neighborhood of $\partial\Omega$ on which $\|f_\tau(x)\|\ge\frac{\varepsilon}{2}$.

Hence, in both cases, one has the homotopy invariance property.

Property 2.3 (Excision). *Let* $f:\bar{\Omega}\to E$ *be a continuous* Γ*-equivariant map such that* $f(x)\neq 0$ *in* $\bar{\Omega}\backslash\Omega_0$, *where* $\Omega_0\subset\Omega$ *is open and* Γ*-invariant. Then*

$$\deg_\Gamma(f;\Omega)=\deg_\Gamma(f|_{\bar{\Omega}_0};\Omega_0).$$

Proof. If $\tilde{f}$, N and φ correspond to $\deg_\Gamma(f;\Omega)$, then $\tilde{f}$ is also an extension of $f|_{\bar{\Omega}_0}$, which never vanishes on the Γ-invariant neighborhood $\bar{N}'=(\bar{\Omega}\backslash\Omega_0)\cup\bar{N}\cup(\bar{\Omega}\cap\bar{N}_0)$ of $\partial\Omega_0$. Since $\bar{\Omega}_0\cup\bar{N}'=\bar{\Omega}\cup\bar{N}$, the Uryson function φ is also a Uryson function associated to $\bar{N}'$. Thus, from Proposition 1.1, one has that $[F]_\Gamma=\deg_\Gamma(f|_{\Omega_0};\Omega_0)$. In particular, if $f(x)\neq 0$ for all x in $\bar{\Omega}$, then $\deg_\Gamma(f;\Omega)=0$, taking $\Omega_0=\phi$. □

Remark 2.2. Using the excision property, we may extend the definition of Γ-degree to the class of Γ-equivariant maps $f : \Omega \to E$, when Ω is not necessarily bounded, provided that $f^{-1}(0)$ is a compact set, by restricting f to a bounded open Γ-invariant set Ω_0 containing $f^{-1}(0)$.

Property 2.4 (Γ-homotopy invariance). *Let $f_\tau : \bar{\Omega} \to E, 0 \leq \tau \leq 1$, be a continuous one-parameter family of Γ-equivariant maps not vanishing on $\partial\Omega$ for all τ in I. Then, the Γ-degree $\deg_\Gamma(f_\tau; \Omega)$ does not depend on τ.*

Property 2.5 (Suspension). *If there is a Γ-extension $\tilde{f}$ to B_R of f, such that $\tilde{f}(x) \neq 0$ on $\bar{B}_R \backslash \Omega$ (in particular, if $\Omega = B_R$), then*

$$\deg_\Gamma(f; \Omega) = \deg_\Gamma(\tilde{f}; B_R) = \Sigma_0[\tilde{f}]_\Gamma,$$

where Σ_0 is the suspension (one-dimensional) homomorphism, by $2t - 1$.

Proof. Since $\deg_\Gamma(f; \Omega) = [2t + 2\varphi(x) - 1, \tilde{f}(x)]_\Gamma$, we may deform $\varphi(x)$ to 0 and obtain the equality with $[2t-1, \tilde{f}(x)]_\Gamma$. Using a radial extension of $\tilde{f}$ to $B_{R'}$, $R' > R$, one obtains similarly that this class is equal to $\deg_\Gamma(\tilde{f}; B_R)$. (One may also use the excision property to get $\deg_\Gamma(\tilde{f}; B_R) = \deg_\Gamma(f; \Omega)$). □

Property 2.6 (Hopf property). *If Ω is a ball and Σ_0 is one-to-one, then $\deg_\Gamma(f; \Omega) = \deg_\Gamma(g; \Gamma)$ if and only if $f|_{\partial\Omega}$ is Γ-homotopic to $g|_{\partial\Omega}$.*

Proof. Follows immediately from Property 2.4. In this case the Γ-degree characterizes completely $\Pi_{S^B}^\Gamma(S^E)$. □

Property 2.7 (Additivity up to one suspension). *If $\Omega = \Omega_1 \cup \Omega_2$, Ω_i open with $\bar{\Omega}_1 \cap \bar{\Omega}_2 = \phi$, then*

$$\Sigma_0 \deg_\Gamma(f; \Omega) = \Sigma_0 \deg_\Gamma(f; \Omega_1) + \Sigma_0 \deg_\Gamma(f; \Omega_2),$$

where Σ_0 is again the suspension by $2t - 1$.

Proof. Take $N = N_1 \cup N_2$, with $\bar{N}_1 \cap \bar{N}_2 = \phi$ and let $\varphi, \varphi_1, \varphi_2$ denote the partition functions associated to N, N_1, N_2. Then

$$F(t, x) = (2t + 2\varphi(x) - 1, \tilde{f}(x))$$

is Γ-deformable to

$$\tilde{F}(t, x) = \begin{cases} (2t + (1 - 2t)(2\varphi(x) - 1), \tilde{f}(x)), & 0 \leq t \leq \frac{1}{2} \\ (1, \tilde{f}(x)), & \frac{1}{2} \leq t \leq 1, \end{cases}$$

by replacing $2t + 2\varphi(x) - 1$ with $2t + (1 - 2t\tau)(2\varphi(x) - 1)$ for $0 \leq t \leq \frac{1}{2}$ and by $\tau + (1 - \tau)(2t + 2\varphi(x) - 1)$ for $\frac{1}{2} \leq t \leq 1$: the only zeros of this homotopy are such that $f(x) = 0$, for x in Ω, and $2t = (1 + \tau)^{-1}$.

Now, if one changes t by $1-t$ in the above formula, one obtains $-[\tilde{F}]_\Gamma = -[F]_\Gamma$. Hence,

$$[F]_\Gamma - [F_1]_\Gamma = \begin{cases} (2t + (1-2t)(2\varphi(x)-1), \tilde{f}(x)), & 0 \le t \le \frac{1}{2} \\ (2(1-t) + (2t-1)(2\varphi(x)-1), \tilde{f}(x)), & \frac{1}{2} \le t \le 1, \end{cases}$$

where $[F_1]_\Gamma = \deg_\Gamma(f; \Omega_1)$.

Note that, since $F(t,x) = \tilde{F}_1(t,x) = (1, \tilde{f}(x))$ for $\frac{1}{2} \le t \le 1$, then

$$[\tilde{F}]_\Gamma - [\tilde{F}_1]_\Gamma = \begin{cases} \tilde{F}(2t, x), & 0 \le t \le \frac{1}{2} \\ \tilde{F}_1(2-2t, x), & \frac{1}{2} \le t \le 1, \end{cases}$$

according to Definition 8.2 of Chapter 1, is effectively the above difference by using the Γ-homotopy

$$H_\tau(t,x) = \begin{cases} \tilde{F}((2-\tau)t, x), & 0 \le t \le \frac{1}{2} \\ \tilde{F}_1((2-\tau)(1-t), x), & \frac{1}{2} \le t \le 1. \end{cases}$$

Consider next the Γ-homotopy

$$\tilde{H}_\tau(t,x) = (h_\tau(t,x), \tilde{f}(x)),$$

where

$$h_\tau(t,x) = \begin{cases} 1, & \text{if } x \notin \bar{\Omega} \cup \bar{N} \\ 2t + (1-2t)(2\varphi_1 - 1), & \text{if } x \in \bar{\Omega}_1 \cup \bar{N}_1 \text{ and } 0 \le 2t \le \tau \\ \tau + (1-\tau)(2\varphi_1 - 1), & \text{if } x \in \bar{\Omega}_1 \cup \bar{N}_1 \text{ and } \tau \le 2t \le 2-\tau \\ 2(1-t) + (2t-1)(2\varphi_1 - 1), & \text{if } x \in \bar{\Omega}_1 \cup \bar{N}_1 \text{ and } 2-\tau \le 2t \le 2 \\ 2t + (1-2t)(2\varphi_2 - 1), & \text{if } x \in \bar{\Omega}_2 \cup \bar{N}_2 \text{ and } 0 \le t \le \frac{1}{2} \\ 1, & \text{if } x \in \bar{\Omega}_2 \cup \bar{N}_2 \text{ and } \frac{1}{2} \le t \le 1 \end{cases}$$

It is easy to check that $\tilde{H}_\tau$ is well defined (recall that $\varphi|_{\bar{\Omega}_i \cup \bar{N}_i} = \varphi_i, i = 1, 2$) and continuous. Clearly, $[\tilde{H}_1]_\Gamma = [F]_\Gamma - [F_1]_\Gamma$, since $\varphi_2|_{\bar{\Omega}_1 \cup \bar{N}_1} = 1$. On the other hand, $\tilde{H}_0(t,x) = (2\varphi_1(x) - 1, \tilde{f}(x))$ if $x \in \bar{\Omega}_1 \cup \bar{N}_1$ (hence, H_0 is non-zero there) and is $\tilde{F}_2(t,x)$ on $\bar{\Omega}_2 \cup \bar{N}_2$. That is, H_0 is an extension of $\tilde{F}_2$ which is not vanishing on $I \times (B_R \backslash (\bar{\Omega}_2 \cup \bar{N}_2))$.

Now, from Properties 2.3 and 2.4,

$$\deg_\Gamma(H_0; I \times B_R) = \deg_\Gamma(\tilde{F}_2; I \times \Omega_2) = \deg_\Gamma(\tilde{F}_2; I \times B_R) = \Sigma_0[F_2].$$

Finally, $\deg_\Gamma(H_0; I \times B_R) = \Sigma_0[H_0]_\Gamma = \Sigma_0[H_1]_\Gamma = \Sigma_0([F]_\Gamma - [F_1]_\Gamma)$. Since Σ_0 is a morphism, this proves the additivity. □

Remark 2.3. In the above proof, if Ω_2 is a ball (hence, by the invariance, centered at the origin), then, since $\tilde{H}_0$ is non-zero on $I \times (B_R \backslash \bar{\Omega}_2 \cup \bar{N}_2)$, the class of $\tilde{H}_0$ on $\partial(I \times B_R)$ is the same, by a radial retraction, as the class of $\tilde{H}_0$ on $\partial(I \times \bar{\Omega}_2)$. That is, $[\tilde{H}_0]_\Gamma = [\tilde{F}_2]_\Gamma = [\tilde{H}_1]_\Gamma = [\tilde{F}]_\Gamma - [\tilde{F}_1]_\Gamma$. Hence, in this case, the addition formula is true without a suspension.

This is not true in general, as the following example shows:

Example 2.1. Let $f(x_1, x_2, x_3) = f_1 + if_2 = (x_1^2 + x_2^2 - 1 + ix_3)((x_1 - 1)^2 + x_3^2 - 1 + ix_2)$, be a map from $\mathbb{R}^3$ into $\mathbb{R}^2$ (here $\Gamma = \{e\}$.) The zeros of f are the two linked circles $S_1 = \{x_1^2 + x_2^2 = 1, x_3 = 0\}$ and $S_2 = \{(x_1 - 1)^2 + x_3^2 = 1, x_2 = 0\}$. Take $B = \{(x_1, x_2, x_3) : x_1^2 + x_2^2 + x_3^2 < 4\}$ and Ω_j be two small disjoint tubular neighborhoods of S_j, $j = 1, 2$.

Then one has, by Property 2.3, that $\deg(f; \Omega) = \deg(f; B) = \Sigma[f]$. But $[f]$ is in $\Pi_2(S^1) = 0$, hence $\deg(f; B) = 0$, see Remark 8.1 in Chapter 1.

On Ω_1, one may perform the deformation

$$(x_1^2 + x_2^2 - 1 + ix_3)[\tau((x_1 - 1)^2 + x_3^2 - 1) - (1 - \tau)x_1 + ix_2];$$

on $\partial\Omega_1$ the first factor is non-zero; thus, a zero would have $x_2 = 0$, x_1 close to ± 1 and x_3 close to 0. If x_1 is close to 1, the deformed term is negative, while if x_1 is close to -1, the deformed term is positive. Hence, $\deg(f; \Omega) = \deg(-\bar{z}(|z|^2 - 1 + ix_3); \Omega_1)$, where $z = x_1 + ix_2$. Now, in $(2t + 2\varphi(x) - 1, \tilde{f}(x))$, one may take for $\tilde{f}(x)$ the above map (since φ is 1 if $z = 0$) and linearly deform $\varphi(x)$ to $(|z|^2 - 1)^2$. Then one may deform $(|z|^2 - 1 + ix_3)$ via $(|z|^2 - 1)[(1 + \tau)(1 - \tau(2t - 1))] + ix_3$, since $1 - \tau(2t - 1) \geq 0$, being 0 only if $\tau = t = 1$, for which $2t - 1 + 2(|z|^2 - 1)^2 \geq 1$. One obtains $\deg(f; \Omega_1) = [2t - 1 + 2(|z|^2 - 1)^2, -\bar{z}(4(|z|^2 - 1)(1 - t) + ix_3)]$. By performing the rotation, on the first component and on the term $4(1 - t)(|z|^2 - 1)$

$$\begin{pmatrix} \tau & -(1 - \tau) + 2\tau(|z|^2 - 1) \\ 1 - \tau - 2\tau(|z|^2 - 1) & 2\tau \end{pmatrix} \begin{pmatrix} 2t - 1 \\ |z|^2 - 1 \end{pmatrix},$$

one arrives at $\deg(f; \Omega_1) = [1 - |z|^2, -\bar{z}(2t - 1 + ix_3)] = \eta$, where η is the Hopf map of Remark 8.1 of Chapter 1. Then, $\deg(f; \Omega_1) = 1$.

Similarly for Ω_2, make the deformation

$$[\tau(x_1^2 + x_2^2 - 1) + (1 - \tau)(x_1 - 1) + ix_3]((x_1 - 1)^2 + x_3^2 - 1 + ix_2).$$

On $\partial\Omega_2$, an eventual zero would be for $x_3 = 0$, (x_1, x_2) close to $(2, 0)$ or to $(0, 0)$. In the first case the deformed term is positive, while in the second case it is negative.

The resulting map $(x_1 - 1 + ix_3)((x_1 - 1)^2 + x_3^2 - 1 + ix_2)$ can be written as $(y_1 - iy_2)(y_1^2 + y_2^2 - 1 + iy_3)$ under the change of variables $y_1 = x_1 - 1$, $y_2 = -x_3$, $y_3 = x_2$, with a positive Jacobian and Ω_2 is sent onto Ω_1. Then, one has $\deg(f; \Omega_2) = 1$: the rotation from $\bar{z}$ to $-\bar{z}$, having Jacobian 1, is a valid deformation.

Then, $\deg(f; \Omega_1 \cup \Omega_2) = 0 \neq \deg(f; \Omega_2) + \deg(f; \Omega_2) = 2$. Clearly, when one suspends, the equality holds since $2\Sigma_0\eta = 0$ (see Remark 8.1 of Chapter 1).

Property 2.8 (Universality). *If $\Delta(f;\Omega)$ is any other Γ-degree with the properties* 2.1–2.3 *and Σ_0 is one-to-one, then if $\Delta(f;\Omega)$ is non-trivial, this is also the case for* $\deg_\Gamma(f;\Omega)$.

Proof. One has $\Delta(F; I\times B_R) = \Delta((2t-1,f); I\times\Omega) = \Sigma_0\Delta(f;\Omega)$, where the first equality comes from the excision property 2.3 and the second is the suspension. Hence, if $\deg_\Gamma(f;\Omega)=0$, then $[F]_\Gamma$ has a non-zero Γ-extension from $\partial(I\times B_R)$ to $I\times B_R$. By Property 2.1, $\Delta(F; I\times B_R)$ must be trivial and, since Σ_0 is one-to-one, $\Delta(f;\Omega)$ is also trivial. □

Remark 2.4. Note that properties 2.3–2.7 are also valid in the infinite dimensional case, since either $\tilde{f}(x)$ is untouched or changed by a scaling. Hence, all the above properties hold in the two cases of infinite dimensional settings: the general one and that of Γ-compact perturbations of the identity.

Property 2.9 (Brouwer topological degree). *If $B=E$ and $\Gamma=\{e\}$, then*

$$\deg_{\{e\}}(f;\Omega) = \deg_B(f;\Omega), \text{ the Brouwer degree of } f.$$

Proof. Since $\deg_{\{e\}}(f;\Omega) = [F] = \deg_B(F;[0,1]\times B_R)$, from Remark 8.1 of Chapter 1, this last degree, by excision, is equal to

$$\deg_B((2t-1,f(x)); I\times\Omega) = \deg_B(2t-1;I)\deg_B(f(x);\Omega) = \deg_B(f;\Omega)$$

recalling that $\varphi(x)=0$ on Ω and using the product formula for the Brouwer degree. □

It is clear that for a compact perturbation of the identity, $B=E$ and $\Gamma=\{e\}$ one recovers the Leray–Schauder degree.

In [I.M.V. 0.], the class of Γ-epi maps has been introduced. Its definition, for the case of a bounded domain runs as follows

Definition 2.1. A continuous Γ-equivariant map $f:\bar{\Omega}\to E$ is called Γ-*epi* provided that

1. $f(x)\neq 0$ on $\partial\Omega$
2. $f(x)=h(x)$ has a solution in Ω, for any h continuous Γ-equivariant compact map with support contained in Ω.

Lemma 2.1. *If $\deg_\Gamma(f;\Omega)\neq\{0\}$ then f is Γ-epi.*

Proof. This follows at once from $\deg_\Gamma(f;\Omega) = \deg_\Gamma(f-h;\Omega)$, since if two Γ-maps, f and g, coincide on $\partial\Omega$, they must have the same Γ-degree: this last fact is an immediate consequence of the homotopy property, by using the deformation $\tau f(x)+(1-\tau)g(x)$ (which is a Γ-compact perturbation of the identity). □

2.3 Approximation of the Γ-degree

We have seen, in the two last sections, that the Γ-degree may be defined in the same way in the case of infinite dimensional spaces B and E, with all the properties 2.1–2.8. In this section, we shall compute $\deg_\Gamma(f;\Omega)$ for f a Γ-compact perturbation of the identity, by appealing to suitable finite dimensional approximations. Thus, we shall consider two Γ-spaces B and E such that

$$B = U \times W$$
$$E = V \times W$$

where U and V are finite dimensional Γ-representations and W is an infinite dimensional Γ-space. The maps and homotopies will be of the following form:

$$f(x) = f(u,w) = (g(u,w), w - h(u,w)),$$

where $g(u,w)$ is in V, h is compact and g and h are Γ-equivariant. Homotopies will affect only g and h.

From Theorem 4.1 of Chapter 1, we know that for any integer n, there is a finite dimensional Γ-subrepresentation M_n of W and a Γ-equivariant map $h_n(u,w) : B \to M_n$, such that

$$\|h(u,w) - h_n(u,w)\| \leq 1/2^n,$$

for any (u,w) in B_R. Define

$$f_n(x) = f_n(u,w) = (g(u,w), w - h_n(u,w)).$$

We have seen, the last time in Remark 2.1, that the compactness of $h(u,w)$, together with the finite dimensionality of U, and the fact that $f(x)$ is non-zero on $\bar{N}$ imply that there is an $\varepsilon > 0$, such that $\|f(x)\| > 2\varepsilon$, for x in $\bar{N}$, where N is the Γ-invariant neighborhood of $\partial\Omega$ used in the definition of the Γ-degree of f.

Hence, it follows that there is an integer n_0 such that for $n \geq n_0$,

$$\|f_n(x)\| > \varepsilon, \quad \text{for } x \text{ in } \bar{N},$$

and that

$$\deg_\Gamma(f;\Omega) = \deg_\Gamma(f_n;\Omega),$$

since the linear deformation $\tau f(x) + (1-\tau)f_n(x)$ is non-zero on $\partial\Omega$ and is a Γ-compact perturbation of the identity.

Furthermore, if one writes $w = w_n \oplus \tilde{w}_n$, with w_n in M_n and $\tilde{w}_n$ in a Γ-invariant complement $\tilde{M}_n$, it is clear that the Γ-homotopy

$$f_n^\tau(x) = (g(u, w_n + \tau\tilde{w}_n), w - h_n(u, w_n + \tau\tilde{w}_n)),$$

is also valid (since a zero of $f_n^\tau(x)$ must have $\tilde{w}_n = 0$ due to the fact that $h_n(x)$ is in M_n). If $x_n = u \oplus w_n$, let

$$\tilde{f}_n(x_n) = (g(u,w_n), w_n - h_n(u,w_n)),$$

and $\Omega_n = \Omega \cap (U \times M_n)$, $\bar{N}_n = N \cap (U \times M_n)$. Since $\partial\Omega_n \subset \bar{N}_n \cap (U \times M_n)$, we have that $\tilde{f}_n(x_n)$ does not vanish on $\bar{N}_n$, for $n \geq n_0$. Therefore, $\deg_\Gamma(\tilde{f}_n; \Omega_n)$ is well defined and, since $f_n^0(x)$ is the suspension of $\tilde{f}_n$ by $\tilde{M}_n$, one has

$$\deg_\Gamma(f; \Omega) = \Sigma^{\tilde{M}_n} \deg_\Gamma(\tilde{f}_n; \Omega_n).$$

It is clear that, if we had not taken care of seeing that the finite dimensional construction extends directly to the infinite dimensional case, we would have here an alternative way of defining the Γ-degree through finite dimensional approximations. This is, of course, nothing else but an adaptation of the classical technique due to Leray and Schauder when constructing the topological degree for compact perturbations of the identity via the Brouwer degree of their finite dimensional approximations, noticing that $\Sigma^{\tilde{M}_n}$ in this case is an isomorphism, due to the product formula.

To be more precise, we would have to proceed by comparing $\deg_\Gamma(\tilde{f}_n; \Omega_n)$ and $\deg_\Gamma(\tilde{f}_m; \Omega_m)$, for $n, m \geq n_0$. To this end, denote by $M_{n,m}$ the Γ-invariant space (M_n, M_m) and let $P_{n,m}$, P_n and P_m be the Γ-projections on $M_{n,m}$, M_n and M_m respectively. Set

$$\tilde{W} = (I - P_n) \circ P_{n,m} W$$

and $\tilde{w} = (I - P_n) \circ P_{n,m} w$. Clearly, $\tilde{W}$ is Γ-invariant and $M_n \oplus \tilde{W} = M_{n,m}$. Put $\Omega_{n,m} = \Omega \cap (U \times M_{n,m})$ and let

$$\tilde{f} : \bar{B} \cap (U \times M_{n,m}) \to V \times M_{n,m}$$

be the Γ-equivariant map defined by

$$\tilde{f}(u, w_n, \tilde{w}) = (g(u, w_n, \tilde{w}), w_n - h_n(u, w_n, \tilde{w}), \tilde{w}).$$

Notice that $\tilde{f}(u, w_n, 0) = \tilde{f}_n(u, w_n)$.

One has the following

Lemma 3.1. $\deg_\Gamma(\tilde{f}; \Omega_{n,m}) = \Sigma^{\tilde{W}} \deg_\Gamma(\tilde{f}_n; \Omega_n)$.

Proof. Note first that, by the excision property of the Γ-degree, we may replace the set $\Omega_{n,m}$ by the set $\Omega_{n,m} \cup (\Omega_n \times \{\tilde{w} \in \tilde{W} : \|\tilde{w}\| < \varepsilon\})$ and, in turn, this set by the set $\Omega_n \times \{\tilde{w} \in \tilde{W} : \|\tilde{w}\| < \varepsilon\}$. We may also deform the map $\tilde{f}$ to $(\tilde{f}_n(u, w_n), \tilde{w})$.

Set $N_n = N \cap (U \times M_n)$ and $\varphi_n = \varphi|_{B \cap (U \times M_n)}$, where φ is the Uryson function associated to N. Obviously, $\partial\Omega_n \subset N_n$ and φ_n is a Γ-invariant Uryson function associated to N_n.

If B_ε denotes $\{\tilde{w} \in \tilde{W} : \|\tilde{w}\| < \varepsilon\}$, then $N_n \times B_{2\varepsilon}$ is a Γ-invariant neighborhood of $\partial(\Omega_n \times B_\varepsilon)$ such that $\tilde{f}(u, w_n, \tilde{w})$ is not zero on it. Let $\psi : \bar{B} \cap (U \times M_{n,m}) \to [0, 1]$ be defined as

$$\psi(u, w_n, \tilde{w}) = \begin{cases} \varphi_n(u, w_n), & \text{if } \|\tilde{w}\| \leq \varepsilon \\ \varphi_n(u, w_n)(2 - \|\tilde{w}\|/\varepsilon) + \|\tilde{w}\|/\varepsilon - 1, & \text{if } \varepsilon < \|\tilde{w}\| \leq 2\varepsilon \\ 1, & \text{if } 2\varepsilon \leq \|\tilde{w}\|. \end{cases}$$

Clearly, ψ is a Γ-invariant Uryson function associated to $\bar{N}_n \times \bar{B}_{2\varepsilon}$. It follows that in the Γ-homotopy class of $\deg_\Gamma(\tilde{f}; \Omega_{n,m})$ we have the map $(2t + 2\psi(u, w_n, \tilde{w}) - 1, \tilde{f}(u, w_n, \tilde{w}))$, which can be deformed (via a convex Γ-homotopy) to the map

$$(2t + 2\varphi_n(u, w_n) - 1, \tilde{f}_n(u, w_n), \tilde{w}),$$

which is the $\tilde{W}$-suspension of $\deg_\Gamma(\tilde{f}_n; \Omega_n)$. □

Clearly, Lemma 3.1 can be equally applied to $\tilde{f}_m$ and to $\hat{f}(x) = (g(u, w_n, \hat{w}), w_m - h_m(u, w_m, \hat{w}), \hat{w})$, where $w_m = P_m w$ and $\hat{w} = (I - P_m) \circ P_{nm} w$. Hence, we have that

$$\deg_\Gamma(\hat{f}; \Omega_{n,m}) = \Sigma^{\hat{W}} \deg_\Gamma(f_m, \Omega_m).$$

Finally, it is clear that $\hat{f}$ and $\tilde{f}$ are Γ-homotopic, via a convex homotopy, on $\partial\Omega_{m,n}$. Therefore, $\Sigma^{\tilde{W}} \deg_\Gamma(f_n; \Omega_n) = \Sigma^{\hat{W}} \deg_\Gamma(f_m; \Omega_m)$, provided $n, m \geq n_0$.

To perform the last step of our construction, we would need that the Γ-suspensions $\Sigma^{\tilde{W}}$ and $\Sigma^{\hat{W}}$ should be one-to-one. We could then define $\deg_\Gamma(f; \Omega)$ as the direct limit of the finite dimensional Γ-degrees, $\deg_\Gamma(f_n; \Omega_n)$ and $\Pi^\Gamma_{S^{U \times W}}(S^{V \times W})$ as the direct limit of $\Pi^\Gamma_{S^{U \times M_n}}(S^{V \times M_n})$.

Remark 3.1. Since in this alternative approach, we are asking for one-to-one suspensions, we would have, in this case, the complete additivity of the Γ-degree.

Furthermore, it is clear that if $U = V$ and $\Gamma = \{e\}$, then $\deg_\Gamma(f; \Omega)$ is the Leray–Schauder degree of f with respect to Ω.

Note that, in order to apply the finite dimensional approximation, one has to keep track of the suspensions used, in particular of the orientation chosen.

2.4 Orthogonal maps

The reader can see easily that one may extend the Γ-degree to other categories of infinite dimensional maps, such as k-set contractions, A-proper or C^1-Fredholm nonlinear maps, as in the case of the Leray–Schauder degree. However, the case of orthogonal maps is more interesting since the invariants which will give this degree are much richer, as we shall see in the next chapter, even in the finite dimensional case.

Let then Γ be abelian, $U = \mathbb{R}^k \times V$ be a finite dimensional representation of Γ (with trivial action on $\mathbb{R}^k$), Ω be an open Γ-invariant subset of U and consider a Γ-orthogonal map $f(\lambda, x)$ from $\bar{\Omega}$ into V such that:

$$\begin{aligned} f(\lambda, \gamma x) &= \gamma f(\lambda, x) & \\ f(\lambda, x) \cdot A_j x &= 0, & j = 1, \dots, n, \\ f(\lambda, x) &\neq 0 & \text{if } (\lambda, x) \in \partial\Omega, \end{aligned}$$

for A_j the infinitesimal generators of the action of the torus part of Γ.

As proved in Theorem 7.1 of Chapter 1, one may extend $f(\lambda, x)$ to a Γ-orthogonal map $\tilde{f}(\lambda, x)$, for (λ, x) in B_R, a large ball, centered at the origin and containing Ω.

It is then clear that one may repeat the construction for the Γ-degree: take an invariant neighborhood N of $\partial\Omega$ where $\tilde{f}(\lambda, x)$ is non-zero, construct a Γ-invariant Uryson function and define

$$F(t, \lambda, x) = (2t + 2\varphi(\lambda, x) - 1, \tilde{f}(\lambda, x))$$

which will be a Γ-orthogonal map on $I \times B_R$ and non-zero on its boundary, thus, defining an element of the abelian group $\Pi^{\Gamma}_{\perp S^U}(S^V)$, see Lemma 8.3 in Chapter 1.

Definition 4.1. Define the *orthogonal degree* of f, $\deg^{\Gamma}_{\perp}(f; \Omega)$, as $[F(t, \lambda, x)]_{\Gamma}$ in $\Pi^{\Gamma}_{\perp S^U}(S^V)$.

It is easy to see that, as in Proposition 1.1, this orthogonal degree is independent of the construction, since all the deformations can be chosen to be Γ-orthogonal.

Theorem 4.1. *The orthogonal degree has all the properties* 2.1–2.7, *i.e., existence, homotopy invariance* (*for* Γ*-orthogonal deformations*), *excision, suspension, the Hopf property, additivity* (*up to one suspension*) *and universality.*

Proof. We invite the reader to check all those properties. If $k = 0$, we shall prove later that the additivity is true without any suspension. □

We leave also to the reader the task of extending this degree to infinite dimensions for Γ-orthogonal and compact perturbations of the identity. The examples we shall be looking at can be studied by a global reduction to finite dimensions, as explained in Remark 8.2 of Chapter 1, *avoiding in this way some of the technicalities necessary for the infinite dimensional setting.*

Remark 4.1. We have seen that gradients of Γ-invariant functionals are orthogonal maps (Example 7.1 of Chapter 1). That is, if $\Phi(\lambda, x)$ is Γ-invariant then $f(\lambda, x) = \nabla_x\Phi(\lambda, x)$ is Γ-orthogonal. In this case, one could reduce the class of maps to gradients and define a degree in the following way: Assume that $f(\lambda, x)$ is non-zero on $\partial\Omega$ and let B_R be the ball containing Ω. Let $\tilde{\Phi}(\lambda, x)$ be an invariant extension of Φ to B_R.

By using mollifiers, one may assume that $\tilde{\Phi}$ is C^1 in x and that $\nabla_x\tilde{\Phi}(\lambda, x) \equiv \tilde{f}(\lambda, x)$ is arbitrarily close to $f(\lambda, x)$. In fact, let $\varphi(\rho) : \mathbb{R}^+ \to \mathbb{R}^+$ be decreasing, C^∞, with values A for $\rho < \rho_0$ and 0 for $\rho \geq 1$, where A is such that $\int_U \varphi(\|z\|)dz = 1$, for $z = (\lambda, x)$. If $\dim U = N$, let

$$\tilde{\Phi}_\varepsilon(\lambda, x) = \varepsilon^{-N} \int_U \varphi(\|z - y\|)\tilde{\Phi}(y)\, dy.$$

Then, $\tilde{\Phi}_\varepsilon(\lambda, x)$ is C^∞ and Γ-invariant (since the action of Γ is an isometry and $\|\gamma z - y\| = \|z - \gamma^T y\|$). Furthermore, since

$$\tilde{\Phi}_\varepsilon(\lambda, x) = \int_U \varphi(\|y\|)\tilde{\Phi}(z + \varepsilon y)\, dy,$$

$\tilde{\Phi}_\varepsilon(\lambda, x)$ approximates $\tilde{\Phi}(\lambda, x)$ uniformly on B_R and its gradient, f_ε, with respect to x, does approximate $f(\lambda, x)$ on $\bar{\Omega}_{\varepsilon_0} \equiv \{(\lambda, x) \in \Omega : \operatorname{dist}(\lambda, x; \partial\Omega) \geq \varepsilon_0\}$, for $\varepsilon \leq \varepsilon_0$. Since $f(\lambda, x)$ is non-zero on $\partial\Omega$, one may choose ε_0 such that $f(\lambda, x) \neq 0$ on $\bar{\Omega}\backslash\Omega_{\varepsilon_0}$ and replace Ω by Ω_{ε_0}.

For the construction of the invariant neighborhood N of $\partial\Omega$, take N to be an ε_1-neighborhood, N_1 and N_2 be $\varepsilon_1/3$ and $2\varepsilon_1/3$ neighborhoods of $\partial\Omega$. Choose then φ_1 to have value 0 in $\Omega \cup N_1$, and 1 outside $\Omega \cup N_2$. Take then mollifiers φ_ε in order to obtain a C^1 invariant function φ, such that φ is 0 in $\bar{\Omega}$ and 1 outside $\Omega \cup N$, by taking $\varepsilon < \varepsilon_1$.

Next, let $\varepsilon > 0$ be such that

$$4\varepsilon\|\nabla_x\varphi(\lambda, x)\| \leq \|\tilde{f}(\lambda, x)\|$$

for all (λ, x) in N.

Define, for t in $[0, 1]$,

$$\hat{\Phi}(t, \lambda, x) = \varepsilon(t^2 + t(2\varphi(\lambda, x) - 1)) + \tilde{\Phi}(\lambda, x).$$

Then,

$$\nabla_{(t,x)}\hat{\Phi}(t, \lambda, x) = \begin{pmatrix} \varepsilon(2t + 2\varphi(\lambda, x) - 1) \\ \tilde{f}(\lambda, x) + 2\varepsilon t\nabla_x\varphi(\lambda, x). \end{pmatrix}$$

Thus, the zeros of this gradient are such that $\tilde{f}(\lambda, x) = 0$ and $t = 1/2$. It is clear that if one has a gradient Γ-homotopy on $\partial\Omega$, the corresponding gradients of $\hat{\Phi}$ will be Γ-homotopic as maps from $\partial(I \times B_R)$ into $\mathbb{R} \times V\backslash\{0\}$.

Definition 4.2. Let $\Pi^\Gamma_{\nabla S^U}(S^V)$ be the set of Γ-homotopic gradients (with respect to t and x) from $S^U = \partial(I \times B_R)$ into $S^V \equiv \mathbb{R} \times V\backslash\{0\}$. Define the *gradient degree* of $\nabla_x\Phi(\lambda, x)$ with respect to Ω as $\deg^\Gamma_\nabla(\nabla_x\Phi(\lambda, x); \Omega) \equiv [\nabla_{(t,x)}\hat{\Phi}(t, \lambda, x)]_\nabla$.

However, at this point, we don't know if $\Pi^\Gamma_{\nabla S^U}(S^V)$ is a group, since it is not clear that the Borsuk extension theorem holds for gradient maps. Thus, we may consider instead the orthogonal degree of $\nabla_x\Phi(\lambda, x)$, which is an easier object to study. Of course, one could also forget the orthogonality and consider only $\deg_\Gamma(\nabla_x\Phi(\lambda, x); \Omega)$, obtaining the following maps:

$$\Pi^\Gamma_{\nabla S^U}(S^V) \xrightarrow{\perp} \Pi^\Gamma_{\perp S^U}(S^V) \xrightarrow{\Pi} \Pi^\Gamma_{S^U}(S^V),$$

where $\perp$ means forgetting the gradient character but retaining the orthogonality and Π corresponds to maintaining only the equivariance. It is clear that Π is a morphism of abelian groups, and one may show (see Chapter 3, § 6) that Π is onto if $k = 0$. On the other hand, one may conjecture, if $k = 0$, that $\perp$ is one-to-one and onto.

2.5 Applications

There are a certain number of classical applications of any degree theory, such as continuation and bifurcation for problems with parameters. In the case of symmetries one may also consider the implication of breaking the symmetry.

In the case of parametrized problems, assume that we have continuous families of Γ-perturbations of the identity, from $B = \mathbb{R}^k \times U \times W$ into $E = V \times W$,

$$f(\lambda, x) = (g(\lambda, u, w), w - h(\lambda, u, w)),$$

where λ is in $\mathbb{R}^k$, $x = (u, w)$ with U and V finite dimensional Γ-representations while W may be an infinite dimensional representation, in which case h is assumed to be compact. Furthermore,

$$f(\lambda, \gamma x) = \tilde{\gamma} f(\lambda, x).$$

Let S be the set of zeros of f and assume we know an invariant closed subset T of S, called *"trivial"* solutions and that we wish to concentrate on an invariant set G of *"good"* or non-trivial solutions. Suppose $S = G \cup T$. Clearly $\bar{G} \backslash G \subset T$. The set $\bar{G} \backslash G$ will be called the *"bifurcation" set.*

Note that if A is a closed bounded subset of B, then $S \cap A$, $\bar{G} \cap A$ and $T \cap A$ are compact, from the finite dimensionality of $\mathbb{R}^k \times U$ and the compactness of h (this argument has already been used several times).

Let (λ_0, x_0) be a point in $\bar{G} \backslash G$ and let $\mathcal{C}$ be the connected component of $\bar{G}$ containing (λ_0, x_0). Assume that $\Gamma\mathcal{C}$ is bounded (hence compact) and let Ω be an open invariant bounded subset containing $\Gamma\mathcal{C}$.

The following result is an adaptation of a well known topological lemma.

Lemma 5.1. *There is an open bounded invariant set* Ω_1 *such that* $\Gamma\mathcal{C} \subset \Omega_1 \subset \Omega$ *and* $\bar{G} \cap \partial\Omega_1 = \phi$.

Proof. Set $G_1 = \bar{G} \cap \partial\Omega$, then G_1 and $\mathcal{C}$ are two disjoint compact subsets of $\bar{G} \cap \bar{\Omega}$ and hence at a positive distance one from the other. Note that G_1 is invariant while $\mathcal{C}$ may not be so.

It is easily seen that there is an $\varepsilon_0 > 0$, such that, if $\varepsilon < \varepsilon_0$, no ε-chain in $\bar{G} \cap \bar{\Omega}$ can join G_1 to $\mathcal{C}$: recall that an *ε-chain* is a finite number of balls, $A_1, \dots, A_m$, with diameter smaller than ε, such that $\tilde{A}_i = A_i \cap (\bar{G} \cap \bar{\Omega})$ has the property that $\tilde{A}_i \cap \tilde{A}_j = \phi$ if and only if $|i - j| > 1$.

In fact, if this is not true, there are a_n in G_1 and b_n in $\mathcal{C}$ and a $(1/n)$-chain joining a_n to b_n. By compactness there is a subsequence (a_{n_j}, b_{n_j}) converging to (a, b) in $(G_1, \mathcal{C})$ and, for all $\varepsilon > 0$, there is an ε-chain joining a to b. Let then $\mathcal{C}_a$ be the set of all x in $\bar{G} \cap \bar{\Omega}$ such that, for all $\varepsilon > 0$, a and x can be joined by an ε-chain. Both a and b are in $\mathcal{C}_a$ and clearly $\mathcal{C}_a$ is closed, hence compact. Furthermore, $\mathcal{C}_a$ is connected since if not there would be two open subsets X and Y with $\mathcal{C}_a \cap X$, $\mathcal{C}_a \cap Y$ disjoint, non-empty and covering $\mathcal{C}_a$.

From the above, it follows that $\mathcal{C}_a \cap \bar{X}$ and $\mathcal{C}_a \cap \bar{Y}$ are disjoint and hence at a positive distance ε_1 one from the other. Note that any two points in $\mathcal{C}_a$ are joined by ε-chains (passing through a and eliminating intersections). Hence, for x in $\mathcal{C}_a \cap \bar{X}$ and y in $\mathcal{C}_a \cap \tilde{Y}$, for any n larger than $4/\varepsilon_1$, there is a $(1/n)$-chain joining x to y and a point c_n on it, with c_n in $\bar{G} \cap \bar{\Omega}$ and distant at least $\varepsilon_1/2$ from $\mathcal{C}_a$. Passing through a subsequence one would get a c, with $\operatorname{dist}(c, \mathcal{C}_a) \geq \varepsilon_1/2$ and such that, for any $\varepsilon > 0$, there is an ε-chain from c to x and to y. But then one would have an ε-chain from c to a, i.e., c should be in $\mathcal{C}_a$, giving a contradiction. Hence $\mathcal{C}_a$ is connected, a and b are in $\mathcal{C}_a$ and $\mathcal{C}$ is a connected component. One would have that $a \in \mathcal{C}_a \subset \mathcal{C}$, contradicting the fact that G_1 and $\mathcal{C}$ are disjoint.

This proves the existence of ε_0 such that, if $\varepsilon \leq \varepsilon_0$, no ε-chain in $\bar{G} \cap \bar{\Omega}$ can join G_1 to $\mathcal{C}$. Choose then

$$G_2 = \{x \in \bar{G} \cap \bar{\Omega} : \text{ there is a } y \text{ in } G_1 \text{ and an } \varepsilon_0\text{-chain from } x \text{ to } y\}.$$

Clearly, $G_1 \subset G_2$, $G_2 \cap \mathcal{C} = \phi$ by construction and G_2 is closed. Furthermore, if x is in G_2, then any point in $(\bar{G} \cap \bar{\Omega}) \cap B(\varepsilon_0, x)$ is also in G_2, hence G_2 is relatively open. Note also that G_2 is invariant, since the action of Γ is an isometry and if one has an ε_0-chain from x to y, its image under γ will join γx to γy.

Let $C_2 = (\bar{G} \cap \bar{\Omega}) \backslash G_2$, then C_2 and G_2 are compact, disjoint, invariant and cover $\bar{G} \cap \bar{\Omega}$. They are at a positive distance ε_2 one from the other. Choose Ω_1 an $\varepsilon_2/2$-neighborhood of C_2. Clearly, $\Gamma\mathcal{C} \subset \Omega_1 \subset \Omega$ and $\bar{G} \cap \partial\Omega_1 = \phi$ and Ω_1 is invariant. □

A first application of this result will be for the *continuation problem:*

Theorem 5.1. *Let Ω be an open bounded Γ-invariant subset of $\mathbb{R}^k \times U \times W$ and set $\Omega_{\lambda_0} = \Omega \cap \{\lambda = \lambda_0\}$. Assume that $f(\lambda, x)$ is non-zero on $\partial\Omega_{\lambda_0}$ and that $\deg_\Gamma(f(\lambda_0, x); \Omega_{\lambda_0})$ is non-trivial. Suppose that the suspension by any trivial representation of Γ is one-to-one. Then, there is a set of solutions Σ of $f(\lambda, x) = 0$ in Ω, such that Σ/Γ is a connected component of orbits and Σ joins λ_0 to $\partial\Omega$. In fact, for each straight line $\mathcal{L}$ in $\mathbb{R}^k$, passing through λ_0, there is such a set of solutions $\Sigma_{\mathcal{L}}$ going from the left part of $\partial\Omega_{\mathcal{L}}$ to its right part, where $\Omega_{\mathcal{L}} = \Omega \cap \{\lambda \in \mathcal{L}\}$ and the left part means λ in $\mathcal{L}$ and to the left of λ_0 (with respect to the given orientation of $\mathcal{L}$).*

Remark 5.1. In fact, using the tool of Γ-epi maps, [I.M.P.V] and [I.M.V.0], one may show that there is a Σ, with $\Sigma_{\mathcal{L}} \subset \Sigma$ for all $\mathcal{L}$, and Σ/Γ has local dimension at least k, where the dimension is the covering dimension.

Proof of Theorem 5.1. Let $H(\lambda, x) = (\lambda - \lambda_0, f(\lambda, x))$, then $\deg_\Gamma(H(\lambda, x); \Omega)$ is defined and non-trivial, since, by excision, one may replace Ω by $B_\varepsilon(\lambda_0) \times \Omega_{\lambda_0}$, where $B_\varepsilon(\lambda_0)$ is the ball, in $\mathbb{R}^k$, of center λ_0 and radius ε. Hence $H(\lambda, x)$ is a suspension. Since the Γ-degree is non-trivial, the equation $f(\lambda_0, x) = 0$ has solutions in Ω_{λ_0} and, as well, $f(\lambda, x)$ has zeros in Ω_λ for λ close to λ_0. Using the fact that $h(\lambda, x)$ is compact and $\mathbb{R}^k \times U$ is finite dimensional, one has that there are points (λ_0, x_0) in Ω_{λ_0} which are limit points of zeros of $f(\lambda, x)$, for $\lambda \neq \lambda_0$.

Let T be such a point (λ_0, x_0) and let $G = S\backslash T$, hence $\bar{G} = S$. If the connected component $\mathcal{C}$ of (λ_0, x_0) in S does not touch $\partial\Omega$, then we may apply Lemma 5.1 and get Ω_1 contained in Ω, containing $\Sigma = \Gamma\mathcal{C}$ and such that $f(\lambda, x)$ is non-zero on $\partial\Omega_1$. Thus, $\deg_\Gamma(H(\lambda, x); \Omega_1)$ is defined and, deforming λ_0 to λ_1 outside Ω_1, this Γ-degree is trivial (since if $\lambda = \lambda_1$, $f(\lambda_1, x)$ has no zeros in Ω_1). By the addition formula (and using the fact that the suspension is one-to-one), one has that $\deg_\Gamma(H(\lambda, x); \Omega\backslash\bar{\Omega}_1)$ is non-trivial.

If all the components of zeros of $f(\lambda_0, x)$ in Ω_{λ_0} do not join $\partial\Omega$, one may repeat the above construction and arrive, through Zorn's lemma, at minimal sets Ω_j on which the Γ-degree of $H(\lambda, x)$ is non-trivial. An application of the above argument would contradict the minimality and prove the first part of the theorem.

For the second part, by restricting λ to $\mathcal{L}$, we may as well assume that k is 1. Let then T be $S \cap \Omega^C$ and $G = S \cap \Omega$. Suppose that none of the connected components of $\bar{G}$ originating on the left part of $\partial\Omega$ crosses all the way, in Ω, to the right part of $\partial\Omega$. From the first part, we know there is at least one of these components which reaches λ_0 (if not, start from the right part of $\partial\Omega$). Let $\mathcal{C}$ be such a component. Since $f(\lambda, x)$ is non-zero on $\partial\Omega$ for $|\lambda - \lambda_0| < \varepsilon$, for some ε (a compactness argument), one may construct Ω_1 in Ω, containing $\Gamma\mathcal{C} \cap \{\lambda \geq \lambda_0 - \varepsilon\}$ and such that $f(\lambda, x)$ is non-zero on $\partial\Omega_1 \cap \{\lambda \geq \lambda_0\}$ (this requires a slight modification of the sets G and S). Hence, $\deg_\Gamma(H(\lambda, x); \Omega_1)$ is well defined and zero, by pushing λ_0 to the right. By repeating the argument of the first part (taking out Ω_1 and doing the same excision on all the components going from the left of $\partial\Omega$ to λ_0) one arrives at the same contradiction. □

The second classical application is that of *bifurcation*. Let $f(\lambda, x)$ be a $C^1 - \Gamma$-compact perturbation of the identity from $\mathbb{R}^k \times B$ into E, with a known branch of solutions $(\lambda, x(\lambda))$, where $x(\lambda)$ is continuous in λ.

If one linearizes $f(\lambda, x)$ around this solution, one has

$$f(\lambda, x) = A(\lambda)(x - x(\lambda)) + g(\lambda, x),$$

where $A(\lambda) = Df_x(\lambda, x(\lambda))$ and $g(\lambda, x) = o(\|x - x(\lambda)\|)$.

Let H_λ be the isotropy subgroup of $x(\lambda)$. If for some λ_0, one has that $A(\lambda_0)$ is invertible, then the orbit $\Gamma x(\lambda_0)$ must be finite since any one parameter subgroup in Γ gives rise to a vector in $\ker A(\lambda_0)$, see Property 3.3. in Chapter 1. Furthermore, from the diagonal structure of $A(\lambda)$, one has that $A(\lambda_0)^{H_{\lambda_0}} = D_x f^{H_{\lambda_0}}(\lambda_0, x(\lambda_0))$ is also invertible. From the implicit function theorem, one has that, near $(\lambda_0, x(\lambda_0))$, the solutions are in $B^{H_{\lambda_0}}$, that is, by uniqueness, that $x(\lambda)$ belongs to that space and $H_{\lambda_0} < H_\lambda$, for λ close to λ_0. Furthermore, since $Df_x(\lambda_0, \gamma x(\lambda_0))$ is also invertible, for any γ in Γ, one has that, by the same implicit function theorem, the order of the orbit of $x(\lambda)$ is the same as that of $x(\lambda_0)$. Thus, $H_{\lambda_0} = H_\lambda$, for λ close to λ_0.

Lemma 5.2. *Let $\mathcal{A} = \{\lambda : A(\lambda)$ is invertible$\}$. Then, $\mathcal{A}$ is open and on any connected component $\mathcal{B}$ of $\mathcal{A}$, H_λ is constant and will be called the isotropy of that component. If λ_0 is a limit point of $\mathcal{B}$, with isotropy H, then $H < H_{\lambda_0}$.*

Proof. For the last part, it is enough to recall that $x(\lambda)$ is in E^H, which is closed. Hence $x(\lambda_0)$ is fixed by H. □

Since $A(\lambda)$ is H-equivariant, we shall assume in most of this book that $H = \Gamma$ and look, except for some examples, at bifurcation from stationary solutions. Hence, the change of variables which takes $x(\lambda)$ into 0 is admissible and one may consider the bifurcation problem

$$f(\lambda, x) = Ax - T(\lambda)x - g(\lambda, x)$$

where $A, T(\lambda)$ and $g(\lambda, x)$ are as in § 9 of Chapter 1, i.e., A being a Γ-compact perturbation of the identity, is a Fredholm operator of index 0, $\|T(\lambda)\| \to 0$, uniformly as λ goes to 0, and $g(\lambda, x) = o(\|x\|)$, uniformly in λ.

Among all the possible hypotheses on $A - T(\lambda)$, see [I.M.P.V] and [I], we shall choose the following:

For λ close to 0, $\mathbb{R}^k$ decomposes as $\mathbb{R}^{k_1} \times \mathbb{R}^{k_2}$, with $\lambda = (\lambda_1, \lambda_2)$, such that $A - T(\lambda_1, 0)$ is invertible for λ_1 in $\{0 < \|\lambda_1\| \leq 2\rho\} = \bar{B}^1_{2\rho}$.

Let G be the set of non-trivial zeros of $f(\lambda, x)$, i.e., with $x \neq 0$. Denote by $G_1 = G \cap \{\lambda_2 = 0\}$.

Theorem 5.2. *Under the above hypothesis, assume that*

$$\deg_\Gamma(\|x\| - \varepsilon\ ,\ (A - T(\lambda_1, 0))x;\ B^1_{2\rho} \times B_{2\varepsilon})$$

is non-trivial, where $B_{2\varepsilon} = \{x : \|x\| < 2\varepsilon\}$, and that any suspension by a trivial representation of Γ is one-to-one, then there is a branch Σ of non-trivial solutions bifurcating from $(0, 0)$, such that the following holds.

1) *If $\Sigma_1 = \Sigma \cap (\lambda_2 = 0)$, then $\bar{\Sigma}_1$ is connected, Σ_1 is either unbounded or returns to $(\lambda_1, 0)$, with $\|\lambda_1\| > 2\rho$.*

2) *Let $\mathfrak{C}_1$ be the connected component of $(0, 0)$ in $\bar{G}_1$ and assume that $\mathfrak{C}_1$ is bounded. Then, if all the return points $(\lambda_1^j, 0, 0)$ of $\mathfrak{C}_1$ satisfy the above hypothesis, one has*

$$\sum_j \deg_\Gamma(\|x\| - \varepsilon, (A - T(\lambda_1, 0))x;\ B^1_{2\rho_j}(\lambda_1^j) \times B_{2\varepsilon})$$

is trivial, where $B^1_{2\rho_j}(\lambda_1^j) = \{\lambda_1 : \|\lambda_1 - \lambda_1^j\| < 2\rho_j\}$.

3) *The local dimension of Σ/Γ is at least $k_2 + 1$.*

4) *If B^Γ has a closed invariant complement B_1 and $f(\lambda, 0, x)$ is not zero on $B^1_{2\rho} \times (B_{2\varepsilon} \backslash \{0\})$, then (1)–(3) is true for a set $\tilde{\Sigma}$ of points which are non-stationary, except for the return points.*

Proof. As in the continuation problem, we shall not prove here (3) since (3) depends on the notion of Γ-epi maps (see [I]). The argument relies on the fact that

$$\deg_\Gamma(\lambda_2, \|x\| - \varepsilon, f(\lambda, x); B^1_{2\rho} \times B^2_{2\rho} \times B_{2\varepsilon})$$

is the suspension by λ_2 of the previous degree, by deforming $g(\lambda, x)$ to 0, hence non-trivial.

For $\lambda_2 = 0$, let $T_1 = (\lambda_1, 0, 0)$ and $\mathcal{C}_1$ be the connected component of $(0, 0)$ in $\bar{G}_1$. Note that, since 0 is stationary, one has that $\Gamma\mathcal{C}_1 = \mathcal{C}_1$. If $\mathcal{C}_1$ is bounded and does not return to T_1, let Ω_1 given by Lemma 5.1, be such that $f(\lambda_1, 0, x) \neq 0$ on $\partial\Omega_1$, unless $x = 0$ and $\|\lambda_1\| \leq 2\rho$. Then,

$$\deg_\Gamma(\|x\| - \varepsilon, f(\lambda_1, 0, x); \Omega_1)$$

is well defined for all positive ε.

In particular, one may use the excision property, to see that for ε small enough, the above degree is equal to

$$\deg_\Gamma(\|x\| - \varepsilon, f(\lambda_1, 0, x); B^1_{2\rho} \times B_{2\varepsilon}).$$

Choose ε so small that $(A - T(\lambda))x = \tau g(\lambda, x)$, for $\lambda = (\lambda_1, 0)$, $\|\lambda_1\| = 2\rho$ and $\|x\| \leq 2\varepsilon$, is true only for $x = 0$: since $A - T(\lambda)$ is invertible, the left hand side dominates the right hand side.

Thus, $\deg_\Gamma(\|x\| - \varepsilon, f(\lambda_1, 0, x); \Omega_1)$ is non-trivial.

However, since Ω_1 is bounded, by taking ε very large, one does not have zeros of $f(\lambda_1, 0, x)$, with $\|x\| = \varepsilon$, in Ω_1. This proves (1).

For (2), it is enough to remark that if $\mathcal{C}_1$ is bounded and the return points satisfy the non-degeneracy hypothesis, then they are isolated and hence in finite number. In this case, $\deg_\Gamma(\|x\| - \varepsilon, f(\lambda_1, 0, x); \Omega_1)$ is the sum given in (2).

For (4), write $x = x_0 \oplus x_1$, with x_0 in B^Γ and x_1 in B_1. It is enough to complement $f(\lambda_1, 0, x)$ by $\|x_1\| - \varepsilon$ instead of $\|x\| - \varepsilon$, taking the set T as the set of stationary zeros, $\{(\lambda, x) : f(\lambda_1, 0, x) = 0, x \text{ in } B^\Gamma\}$. □

Remark 5.2. Using the arguments of [IMV1] and [I] one may characterize the set of points λ such that $A - T(\lambda)$ is not invertible.

Corollary 5.1. *If* $k_1 = 1$, *then* $\deg_\Gamma(\|x\| - \varepsilon, (A - T(\lambda_1, 0))x; B^1_{2\rho} \times B_{2\varepsilon})$ *is non-trivial if and only if*

$$\deg_\Gamma((A - T(-\rho, 0))x; B_{2\varepsilon}) \neq \deg_\Gamma((A - T(\rho, 0))x; B_{2\varepsilon}).$$

Proof. In fact, the homotopy $\tau(\|x\| - \varepsilon) + (1 - \tau)(\rho - |\lambda_1|)$ is valid, since on $\partial(B^1_{2\rho} \times B_{2\varepsilon})$, either $\|x\| = 2\varepsilon$ and $\lambda_1 = 0$, since $A - T(\lambda_1, 0)$ is invertible for $\lambda_1 \neq 0$, in which case the deformed term is positive, or $|\lambda_1| = 2\rho$ and $x = 0$, in which case the deformed term is negative.

Thus, the first degree is equal to $\deg_\Gamma(\rho - |\lambda_1|, (A - T(\lambda_1, 0))x;\ B^1_{2\rho} \times B_{2\varepsilon}) = \deg_\Gamma(\rho-\lambda_1, (A-T(\lambda_1, 0))x;\ B_+ \times B_{2\varepsilon}) + \deg_\Gamma(\rho+\lambda_1, (A-T(\lambda_1, 0))x;\ B_- \times B_{2\varepsilon})$, where $B_\pm = \{\lambda_1 : |\lambda_1 \mp \rho| < \rho/2\}$: by Remark 2.3, the addition formula holds without suspension. On $B_+ \times B_{2\varepsilon}$, one may deform $A - T(\lambda_1, 0)$ to $A - T(\rho, 0)$, while, on $B_- \times B_{2\varepsilon}$, one deforms to $A - T(-\rho, 0)$. Furthermore, from the definition of $\Pi^\Gamma_{S^B}(S^E)$, it is clear that $\deg_\Gamma(\rho-\lambda_1, (A-T(-\rho, 0)x;\ B_+ \times B_{2\varepsilon}) = -\Sigma_0 \deg_\Gamma((A-T(-\rho))x;\ B_{2\varepsilon})$. Hence,

$$\begin{aligned}&\deg_\Gamma(\|x\| - \varepsilon, (A - T(\lambda_1, 0))x;\ B^1_{2\rho} \times B_{2\varepsilon})\\ &\quad = \Sigma_0(\deg_\Gamma((A - T(-\rho, 0))x;\ B_{2\varepsilon}) - \deg_\Gamma((A - T(\rho, 0)x;\ B_{2\varepsilon})).\end{aligned}$$

Since Σ_0 is one-to-one, one gets the result. □

Recall that, under the above hypothesis, the equation $f(\lambda, x) = 0$ is equivalent to the bifurcation equation (see § 9 of Chapter 1)

$$B(\lambda)x_1 + G(\lambda, x_1) = 0,$$

where x_1 is in $\ker A$, of dimension d, and $B(\lambda)$ is a $d \times d$ equivariant matrix with $G(\lambda, x_1) = o(\|x_1\|)$. Furthermore $B = \ker A \oplus B_2$, where B_2 is a subrepresentation of Γ.

Corollary 5.2.

$$\begin{aligned}&\deg_\Gamma(\|x\| - \varepsilon, (A - T(\lambda_1, 0))x;\ B^1_{2\rho} \times B_{2\varepsilon})\\ &\quad = \deg_\Gamma(\|x_1\| - \varepsilon, B(\lambda_1)x_1, Ax_2;\ B^1_{2\rho} \times B_{2\varepsilon})\end{aligned}$$

and is non-trivial if and only if $\Sigma^{B_2}_A J^\Gamma(B(\lambda_1))$ is non-trivial, where

$$J^\Gamma : \Pi_{k_1-1}(\mathrm{GL}_\Gamma(\ker A)) \to \Pi^\Gamma_{S^{\mathbb{R}^{k_1}\times \ker A}}(S^{\ker A})$$

is the equivariant Whitehead map, and $\Sigma^{B_2}_A$ is the suspension by Ax_2.

Proof. It is enough to deform $(A - T(\lambda_1, 0)x$ to $B(\lambda_1)x_1 \oplus Ax_2$, as it follows from § 9 of Chapter 1. The exact effect of the suspension by Ax_2 depends on Γ and will be given in the next chapter. □

Remark 5.3. The above considerations hold also for Γ-orthogonal problems.

2.6 Operations

In classical degree theory one has formulae for the products and compositions of maps. This is also the case when one has symmetries. Furthermore, for the equivariant problem, one may also consider the effect of changing the group of symmetries. In this last section we shall have a first visit to these operations, by relating them to the corresponding homotopy groups of spheres. In the next chapter, we shall compute these groups and give results for these operations.

2.6.1 Symmetry breaking

The first type of general operation is that of *symmetry breaking or forcing*. More specifically, assume that $f(\lambda, x)$ is Γ-equivariant and has a well-defined Γ-degree with respect to some open, bounded and Γ-invariant subset Ω of $\mathbb{R}^k \times B$.

Let $g(\lambda, x)$ be a Γ_0-perturbation of the identity, where Γ_0 is a subgroup of Γ.

Then, for ε small enough, the map $f(\lambda, x) + \varepsilon g(\lambda, x)$ is non-zero on $\partial\Omega$ (using the fact that $\|f(\lambda, x)\| \geq \eta > 0$, for some η, on $\partial\Omega$ and that g is bounded on $\bar{\Omega}$) and thus $\deg_{\Gamma_0}(f(\lambda, x) + \varepsilon g(\lambda, x); \Omega)$ is also well defined.

Since any Γ-map can be considered as a Γ_0-map, one has a natural morphism

$$P_* : \Pi^{\Gamma}_{S^{\mathbb{R}^k \times B}}(S^E) \to \Pi^{\Gamma_0}_{S^{\mathbb{R}^k \times B}}(S^E)$$

and $\deg_{\Gamma_0}(f(\lambda, x) + \varepsilon g(\lambda, x); \Omega) = P_* \deg_{\Gamma}(f(\lambda, x); \Omega)$.

Example 6.1. Consider the problem of finding 2π-periodic solutions to

$$\nu \frac{dX}{dt} = f(\lambda, X),$$

for X in $\mathbb{R}^N$ and f autonomous. As we have seen, this gives rise to an S^1-equivariant problem.

If one perturbs $f(\lambda, x)$ by $\varepsilon g(t, \lambda, X)$, where $g(t, \lambda, X) = g(t + \frac{2\pi}{p}, \lambda, X)$, then the S^1-equivariance is broken to a $\mathbb{Z}_p$-equivariance: see § 3 and § 9 of Chapter 1. This sort of example will be studied in Chapter 1V.

Note also that if a map $f(\lambda, x)$ is Γ-orthogonal, it will be Γ_0-orthogonal, since the torus part of Γ_0 is a subgroup of the torus part of Γ. One will have the morphism

$$P_\perp : \Pi^{\Gamma}_{\perp S^{\mathbb{R}^k \times B}}(S^E) \longrightarrow \Pi^{\Gamma_0}_{\perp S^{\mathbb{R}^k \times B}}(S^E)$$

and the Γ_0-orthogonal degree of $f(\lambda, x) + \varepsilon g(\lambda, x)$ will be the image, under $P_\perp$, of the Γ-orthogonal degree of $f(\lambda, x)$.

The properties of P_* and $P_\perp$ will be studied in §7 of next chapter.

2.6.2 Products

Consider the classical problem of a product of maps $(f_1(x_1), f_2(x_2))$ defined on a product $\Omega = \Omega_1 \times \Omega_2$ from $V_1 \times V_2$ into $W_1 \times W_2$, where f_1 and f_2 are Γ-equivariant and Ω_i are Γ-invariant, open and bounded. The associated maps, which define the Γ-degree, are

$$F_i(t_i, x_i) = (2t_i + 2\varphi_i(x_i) - 1, \tilde{f}_i(x_i)).$$

One may consider the pair $(F_1(t_1, x_1), F_2(t_2, x_2))$ from $\mathbb{R} \times V_1 \times \mathbb{R} \times V_2$ into $\mathbb{R} \times W_1 \times \mathbb{R} \times W_2$. Let

$$\varphi(x_1, x_2) = \varphi_1(x_1) + \varphi_2(x_2) - \varphi_1(x_1)\varphi_2(x_2) = \varphi_2(1 - \varphi_1) + \varphi_1.$$

Then clearly, $0 \leq \varphi \leq 1$, $\varphi \equiv 0$ on $\Omega_1 \times \Omega_2$ and $\varphi \equiv 1$ on the complement of $(\Omega_1 \cup N_1) \times (\Omega_2 \cup N_2)$. Furthermore, (F_1, F_2) is linearly deformable to $(2t_1 + 2\varphi - 1, \tilde{f}_1, F_2)$, since $\tilde{f}_i(x_i) \neq 0$ on N_i. This last map is in turn deformable to $(2t_1 + 2\varphi - 1, \tilde{f}_1, 2t_2 - 1, \tilde{f}_2)$. Hence,

Lemma 6.1. *If Σ_0 is the suspension by $2t_2 - 1$, one has*

$$[F_1, F_2] = \Sigma_0 \deg_\Gamma((f_1, f_2); \Omega_1 \times \Omega_2).$$

Note that if f_1 and f_2 are Γ-orthogonal, this will be also the case for (f_1, f_2) and the same relation holds for the Γ-orthogonal classes.

Furthermore, it is easy to see that, since $[F_i]$ belongs to $\Pi^\Gamma_{S^{V_i}}(S^{W_i})$, then $[F_1, F_2]$ is in $\Pi^\Gamma_{S^{V_1 \times \mathbb{R} \times V_2}}(S^{W_1 \times \mathbb{R} \times W_2})$ and that one has a morphism of groups, i.e.,

$$\begin{aligned} [F_1 + G_1, F_2] &= [F_1, F_2] + [G_1, F_2] \\ [F_1, F_2 + G_2] &= [F_1, F_2] + [F_1, G_2], \end{aligned}$$

where, for this last operation, with the sum defined on t_2, one has to translate this sum to t_1 (see §7 of next chapter).

Example 6.2. If $V_1 = W_1 = \mathbb{R}^n$, $V_2 = W_2 = \mathbb{R}^m$ and $\Gamma = \{e\}$, then $[F_i] = \deg(f_i; \Omega_i)[\mathrm{Id}]$, hence, from the above morphism, one obtains that, for the Brouwer degree,

$$\deg((f_1, f_2); \Omega_1 \times \Omega_2) = \deg(f_1; \Omega_1)\deg(f_2; \Omega_2).$$

The situation for an abelian group is more complicated but several results will be given in §7 of next chapter.

2.6.3 Composition

The last operation which we shall study is that of composition of maps. Consider three representations V, W and U of the group Γ and assume $f : V \to W$ and $g : W \to U$ are Γ-equivariant maps. Then $g \circ f$ is also Γ-equivariant. Let Ω be a bounded open invariant subset of V.

Assume $f : \bar{\Omega} \to W$ is non-zero on $\partial\Omega$. Let $\Omega_1 = f(\Omega)$. Assume Ω_1 is open and that g is non-zero on $\partial\Omega_1$. It is easy to see that Ω_1 is invariant and bounded (in infinite dimensions this is due to the appropriate compactness), that $f(\partial\Omega) \subset \partial\Omega_1$ and that 0 is away from $\partial\Omega_1$.

Let B be the ball used in the definition of the Γ-degree of f, with the associated extension $\tilde{f}$ of f. Then $\tilde{f}(B) \subset B_1$ for some ball B_1 centered at the origin. If $\tilde{g}$ is the extension of g to B_1, then $\tilde{g} \circ \tilde{f}$ will be an equivariant extension of $g \circ f$ to B. If N_1 is a neighborhood of $\partial\Omega_1$, where $\tilde{g}$ is non-zero and not containing 0, then one may choose a neighborhood N of $\partial\Omega$ contained in $\tilde{f}^{-1}(N_1)$, with its associated φ. Thus,

$$\begin{aligned}
[F] &= [2t + 2\varphi(x) - 1, \tilde{f}(x)] = \deg_\Gamma(f; \Omega) \\
[H] &= [2t + 2\varphi(x) - 1, \tilde{g}(\tilde{f}(x))] = \deg_\Gamma(g \circ f; \Omega) \\
[G] &= [2t_1 + 2\varphi_1(y) - 1, \tilde{g}(y)] = \deg_\Gamma(g; \Omega_1)
\end{aligned}$$

are well defined.

In order to be able to compare these Γ-homotopy classes, let us replace $2t - 1$ by s and $2t_1 - 1$ by s_1, hence s and s_1 belong to the interval $[-1, 1]$, and replace the component $2t + 2\varphi(x) - 1$ by $(s + 2\varphi(x))/3$, which belongs to the interval $[-1, 1]$. Thus F is Γ-homotopic on $\partial([-1, 1] \times B)$ to (s_1, y), with $s_1 = (s + 2\varphi(x))/3$ and $y = \tilde{f}(x)$. Then,

$$[G(F(s, x))] = [(s + 2\varphi(x))/9 + 2\varphi_1(\tilde{f}(x))/3, \tilde{g}(\tilde{f}(x))].$$

One may deform $\varphi_1(\tilde{f}(x))$ to 0, since $\tilde{g}(\tilde{f}(x)) \neq 0$ on N and $\varphi(x) = 1$ outside $\Omega \cup N$, that is the first component is larger than 1/9 and the deformation is valid. We have proved

Lemma 6.2. *Under the above hypothesis one has*

$$\deg_\Gamma(g \circ f; \Omega) = [G \circ F]_\Gamma,$$

where $[F]_\Gamma = \deg_\Gamma(f; \Omega)$ *and* $[G]_\Gamma = \deg_\Gamma(g; f(\Omega))$.

Remark 6.1. If $V = W = U = \mathbb{R}^n$ and $\Gamma = \{e\}$, the classical formula for the degree of a composition runs as follows.

Proposition 6.1. *Let* Ω_i *be the bounded components of* $\mathbb{R}^n \setminus f(\partial\Omega)$ *and suppose that* $g \circ f$ *is non-zero on* $\partial\Omega$. *Then*

$$\deg(g \circ f; \Omega) = \sum \deg(g; \Omega_i) \deg(f - p_i; \Omega),$$

where p_i *is any point in* Ω_i.

Proof. Assuming, from Sard's lemma, that f and g are C^1 and that 0 is a regular value for $g \circ f$, one has

$$\deg(g \circ f; \Omega) = \sum_{g \circ f(x) = 0} \operatorname{Sign} \det D_x(g \circ f(x)) =$$

$$= \sum_{\substack{y=f(x)\\ g\circ f(x)=0}} \text{Sign det } D_y g(y) \text{ Sign det } D_x f(x)$$

$$= \sum_{\substack{y\in\mathbb{R}^n\setminus f(\partial\Omega)\\ g(y)=0}} \text{Sign det } Dg(y) \deg(f(x) - y; \Omega).$$

But $\mathbb{R}^n \setminus f(\partial\Omega)$ is the union of the disjoint connected sets Ω_i, so that $\deg(f(x) - y; \Omega) = \deg(f(x) - p_i; \Omega)$ for y in Ω_i Hence

$$\deg(g \circ f; \Omega) = \sum_i \deg(f(x) - p_i; \Omega)\Big(\sum_{\substack{y\in\Omega_i\\ g(y)=0}} \text{Sign det } Dg(y)\Big)$$

$$= \sum_i \deg(f(x) - p_i; \Omega) \deg(g; \Omega_i). \qquad \square$$

In the equivariant case, it is clear that Ω_i are open, connected and invariant. Furthermore,

$$\deg_\Gamma(g \circ f; \Omega) = \sum_i \deg_\Gamma(g \circ f; f^{-1}(\Omega_i) \cap \Omega)$$

up to one suspension. For Ω_i which contains 0, one may apply Lemma 6.2 (taking $p_i = 0$), however, in general $f(x) - p_j$ will not be equivariant. This explains the hypothesis of Lemma 6.2.

Now, in order to use the algebraic structure of the homotopy groups of spheres, one needs to look at maps from S^V into S^W, i.e., to normalize F to $\hat{F}(s, x) = F(s, x)/\|F(s, x)\|$ (one may assimilate the radii R and R_1 to a change of scale). In that case $\hat{F}(s, x)$ sends the boundary of the cylinder $[-1, 1] \times \{x : \|x\| \le 1\}$ into the boundary of the cylinder $[-1, 1] \times \{y : \|y\| \le 1\}$ in W. One may then take the composition with a Γ-map G and obtain a pairing

$$\Pi^\Gamma_{S^V}(S^W) \times \Pi^\Gamma_{S^W}(S^U) \to \Pi^\Gamma_{S^V}(S^U)$$

$$([\hat{F}]_\Gamma, [G]_\Gamma) \to [G \circ \hat{F}]_\Gamma$$

which is well defined on homotopy classes. We shall see, in §7 of next chapter, that the pairing is in fact a morphism.

Remark 6.2. As maps from S^V into $W \setminus \{0\}$, it is clear that $F(s, x)$ and $\hat{F}(s, x)$ are Γ-homotopic. However, this homotopy is not true for $G(F)$ and $G(\hat{F})$. For instance, if $V = W = V = \mathbb{R}^n$ and $\Gamma = \{e\}$, then any map $F : B \to \mathbb{R}^n \setminus \{0\}$, is such that

$$[F] = \deg(F; B)[\text{Id}].$$

The morphism property of the pairing implies that

$$[G \circ \hat{F}] = \deg(G \circ \hat{F}; B)[\text{Id}] = \deg(G; B) \deg[\hat{F}; B)[\text{Id}]$$

and $\deg(\hat{F}; B) = \deg(F; B)$. The first equality follows also from Proposition 6.1, since $\hat{F}(\partial B) \subset \partial B$ and $\Omega_i = B$ contains the origin. However, $\deg(G \circ F; B)$ may be quite different, as the following example shows.

Example 6.3. On $\mathbb{R}^2$, let $f(x, y) = (x, (x - \varepsilon)y)$, where $0 < \varepsilon < 1/2$. Then f sends the unit disk B into the interior of a lemniscate, with two components, Ω_- containing the origin and Ω_+. By computing the Jacobian of f at the origin, one has that $\deg(f; B) = -1$, while $\deg(f(x, y) - (2\varepsilon, 0); B) = 1$. Hence, from Proposition 6.1., one has

$$\deg(g \circ f; B) = \deg(g; \Omega_+) - \deg(g; \Omega_-).$$

For instance, if $g(x_1, y_1) = (x_1 - 2\varepsilon, y_1)$, then $\deg(g; \Omega_+) = 1$ and $\deg(g; \Omega_-) = 0$, hence $\deg(g \circ f; B) = \deg((x - 2\varepsilon, (x - \varepsilon)y); B) = 1$, which is not the product of $\deg(g; B)$, which is 1, by $\deg(f; B)$.

However, if one considers $f/\|f\|$ on ∂B, then

$$g(f/\|f\|) = (x/\|f\| - 2\varepsilon, (x - \varepsilon)y/\|f\|).$$

This map, which is non-zero on ∂B, is homotopic to $\|f\|g(f/\|f\|)$, which is $(x - 2\varepsilon\|f\|, (x - \varepsilon)y)$, a continuous map on B. For $x = \varepsilon$ or $y = 0$ then $\|f\| = |x|$, thus the only zero of this map is $(0, 0)$. Near the origin, one may deform the map to $(x, -\varepsilon y)$, with degree equal to -1.

Example 6.4. Let Ω be the annulus $\{z \in \mathbb{C}; 1/4 < |z| < 1\}$ and let $f(z) = z^n$, for some integer $n \geq 1$. Let $\Omega_1 = f(\Omega)$ and $g(y) = y - 1/2$ be defined on Ω_1. Then

$$[F] = [(s + 2\varphi(z))/3, z^n] = 0, [G] = [s_1 + 2\varphi_1(y), y - 1/2] = 1 \times [\mathrm{Id}],$$

since $F(s, z) \neq 0$ on $B = \{(s, z) : |s| \leq 1, |z| \leq 1\}$ and G has degree 1 with respect to $B_1 = \{(s_1, y) : |s_1| \leq 1, |y| \leq 1\}$. On the other hand

$$[G \circ F] = [(s + 2\varphi(z))/3 + 2\varphi_1(z^n), z^n - 1/2] = [s, z^n - 1/2] = n[\mathrm{Id}],$$

since $\varphi_1(1/2) = \varphi((1/2)^{1/n}) = 0$.

However, since $F(s, z) \neq 0$ and $G(s_1, y) \neq 0$ on ∂B_1, one may perform the deformation $G(\hat{F}(\tau s, \tau z))$, where $\hat{F}(s, z) = F(s, z)/\|F(s, z)\|$, and have $[G \circ \hat{F}] = [G(1, 0)] = 0$.

There is however one case where $[G \circ F] = [G \circ \hat{F}]$, which we shall phrase in terms of the hypothesis of Lemma 6.2.

Lemma 6.3. *Assume $\Omega_1 = f(\Omega)$ is open and has the following property: If y is non-zero and in $\Omega_1 \cap g^{-1}(0)$, then the segment $\{ty, 0 \leq t \leq 1\}$ is entirely in Ω_1. Then* $\deg_\Gamma(g \circ f; \Omega) = [G \circ \hat{F}]_\Gamma$.

Proof. One has that $F(s, x) = ((s + 2\varphi(x))/3, \tilde{f}(x))$ and, for (s, x) on $\partial(I \times B_R)$, $\hat{F}(s, x) = (\alpha(s + 2\varphi(x))/3, \alpha\tilde{f}(x))$, where $\alpha(s_1, y) = (\alpha s_1, \alpha y)$ belongs to $\partial(I \times B_1)$, i.e., is the intersection of the line segment, going from the origin to (s_1, y), with the boundary of the cylinder. Then,

$$G(\hat{F}(s,x)) = (\alpha(s+2\varphi(x))/3 + 2\varphi_1(\alpha\tilde{f}(x)), \tilde{g}(\alpha\tilde{f}(x))).$$

Replacing α by τ, going from 1 to $\alpha(s_1, y)$, one would obtain a Γ-homotopy from $G\circ F$ to $G \circ \hat{F}$, provided that, on $\partial(I \times B_R)$, this homotopy is valid. Now if one has a zero, then $\tau\tilde{f}(x)$ cannot belong to $(\Omega_1 \cup N_1)^c$, since there φ_1 is 1 and $|\alpha(s+2\varphi(x))/3| \leq 1$. Thus, for a zero, one would need that $\tau\tilde{f}(x)$ should be in Ω_1 and $s+2\varphi(x) = 0$. This last equality, on $\partial(I \times B_R)$, is possible only if $s = -1$ and $\varphi(x) = 1/2$, i.e., x is in N, $\tilde{f}(x)$ in N_1 (and non-zero), and $g(\tau\tilde{f}(x)) = 0$. If $y = \tau\tilde{f}(x)$, one has a contradiction with the assumption on Ω_1. □

Note that the above condition is violated in the examples above and that, on the contrary, it is fulfilled if Ω_1 is starshaped about the origin, or if $g^{-1}(0) \cap \Omega_1 \subset B(0,\rho) \subset \Omega_1$, where $B(0,\rho)$ is a ball of radius ρ and center at the origin.

Remark 6.3 (Orthogonal maps). Let $V = W = U$ and Γ be abelian. If f and g are Γ-orthogonal, i.e., $f(x) \cdot A_j x = 0$, $g(x) \cdot A_j x = 0$, where A_j are the infinitesimal generators for the torus part of Γ, it is easy to see that the composition of f and g is not necessarily orthogonal. However, if one follows the case of gradients,

$$g(y) = \nabla\Phi(y), \quad h(x) = \nabla_x(\Phi(f(x)),$$

then

$$h(x) = Df(x)^T g(f(x))$$

where $g(y)$ is Γ-orthogonal, then one may ask that $f(x)$ is C^1 and Γ-equivariant. From the relations $Df(\gamma x)\gamma = \gamma Df(x)$ and $Df(x)A_j x = A_j f(x)$ (obtained by differentiating $f(\gamma x) = \gamma f(x)$), one sees that $h(x)$ is Γ-orthogonal.

If $f(\partial\Omega) \subset \partial\Omega_1$, $0 \notin \partial\Omega_1$ and $g(y)$ is non-zero on $\partial\Omega_1$, then one may look at $\deg_\Gamma(f;\Omega)$, $\deg_\perp(g;\Omega_1)$ and $\deg_\perp(h;\Omega)$ provided h is non-zero on $\partial\Omega$, for instance if $Df(x)$ is invertible on $\partial\Omega$.

As in Lemma 6.2, one has, for $|s|, |s_1| \leq 1$,

$$\begin{aligned}
F(s,x) &= (s+2\varphi(x))/3, \tilde{f}(x)) \\
G(s_1,y) &= (s_1 + 2\varphi_1(y), \tilde{g}(y)) \\
[H(s,x)]_\perp &= [s+2\varphi(x), D\tilde{f}(x)^T\tilde{g}(\tilde{f}(x))]_\perp = [DF(s,x)^T G(F(s,x))]_\perp
\end{aligned}$$

assuming $\tilde{f}$ is C^1 and $D\tilde{f}$ is invertible in $N \subset f^{-1}(N_1)$.

In order to get a result similar to Lemma 6.3, i.e., working with $\hat{F}$, one needs to make cumbersome computations. Instead, we shall consider the following particular case:

Proposition 6.2. *Assume $\Omega = B(0, R)$, $f(0) = 0$, $Df(x)$ invertible in Ω, $g(y) \neq 0$ for $\|y\| \geq R_1$ and $\|f(x)\| \geq R_1$ if $\|x\| = R$. Then*

$$\deg_\perp((Df)^T g(f(x)); B(0, R)) = \deg_\perp(g(y); B(0, R_1)).$$

Proof. Since $\Omega = B(0, R)$ and one may choose $\Omega_1 = B(0, R_1)$, then the construction of F and G is not necessary: one may compute directly the classes of $h(x)$ and of $g(y)$. Note that the invertibility of $Df(x)$ implies that 0 is the only zero of $f(x)$. Note also that $Dh(x) = Df(x)^T Dg(y) Df(x)$, whenever $g(f(x)) = 0$. Hence, if 0 is a regular value of h, then the Brouwer degree of h is that of g.

Now, on $\partial\Omega$, one may deform orthogonally $h(x)$ to the map

$$Df(x)^T \|f(x)\|^2 g(R_1 f(x)/\|f(x)\|)$$

via $(1 - \tau + \tau\|f(x)\|^2)h(x)$ first and then via

$$Df(x)^T \|f(x)\|^2 g(f(x)(1 - \tau + \tau R_1/\|f(x)\|)).$$

The new map has its only zero at $x = 0$, hence one may deform x in $\partial\Omega$ to εx, for ε small and use the homotopy where $f(\varepsilon x)$ is replaced by $(1 - \tau)f(\varepsilon x) + \tau Df(0)\varepsilon x$ and $Df(\varepsilon x)$ by $(1 - \tau)Df(\varepsilon x) + \tau Df(0)$: since $Df(0)$ commutes with any γ in Γ (and hence with A_j) the deformation is clearly Γ-orthogonal and, for ε small enough, the path from $Df(0)$ to $Df(\varepsilon x)$ consists of invertible matrices, that is the only zero of the deformation is at $x = 0$.

At this stage, one has that

$$[h]_\perp = [Df(0)^T \|Df(0)x\|^2 g(R_1 Df(0)x/\|Df(0)x\|)]_\perp.$$

Now, in $\mathrm{GL}_\Gamma(V)$, one has that $Df(0)$ is Γ-deformable to

$$A = \mathrm{diag}(\varepsilon_\Gamma, \varepsilon_{\mathbb{Z}_2}, \dots, I),$$

where $\varepsilon_\Gamma = \mathrm{diag}(\mathrm{Sign}\det Df(0)^\Gamma, I)$ on V^Γ and $\varepsilon_{\mathbb{Z}_2}$ is a similar matrix on $V^H \cap (V^\Gamma)^\perp$, for each H with $\Gamma/H \cong \mathbb{Z}_2$ and the last I is on the other irreducible representations (see Theorem 8.3 in Chapter 1). Hence, by replacing $Df(0)$ by its deformation to A, one obtains an orthogonal deformation for h and

$$[h]_\perp = [A\|x\|^2 g(R_1 A(x/\|x\|))]_\perp = [Ag(Ay)]_\perp,$$

where y is on $\partial B(0, R_1)$.

Now, we shall see, in Theorem 6.1 of Chapter 3, that

$$[g(y)]_\perp = \sum d_H [F_H(y)]_\perp,$$

with one generator F_H for each isotropy subgroup H of Γ. In particular, each F_H has the form $(\dots, X_0, (y^2 - y)y, \dots)$, where y is a real coordinate where Γ acts as $\mathbb{Z}_2$ and X_0 corresponds to V^Γ. It is then clear that

$$[Ag(Ay)]_\perp = \sum d_H [AF_H(Ay)]_\perp = \sum d_H [F_H(y)]_\perp = [g(y)]_\perp,$$

from the form of F_H, since $\varepsilon_\Gamma^2 X_0 = X_0$, $-((-y)^2 - 1)(-y) = (y^2 - 1)y$. □

2.7 Bibliographical remarks

The literature on the "classical" degree theory and its extensions (to k-set contractions, A-proper maps, etc.) is very extensive. For the reader with interest in analysis, the most accessible texts are the books by Nirenberg, Berger and Krasnosel'skii–Zabrejko. For a survey of the Russian literature, the reader may consult the paper [Z], by Zabrejko.

On the equivariant side, the situation is scarcer. There are indices coming from Algebraic Topology, with the inconvenient that, having to assume that the orbit space is a nice manifold, the action has to be free.

For the case of autonomous differential equation, Fuller has introduced in [F], a degree which is a rational. The relation between the Fuller degree and ours has been shown in [I.M.V. 2]. Dancer, in [Da], has defined a degree for S^1-gradient maps, which is also a rational, and can be shown to follow from the S^1-degree with a "Lagrange multiplier", see [I.M.V. 2].

Geba et al. have defined an S^1-degree in [D.G.J.M.] and then a degree for a general Lie group in [GKW.], which corresponds to the "free part" of our degree. Their definition, using the "normal map" approach will be related to ours in the next chapter. Finally, Rybicki has also defined a degree for S^1-orthogonal maps in [R] and Geba for Γ-gradient maps, for a general Γ, in [G].

The material in this chapter is taken from [IMV1] for § 1–3, from [I.V. 3] for § 4. Lemma 5.1 is known as "Whyburn lemma" and has been widely used in the bifurcation literature.

Chapter 3

Equivariant Homotopy Groups of Spheres

As we have seen in the preceding chapter, the equivariant degree in an element of $\Pi^{\Gamma}_{S^B}(S^E)$, the group of all equivariant homotopy classes of Γ-maps from S^B into S^E. Thus, it is necessary to compute these groups, to know their generators and to understand the effect of some operations, like changing the group Γ, taking products or composition of maps. This chapter is devoted to these topological considerations. Our tools have been completely expounded in § 8 of Chapter 1 and are based on the idea of extension of maps, what is known as obstruction theory. However, we have avoided, as much as possible, most of the abstract scaffolding of Algebraic Topology so that any reader should be able to follow our constructions. The price we have to pay is maybe some long formulae and the restriction to abelian groups and to spaces which satisfy hypothesis (H) of § 2 of Chapter 1. We refer to the section on bibliographical remarks for the very few results for the non-abelian case and some other abstract results. If, nevertheless, the reader wishes to go quickly to applications, then he has only to see the main results of this chapter and go on to the next one.

3.1 The extension problem

Let the abelian group Γ act on the finite dimensional representations B and E and consider an element F of $\Pi^{\Gamma}_{S^B}(S^E)$, i.e., F is an equivariant map from $\partial(I \times B)$ into $\mathbb{R} \times E\backslash\{0\}$. Let V and W stand for $I \times B$ and $\mathbb{R} \times E$ respectively.

For any isotropy subgroup K denote by $B^K = (I \times B)^K$.

The problem we shall be considering in this section is the following: Let H be a fixed isotropy subgroup and assume that F has a Γ-equivariant extension: $\tilde{F} : \bigcup_{K>H} B^K \to W\backslash\{0\}$.

Under which condition F on ∂B^H (and $\tilde{F}$ on the union of the balls B^K) will have a non-zero Γ-extension from B^H into $W^H\backslash\{0\}$?

From Gleason's lemma (Lemma 4.1 of Chapter 1) one has a Γ-extension from

$$\partial B^H \bigcup_{K>H} B_\varepsilon(K) \quad \text{into} \quad W^H\backslash\{0\}$$

which will be non-zero in an ε-neighborhood $B_\varepsilon(K)$ of B^K.

Note that all points in $B^H\backslash\bigcup_{K>H} B_\varepsilon(K)$ have H as isotropy subgroup and that Γ/H acts freely on them.

Theorem 1.1. (a) *If* $\dim V^H - \dim \Gamma/H < \dim W^H$, *then there is a non-zero* Γ-*extension to* B^H.

(b) *If instead one has equality of the dimensions, then there in an integer which is an obstruction to the* Γ-*extension.*

Proof. Write V^H componentwise, in real and complex representations as

$$(x_1, \ldots, x_m) = (y_1, \ldots, y_r, z_1, \ldots, z_s)$$

with y_j real and z_j complex. Then $H = \bigcap H_j$, where $H_j = \Gamma_{x_j}$. Let $k = \dim \Gamma/H$, define $\tilde{H}_j = H_1 \cap \cdots \cap H_j$ and let k_j be the order of $\tilde{H}_{j-1}/\tilde{H}_j$. As seen in § 2 of Chapter 1, there are exactly k complex coordinates for which $k_j = \infty$, i.e., where $\tilde{H}_{j-1}/\tilde{H}_j$ acts as S^1. Let

$$\mathcal{C}_\varepsilon = \{\varepsilon(1 - 1/k_j) \le |x_j| \le R, 0 \le \operatorname{Arg} x_j < 2\pi/k_j\}$$

be a fundamental cell corresponding to H.

Hence, if $k_j = 1$, there are no limitations on x_j, while on y_j with $k_j = 2$, one has $y_j \ge \varepsilon/2$. Also, if $k_j = \infty$, then x_j is real and positive.

Then, $\bar{\mathcal{C}}_\varepsilon$ is a $(r + 2s - k)$- dimensional ball and, from the hypothesis of the theorem, one has $\dim W^H \ge \dim \mathcal{C}_\varepsilon$. From the fundamental cell lemma (Lemma 2.2 in Chapter 1), $B^H \backslash \bigcup_{K>H} B_{\varepsilon_K}(K)$ is covered properly by the images of $\mathcal{C}_\varepsilon$ under Γ/H, where ε_K is chosen to fit $\varepsilon(1 - 1/k_j)$.

In particular, if X belongs to $\bar{\mathcal{C}}_\varepsilon \backslash \mathcal{C}_\varepsilon$, i.e., for some j, with x_j complex and $k_j > 1$, one has $\operatorname{Arg} x_j \equiv \varphi_j = 2\pi/k_j$, then there is a unique point $\gamma_j X$, with γ_j in $\tilde{H}_{j-1}/\tilde{H}_j$, such that $(\gamma_j X)_i = x_i$ for $i < j$ and the argument of $(\gamma_j X)_j$ is 0, i.e., $\gamma_j X$ belongs to $\partial \mathcal{C}_\varepsilon$. In particular, y_i, with y_i real and positive, i.e., with $k_i = 2$, and z_i, with $k_i = \infty$, are left unchanged under γ_j. For such an X, let j_1 be the first index such that $\varphi_{j_1} = 2\pi/k_{j_1}$, then, if $X = (x_1, \ldots, x_m)$, there is a unique γ_{j_1} in $\tilde{H}_{j_1-1}/\tilde{H}_{j_1}$ such that $\gamma_{j_1} X = (x_1, \ldots, x_{j_1-1}, |x_{j_1}|, \gamma_{j_1} x_{j_1+1}, \ldots, \gamma_{j_1} x_m)$. If j_2 is the first index for $\gamma_{j_1} X$ such that $\operatorname{Arg}(\gamma_{j_1} x_{j_2}) = 2\pi/k_2$, one obtains a γ_{j_2} such that $\gamma_{j_2}\gamma_{j_1} X = (x_1, \ldots, |x_{j_1}|, \ldots, |x_{j_2}|, \ldots, \gamma_{j_2}\gamma_{j_1} x_m)$. In a finite number of steps one gets the unique γ in Γ/H, given in the fundamental cell lemma, such that γX belongs to $\mathcal{C}_\varepsilon \cap \partial \mathcal{C}_\varepsilon$.

Note that any equivariant Γ-extension of F must satisfy $F(X) = \tilde{\gamma}^{-1} F(\gamma X)$, i.e., F on $\bar{\mathcal{C}}_\varepsilon \backslash \mathcal{C}_\varepsilon$ is determined by F restricted to $\mathcal{C}_\varepsilon \cap \partial \mathcal{C}_\varepsilon$, while on interior points of $\mathcal{C}_\varepsilon$, F may be arbitrary.

Let $\tilde{s}$ be the number of complex z_j's with $k_j < \infty$. Then $s = \tilde{s} + k$ and $\dim \mathcal{C}_\varepsilon = r + 2\tilde{s} + k \le \dim W^H$. Let $\hat{s}$ be the number of complex z_j's, with $1 < k_j < \infty$.

The proof of the theorem, by induction on $\dim V^H$, will consist in showing that, on $\partial \mathcal{C}_\varepsilon$, there is a non-zero extension of F (and of $\tilde{F}$). This extension needs to have the necessary symmetry properties allowing, under the action of Γ/H, a consistent extension, i.e., Γ-equivariant, to B^H of any continuous extension inside $\mathcal{C}_\varepsilon$.

The minimal dimension possible, for the starting point of the induction, will be that for the case when $H = \bigcap H_j$, while any intersection without one group will give a group strictly bigger than H. In particular, $k_j > 1$.

(a) *The simplest case*. Points on $\partial\mathcal{C}_\varepsilon$ are then either with $|x_j| = R$, where F is given or $y_j = \varepsilon/2$ real or $|z_j| = \varepsilon(1-1/k_j)$ with $k_j > 1$ and z_j complex, or $\operatorname{Arg} z_j = 0$ or $2\pi/k_j$ when $1 < k_j < \infty$. Now, if $x_j = 0$, from the minimality of the intersection, the corresponding isotropy subgroup contains strictly H and for $|x_j| \le \varepsilon(1 - 1/k_j)$ one has the given equivariant extension $\tilde{F}$ of F. In particular, one does not have to worry about y_j real or z_j with $k_j = \infty$.

The rest of $\partial\mathcal{C}$ corresponds to points with some phase φ_j on ∂I_j, where $I_j = [0, 2\pi/k_j]$, for $1 < k_j < \infty$ and z_j complex (there are $\hat{s}$ such j's). Let $\Delta_{\hat{s}-i}$ be the $(\hat{s} - i)$-torus

$$\Delta_{\hat{s}-i} \equiv \{\varphi_j = 0 \text{ for } j = 1, \dots, i, \varphi_j \in I_j \text{ for } j = i + 1, \dots, \hat{s}\}.$$

(There is a slight abuse of notation here: x_j are not necessarily consecutive variables since we are taking out the real variables and those for which $k_j = \infty$).

Then, $\Delta_0 = \{\varphi_j = 0 \text{ for } j = 1, \dots, \hat{s}\}$ gives a piece of $\partial\mathcal{C}$ which has dimension $r + 2\tilde{s} - \hat{s} + k < \dim W^H$, since $\hat{s} \ge 1$. From Remark 8.1 in Chapter 1, one has a continuous non-zero extension on it: that is, any map from $\partial(B^{n+1})$ into $\mathbb{R}^{m+1}\backslash\{0\}$ has a non-zero extension to B^{n+1}, provided $n < m$.

For $\Delta_1 = \{\varphi_j = 0 \text{ for } j < \hat{s} \text{ and } \varphi_{\hat{s}} \text{ in } I_{\hat{s}}\}$, one has already an extension for $\varphi_{\hat{s}} = 0$ (i.e., on Δ_0). Furthermore, there is a unique $\gamma_{\hat{s}}$ in $\tilde{H}_{\hat{s}-1}/\tilde{H}_{\hat{s}}$ such that $\operatorname{Arg}(\gamma_{\hat{s}}|x_{\hat{s}}|e^{2\pi i/k_{\hat{s}}}) = 0$, hence $\gamma_{\hat{s}}$ leaves untouched the preceding x_j's while, for an eventual $j > \hat{s}$, one would have $k_j = \infty$ and $\gamma_{\hat{s}}x_j = x_j$: by the minimality of V^H, there are no x_j with $k_j = 1$. Define

$$F(x_1, \dots, |z_{\hat{s}}|e^{2\pi i/k_{\hat{s}}}, \dots) = \tilde{\gamma}_{\hat{s}}^{-1} F(x_1, \dots, |z_{\hat{s}}|, \dots)$$

which gives an extension to the front face, $\varphi_{\hat{s}} = 2\pi/k_{\hat{s}}$, of Δ_1 from the back face $\varphi_{\hat{s}} = 0$, which is compatible with the action of Γ/H (leaving fixed φ_j for $j < \hat{s}$). If $\hat{s} > 1$, i.e., if Δ_1 corresponds to a piece of $\partial\mathcal{C}_\varepsilon$ ($\hat{s} = 1$ would give $\mathcal{C}_\varepsilon$), then this piece is a ball of dimension $r + 2\tilde{s} - \hat{s} + k + 1 < \dim W^H$. Hence, again from Remark 8.1 of Chapter 1, one obtains a continuous non-zero extension on it.

Assume now that one has obtained an extension for the last $\hat{s} - i$ phases, that is for $\Delta_{\hat{s}-i}$. Consider now $\Delta_{\hat{s}-i+1}$ which has a back face $\varphi_i = 0$, i.e., $\Delta_{\hat{s}-i}$, and a front face for $\varphi_i = 2\pi/k_i$. For a point X on that front face, let γ_i be the unique element of $\tilde{H}_{i-1}/\tilde{H}_i$ such that

$$\gamma_i(x_1, \dots, x_{i-1}, |x_i|e^{2\pi i/k_i}, x_{i+1}, \dots, x_{\hat{s}}) = (x_1, \dots, x_{i-1}, |x_i|, \gamma_i x_{i+1}, \dots, \gamma_i x_{\hat{s}})$$

in $\mathcal{C}_\varepsilon$, i.e, with φ_j in $[0, 2\pi/k_j)$ for $j > i$. Define then

$$F(X) = \tilde{\gamma}_i^{-1} F(\gamma_i X)$$

which will preserve the symmetry on that face. It remains to extend F to $I_i = \{\varphi_i \in [0, 2\pi/k_i]\}$. This is done in the following sequence:

1. Extend to $I_i \times \{\varphi_j = 0, \text{ for } j > i\}$ by the dimension argument.

2. Extend to $I_i \times \{\varphi_j = 0, \text{ for } i < j < \hat{s}\} \times \{\varphi_{\hat{s}} = 2\pi/k_{\hat{s}}\}$ by the action of $\gamma_{\hat{s}}$.

3. Extend to $I_i \times \{\varphi_j = 0, \text{ for } i < j < \hat{s}\} \times I_{\hat{s}}$ by the dimension argument.

4. Extend to $I_i \times \{\varphi_j = 0, \text{ for } i < j < \hat{s} - 1\} \times \{\varphi_{\hat{s}-1} = 2\pi/k_{\hat{s}-1}\} \times I_{\hat{s}}$ by the action of $\gamma_{\hat{s}-1}$.

5. Extend to $I_i \times \{\varphi_j = 0, \text{ for } i < j < \hat{s} - 1\} \times I_{\hat{s}-1} \times \{\varphi_{\hat{s}} = 0\}$ by the dimension argument.

6. Extend to $I_i \times \{\varphi_j = 0 \text{ for } i < j < \hat{s} - 1\} \times I_{\hat{s}-1} \times \{\varphi_{\hat{s}} = 2\pi/k_{\hat{s}}\}$ by the action of $\gamma_{\hat{s}}$.

7. Extend to $I_i \times \{\varphi_i = 0, \text{ for } i < j < \hat{s} - 1\} \times I_{\hat{s}-1} \times I_{\hat{s}}$ by the dimension argument.

8. One continues with $I_{\hat{s}-2}$, first with $\varphi_{\hat{s}} = \varphi_{\hat{s}-1} = 0$ and so on ..., i.e., repeating all the constructions which lead to the extension to $\Delta_{\hat{s}-i}$ but now with I_i instead of $\varphi_i = 0$. Each time one makes an extension, one has to be sure that all the symmetries affecting the phases, which are placed later in the sequence, are taken care of.

Example 1.1. We invite the reader to make a pause and to see in simple examples what is the mechanics of the construction. We propose the following case. Take $\Gamma = \mathbb{Z}_{30}$, acting on (z_1, z_2, z_3) via $(e^{k\pi i}, e^{k\pi i/3}, e^{2k\pi i/5})$ for $k = 0, \dots, 29$. Then, $H_1 = \mathbb{Z}_{15}$, corresponding to even k's and $k_1 = 2$. Also $H_2 \cong \mathbb{Z}_5$, corresponding to multiples of 6, with $H_2 < H_1$, that is $H_1 \cap H_2 = H_2$ and $k_2 = 3$. One has $H_3 \cong \mathbb{Z}_6$, corresponding to multiples of 5, with $k_3 = 5$, $H_1 \cap H_3 = \mathbb{Z}_3$, with multiples of 10 and $H_2 \cap H_3 = \{e\}$.

Then, $0 \le \varphi_1 < \pi, 0 \le \varphi_2 < 2\pi/3, 0 \le \varphi_3 < 2\pi/5$ is the fundamental cell. The γ_3^{-1} which sends $(\varphi_1, \varphi_2, \varphi_3) = (0, 0, 0)$ into $(0, 0, 2\pi/5)$ corresponds to $k = 6$, while $\gamma_2^{-1}(0, 0, 0) = (0, 2\pi/3, 0)$ for $k = 20$ and $\gamma_1^{-1}(0, 0, 0) = (\pi, \pi/3, 0)$ for $k = 25$.

Consider now the following equivariant map on $[0, 1] \times \{(z_1, z_2, z_3) : |z_i| \le 2, i = 1, 2, 3\}$

$$(2t + 1 - 2|z_1 z_2 z_3|, (z_1^2 + 1)z_1, (\bar{z}_1 z_2^3 - 1)z_2, (z_3^5 + 1)z_3),$$

which is non-zero if one z_i is 0. One recognizes in z_1^2, $\bar{z}_1 z_2^3$ and z_3^5 the invariant monomials of Lemma 6.3 in Chapter 1. The zeros of the map are for $z_1 = \pm i$, $z_3 =$

$e^{\pi i/5+2k\pi i/5}$, $z_2 = e^{\pi i/6+2k\pi i/3}$ or $e^{\pi i/2+2k\pi i/3}$ and $t = 1/2$, i.e., 30 in all, but only one in $\mathcal{C}$: $(i, e^{\pi i/6}, e^{\pi i/5})$, for $t = 1/2$. Now, on $\partial\mathcal{C}$, one may deform the map to

$$(2t-1, z_1 - z_1^0, z_2 - z_2^0, z_3 - z_3^0),$$

where (z_1^0, z_2^0, z_3^0) is the unique zero, which has an index 1. We leave to the reader the details of the deformation, as well as the computation of the ordinary degree (i.e., non-equivariant) of the map, which is 30: make a deformation to $(2t-1, z_1^2, z_2^3, z_3^5)$ or use the fact that each zero has the same index (see Property 3.3 in Chapter 1).

A final note on this example: if one takes the order (z_2, z_1, z_3), then $\tilde{k}_1 = 6$, $\tilde{k}_2 = 1$, $\tilde{k}_3 = 5$, that is, the fundamental cell depends on the order of the coordinates. Of course, if Γ/H is finite, then $|\Gamma/H| = \Pi k_j$ is independent of the order.

(b) *End of the proof of the theorem.* In order to complete the induction argument, one needs to see what happens when adding a new variable, without changing H. Thus, one may assume that this new variable is the last one, x_{n+1}, such that $\tilde{H}_n = H = \tilde{H}_{n+1}$, that is $k_{n+1} = 1$. Hence, $\mathcal{C}_{n+1}$, the new fundamental cell, is $\mathcal{C}_n \times \{|x_{n+1}| \leq R\}$. On $\partial\mathcal{C}_{n+1}$, one has either $|x_{n+1}| = R$, with the original map F, or X in $\partial\mathcal{C}_n$ and $|x_{n+1}| \leq R$. The first step is the extension to $\partial\mathcal{C}_{n+1} \cap \{x_j = 0\}$, where the minimality argument is replaced by the induction hypothesis.

The next step is the construction on $\Delta_{\hat{s}}$, where one allows some k_j's to be 1, i.e., that there is no restriction on φ_j. One may perform the same steps by either ignoring these phases or by taking $\gamma_j = \text{Id}$.

Thus, if $\dim V^H < \dim W^H + k$, one may go all the way and obtain an extension to $\mathcal{C}$ which respects the action of Γ/H on $\partial\mathcal{C}$. Then this extension is reproduced by Γ/H to give a Γ-equivariant map on B^H. While if $\dim V^H = \dim W^H + k$, one has a non-zero extension to $\partial\mathcal{C}$ and, given any continuous extension to $\bar{\mathcal{C}}$, with possible zeros, one obtains a Γ-equivariant map on B^H which is non-zero on $\Gamma(\partial\mathcal{C})$. The possibility of a non-zero continuous extension to $\mathcal{C}$ will be determined by the Brouwer degree of this map from $\partial\mathcal{C}$ into $W^H \backslash \{0\}$: see again Remark 8.1 of Chapter 1: such a map has a non-zero extension if and only if its degree is 0. □

Corollary 1.1. *If for all isotropy subgroups H for the action of Γ on V, one has $\dim V^H < \dim W^H + \dim \Gamma/H$, then $\Pi^{\Gamma}_{S^V}(S^W) = 0$.*

Proof. This is clear, since one may extend any $F : S^V \to \mathbb{R} \times W\backslash\{0\}$ to a non-zero equivariant map on $I \times B$, starting with $H = \Gamma$, $I \times B^{\Gamma}$ and the invariant part of F, then on all maximal isotropy subgroups (which correspond to some of the coordinates) and then, by stages, for a given H, having first extended for all K's, with $H < K$ □

Corollary 1.2. *Let J be the subset of* Iso(V) *consisting of all isotropy subgroups H with the property that for any $K \leq H$, one has* $\dim V^K < \dim W^K + \dim \Gamma/K$. *Denote by S_{J^C} the union of S^H's for H in J^C. Then the following holds.*

(a) *If F in $\Pi^{\Gamma}_{S^V}(S^W)$ is such that F restricted to S_{J^C} has a non-zero Γ-equivariant extension to the union of the B^H's, for H in J^c, then $[F]_\Gamma = 0$.*

(b) *If F and G are Γ-homotopic on S_{J^c}, then $[F]_\Gamma = [G]_\Gamma$.*

(c) *If $\tilde{J}$ is the subset of $Iso(V)$ defined as J but with* $\dim V^K \leq \dim W^K + \dim \Gamma/K$ *instead of a strict inequality, then if there is a non-zero Γ-equivariant map $\tilde{F}$ defined on $S_{\tilde{J}^c}$, then $\tilde{F}$ extends to an element of $\Pi^{\Gamma}_{S^V}(S^W)$. In particular, if $\tilde{J}$ is all of* $\mathrm{Iso}(V)$, *there is an equivariant map from S^V into S^W, even if $V^\Gamma = \{0\}$.*

Proof. (a) Starting from maximal elements in J, one extends F, step by step, to a non-zero equivariant map from S^V into $\mathbb{R} \times W\backslash\{0\}$, thus, $[F]_\Gamma = 0$.

(b) Either replace V by $I \times V$, defining a new map, on $I \times S^V$, by F for $\tau = 0$ and G for $\tau = 1$ and the Γ-homotopy on $I \times S_{J^c}$, or use the algebraic structure of $\Pi^{\Gamma}_{S^V}(S^W)$ by considering $[F]_\Gamma - [G]_\Gamma$, where the sum is defined on the first variable. This map, being trivial on S_{J^c}, has a Γ-equivariant extension to the union of the B^H's, for H in J^c. By (a), one obtains $[F]_\Gamma - [G]_\Gamma = 0$.

(c) It is enough to follow the extension procedure given in Theorem 1.1, but now on $\mathcal{C} \cap S^V$, which has one dimension less. Since this construction does not involve the group structure, one obtains the result. □

Returning to the case of a single H with $\dim V^H = \dim W^H + \dim \Gamma/H$, we have seen that the Brouwer degree of the extension to $\partial\mathcal{C}$ is the obstruction for the extension to B^H. A priori, this degree may depend on the extensions to $\partial\mathcal{C}$ and on the choice of $\mathcal{C}$, i.e., on the decomposition of Γ/H. We shall give below several conditions under which this degree is independent of these factors.

We shall first complete some of our results on the fundamental cell.

Definition 1.1. Let $z_1, \ldots, z_k$ be the complex coordinates with $k_j = \infty$, in the decomposition of the fundamental cell $\mathcal{C}$, ($z_1, \ldots, z_k$ are not necessarily consecutive). The ball $B_k = \{x \in B^H, z_j$ real and non-negative, $j = 1, \ldots, k\}$ will be called the *global Poincaré section*. Note that B_k has dimension $\dim V^H - \dim \Gamma/H$. Let $H_0 = H_1 \cap \cdots \cap H_k$, with H_j the isotropy of the coordinate z_j, $j = 1, \ldots, k$, then H_0, which leaves B_k globally invariant, will be called the isotropy of the Poincaré section B_k.

Lemma 1.1. (a) *H_0 acts as a finite group on B_k and $|H_0/H| = \Pi k_j$, for those x_j with $k_j < \infty$. The fundamental cell for this action of H_0 on B_k is $\mathcal{C}$.*

(b) *Any Γ-equivariant map on B^H induces, by restriction, an H_0-equivariant map on B_k and, conversely, any H_0-equivariant map on B_k can be extended to a Γ-equivariant map on B^H. These two operations are the inverse of one another.*

(c) *If F_0 is a non-zero H_0-equivariant map on ∂B_k which has a non-zero H_0-extension $\tilde{F}_0$ to $\bigcup_{K>H}(B^K \cap B_k)$, then one obtains a Γ-equivariant map F on S^H and a non-zero Γ-extension $\tilde{F}$ to $\bigcup_{K>H}(B^K)$. If $(F_0, \tilde{F}_0)$ is H_0-homotopic to $(G_0, \tilde{G}_0)$,* *then* $(F, \tilde{F})$ *is* Γ*-homotopic to* $(G, \tilde{G})$*. Conversely, if* $(F, \tilde{F})$ *is* Γ*-equivariant on* $S^H \bigcup_{K>H}(B^K)$ *and* $\dim V^H \leq \dim W^H + \dim \Gamma/H$*, then one may build* $(F_0, \tilde{F}_0)$ *on* $\partial B_k \bigcup_{K>H}(B^K \cap B_k)$ *a non-zero* H_0*-map which depends only on the* Γ*-homotopy class of* $(F, \tilde{F})$*.*

Proof. (a) We have seen, in Lemma 2.4 (a) of Chapter 1, that the matrix (n_i^j) giving the action of T^n has rank $k = \dim \Gamma/H$ and an invertible submatrix corresponding to $z_1, \ldots, z_k$. Hence $\dim \Gamma/H_0 = k$ and, since $\Gamma/H = (\Gamma/H_0)(H_0/H)$, one gets that H_0/H is a finite group.

Write now Γ/H as $(\Gamma/H_1) \times \cdots \times (\tilde{H}_{i-1}/\tilde{H}_i) \times \cdots \times (\tilde{H}_{m-1}/H)$, where $k_j = |\tilde{H}_{j-1}/\tilde{H}_j|$. For a complex coordinate, if $k_j < \infty$, then $\tilde{H}_{j-1}/\tilde{H}_j \cong \mathbb{Z}_{k_j}$, from Lemma 1.1 in Chapter 1, and one may choose as a generator the γ_j, given in the proof of Theorem 1.1, such that $\gamma_j(|z_j|e^{2\pi i/k_j}) = |z_j|$ and sends $\bar{\mathfrak{C}}$ onto itself. In particular, γ_j leaves invariant the argument of the coordinates $z_1, \ldots, z_k$ with $k_i = \infty$, that is γ_j belongs in fact to H_0. For a real coordinate y_i with isotropy H_i and $k_i = 2$, then we have seen that all the preceding γ_j's, corresponding to complex coordinates, belong to H_i. Furthermore, H_i contains T^n and any subgroup of odd order. If γ_i generates $\tilde{H}_{i-1}/\tilde{H}_i \cong \mathbb{Z}_2$, then the action of Γ on y_i is by $\exp(2\pi i\langle K, L^i/M\rangle)$ and γ_i corresponds to a choice K_0 of K such that $\langle K_0, L^i/M\rangle = 1/2$. Since the action on $z_1, \ldots, z_k$ is by $\exp(i\langle N^j, \Phi\rangle + 2\pi i\langle K, L^j/M\rangle)$, where $N_1, \ldots, N_k$ are linearly independent, there is a Φ_0 such that $\langle N^j, \Phi_0\rangle + 2\pi\langle K_0, L^j/M\rangle$ is a multiple of 2π for $j = 1, \ldots, k$. That is γ_i is in H_0.

Note that γ_j is in $\tilde{H}_{j-1}$, i.e., leaves invariant x_i for $i < j$, and that $\gamma_j^{k_j}$ is in $\tilde{H}_j$. Following the decomposition of Γ/H, one may write any γ in Γ/H as $\gamma_1^{\alpha_1} \ldots \gamma_{m-k}^{\alpha_{m-k}} \delta_1 \ldots \delta_k$, where $0 \leq \alpha_j < k_j$ and $\delta_1, \ldots, \delta_k$ correspond to the coordinates with $k_j = \infty$. Thus, $\gamma = \gamma_0\delta$, where $\gamma_0 = \prod \gamma_j^{\alpha_j}$ is in H_0/H and $\delta = \delta_1 \ldots \delta_k$. From the fact that $\gamma_j^{k_j}$ acts trivially on x_j, it is easy to see that the set of all possible γ_0's gives a subgroup of H_0/H of order $\prod k_j$. Furthermore, if γ is in H_0 then, from $\delta = \gamma_0^{-1}\gamma$, one would have $\delta_1 \ldots \delta_k = \tilde{\gamma}_0$ is in H_0. But then $\delta_1 = \tilde{\gamma}_0\delta_2^{-1} \ldots \delta_k^{-1}$ would be in H_1, since δ_j fixes z_1 for $j > 1$, that is δ_1 would be in $\tilde{H}_1$, hence trivial. Continuing this argument one gets that δ_j are all trivial and $\gamma = \gamma_0$, that is H_0/H has order $\prod k_j$.

Now, for fixed j with $k_j < \infty$, one has that $\gamma_j^p(\mathfrak{C})$, for $p = 0, \ldots, k_j - 1$, are k_j disjoint cells contained in B_k, since γ_j is in H_0. Moreover, $\bar{\mathfrak{C}}$ and $\gamma_j\bar{\mathfrak{C}}$ have only the face $\operatorname{Arg} x_j = 0$ in common. Note also that if $0 \leq \operatorname{Arg} x_i < 2\pi/k_i$, for $i \neq j$, then $\operatorname{Arg} \gamma_j x_i$ belongs to an interval of length $2\pi/k_i$ which intersects the previous one, since γ_j was defined from a point X of $\bar{\mathfrak{C}}\backslash\mathfrak{C}$ such that $\gamma_j X$ was in $\mathfrak{C}$. Furthermore, if x_j is complex, then γ_j preserves the argument of any y_i real with $k_i = 2$.

Now, if there is an X in $\gamma_i^p\mathfrak{C}^o \cap \gamma_j^q\mathfrak{C}^o$, where $\mathfrak{C}^o$ is the interior of $\mathfrak{C}$ and $0 \leq p < k_i$, $0 \leq q < k_j$, then if γ_i corresponds to y_i real and $k_i = 2$, then $\gamma_j^q(y_i) > 0$ and $\gamma_i(y_i) < 0$, which is impossible. Thus, the only possibility is for complex z_i and z_j

with $\operatorname{Arg} z_i$ in $(-2\pi p/k_i, -2\pi(p-1)/k_i)$ and $\operatorname{Arg} z_j$ in $(-2\pi q/k_j, -2\pi(q-1)/k_j)$. Assuming $i < j$, one has that γ_j fixes x_i, hence $\operatorname{Arg} z_i$ must belong to $(0, 2\pi/k_i)$ and $p = 0$. But then $\mathcal{C}^o \cap \gamma_j^q \mathcal{C}^o = \phi$ unless $q = 0$. Similarly, if $\gamma_0 = \prod \gamma_j^{\alpha_j}$ and $\tilde{\gamma}_0 = \prod \gamma_j^{\tilde{\alpha}_j}$ and one has an X in $\gamma_0\mathcal{C}^o \cap \tilde{\gamma}_0\mathcal{C}^o$, then $X = \gamma_0 X_0 = \tilde{\gamma}_0 \tilde{X}_0$ and $X_0 = \gamma_0^{-1}\tilde{\gamma}_0\tilde{X}_0$. Since $\gamma_1^{\tilde{\alpha}_1-\alpha_1}$ will move the argument of x_1 out of $\mathcal{C}^o$, where $x_j \neq 0$, unless $\alpha_1 = \tilde{\alpha}_1$, one gets that this equality is possible only if $\gamma_0 = \tilde{\gamma}_0$. Thus, the $\prod k_j$ images of $\mathcal{C}^o$ do not intersect and, since B_k can also be decomposed in $|H_0/H|$ cells of equal volume, one has that the images of $\mathcal{C}$ cover properly B_k. Furthermore, from the decomposition of any γ_0 in H_0/H as $\prod \gamma_j^{\alpha_j}$, one has that if X belongs to $\mathcal{C}$, then $\gamma_0 X$ belongs to the corresponding image of $\mathcal{C}$ and H_0/H acts freely on $\mathcal{C}$, that is $\mathcal{C}$ is a fundamental cell for H_0 acting on B_k.

(b) If F is Γ-equivariant on B^H, then F restricted to B_k is H_0-equivariant. Conversely, if F_0 is H_0-equivariant on B_k, take any X in B^H. Then, there is a δ in Γ/H_0 such that δX is in B_k. Define

$$F(X) = \tilde{\delta}^{-1} F_0(\delta X).$$

Recall that δ is given by the solution of the system

$$\langle N^j, \Phi\rangle + 2\pi\langle K, L^j/M\rangle = \operatorname{Arg} z_j, \quad j = 1, \dots, k,$$

see Lemma 2.4 in Chapter 1, where the vectors $N_1, \dots, N_k$ are linearly independent. Thus, one may solve for $\Psi = (\varphi_1, \dots, \varphi_k)$ (for instance) and some K so that φ_j are in $[0, 2\pi)$. It is then clear that if δ' solves also this system, then $\delta^{-1}\delta'$ will fix the argument of z_j, i.e., will belong to H_0. But then, $F_0(\delta' X) = F_0(\delta^{-1}\delta'\delta X) = \tilde{\delta}^{-1}\tilde{\delta}' F_0(\delta X)$, since F_0 is H_0-equivariant. Thus, $F(X)$ is well defined and $F|_{B_k} = F_0$.

Let $\gamma = \gamma_0\delta_0$ be in Γ, with γ_0 in H_0 and δ_0 in Γ/H_0. Then, if δX is in B_k, one has that $\delta\delta_0^{-1}(\gamma X)$ is in B_k, since γ_0 is in H_0 and Γ is abelian. Thus,

$$F(\gamma X) = \tilde{\delta}_0\tilde{\delta}^{-1} F_0(\delta\delta_0^{-1}\gamma X) = \tilde{\delta}_0\tilde{\delta}^{-1} F_0(\gamma_0\delta X) = \tilde{\delta}_0\tilde{\gamma}_0\tilde{\delta}^{-1} F_0(\delta X) = \tilde{\gamma} F(X),$$

where one has used the H_0-equivariance of F_0. Hence, F is Γ-equivariant.

(c) Let $(F_0, \tilde{F}_0)$ be H_0-equivariant and non-zero, on ∂B_k for F_0 and on $B^K \cap B_k$ for $\tilde{F}_0$. The above construction gives the extension, after noticing that if X is in B^K then δX is also in B^K and that $B^K \cap B_k = B_k^{K\cap H_0}$. Furthermore, any H_0-homotopy for $(F_0, \tilde{F}_0)$ will generate a Γ-homotopy for $(F, \tilde{F})$.

Conversely, consider z_i, with $k_i = \infty$, and set $V_i^H = V^H \cap \{z_i = 0\}$ and let B_i^H be the corresponding ball with dimension equal to $\dim V^H - 2$. If the isotropy of B_i^H is bigger than H, then F has the extension $\tilde{F}$ on B_i^H. However, if this isotropy is H, the dimension hypothesis implies that one has a Γ-extension to B_i^H. For two different Γ-extensions F_1 and F_2, define, on the boundary of $[0, 1] \times B_i^H$, a Γ-equivariant map defined as F_1 for $\tau = 0$, F_2 for $\tau = 1$ and F for $[0, 1] \times \partial B_i^H$ (and of course $\tilde{F}$ on B_i^K). Applying Theorem 1.1 to $I \times B_i^H$, with dimension $\dim V^H - 1 < \dim W^H + \dim \Gamma/H$, one obtains a Γ-equivariant extension to $I \times B_i^H$, i.e., a Γ-equivariant homotopy from F_1 to F_2.

It is clear that, starting from $\bigcap B_i^H$ and going up in dimension, one may extend this homotopy to a Γ-homotopy to $\bigcup B_i^H$ and, by restriction, to an H_0-homotopy on B_k. Thus, this construction of F_0 on B_k is independent of the extensions to B_i^H.

Furthermore, if one has a Γ-homotopy of $(F, \tilde{F})$ on $(S^H, \bigcup B^K)$ one may extend it, using the arguments of Corollary 1.2 (c), to a Γ-homotopy on $(I \times \bigcup S_i^H, I \times \bigcup B^K)$ and, by gluing the Γ-homotopy of $(F, \tilde{F})$ and by restricting to ∂B_k, one obtains an H_0-homotopy for $(F_0, \tilde{F}_0)$, since $\dim I \times S_i^H \leq \dim W^H + \dim \Gamma/H$. Thus, F_0 depends only on the Γ-homotopy class of $(F, \tilde{F})$. □

Definition 1.2. If F is a non-zero Γ-map on S^H which extends to a non-zero Γ-map $\tilde{F}$ on $\bigcup_{K>H} B^K$ and $\dim V^H = \dim W^H + \dim \Gamma/H$, the obstruction for the extension to B^H will be called the *extension degree* and denoted by $\deg_E(F, \tilde{F})$.

Theorem 1.2. *Let the following condition hold:*

$$(\tilde{\text{H}}) \quad \textit{For all } \gamma \textit{ in } \Gamma \textit{ one has } \det \gamma \det \tilde{\gamma} > 0.$$

Then $\deg_E(F, \tilde{F})$ *depends only on the Γ-homotopy class of $(F, \tilde{F})$ and on H_0. In fact, if F_0 is any H_0-equivariant extension to B_k, one has*

$$\deg(F_0; B_k) = |H_0/H| \deg_E(F, \tilde{F}).$$

Proof. From the preceding lemma, one has that F_0 depends only on the Γ-homotopy class of $(F, \tilde{F})$. One may also perform the construction of Theorem 1.1, by choosing the first k coordinates to be $z_1, \ldots, z_k$, and get a non-zero H_0-equivariant map on $\partial \mathcal{C}$.

Then, if $\gamma_0 = \prod \gamma_i^{\alpha_i}$ is in H_0, one defines on $\gamma_0(\partial \mathcal{C})$

$$F_{\gamma_0}(X) = \tilde{\gamma}_0 F_0(\gamma_0^{-1} X).$$

Then, whenever $X = \gamma_0 X_0 = \gamma_1 X_1$, with X_0 and X_1 in $\partial \mathcal{C}$, one has $X_0 = \gamma_0^{-1} \gamma_1 X_1$ and $F_0(X_0) = \tilde{\gamma}_0^{-1} \tilde{\gamma}_1 F_0(X_1)$, by the equivariance of F_0 on $\partial \mathcal{C}$. Thus, $F_{\gamma_0}(X) = F_{\gamma_1}(X)$. Furthermore, the new map is clearly H_0-equivariant on $H_0(\partial \mathcal{C})$, which contains ∂B_k.

This implies that

$$\deg(F_0; B_k) = \sum_{\gamma_0 \in H_0} \deg(F_{\gamma_0}; \gamma_0(\mathcal{C})).$$

Now, from the property of the composition for the Brouwer degree (this is easy to prove for the case where the zeros are non-degenerate), one has

$$\deg(F_{\gamma_0}; \gamma_0(\mathcal{C})) = \text{Sign} \det \tilde{\gamma}_0 \ \text{Sign} \det \gamma_0 \deg(F_0; \mathcal{C}).$$

From $(\tilde{\mathrm{H}})$, one has that

$$\deg(F_{\gamma_0}; \gamma_0(\mathcal{C})) = \deg(F_0; \mathcal{C})$$

and

$$\deg(F_0; B_k) = \prod k_i \deg(F_0; \mathcal{C}) = |H_0/H| \deg(F_0; \mathcal{C}). \qquad \square$$

Remark 1.1. Condition $(\tilde{\mathrm{H}})$ affects only the real variables $y_1, \dots, y_r$, since on any complex variable, the real determinant is always positive. Thus, if $\det \gamma$ and $\det \tilde{\gamma}$ have opposite sign, this must be on the real variables, where the generators of the action of Γ may be chosen to be in H_0, as we have seen in the proof of Lemma 1.1. Hence, for such a γ_0 one would have:

$$\deg(F_0(\gamma_0 X); B_k) = \operatorname{Sign}\det \gamma_0 \deg(F_0(X); B_k) = \operatorname{Sign}\det \tilde{\gamma}_0 \deg(F_0(X); B_k),$$

by using again the composition property. Thus, if $(\tilde{\mathrm{H}})$ is not valid, one has

$$\deg(F_0(X); B_k) = 0.$$

Remark 1.2. The independence of the extension degree on the extension process includes the fact that one may change the order of the special variables $z_1, \dots, z_k$ which give H_0 and B_k. However, the extension degree could be different for a different choice of H_0. For instance, if S^1 acts on $\mathbb{C}^2$ via $(e^{2i\varphi}, e^{3i\varphi})$, then $H_1 = \{\varphi = 0 \text{ or } \pi\}$, $H_2 = \{\varphi = 0, 2\pi/3 \text{ or } 4\pi/3\}$. If one takes for H_0 the first coordinate, then $k_1 = 2$ and $\mathcal{C}_1 = \{z_1 \in \mathbb{R}^+, 0 \le \operatorname{Arg} z_2 < \pi\}$, while if one takes for H_0 the second coordinate, then $k_2 = 3$ and $\mathcal{C}_2 = \{0 \le \operatorname{Arg} z_1 < 2\pi/3, z_2 \in \mathbb{R}^+\}$. The S^1-map, from $\mathbb{R}^2 \times \mathbb{C}^2$ into $\mathbb{R} \times \mathbb{C}^2$, defined by

$$F(t, \lambda, z_1, z_2) = (2t + 1 - 2|z_1^2 z_2|, (\lambda + i(2t - 1))z_1, (\bar{z}_1^3 z_2^2 + 1)z_2)$$

has its zeros, on $[0, 1] \times \mathbb{R} \times \mathbb{C}^2$, for $t = 1/2$, $\lambda = 0$, $|z_1| = |z_2| = 1$ with $\bar{z}_1^3 z_2^2 = -1$, which is a pair of circles ($z_1 = e^{i\theta}$, $z_2 = \pm i\, e^{3i\theta/2}$).

On $\partial\mathcal{C}_1$, one may perform the following deformations:

1. $(1 - \tau)z_1 + \tau$ in the second component.
2. Replace z_2 in the third component by $|z_2|e^{i(1-\tau)\theta}$, with $0 \le \theta \le \pi$ and $|z_2|$ by $(1 - \tau)|z_2| + \tau$.
3. Replace $2t$, in the first component, by $2(1 - \tau)t + \tau$.

   At this stage, one has deformed F, on $\partial\mathcal{C}$, to the map

$$(2 - 2|z_1|^2 |z_2|, \lambda + i(2t - 1), \bar{z}_1^3 z_2^2 + 1).$$

4. Replace $|z_2|$, in the first component, by $(1-\tau)|z_2| + \tau|z_1|^{-3/2}$, arriving at

$$(2(1 - z_1^{1/2}), \lambda + i(2t-1), z_1^3 z_2^2 + 1), \text{ since } z_1 \geq 0.$$

5. Deform linearly z_1^3 to 1.

One may then linearize the map at $(t = 1/2, \lambda = 0, z_1 = 1, z_2 = i)$, obtaining that the sign of the Jacobian is 1, i.e., the extension degree for $\mathcal{C}_1$ is 1 (and $\deg(F_0; B_1) = 2$).

On the other hand, on $\partial\mathcal{C}_2$, one may perform the following deformations:

1. z_2 to 1 in the last component.
2. z_1 to 1 in the second component, via a rotation.
3. $2t + 1$ to 2 in the first component.
4. After reducing $\mathcal{C}_2$ to the set $\{z_2 > \varepsilon, 0 < \operatorname{Arg} z_1 < 2\pi/3\}$, deform linearly $|z_1|^2$, in the first component, via $(1-\tau)|z_1|^2 + \tau|z_2|^{-4/3}$.
5. Deform linearly z_2^2 in the last component to 1, arriving at

$$\left(2\left(1 - z_2^{-1/3}\right), \lambda + 1(2t-1), \bar{z}_1^3 + 1\right).$$

The linearization, at the only zero in $\mathcal{C}_2$, has a positive determinant, i.e., the extension degree for $\mathcal{C}_2$ is 1 (and $\deg(F_0; B_2) = 3$).

We shall see, in the next result, that the extension degree is independent of H_0.

Let us now continue with the extension problem. Denote by $\bar{V}$ and $\bar{W}$ the subspaces V^H and W^H respectively.

Definition 1.3. (a) Let $\Pi(H)$ denote the subset of $\Pi^{\Gamma}_{S^{\bar{V}}}(S^{\bar{W}})$ consisting of those elements F which have a non-zero Γ-extension to all B^K with $K > H$.

(b) Denote by $\Pi(H, K)$ the set of Γ-homotopy classes of maps $[F, \tilde{F}]$, with $F : \partial B^H \to W^H \backslash \{0\}$, $\tilde{F} : \bigcup B^K \to W \backslash \{0\}$, for $K > H$ and $\tilde{F}$ a Γ-extension of F.

(c) Let Π be the assignment $[F, \tilde{F}] \to [F]$, from $\Pi(H, K)$ into $\Pi(H)$.

Note that if F is in $\Pi(H)$, with extension $\tilde{F}$, and F is Γ-homotopic to G on S^H, then, from the equivariant Borsuk extension theorem (Theorem 6.2 of Chapter 1), G has a non-zero Γ-extension $\tilde{G}$ to $\bigcup B^K$ and $(F, \tilde{F})$ is Γ-homotopic to $(G, \tilde{G})$. This implies that $\Pi(H)$, $\Pi(H, K)$ and Π depend only on Γ-homotopy classes. We have the following result.

Theorem 1.3. *Assume that* $\dim \bar{V} = \dim \bar{W} + \dim \Gamma/H$ *and that* $(\tilde{\mathrm{H}})$ *hold. Then,* $\Pi(H)$ *is a subgroup of* $\Pi^{\Gamma}_{S^{\bar{W}}}(S^{\bar{W}})$. *Furthermore,* $\Pi(H, K)$ *is an abelian group which is isomorphic to* $\mathbb{Z}$ *via the extension degree. The map* Π *is a morphism onto* $\Pi(H)$, *with* $\ker \Pi = \{[(1, 0), \tilde{F}]\}$, *for all possible extensions* $\tilde{F}$ *of the map* $(1, 0)$: *recall that*

$\bar{W} = \mathbb{R} \times E^H$ and $(1, 0)$ corresponds to a map with value 1 on $\mathbb{R}$ and 0 on E^H. In particular, the extension degree is independent of H_0, up to conjugations, and any extension degree is achieved.

Proof. Recall that B^H is a cylinder $I \times B_R$, for t in $I = [0, 1]$ and $\|X\| \leq R$. Let

$$A = \{(t, X) \text{ with either } t = 0 \text{ or } 1 \text{ or } X \text{ in } B^K, \text{ for } K > H\}.$$

If $[F, \tilde{F}]$ is in $\Pi(H, K)$, then $\tilde{F}$ is defined in A and F is non-zero there. Furthermore, the Γ-homotopy $(F(t, \tau X), \tilde{F}(t, \tau X))$ is admissible on A, since if (t, X) is in A then $(t, \tau X)$ is also in A and these maps are non-zero on A. Since $(t, 0)$ is in B^Γ, one has that $\tilde{F}(t, 0) \neq 0$ and $(F(t, 0), \tilde{F}(t, 0))$ is deformable to $(F(0, 0), F(0, 0))$, since H is a proper subgroup of Γ (if $H = \Gamma$ there is nothing to prove). This last map is in turn deformable to $((1, 0), (1, 0))$, since, if dim $W^\Gamma = 1$, the admissibility of F requires that $F(0, 0) > 0$: see §8 of Chapter 1.

Thus, $(F, \tilde{F})$ is Γ-homotopic on A to $((1, 0), (1, 0))$. The Borsuk equivariant extension theorem implies that $(F, \tilde{F})$ is Γ-homotopic, on $\partial B^H \bigcup B^K$, to a map $(F_1, (1, 0))$. Hence one may assume that $(F, \tilde{F})$ is of the form $(F_1, (1, 0))$ on A. This implies, as in §8 of Chapter 1, that one may define a group structure on $\Pi(H, K)$. Furthermore, if dim $V^\Gamma > 1$, one has that $\Pi(H, K)$ is abelian. If V^Γ is reduced to t, the commutativity will follow from the rest of the proof.

Note that, by reducing A to $\{t = 0 \text{ or } 1\}$, one sees that $\Pi(H)$ is a subgroup of $\Pi^\Gamma_{S^{\bar{V}}}(S^{\bar{W}})$, abelian if dim $V^\Gamma > 1$. Furthermore, it is clear that Π is a morphism, onto $\Pi(H)$ and with ker $\Pi = \{((1, 0), \tilde{F})\}$.

Note also that, up to here, we have not used any of the two hypotheses.

However, if these hold, then $\deg_E(F, \tilde{F})$ depends only on $[F, \tilde{F}]$ and on H_0. Hence, one has a map from $\Pi(H, K)$ into $\mathbb{Z}$, given by $\deg_E(F, \tilde{F}) = \deg_E(F_1, (1, 0))$. From Theorem 1.1, this assignment, which is clearly a morphism, is one-to-one, since if the extension degree is 0, one has a Γ-extension to B^H, which is Γ-homotopic, radially and together with $\tilde{F}$, to $(F(0), \tilde{F}(0))$, i.e., to $((1, 0), (1, 0))$.

Thus, $\Pi(H, K)$ is isomorphic to a subgroup of $\mathbb{Z}$, hence abelian.

Finally, define a map F_0 on $\mathcal{C} \cup B^K$ with value $(1, 0)$ on $\bigcup B^K, t = 0$ or $1, x_j = 0$ if $k_j > 1$ (including $z_1, \ldots, z_k$) and on Arg $x_j = 0$ or $2\pi/k_j$ (if $1 < k_j < \infty$) and, on the rest of $\partial\mathcal{C}$, which defines a continuous map of degree 1: one may always localize a map of any degree in a neighborhood of a point on a sphere, with a constant value outside the neighborhood. One may either use an explicit construction or use the fact that the complement of the neighborhood is contractible and appeal to the classical Borsuk extension theorem. This map is invariant on $\bar{\mathcal{C}} \backslash \mathcal{C}$, hence one may extend it, by the action of Γ/H to a Γ-equivariant map F_0 on S^H, which has an extension degree equal to 1. Thus, the morphism

$$\Pi(H, K) \xrightarrow{\deg_E} \mathbb{Z}$$

is an isomorphism, since one may achieve any extension degree, and $\Pi(H, K) \cong \mathbb{Z}$. Any $[F, \tilde{F}]$ in $\Pi(H, K)$ can be written as

$$[F, \tilde{F}] = \deg_E(F, \tilde{F})[F_0, (1, 0)],$$

where F_0 is the above map.

Now, if $\tilde{z}_1, \ldots, \tilde{z}_k$ and $\tilde{H}_0$ correspond to another choice of fundamental cell, with generator $[\tilde{F}_0, (1, 0)]$ and extension degree $\deg_{\tilde{H}_0}$, one has

$$\begin{aligned}[F_0, (1,0)] &= \deg_{\tilde{H}_0}(F_0)[\tilde{F}_0, (1,0)] \\ [\tilde{F}_0, (1,0)] &= \deg_{H_0}(\tilde{F}_0)[F_0, (1,0)],\end{aligned}$$

which means that $\deg_{\tilde{H}_0}(F_0)\deg_{H_0}(\tilde{F}_0) = 1$, since $[F_0, (1, 0)]$ is not trivial, having $\deg_E(F_0, (1, 0)) = 1$, hence $(F_0, (1, 0))$ has no non-zero Γ-extension to B^H. Thus, $\deg_{\tilde{H}_0}(F_0) = \deg_{H_0}(\tilde{F}_0) = \pm 1$.

Moreover, one may construct $\tilde{F}_0$, by a change of variables, from F_0, leaving untouched the variables which are different from $z_1, \ldots, z_k, \tilde{z}_1, \ldots, \tilde{z}_k$, in particular the real variables: one may localize the map in the intersection of the sectors for all these variables. Since one does not alter the order of the variables, the Jacobian of the change of variables is 1 and one has the same extension degree. This last argument is valid only on complex representations. Hence, if Γ acts on (z_1, z_2) in a complex conjugate way, then either one changes z_2 to $\bar{z}_2$ (and the same action) or one has a Jacobian which is -1: see Remark 5.3 in Chapter 1. □

Remark 1.3. If H belongs to $\tilde{J}$, defined in Corollary 1.2, then one may extend F_0 to an equivariant map on V, with value $(1, 0)$ on any B^K whenever K is not a subgroup of H: if $H < K$ this is how F_0 was constructed and for other K's, which are not subgroups of H, define F_0 as $(1, 0)$ and use the dimension argument of Corollary 1.2 (c).

One may have examples where Π is not one-to-one, although most of our applications will be for the case where $\ker \Pi = \{0\}$.

Example 1.2. Let $V = [0, 1] \times \mathbb{C}$ and $W = \mathbb{R}^2$, with $\Gamma = S^1$, acting as $e^{i\varphi}$ on $\mathbb{C}$ and with a trivial action on W. Consider the S^1-map, on $[0, 1] \times \{z : |z| \le 2\} = B$:

$$F(t, z) = (1 - 4t(1-t)(|z| - 2)^2, (2t - 1)t(1-t)(|z| - 2)).$$

Then, $F(t, z) = (1, 0)$ on ∂B and $F(t, z) = 0$ only for $t = 1/2, |z| = 1$ in B. In particular, $F^{S^1}(t, 0) \neq 0$. For $z > 0$, one sees that the Jacobian of the map at $(1/2, 1)$ is positive, i.e., $\deg_E((1, 0), F^{S^1}) = 1$. In fact, by using the addition of homotopy classes, one obtains that $\ker \Pi \cong \mathbb{Z}$ and $\Pi(\{e\}) = 0$.

This case, for $\Gamma = S^1$, $\dim V^\Gamma = \dim W^\Gamma + 1 - 2p$, $\dim V = \dim W + 1$, was studied in [I.M.V. 1 Appendix D] and [I.M.V. 2, Lemmas 2.2 and 2.3 and Theorem 3.1], where it is shown that, if $p \neq 1$, then Π is one-to-one, while if $p = 1$, $\ker \Pi \cong$

$\mathbb{Z}_{m_0(\Pi n_j)/(\Pi m_j)}$, where the action on $(V^\Gamma)^\perp$ is by $e^{im_j\varphi}$, $j = 1, \dots, n$, and on $(W^\Gamma)^\perp$ by $e^{in_j\varphi}$, $j = 1, \dots, n-1$, $n_j = k_j m_j = \tilde{k}_j m_n$, $j = 1, \dots, n-1$, and m_0 is the largest common divisor of the m_j's. Note that $\big(\prod k_j\big)m_0/m_n$ is an integer, since there are integers a_j, $j = 1, \dots n$, such that

$$\sum a_j m_j/m_0 = 1,$$

then, one obtains, using $m_j = \tilde{k}_j m_n/k_j$,

$$(m_n/m_0)\Big(a_n \prod k_j + \sum \Big(a_j \tilde{k}_j \prod_{i \neq j} k_i\Big)\Big) = \prod k_j,$$

that is m_n/m_0 divides $\prod k_j$.

As in the above example, this result depends on an explicit construction of ker Π: Let $V = [0,1] \times \mathbb{R}^l \times \mathbb{C}^n$, $W = \mathbb{R}^{l+2} \times \mathbb{C}^{n-1}$, with $x = (X_0, Z)$ in $\mathbb{R}^l \times \mathbb{C}^n$, and $\|x\| = \max |x_j|$. The ball B will be $[0,1] \times \{x : \|x\| \leq 2\}$. With the above actions one has that $H = \mathbb{Z}_{m_0}$ and taking for $H_0 \cong \mathbb{Z}_{m_n}$ with the last variable, one has $B_1 = B \cap \{z_n \geq 0\}$. Consider the equivariant map

$$F(t,x) = (1,0) + t(1-t)(\|x\| - 2)\big(-4(\|x\| - 2), 2t - 1, \\ X_0, z_1^{k_1} - z_n^{\tilde{k}_1}, \dots, z_{n-1}^{k_{n-1}} - z_n^{\tilde{k}_{n-1}}\big).$$

The zeros of F are for $t = 1/2$, $X_0 = 0$, $\|Z\| = 1$ and hence $|z_j| = 1$ for $j = 1, \dots, n$. One easily checks that

$$\deg(F; B_1) = \prod k_j.$$

Hence, the extension degree $\deg_E((1,0), F) = \big(\prod k_j\big)(m_0/m_n)$.

Furthermore, let $\tilde{F}(t, X_0, z_1, \dots, z_{n-1})$ be a non-zero S^1-extension of $(1,0)$ to $B \cap \{z_n = 0\}$, then the map $(1 - z_n/2)\tilde{F} + (z_n/2, 0)$ is an extension of $(1,0)$ to ∂B_1 and, with a trivial action on $z_n \geq 0$, is an S^1-map. We shall show, when treating Borsuk–Ulam results, that such a map has a degree which is a multiple of $\prod k_j$ (in fact the multiple is $\deg((1 - z_n/2)\tilde{F}^{S^1} + (z_n/2, 0); B_1^{S^1})$. Thus, any element in ker Π has an extension degree which is a multiple of $\big(\prod k_j\big)(m_0/m_n)$.

Example 1.3. Let $V = [0,1] \times \mathbb{R}^l \times \mathbb{C}^n$, $W = \mathbb{R}^{l+2p} \times \mathbb{C}^{n-p}$, $p > 1$ and actions of S^1 of the following form: On z_j in $\mathbb{C}^n$ as $e^{im_j\varphi}$, on ξ_j in $\mathbb{C}^{n-p}$ as $e^{in_j\varphi}$, with $n_j = k_j m_j$, for $j = 1, \dots, n-p$, and n_j are multiples of $m_{n-p+1}, \dots, m_n$. As above m_0 is the largest common divisor of the m_j's, hence $H \cong \mathbb{Z}_{m_0}$. Furthermore, since $p > 1$, one has $\dim V^{S^1} \leq \dim W^{S^1} - 2$ and if $K > H$, with $K \cong \mathbb{Z}_m$, then z_j is in V^K if m_j is a multiple of m and one has

$$(\mathrm{H}_1) \quad \dim V^K \leq \dim W^K + \dim \Gamma/K - 2.$$

Lemma 1.2. *For a general abelian group, if* (H_1) *holds, then* $\ker \Pi = \{0\}$, $\Pi(H) = \Pi^{\Gamma}_{S^{\tilde{V}}}(S^{\tilde{W}})$, *which is* $\mathbb{Z}$ *if* $(\tilde{\mathrm{H}})$ *is true and* $\dim V^H = \dim W^H + \dim \Gamma/H$.

Proof. Use the arguments of Corollary 1.2 to show that F in $\Pi^{\Gamma}_{S^{\tilde{V}}}(S^{\tilde{W}})$ has a non-zero Γ-equivariant extension to $\bigcup B^K$ for $K > H$, and, replacing B^H by $I \times B^H$, that any two extensions $\tilde{F}_1$ and $\tilde{F}_2$, will give rise to pairs $(F, \tilde{F}_1)$, $(F, \tilde{F}_2)$ which are Γ-homotopic on $B^H \bigcup B^K$. Thus, $\deg_E(F, \tilde{F})$ is independent of $\tilde{F}$ and, if $F = (1, 0)$, then $\ker \Pi = \{0\}$. Hence, $\Pi(H, K) = \Pi(H) = \Pi^{\Gamma}_{S^{\tilde{V}}}(S^{\tilde{W}})$. □

Example 1.4. If $\dim V^H = \dim W^H$ and Γ/H is finite, then if $(\tilde{\mathrm{H}})$ holds, one has that $\deg(F, B^H) = |\Gamma/H| \deg_E(F, \tilde{F})$, hence the extension degree depends only on F and $\ker \Pi = \{0\}$. In particular, $[F] = d[F_0]$, where d is the extension degree and $[F_0]$ is the generator constructed in Theorem 1.3. Note that if H is in $\tilde{J}$ of Corollary 1.2 (c), then F_0 can be extended to V, such that $F_0 = (1, 0)$ on B^K, for any K which is not a subgroup of H (see Remark 1.3). While, if $K < H$ and $\dim V^K = \dim W^K$, one has that the fundamental cell for V^K is of the form

$$\mathcal{C}_K = \mathcal{C}_H \times \mathcal{C}_\perp$$

where $\mathcal{C}_\perp$ is the fundamental cell on $(V^H)^\perp \cap V^K$ for the action of H. Then, if $B_\perp$ is the ball in this space, one has that B^K is made of $|\Gamma/H|$ images of $\mathcal{C}_H \times B_\perp$. Furthermore, since $F_0 = (1, 0)$ on $\partial\mathcal{C}_H$, one may extend F_0 as $(1, 0)$ on $\partial\mathcal{C}_H \times B_\perp$. Then $\deg(F_0^K; B^K)$ is the sum of the degrees on the $|\Gamma/H|$ images of $\partial\mathcal{C}_H \times B_\perp$, which are all equal, due to the action of this group and hypothesis $(\tilde{\mathrm{H}})$. Then,

$$\deg(F_0^H; B^K) = |\Gamma/H| \deg(F_0^K; \mathcal{C}_H \times B_\perp).$$

This result will be used when studying the Borsuk–Ulam theorems.

In general, a hypothesis which will enable us to compute $\Pi(H)$ and, from there, $\Pi^{\Gamma}_{S^V}(S^W)$ is based on the following:

Definition 1.4. $K > H$ has a *complementing map* in V^H if there is a non-zero equivariant map $F_\perp$ from $V^H \cap (V^K)^\perp \backslash \{0\}$ into $W^H \cap (W^K)^\perp \backslash \{0\}$, with $F_\perp(0) = 0$.

The existence of complementing maps is a non-trivial question: for instance if Γ acts on the above spaces as $\mathbb{Z}_2$, then the Borsuk–Ulam theorem implies that $\dim V^H \cap (V^K)^\perp \leq \dim W^H \cap (W^K)^\perp$. We shall elaborate further on this type of results in Section 4.

Note that hypothesis (H) of §6 of Chapter 1, applied to the two spaces above, implies the existence of an explicit complementing map: see Lemma 6.2 of Chapter 1. In this case $(\tilde{\mathrm{H}})$ holds on these spaces.

Theorem 1.4. *Assume the following hypothesis:*

(K) *Any minimal $K > H$ has a complementing map in V^H*

Here minimal means that adding a variable to V^K, the isotropy of the new space is H. Then:

(a) $\Pi(H, K) \cong \Pi(H)$.

(b) *If furthermore,* ($\tilde{\text{H}}$) *holds and* $\dim V^H = \dim W^H + \dim \Gamma/H$, *then the extension degree is independent of $\tilde{F}$, extension of F to $\bigcup B^K$, and $\Pi(H) \cong \mathbb{Z}$.*

Proof. Consider $((1, 0), \tilde{F})$, an element of $\ker \Pi$, that is, $\tilde{F}$ is a non-zero Γ-extension of $(1, 0)$ defined on ∂B^H to $\bigcup B^K$. Take a minimal K (hence if $\tilde{K} > K$ one has $B^{\tilde{K}} \subset B^K$) and its associated complementing map $F_\perp$. Write X in V^H as $(X_K, X_\perp)$ and define, on the cylinder $B^H = I \times B = \{(t, x), 0 \le t \le 1, \|X\| = \max |x_j| \le 1\}$, the map

$$\hat{F}(t, x) = ((1 - \|X_\perp\|)\tilde{F}(t, X_K) + \|X_\perp\|(1, 0), (1 - \|X\|)t(1 - t)F_\perp(X_\perp)).$$

It is easy to see that

1. $\hat{F}$ is Γ-equivariant,
2. $\hat{F}$ and $\tilde{F}$ coincide on B^K, i.e., if $X_\perp = 0$,
3. $\hat{F}(t, X) = (1, 0)$ if (t, X) is on ∂B^H,
4. $\hat{F}(t, X) \ne 0$ on B^H.

The last property implies that $[(1, 0), \hat{F}] = 0$ in $\Pi(H, K)$, or else

$$[(1, 0), \tilde{F}] - [(1, 0), \hat{F}] = [(1, 0), G] = [(1, 0), \tilde{F}],$$

where $G(t, X)$ is given by the homotopy difference

$$G(t, X) = \begin{cases} \tilde{F}(2t, X), & 0 \le t \le 1/2 \\ \hat{F}(2 - 2t, X), & 1/2 \le t \le 1 \end{cases}$$

(recall that all these maps have value $(1, 0)$ for $t = 0$ and $t = 1$).

Then $G(t, X)$ is Γ-homotopic on B^K (relative to its boundary, i.e., on $\partial B^K \subset \partial B^H$, the homotopy is fixed and equal to (1,0)), to $(1, 0)$. From the equivariant Borsuk extension theorem applied to $\partial B^H \cup B^K$, $((1, 0), G)$ is Γ-homotopic to a map $((1, 0), \tilde{F}_0)$ with value $(1, 0)$ on B^K, or else, one may assume that $\tilde{F}(t, X) = (1, 0)$ on B^K.

Now, if for some other minimal $\tilde{K}$, one has already that $\tilde{F}(t, X) = (1, 0)$ on $B^{\tilde{K}}$, then $\hat{F}(t, X_{\tilde{K}}) = ((1, 0)(1 - \|X_{\tilde{K}}\|)t(1 - t)F_\perp(X_\perp))$, with $X_{\tilde{K}} = X_K \oplus X_\perp$ and X_K

is in $B^K \cap B^{\tilde{K}}$. Then, $G(t, X)$ is Γ-homotopic to $(1, 0)$ on $B^K \cup B^{\tilde{K}}$ (relative to their boundary), by deforming linearly the second part of $\hat{F}(t, X_{\tilde{K}})$ to 0. Thus, $((1, 0), G)$ is Γ-homotopic to $((1, 0), \tilde{F}_0)$, with $\tilde{F}_0 = (1, 0)$ on $B^K \cup B^{\tilde{K}}$.

By induction, one finds that $\tilde{F}$ is Γ-homotopic, relative to ∂B^H, to $(1, 0)$ on $\bigcup B^K$, that is $\ker \Pi = \{0\}$. Part (b) is then a consequence of Theorem 1.3. □

3.2 Homotopy groups of Γ-maps

In this section, we shall continue our computations of the Γ-homotopy groups of spheres. In §1, we have considered a fixed isotropy subgroup H, with $\dim \Gamma/H = k$. We shall now study, for a fixed k, the set of all isotropy subgroups whose Weyl group has dimension k.

Definition 2.1. Denote by $\Pi(k)$ the set of all Γ-homotopy classes of maps

$$F : \bigcup \partial B^H \to W \backslash \{0\},$$

for isotropy subgroups H with $\dim \Gamma/H = k$, which have Γ-extensions

$$\tilde{F} : \bigcup B^K \to W \backslash \{0\},$$

for all K with $\dim \Gamma/K < k$. Define also by $\Pi(k, k-1)$ the set of Γ-homotopy classes $[F, \tilde{F}]_\Gamma$.

Note that if F is in $\Pi(k)$ and F is Γ-homotopic to G on $\bigcup \partial B^H$, then G has also a Γ-extension $\tilde{G}$ to $\bigcup B^K$, from the Borsuk extension theorem, with $(F, \tilde{F})$ being Γ-homotopic to $(G, \tilde{G})$ on $\bigcup \partial B^H \bigcup B^K$. Thus, $\Pi(k)$ and $\Pi(k, k-1)$ depend only on homotopy classes.

As before, one may deform F on $\{t = 0 \text{ or } 1\} \bigcup B^K$ and assume that it has value $(1, 0)$ there. Hence, one may define group structures on $\Pi(k)$ and $\Pi(k, k-1)$ which are abelian if $\dim V^\Gamma > 1$.

Let $\Pi : \Pi(k, k-1) \to \Pi(k)$ be the restriction. Then, Π is a morphism. As in Theorems 1.3 and 1.4, one has

Lemma 2.1. (a) $\Pi(k, k-1)$ *and* $\Pi(k)$ *are groups (abelian if* $\dim V^\Gamma > 1$*). The morphism* Π *is onto and* $\ker \Pi = \{((1, 0), \tilde{F})\}$*, where* $\tilde{F}$ *is any extension of* $(1, 0)$ *to* $\bigcup B^K$*, with* $\dim \Gamma/K \leq k-1$.

(b) *If every* H *with* $\dim \Gamma/H = k$ *satisfies*

$$(\tilde{\mathrm{K}}) \quad \begin{cases} \text{a) Any minimal } K > H \text{ has a complementing map in} V^H, \\ \text{b) } H \text{ has a complementing map } F_\perp \text{ in } V, \end{cases}$$

then $\ker \Pi = \{0\}$.

Proof. (a) is similar to Theorem 1.3. For (b), one starts with H such that any $K > H$ satisfies $\dim \Gamma/K \leq k-1$. If $((1,0), \tilde{F})$ belongs to $\ker \Pi$, then, from Theorem 1.4, one has that $\tilde{F}^H$ is homotopic to $(1,0)$ on B^H. However, in order to continue this homotopy for other $\tilde{H}$'s, one needs to extend the map $\hat{F}(t, X_H)$ from B^H to V. This is where part (b) of ($\tilde{\text{K}}$) is used: replace the above map by

$$\hat{F}(t, X_H, X_\perp) = ((1-\|X_\perp\|)\hat{F}(t, X_H) + \|X_\perp\|(1,0), (1-\|X\|)t(1-t)F_\perp(X_\perp)).$$

Then, the induction argument on H, so that one has compatible extensions on intersections of B^H's, is similar to the proof of Theorem 1.4. □

Definition 2.2. If H has a complementing map $F_\perp$ in V, then for $[F]$ in $\Pi(H)$ (or in $\Pi^\Gamma_{S^{\bar{V}}}(S^{\bar{W}})$, $\bar{V} = V^H$, $\bar{W} = W^H$), the map

$$\tilde{F}(t, X) = (F(t, X_H), t(1-t)F_\perp(X_\perp)),$$

where $X = X_H \oplus X_\perp$, is called the *suspension of F by the complementing map.* The image of $\Pi(H)$ under this construction, which is a morphism, is a subgroup of $\Pi^\Gamma_{S^V}(S^W)$ and is denoted by $\tilde{\Pi}(H)$.

Note that the factor $t(1-t)$ is there only to facilitate the addition property $(F(t, X_H) = (1,0)$ if $t = 0$ or $1)$ and may be deformed to 1 when considering only the equivariant homotopy class of $\tilde{F}$.

Lemma 2.2. (a) *The suspension by complementing maps is one-to-one.*

(b) *If* $\sum[\tilde{F}_j]_\Gamma = 0$, *where* $[F_j]_\Gamma$ *is in* $\Pi(H_j)$, *then* $[F_j]_\Gamma = 0$ *for all j's.*

Proof. If $[\tilde{F}]_\Gamma = 0$, then $\tilde{F}$ is extendable to a non-zero Γ-map on $I \times B_R$. Thus, $\tilde{F}^H = F$ is also extendable, by restriction, on B^H. This proves (a).

For (b), if H_1 is maximal among the isotropy subgroups of the sum, with $[F_1] \neq 0$ and $[\tilde{F}_1]_\Gamma = \sum -[\tilde{F}_j]_\Gamma$, then, recalling that the homotopy sum is by superposition of the maps by rescaling t, one has, for X in V^{H_1},

$$\tilde{F}_j^{H_1}(t, X) = (F_j^{H_1}(t, X_{H_j}), t(1-t)F_{\perp_j}^{H_1}(X_{\perp j})),$$

that is X_{H_j} is in $V^{H_1} \cap V^{H_j}$. If this intersection is strictly contained in V^{H_j}, then its isotropy is larger than H_j and $F_j^{H_1}$ extends as a non-zero Γ-map. While, if the intersection is V^{H_j}, then $V^{H_j} \subset V^{H_1}$ and $H_1 < H_j$ which contradicts the maximality of H_1. By superposing the extensions on B^{H_1}, one would have that, by Borsuk extension theorem, F_1 would have a non-zero Γ-extension, that is $[F_1] = 0$, a contradiction. □

Theorem 2.1. *If* ($\tilde{\text{K}}$) *holds for all H with* $\dim \Gamma/H = k$, *then*

$$\Pi(k) \cong \bigoplus_{\dim \Gamma/H = k} \Pi(H).$$

If furthermore, ($\tilde{\text{H}}$) *holds and* $\dim V^H \le \dim W^H + \dim \Gamma/H$ *for all* H *with* $\dim \Gamma/H = k$*, then* $\Pi(k) \cong \mathbb{Z} \times \cdots \times \mathbb{Z}$*, where there is one* $\mathbb{Z}$ *for each* H_j *such that* $\dim \Gamma/H_j = k$ *and* $\dim V^{H_j} = \dim W^{H_j} + k$.

Proof. Let $[F]_\Gamma$ be an element of $\Pi(k)$ and H_j be a maximal isotropy subgroup, that is, $\dim \Gamma/H_j = k$ but $\dim \Gamma/K \le k - 1$, for all $K > H_j$. Then, $F^{H_j} : \partial B^{H_j} \to W\backslash\{0\}$ is an element of $\Pi(H_j)$, with a well-defined extension degree if ($\tilde{\text{H}}$) holds and $\dim V^{H_j} = \dim W^{H_j} + k$: in that case $[F^{H_j}] = d_j[F_j]$, where F_j is a generator for $\Pi(H_j)$ with extension degree 1. Note that one may assume that F and F_j have value $(1, 0)$ on $\bigcup B^K$, $K > H_j$ and for $t = 0$ or 1.

Now, for any element G in $\Pi(H_j)$, consider the suspension operation defined by

$$\tilde{G}(t, X) = (G(t, X_j), t(1-t)F_{\perp j}(X_{\perp j}))$$

where $X = X_j \oplus X_{\perp j}$, with X_j in V^{H_j}, and $F_{\perp j}$ is the complementing map for H_j.

Then, $[F] - [\tilde{F}^{H_j}]$ has a non-zero extension to B^{H_j}, where $[\tilde{F}]$ is the restriction from $\Pi^\Gamma_{S^V}(S^W)$ to $\Pi(k)$.

One may do the same procedure for each such maximal H_j since we know that, on $V^{H_i} \cap V^{H_j}$, the isotropy subgroups K have $\dim \Gamma/K < k$ and there all maps are assumed to be $(1, 0)$. Let

$$[F_1] = [F] - \sum [\tilde{F}^{H_j}]$$

where the sum is taken over all maximal H_j's.

Note that, at this stage, d_j are uniquely determined by $F^{H_j} = F|_{B^{H_j}}$. Note also that, from the analogue of Lemma 2.1, one may assume $F_1 = (1, 0)$ on $\bigcup B^{H_j}$ and that the homotopy type on $\bigcup \partial B^H \bigcup B^K \bigcup B^{H_i}$ is unchanged by this assumption.

Take then H, with $\dim \Gamma/H = k$ and for all $K > H$, either $\dim \Gamma/K < k$ or $K = H_j$ for some of the preceding H_j's. Then, the map F_1^H defines an element of $\Pi(H)$, which is $d_H[\tilde{F}_H]$ in the particular case of the theorem, where F_H is the generator for $\Pi(H)$ and $\tilde{F}_H$ its suspension by $F_{\perp H}$. Clearly, $[F_1] - [\tilde{F}_1^H]$ is extendable to B^H.

One may perform the same construction for all H's with these characteristics and conclude that $[F_1] - \sum_H [\tilde{F}_1^H]$ is extendable to $\bigcup B^H$. Note again that $[\tilde{F}_1^H]$ is completely and uniquely determined by $[F]$. In the particular case, $[\tilde{F}_1^H] = d_H[\tilde{F}_H]$ has a unique extension degree d_H.

One may go on to the next stages of isotropy subgroups, arriving finally at

$$[F] - \sum_{j,H} [\tilde{F}_j^H] = 0$$

in $\Pi(k)$, with $[\tilde{F}_j^H] = d_H[\tilde{F}_H]$ in the particular case of ($\tilde{\text{H}}$) and a unique extension degree d_H. The set of $[\tilde{F}_j^H]$'s is uniquely determined by $[F]$ and the step by step construction. Note that, from the construction, F_j is (1,0) on the previous stage of isotropy subspaces, in particular $[\tilde{F}_j^H]$ belongs to $\Pi(k)$.

Conversely, if $[F]$ is in $\Pi(H)$, with $\tilde{F}(t, X) = (F(t, X_H), t(1-t)F_\perp(X_\perp))$, then $\tilde{F}^K(t, X_K) = (F^K(t, X_H), t(1-t)F_\perp^K(X_\perp))$. Then, if $V^H \cap V^K$ is strictly contained in V^H, the map F^K is non-zero on $B^K \cap B^H$ and $\tilde{F}^K$ is non-zero on B^K.

While, if $V^H \cap V^K = V^H$ and $K \neq H$, one has $K < H$ and $\dim \Gamma/K \geq \dim \Gamma/H$. Thus, in this case, one cannot have $\dim \Gamma/K < k$ and $\dim \Gamma/H = k$. That, is $[\tilde{F}]$ is in $\Pi(k)$.

Let now M be the morphism from $\bigoplus \Pi(H)$ into $\Pi(k)$ given by

$$M([F^{H_1}], \ldots, [F^H], \ldots) = \sum [\tilde{F}^{H_j}],$$

which is well defined from the previous argument. Furthermore, from Lemma 2.2, M is one-to-one and onto, due to the construction. In the particular case, recall, from Theorem 1.1, that $\Pi(H) = 0$ if $\dim V^H < \dim W^H + k$ and $\Pi(H) \cong \mathbb{Z}$ if one has equality of dimensions (Theorem 1.4 (b)). □

In order to continue with the study of $\Pi_{S^V}^\Gamma(S^W)$ it is natural to keep up with the ordering begun with $\Pi(k)$.

Definition 2.3. Let Π_k be the set of all Γ-homotopy classes of maps $F : \bigcup \partial B^H \to W \backslash \{0\}$, for all H with $\dim \Gamma/H \leq k$.

It is clear that Π_k is a group (abelian if $\dim V^\Gamma > 1$) and that $\Pi(k)$ is a subgroup of Π_k.

Theorem 2.2. *If* ($\tilde{\mathrm{K}}$) *holds for all* H *with* $\dim \Gamma/H = k$, *then*

(a) $\Pi_k \cong \Pi_{k-1} \times \Pi(k)$.

(b) *If moreover,* $\dim V^L < \dim W^L + \dim \Gamma/L$, *for all* L *with* $\dim \Gamma/L > k$, *then* $\Pi_{S^V}^\Gamma(S^W) \cong \Pi_k$.

Proof. Let $P_* : \Pi_k \to \Pi_{k-1}$ be the restriction map. We shall show that P_* is onto and that $\ker P_* = \Pi(k)$.

Let then $[F]$ be an element of Π_{k-1}. Take a minimal K, i.e., with $\dim \Gamma/K = k-1$ and $\dim \Gamma/H = k$ for any $H < K$. Consider the suspension $\tilde{F}_K$ of F^K (first to some H with $\dim \Gamma/H = k$ and then by $F_\perp$). Clearly, $[F] - P_*[\tilde{F}_K]$ is deformable to $(1, 0)$ on ∂B^K. Hence, by the equivariant Borsuk theorem, the above difference is Γ-homotopic in Π_{k-1} to a map $\hat{F}$ which has value $(1, 0)$ on ∂B^K and can be extended as $(1, 0)$ on B^K.

Let $\tilde{K}$ be another minimal isotropy subgroup and consider the suspension $\tilde{F}_{\tilde{K}}$ of $\hat{F}$. Then, $\tilde{F}_{\tilde{K}}|B^K = ((1, 0), t(1-t)F_\perp(X_{\perp,\tilde{K}}))$, thus, $[\hat{F}] - P_*[\tilde{F}_{\tilde{K}}]$ is deformable to $(1, 0)$ on $\partial B^{\tilde{K}} \cup \partial B^K$ and the difference may be replaced by a map with this value on these two spheres.

By performing this operation on all minimal K's, we arrive at $[F] - \sum P_*[\tilde{F}_K]$, which is deformable to $(1, 0)$ on $\bigcup \partial B^K$, for all K's, hence zero in Π_{k-1}. That is,

$$[F] = \sum P_*[\tilde{F}_K].$$

Or else, from the equivariant Borsuk theorem, F has an extension $\tilde{F}$ with $[\tilde{F}] = \sum[\tilde{F}_K]$ and $[F] = P_*[\tilde{F}]$. Note that $\tilde{F}$ depends on the chosen order for the minimal K's (and on the complementing maps) however, for a given choice, it is easy to see that if F is Γ-homotopic to G then $\tilde{F}$ and $\tilde{G}$ are Γ-homotopic, that $[\tilde{F}_K]$ are uniquely determined by this choice and that this construction sends sums into sums, i.e., that it is a morphism.

Let now $[F]$ in Π_k be such that $P_*[F] = 0$. Then, F is extendable to a non zero Γ-map on $\bigcup B^K$ for K with $\dim \Gamma/K \leq k-1$. That is, $[F]$ belongs to $\Pi(k)$. Conversely, if $[F]$ belongs to $\Pi(k)$, then one may assume, from Lemma 2.1, that F restricted to $\bigcup B^K$, $\dim \Gamma/K \leq k-1$, is $(1, 0)$. Thus, $P_*[F] = 0$ and $\ker P_* = \Pi(k)$.

In general, if $[F]$ is an element of Π_k, let $P_*[F] = \sum P_*[\tilde{F}_K]$ and the difference $[F] - \sum[\tilde{F}_K]$ belongs to $\ker P_*$, i.e., is of the form $\sum[\tilde{F}_j^H]$, from Theorem 2.1. Thus,

$$[F] = \sum[\tilde{F}_K] + \sum[\tilde{F}_j^H],$$

where the first sum is on the minimal K's with $\dim \Gamma/K = k-1$ and $\tilde{F}_j^H$ are the generators for $\Pi(k)$. Note that all these maps are defined in $\Pi_{S^V}^{\Gamma}(S^W)$, although the equality is in Π_k.

Under the hypothesis of (b), let $[F]$ be an element of $\Pi_{S^V}^{\Gamma}(S^W)$ and let $P_k[F]$ be the class of its restriction on Π_k, where P_k is the map induced by this restriction. Then $[F] - \sum[\tilde{F}_K] - \sum[\tilde{F}_j^H] \equiv [G]$ is such that $P_k[G] = 0$, that is $P_{k+1}[G]$ belongs to $\Pi(k+1)$. But, from the dimension hypothesis, $\Pi(k+1) = 0 = \Pi(k+l)$ for any $l \geq 1$ by Corollary 1.2. Hence, the Γ-homotopy of $P_k[G]$ extends to a Γ-homotopy of G on $I \times B_R$. □

Remark 2.1. Under the hypothesis of Theorem 2.2, consider the set of H's, with $\dim \Gamma/H = k$, which are minimal. As in the proof of Theorem 2.2, one obtains that $[F] = \sum P_k[\tilde{F}_H]$, for any F in Π_k and P_k is the above morphism induced by restriction to Π_k. Hence, P_k is onto.

On the other hand, if $[F]$ belongs to $\Pi_{S^V}^{\Gamma}(S^W)$, then $P_k[F] = \sum P_k[\tilde{F}_H] = \sum P_k[\tilde{F}_K] + \sum P_k[\tilde{F}_j^H]$, from Theorem 2.2. An easy induction argument leads to

Theorem 2.3. *If* ($\tilde{\text{K}}$) *holds for all isotropy subgroups, then*

$$\Pi_{S^V}^{\Gamma}(S^W) \cong \bigoplus_H \tilde{\Pi}(H),$$

where $\tilde{\Pi}(H)$ stands for the suspension by the corresponding complementing map.

Corollary 2.1. *Let $\tilde{\Gamma} = \Gamma/T^n$, $\tilde{V} = V^{T^n}$, $\tilde{W} = W^{T^n}$, then the following holds.*

(a) *If ($\tilde{\text{H}}$) and ($\tilde{\text{K}}$) hold for all H's with* $\dim \Gamma/H = 0$ *and if, for all H, one has* $\dim V^H \leq \dim W^H$, *then*

$$\Pi^{\Gamma}_{S^V}(S^W) \cong \Pi^{\tilde{\Gamma}}_{S^{\tilde{V}}}(S^{\tilde{W}}) \cong \mathbb{Z} \times \cdots \times \mathbb{Z},$$

with one $\mathbb{Z}$ for each H with $\dim \Gamma/H = 0$ *and* $\dim V^H = \dim W^H$. *One has $[F] = \sum d_H[\tilde{F}_H]$, where d_H is the extension degree and $\tilde{F}_H$ is the generator suspended by its complementing map.*

(b) *If ($\tilde{\text{H}}$) and ($\tilde{\text{K}}$) hold for all H's with* $\dim \Gamma/H = 1$ *and if, for all H, one has* $\dim V^H \leq \dim W^H + 1$, *then*

$$\Pi^{\Gamma}_{S^V}(S^W) \cong \Pi^{\tilde{\Gamma}}_{S^{\tilde{V}}}(S^{\tilde{W}}) \times \mathbb{Z} \times \cdots \times \mathbb{Z},$$

with one $\mathbb{Z}$ for each H with $\dim \Gamma/H = 1$ *and* $\dim V^H = \dim W^H + 1$. *One has $[F] = [\tilde{F}] + \sum d_H[\tilde{F}_H]$, where $\tilde{F}$ is the suspension of F^{T^n}.*

Proof. (a) is an immediate consequence of Theorems 2.2 and 1.4 (b), while, for (b), one needs to recall, from Lemma 2.1 of Chapter 1, that $\tilde{V}$ corresponds to all points with isotropy H with $\dim \Gamma/H = 0$, hence $\Pi_0 = \Pi^{\tilde{\Gamma}}_{S^{\tilde{V}}}(S^{\tilde{W}})$. Since $[F] - [\tilde{F}]$, where $\tilde{F}$ is the suspension of F^{T^n} by its complementing map, is in $\Pi(1)$, one obtains the result. □

Example 2.1. If $V = \mathbb{R}^k \times W$, then ($\tilde{\text{K}}$) is clearly satisfied, with complementing maps which are the identity on $(V^H)^{\perp}$. The hypothesis ($\tilde{\text{H}}$) is also satisfied.

Hence, one may apply Theorems 2.2 and 2.3. In particular, since $\dim V^H = k + \dim W^H < \dim W^H + \dim \Gamma/H$, provided $\dim \Gamma/H > k$. Then, $\Pi^{\Gamma}_{S^V}(S^W) = \bigoplus \tilde{\Pi}(H)$ for H with $\dim \Gamma/H \leq k$. Thus, *only the orbits of dimension less than or equal to k count topologically*. Furthermore, $\Pi^{\Gamma}_{S^V}(S^W) = \Pi_{k-1} \times \mathbb{Z} \times \cdots \times \mathbb{Z}$, with one $\mathbb{Z}$ for each H with $\dim \Gamma/H = k$.

Our last result in this section relates $\Pi^{\Gamma}_{S^V}(S^W)$ to $\Pi^{\Gamma}_{S^{V_0}}(S^{W_0})$, where $V_0 = V^{H_0}$, $W_0 = W^{H_0}$ for some isotropy subgroup H_0 of Γ. This point will be important for symmetry breaking.

Theorem 2.4. *Assume ($\tilde{\text{K}}$) holds for all isotropy subgroups. Let H_0 be an isotropy subgroup of Γ, let $V_0 = V^{H_0}$, $W_0 = W^{H_0}$ and denote by P_0 the morphism from $\Pi^{\Gamma}_{S^V}(S^W)$ into $\Pi^{\Gamma}_{S^{V_0}}(S^{W_0})$ induced by restricting the Γ-maps to V_0. Then P_0 is onto and*

$$P_0\Big(\sum_H [\tilde{F}_H]_\Gamma\Big) = \sum_{H > H_0} [\tilde{F}_H^{H_0}]_\Gamma,$$

where $\tilde{F}_H$ is in $\tilde{\Pi}(H)$.

Proof. Let $F_\perp^H(X_{\perp H})$ be the complementing map for V^H. Then, if $[F_0]$ is in $\Pi^\Gamma_{SV_0}(S^{W_0})$, one has that $[F_0, F_\perp^{H_0}]$ is in $\Pi^\Gamma_{SV}(S^W)$ and P_0 is onto. Furthermore, $F_\perp^H$ is a non-zero Γ-map from $(V^H)^\perp \cap V_0\backslash\{0\}$ into $(W^H)^\perp \cap W_0$, for any $H > H_0$, by Property 3.2 of Chapter 1. Thus, $(F_\perp^H)^{H_0}$ is a complementing map on V_0 and one has hypothesis ($\tilde{\text{K}}$) on V_0.

From Theorem 2.3, one has

$$[F] = \sum_H [\tilde{F}_H],$$

where $[\tilde{F}_H] = [F_H, F_\perp^H]$ and $[F_H]$ in $\Pi(H)$, i.e., $F_H(X)$ is non-zero on $\bigcup B^K$ for $K > H$. From the definition of the homotopy sums and the above Property 3.2 of Chapter 1, one obtains

$$[F^{H_0}] = \sum_H [F_H^{H_0}, (F_\perp^H)^{H_0}].$$

If H_0 is not a subgroup of H, then $V^{H_0} \cap V^H$ is a strict subspace of V^H, with isotropy strictly larger than H, hence $F_H^{H_0} \neq 0$ on B^{H_0} and $[F_H^{H_0}, (F_\perp^H)^{H_0}] = 0$.

The other H's, with $H_0 < H$, will give the result. □

3.3 Computation of Γ-classes

Although the preceding results may be appealing, the construction of the isomorphisms is involved and requires a step by step extension process on the subspaces V^H, for decreasing H's. So the problem is the following: given a Γ-map F, how does one compute its decomposition on $\bigoplus \Pi(H)$?

On one hand F_H, in $\Pi(H)$, is not the restriction of F to V^H, except for the first steps in the construction and, on the other hand, we have given formulae only for the extension degree, i.e., when $\dim V^H = \dim W^H + \dim \Gamma/H$. In this section, we shall give partial answers to these two problems: constructing a new map, a "normal map" for which the restriction argument is valid and, when hypothesis (H) holds, explicit generators for the "free" part of $\Pi^\Gamma_{SV}(S^W)$ with an explicit way of computing the extension degrees via Poincaré sections.

Definition 3.1. We shall define an order on the set, Iso(V), of isotropy subgroups of Γ on V, by denoting them by $H_1, \ldots, H_m$ in such a way that if $H_i > H_j$, then $i < j$, i.e., in decreasing order.

Thus, $H_1 = \Gamma$, H_m is the isotropy of V and the elements of $\Pi(k-1)$ come before those of $\Pi(k)$.

In this section we shall assume that *hypothesis* ($\tilde{\text{K}}$) *holds for all H in* Iso(V), that is, decomposing V as $V^H \oplus V_{\perp H}$, W as $W^H \oplus W_{\perp H}$, one has a complementing Γ-map $F_H^\perp$, from $V_{\perp H}$ into $W_{\perp H}$ with its only zero at 0. We shall assume that these *complementing maps are compatible*, i.e., that $F_H^\perp|V_{\perp K} = F_K^\perp|V_{\perp H}$.

This will be the case for $V = \mathbb{R}^k \times W$, since $F_H^\perp$ is the identity, or when one has hypothesis (H) for U and W with $V = \mathbb{R}^k \times U$ (see below).

Write $X = X_H \oplus X_{\perp H}$ and $F = (F^H, F_{\perp H})$.

Lemma 3.1. *For a fixed H, any map F in $\Pi_{S^V}^\Gamma(S^W)$ is Γ-homotopic to a map $\tilde{F}(t, X) = (F^H(t, X), \tilde{F}_{\perp H}(t, X))$, such that $\tilde{F}_{\perp H}(t, X) = F_H^\perp(X_{\perp H})$ if $\|X_{\perp H}\| \leq \varepsilon$, for ε small enough.*

Proof. Let $\Psi_H : V_{\perp H} \to \mathbb{R}^+$ be defined as a non-increasing function of $\|X_{\perp H}\|$, with value 1 if $\|X_{\perp H}\| \leq \varepsilon$ and value 0 if $\|X_{\perp H}\| \geq 2\varepsilon$. Let

$$\tilde{F}(t, X) = \big(F^H(t, X), (1 - \Psi_H(X_{\perp H}))F_{\perp H} + \Psi_H(X_{\perp H})F_H^\perp(X_{\perp H})\big).$$

Since $F_{\perp H} = 0$ if $X_{\perp H} = 0$ (Property 3.2 of Chapter 1), one has an ε such that $F^H(t, X) \neq 0$ if (t, X) is in $\partial(I \times B_R)$ and $\|X_{\perp H}\| \leq 2\varepsilon$. Hence, $\tilde{F}$ is non-zero on $\partial(I \times B_R)$ and, replacing Ψ_H by $\tau\Psi_H$, it is easy to see that $\tilde{F}$ is Γ-homotopic to F. □

Definition 3.2. A map $\tilde{F}$ in $\Pi_{S^V}^\Gamma(S^W)$ will be called a *normal map* if for all H's, one has $\tilde{F}_{\perp H}(t, X) = F_H^\perp(X_{\perp H})$, if $\|X_{\perp H}\| \leq \varepsilon$.

Lemma 3.2. *Any F in $\Pi_{S^V}^\Gamma(S^W)$ is Γ-homotopic to a normal map $\tilde{F}$.*

Proof. Arrange the isotropy subgroups in decreasing order : $H_1 = \Gamma, \dots, H_m$ is the isotropy of V. For W, decomposed as $W^{H_j} \oplus W_{\perp H_j}$, we shall write any map G as $(G^j, G^{\perp j})$. Starting from $F_0 = F$, define the sequence of maps

$$F_{j+1} = \big(F_j^{j+1}, (1 - \Psi_{j+1})F_j^{\perp j+1} + \Psi_{j+1}\, F_{j+1}^\perp\big),$$

where $\Psi_{j+1} = \Psi_{H_{j+1}}$ and $F_{j+1}^\perp$ is the complementing map for H_{j+1}. From Lemma 3.1, F_{j+1} is Γ-homotopic to F_j and, by induction, to F in $\Pi_{S^V}^\Gamma(S^W)$. The last map, for $j + 1 = m$, will be the map $\tilde{F}$.

Assume, by induction, that if $i \leq j$, then $F_j^{\perp i} = F_i^\perp$ whenever $\|X_{\perp i}\| \leq \varepsilon$ (this is clearly true for $i = j$, since $F_j^{\perp j} = (1 - \Psi_j)F_{j-1}^{\perp j} + \Psi_j F_j^\perp$).

Notice that the compatibility conditions on the complementing maps say that $(F_i^\perp)^{\perp j} = (F_j^\perp)^{\perp i}$. Furthermore, from the projections, for any map G, one has $(G^i)^{\perp j} = (G^{\perp j})^i$ and $(G^{\perp i})^{\perp j} = (G^{\perp j})^{\perp i}$. Then,

$$\begin{aligned} F_{j+1}^{\perp i} &= \big((F_j^{j+1})^{\perp i}, (1 - \Psi_{j+1})(F_j^{\perp j+1})^{\perp i} + \Psi_{j+1}(F_{j+1}^\perp)^{\perp i}\big) \\ &= \big((F_j^{\perp i})^{j+1}, (1 - \Psi_{j+1})(F_j^{\perp i})^{\perp j+1} + \Psi_{j+1}(F_i^\perp)^{\perp j+1}\big). \end{aligned}$$

Hence, using the induction hypothesis, one has, for $\|X_{\perp i}\| \le \varepsilon$:

$$\begin{aligned} F_{j+1}^{\perp i} &= \big((F_i^{\perp})^{j+1}, (1-\Psi_{j+1})(F_i^{\perp})^{\perp j+1} + \Psi_{j+1}(F_i^{\perp})^{\perp j+1}\big) \\ &= \big((F_i^{\perp})^{j+1}, (F_i^{\perp})^{\perp j+1}\big) = F_i^{\perp}. \end{aligned}$$

Thus, for $j = m$, one has $\tilde{F}^{\perp i} = F_i^{\perp}$ provided $\|X_{\perp i}\| \le \varepsilon$, i.e., $\tilde{F}$ is a normal map. □

For a normal map one may compute the decomposition of $[\tilde{F}]$ onto $\bigoplus \Pi(H)$, from restrictions to V^H, in the following way:

Theorem 3.1. *Assume that* ($\tilde{\mathrm{K}}$) *holds for all H, with the compatibility conditions on the complementing maps. Let F, in $\Pi_{SV}^{\Gamma}(S^W)$, be Γ-homotopic to a normal map $\tilde{F}$, with an associated ε. Let B_{ε}^K be an ε-neighborhood of B^K in V and let φ_H be an invariant Uryson function on B^H with value 0 in $B^H \setminus \bigcup_{K>H} B_{\varepsilon}^K$ and value 1 on $\bigcup_{K>H} B^K$. Let s be in* $[0, 1]$, *then, up to one suspension, one has*

$$[2s-1, F(t, X)] = \Sigma[2s + 2\varphi_H(t, X_H) - 1, \tilde{F}^H(t, X_H), F_H^{\perp}(X_{\perp H})],$$

independently of $\tilde{F}$.

Proof. Note first that the left hand side is $\deg_{\Gamma}(F; I \times B_R)$, from the suspension property 2.4 of Chapter 2. Furthermore, each term on the right hand side is in $\tilde{\Pi}(H)$, where V has been replaced by $I \times V$, with s in $I = [0, 1]$. Then, if one has two normal maps homotopic to F, one may choose a common ε, their restriction to B^H will be Γ-homotopic and, from Lemma 2.2 (b), the decomposition will be unique. Note that, since (s, t) is in $(I \times V)^{\Gamma}$, the sum is commutative.

Now, the sets $\big(B_{\varepsilon}^H \setminus \bigcup_{K>H} B_{\varepsilon}^K\big) = A_{\varepsilon,H}$ have disjoint interiors and cover all of $I \times B_R$, as it is easily seen. Furthermore, $\tilde{F}$ is non-zero on their boundary. Thus, up to one suspension, one has

$$\deg_{\Gamma}(\tilde{F}; I \times B_R) = \sum \deg_{\Gamma}(\tilde{F}; A_{\varepsilon H}).$$

On $A_{\varepsilon H}$, $\tilde{F}(t, X) = (\tilde{F}^H(t, X), F_H^{\perp}(X_{\perp H}))$, where $F_H^{\perp}$ has its only zero at $X_{\perp H} = 0$. Hence, by excision

$$\deg_{\Gamma}(\tilde{F}; A_{\varepsilon H}) = \deg_{\Gamma}((\tilde{F}^H, F_H^{\perp}); I \times B_R).$$

In $\tilde{F}^H(t, X)$, one may deform X to X_H, as well as in φ_H and one gets the result. □

We shall continue our more detailed description of $\Pi_{SV}^{\Gamma}(S^W)$ by recalling hypothesis (H): see Section 6 of Chapter 1. We shall assume that

$$V = \mathbb{R}^k \times U \text{ and that } U \text{ and } W \text{ satisfy (H),}$$

i.e.:

(H) *For all isotropy subgroup H and K for U, one has*

$$\dim U^H \cap U^K = \dim W^H \cap W^K.$$

Or equivalently (Lemma 6.2 of Chapter 1):

(a) $\dim U^H = \dim W^H$

(b) *There is a* Γ*-equivariant map:* $(x_1, \ldots, x_s) \to (x_1^{l_1}, \ldots, x_s^{l_s})$, *from U into W, where l_j are integers and x^l, for negative l, means $\bar{x}^{|l|}$ ($l_j = 1$ on U^Γ and on the real representations of* Γ).

Furthermore, ($\tilde{\text{H}}$) holds. From the dimension hypothesis, if a coordinate x_j is not in U^H, then $x_j^{l_j}$ is not in W^H and is a piece of $F_H^\perp$. Thus, ($\tilde{\text{K}}$) holds for all H and the complementing maps are compatible.

Theorem 3.2. *If $V = \mathbb{R}^k \times U$, where U and W satisfy* (H), *then*

$$\Pi^\Gamma_{S^V}(S^W) \cong \Pi_{k-1} \times \mathbb{Z} \times \cdots \times \mathbb{Z},$$

with one $\mathbb{Z}$ for each H with $\dim \Gamma/H = k$. *Moreover,*

$$[F]_\Gamma = [\tilde{F}]_\Gamma + \sum d_H [\tilde{F}_H]_\Gamma,$$

where $[\tilde{F}] = \sum [\tilde{F}_K]$ is constructed from $P_{k-1}[F]$, the restriction of $[F]$ to Π_{k-1} and from the suspensions of Theorem 2.2, *and where $[\tilde{F}_H]$ is in $\tilde{\Pi}(H)$ and the suspension of F_H with extension degree* 1.

Proof. It is enough to use Theorems 2.2, 2.3, Remark 2.1 and Example 2.1 □

The generator F_H was proved to exist in Theorem 1.3. We shall give an explicit form, in case (H) holds.

Let H be an isotropy subgroup with $\dim \Gamma/H = k$. Let $V^\Gamma \cong \mathbb{R}^k \times U^\Gamma$ be generated by $(t, \lambda_1, \ldots, \lambda_k, X_0)$, with t in $[0, 1]$. On $(V^\Gamma)^\perp \cap V^H$, we shall build the fundamental cell $\mathcal{C}_H$, by choosing first $z_1, \ldots z_k$, with $k_j = \infty$ and isotropy H_0, then $x_{k+1}, \ldots, x_m$, with x_j complex and k_j finite, and finally $y_1, \ldots, y_r$ with y_j in $\mathbb{R}$ and $k_j = 1$ or 2. Define the following invariant polynomials:

(a) For $k + 1 \leq j \leq m$: $P_j = P_j(x_1, \ldots, x_j) = x_1^{\alpha_1} \ldots x_j^{k_j}$, as given in Lemma 6.3 of Chapter 1, and $x_1 = z_1, \ldots, x_k = z_k$.

(b) For $1 \leq i \leq r$, $Q_i = Q_i(y_i) = y_i^2$ if $k_i = 2$, or 2 if $k_i = 1$.

Since Γ acts on y_i as $\mathbb{Z}_2$, Q_i is invariant. Define, on $I \times B_R$, with $R > 1$:

$$F_H(t, X_H) = \Big(2t + 1 - 2\prod |x_j|^2 \prod y_j^2, X_0, (\lambda_1 + i(|z_1|^2 - 1))z_l^{l_1}, (\lambda_2 + i(|z_2|^2 - 1))z_2^{l_2}, \dots, (\lambda_k + i\varepsilon(|z_k|^2 - 1))z_k^{l_k}, (P_{k+1} + 1)x_{k+1}^{l_{k+1}}, \dots, (P_m + 1)x_m^{l_m}, (Q_1 - 1)y_1, \dots, (Q_r - 1)y_r\Big).$$

In $\prod |x_j|^2$, one has all j's between 1 and m, while in $\prod y_j^2$ one has only those y_j with $k_j = 2$. The factor ε is $(-1)^{k(k-1)/2+k \dim X_0}$. The order of the components has been taken to be that of the fundamental cell so that the notation is lighter. In fact, they should appear in their natural place. Note that, if H_m is the isotropy of V, one has $\dim U = \dim W^{H_m}$, but W could be larger. However, any Γ map on V will have its range in W^{H_m}.

The map $\tilde{F}_H$ is given by suspending F_H by $x_j^{l_j}$ for the remaining x_j's.

Theorem 3.3. *The map F_H generates $\Pi(H)$, i.e., it has extension degree* 1. *For any integer d_H, one may give a map F in $\Pi(H)$ with extension degree d_H.*

Proof. The zeros of F_H in $I \times B_R$ are for $X_0 = 0, \lambda_1 = \dots = \lambda_k = 0, |z_1| = \dots = |z_k| = 1, y_j^2 = 1$ if $k_j = 2$ and $y_j = 0$ if $k_j = 1$, and $|P_j| = 1$. Since P_j ends with $x_j^{k_j}$, one may solve iteratively the relations $|P_j| = 1$ for $|x_j| = 1$, since $|z_1| = \dots = |z_k| = 1$. Then, $t = 1/2$.

Then, on a zero of F_H, one has $|z_j| = 1$, and, for all j's, $|x_j| = 1$. In particular, any zero, in $I \times B_R$, has isotropy H and $F_H|B^K \neq 0$ for any $K > H$, that is F_H defines an element of $\Pi(H)$.

Furthermore, on $B_k = B^H \cap \{z_j > 0$ for $j = 1, \dots, k\}$, there are exactly $\prod k_j$ zeros, since for $x_1, \dots, x_{j-1}$ fixed, the relation $P_j + 1 = 0$ is solvable for k_j values of x_j. By changing $P_j + 1$ to $P_j + \eta_j$, with $|\eta_j| = 1$, one may choose the phases of η_j iteratively so that none of the zeros is on $\partial\mathcal{C}_H$, that is, there is only one zero X^0 inside $\mathcal{C}_H$, with a well defined extension degree.

In order to compute the extension degree, recall that for $z_l = \dots = z_k = 1$, the only solution in $\mathcal{C}_H$ of $P_j + \eta_j = 0$ is for $x_j = x_j^0$. Perform then the following sequence of deformations:

1. Replace z_j by $\tau + (1 - \tau)z_j$ in $z_j^{l_j}$ and in $|z_j|$ in the first component, for $j = 1, \dots, k$. Replace x_j by $\tau x_j^0 + (1-\tau)x_j$ in $x_j^{l_j}$ and in $|x_j|$ in the first component, for $j = k+1, \dots, m$. For $k_j = 2$, replace y_j by $\tau + (1 - \tau)y_j$ in the term $(y_j + 1)y_j$ and in y_j^2 in the first component. One arrives at the map, deforming $z_j^2 - 1$ to $z_j - 1$,

$$(2t - 1, X_0, \lambda_1 + i(z_1 - 1), \dots, \lambda_k + i\varepsilon(z_k - 1), P_j + \eta_j, y_j - y_j^0).$$

2. In $P_j = x_1^{\alpha_1} \dots x_{j-1}^{\alpha_{j-1}} x_j^{k_j}$, one may deform, linearly in $\mathcal{C}_H$, x_i to x_i^0 for $i < j$, arriving at $(x_j - x_j^0)R_j(x_j)$, where $R_j(x_j)$ is a polynomial of degree $k_j - 1$ with no zeros in $\mathcal{C}_H$, hence deformable, via $R_j((1-\tau)x_j + \tau x_j^0)$, to a constant complex number, which can be deformed to 1.

By the product theorem for the Brouwer degree, the degree of the part $(2t-1, x_j - x_j^0, y_j - y_j^0)$ being 1, one has to compute the degree of the map

$$(\lambda_1, \dots, \lambda_k, X_0, z_1, \dots, z_k) \to (X_0, \lambda_1, z_1 - 1, \lambda_2, z_2 - 1, \dots, \lambda_k, \varepsilon(z_k - 1)).$$

The number of necessary permutations to bring $(\lambda_1, \dots, \lambda_k, X_0, z_1, \dots, z_k)$ into $(X_0, \lambda_1, z_1, \lambda_2, z_2, \dots, \lambda_k, z_k)$ is $k \dim X_0 + k(k-1)/2$. This proves that the extension degree is 1.

For the second part of the theorem, replace $\lambda_k + i\varepsilon(|z_k|^2 - 1)$ by $(\lambda_k + i\varepsilon(|z_k|^2 - 1))^d$, where z^d, with d negative, means $\bar{z}^{|d|}$. One may also replace P_{k+1} by P_{k+1}^d, with the same convention on negative powers meaning conjugation.

While, if there are no complex coordinates (hence $k = 0$) and $r \geq 2$ then, if $|\Gamma/H| > 2$, take two y's say y_1 and y_2, with $k_1 = k_2 = 2$, and replace $(Q_1 - 1)$ and $(Q_2 - 1)$ by the real and imaginary parts of $(y_1^2 - 1 + i(y_2^2 - 1))^d$ respectively. On the other hand, if $|\Gamma/H| = 2$ and $k_1 = 2$, with $k_j = 1$ for $j > 1$, replace the first three components by $(2t + 1 - 2y_1^2 y_2^2, y_1 \operatorname{Re}(y_1^2 - 1 + i(y_1 y_2 - 1))^d, y_2 \operatorname{Im}(y_1^2 - 1 + i(y_1 y_2 - 1))^d)$, which gives an extension degree d, or replace the first two components by $(\operatorname{Re}(2t + 1 - 2y^2 + i(y^2 - 1))^d, y \operatorname{Im}(2t + 1 - 2y^2 + i(y^2 - 1))^d)$. In all cases, it is easy to see that the maps are equivariant and of degree d on $\mathcal{C}_H$. □

For the case $\Gamma = S^1$, several other hypothesis were given in [IMV2, Chapter 3].

Remark 3.1. In Theorem 3.1, we have seen that the generator is unique up to conjugations. Let us make this dependence more precise: assume that, in V^H, one has $z_1, \dots, z_s$ and $z'_1, \dots, z'_s$ such that the action on z'_j is the conjugate of that of Γ on z_j. Assume that one has constructed two fundamental cells $\mathcal{C}$ and $\mathcal{C}'$, where $z_j > 0$ for $\mathcal{C}$ and $z'_j > 0$ for $\mathcal{C}'$. Then, one has two generators F and F'. The map F will have terms of the form $((\lambda_j + i\varepsilon_j(|z_j|^2 - 1))z_j^{l_j}, (z_j z'_j + 1)z_j'^{l'_j})$, while F' will have z_j and z'_j interchanged. The components which do not concern $z_1, \dots, z_s, z'_1, \dots, z'_s$ may be chosen equal for both maps. On $\mathcal{C}$ the map F, as a map from $(\lambda_j, \operatorname{Re} z_j, \operatorname{Re} z'_j, \operatorname{Im} z'_j)$ has degree 1 (from the choice of ε_j: this piece of the map contributes ε_j). While on $\mathcal{C}'$, the map F, as a map from $(\lambda_j, \operatorname{Re} z_j, \operatorname{Im} z_j, \operatorname{Re} z'_j)$, is deformable to $(\lambda_j, \varepsilon_j(|z_j|^2 - 1), \operatorname{Re} z_j \operatorname{Re} z'_j + 1, \operatorname{Re} z'_j \operatorname{Im} z_j)$, near the zero $z_j = -1, z'_j = 1$, and to $(\lambda_j, -\varepsilon_j(\operatorname{Re} z_j + 1), 1 - \operatorname{Re} z'_j, \operatorname{Im} z_j)$, with a contribution of $-\varepsilon_j$ to the degree. Thus, on $\mathcal{C}'$, the map F has extension degree $(-1)^s$ and

$$[F]_\Gamma = (-1)^s [F']_\Gamma.$$

It is easy to see that, if a pair (z_j, z'_j) has $k_j < \infty$, then the generator is independent of the order one has taken for $\mathcal{C}$.

The last part of this section concerns Poincaré sections, as defined in Definition 1.1, which will enable us to compute the d_H's for a certain class of maps, by relating them to usual Brouwer degrees.

Theorem 3.4. *Assume $V = \mathbb{R}^k \times U$, where U and W satisfy* (H). *Let $z_1, \dots, z_k$ with isotropy H_0, with* $\dim \Gamma/H_0 = k$ *and global Poincaré section $B_k = \{(t, X)$ in $I \times B_R$, with $z_j \geq 0$, for $j = 1, \dots, k\}$. Then, if $F : I \times B_R \to \mathbb{R} \times W$, is a Γ-map which is non-zero on $\partial(I \times B_R) \cup \partial B_k$, one has*

$$[F]_\Gamma = \sum_{\underline{H} \leq H \leq H_0} d_H[\tilde{F}_H]_\Gamma,$$

where $\underline{H}$ is the torus part of H_0 (see Lemma 2.6 of Chapter 1, hence $\dim \Gamma/H = k$). *Furthermore, for each H in the above sum, one has*

$$\deg(F^H; B_k^H) = \sum_{H \leq K \leq H_0} \beta_{HK} d_K |H_0/K|,$$

where, if $(x_1^{l_1}, \dots, x_s^{l_s})$ is the complementing map of V^K in V^H (i.e., an equivariant map from $(V^K)^\perp \cap V^H$ into $(W^K)^\perp \cap W^H$), then $\beta_{HK} = \prod l_j$. In particular, $\beta_{HH} = 1$.

Proof. Note first that if F is non-zero on ∂B_k, it is also non-zero on $\partial(I \times B_R)$ due to the action of Γ/H_0 (see Lemma 1.1). Furthermore, for any isotropy subgroup H, F^H is a non-zero map from ∂B_k^H into W^H. If $\underline{H} < H < H_0$, then V^H contains $z_1, \dots, z_k$ and since $\dim \Gamma/H = k$, the spaces B_k^H and W^H have the same dimension and $\deg(F^H; B_k^H)$ is well defined.

Now, if K is not a subgroup of H_0, in particular, if $\dim \Gamma/K < k$, then $z_j = 0$ for some $j = 1, \dots, k$, in V^K. This implies that $F^K \neq 0$, in particular, $[F]_\Gamma$ is in $\Pi(k)$ and $[\tilde{F}]_\Gamma = 0$, as given in Theorem 3.2. Then, $[F] = \sum d_H[\tilde{F}_H]$.

For such a K one has, from Theorem 2.4,

$$0 = [F^K] = \sum_{K < H} d_H[\tilde{F}_H^K].$$

From Lemma 2.2 and Theorem 2.4, one gets that $d_H[\tilde{F}_H^K] = 0 = d_H[F_H, F_H^{\perp K}]$, since $V^H \subset V^K$. Since $F_H^{\perp K}$ is a complementing map for V^H in V^K and this suspension is one-to-one (Lemma 2.2 (a)), one has $d_H = 0$ for all $H \geq K$. Since $\underline{H}$ is the unique smallest isotropy subgroup, with a Weyl group of dimension k contained in H_0, the above sum is reduced to those H's between $\underline{H}$ and H_0:

$$[F]_\Gamma = \sum_{\underline{H} < H < H_0} d_H[\tilde{F}_H]_\Gamma.$$

Let $V_0 = \mathbb{R}^k \times V^{\underline{H}} \cap B_k$, $W_0 = W^{\underline{H}}$. Then, from Lemma 1.1, we know that H_0 acts as a finite group on V_0. The isotropy subgroups for that action are exactly those H's with $\underline{H} < H < H_0$, since $H_{0X} = \Gamma_X \cap H_0$. Furthermore, one has dim $V_0^H =$ dim W^H and $\{x_j^{l_j}\}$ gives, for x_j different from $z_1, \dots, z_k$, complementing maps. Thus, one has property (H) for V_0. Moreover, $\mathcal{C}_H$ is also the fundamental cell for the action of H_0 on V_0^H (see Lemma 1.1) and, the generators F_H of Theorem 3.3 have extension degree equal to 1, one may choose $F_H|B_k$ as the generators for $\Pi^{H_0}_{S^{V_0}}(S^{W_0})$. Applying the above argument to H_0 and $F|_{V_0}$, which gives an element of $\Pi^{H_0}_{S^{V_0}}(S^{W_0})$, one has

$$[F^0]_{H_0} = \sum_{\underline{H} \le H \le H_0} d'_H [\tilde{F}^0_H]_{H_0},$$

where F^0 stands for $F|_{V_0}$. This equality means that one has an H_0-homotopy $F^0(\tau, X)$ on $\partial(B_k \cap V_0)$ from the left-hand side to the right-hand side. From Lemma 1.1, by the action of Γ/H_0, one may lift this homotopy between $F(X)$ and $\sum_{\underline{H}<H<H_0} d'_H[\tilde{F}_H]_\Gamma$: again use Lemma 1.1 (b) to see that the lifting of F^0 and $\tilde{F}^0_H$ is $F^{\underline{H}}$ and $\tilde{F}^{\underline{H}}_H$, respectively, on $V^{\underline{H}}$. Hence,

$$[F^{\underline{H}}]_\Gamma = \sum_{\underline{H} \le H \le H_0} d'_H [\tilde{F}^{\underline{H}}_H]_\Gamma,$$

but, from Theorem 2.4, $[F^{\underline{H}}]_\Gamma$ has the same decomposition, with d_H instead. From Lemma 2.2, one has $d'_H = d_H$. Since,

$$[F^{0H}]_{H_0} = \sum_{H \le K \le H_0} d_K [\tilde{F}^{0H}_K]_{H_0},$$

which is a homotopy on $\partial(B_k \cap V_0)$, the two sides have the same Brouwer degree, for which the sum operation is an isomorphism, i.e., the degree of a topological sum is the sum of the degrees. Since $\tilde{F}^{0H}_K = (F^0_K, F^{\perp H}_K)$, one has, from Theorem 3.3, that $\deg(\tilde{F}^{0H}_K; B^H_k) = \left(\prod k_j\right)\left(\prod l_j\right) = |H_0/K|\beta_{HK}$, using Theorem 1.2. This finishes the proof of the theorem. □

Remark 3.2. The passage through H_0 may seem, at first sight, unnecessary. The point is that a Γ-homotopy on S^V does *not* imply an H_0-homotopy on ∂B_k, since, even if the two maps are non-zero on ∂B_k, the Γ-homotopy may have zeros, when one z_j is 0. In Lemma 1.1, we have proved that this can be fixed for $\Pi(H)$, and the summation formulae, for $[F^0]_{H_0}$ and $[F^{\underline{H}}]_\Gamma$, extend this property to $\bigoplus \tilde{\Pi}(H)$.

Now, if $H < H_0$, one has dim $V^H \cap B_k =$ dim W^H, and if $\underline{H}$ is not a subgroup of H, then in V^H one has coordinates of $(V^{\underline{H}})^\perp$, if $\underline{H}$ is not the isotropy of V. From the definition of the torus part of H_0, this implies that dim $H_0/H > 0$. From Theorem 2.2 (b), one has

$$\Pi^{H_0}_{\partial B_k}(S^W) \cong \Pi_0 = \Pi^{H_0}_{S^{V_0}}(S^{W_0}).$$

In particular,

$$[F|B_k]_{H_0} = [F^0]_{H_0} = \sum_{\underline{H} \le H \le H_0} d_H [\tilde{F}^0_H]_{H_0}.$$

Remark 3.3. Consider the map $\tilde{F} = (F^{\underline{H}}, F^{\perp}_{\underline{H}})$, where $F^{\perp}_{\underline{H}}$, from $(V^{\underline{H}})^{\perp}$ into $(W^{\underline{H}})^{\perp}$, has degree $\prod l_j$, coming from the complementing map. Then $\tilde{F}$ is non-zero on ∂B_k, $\tilde{F}^H = F^H$ for $\underline{H} < H < H_0$, thus, $\deg(\tilde{F}^H; B_k^H) = \deg(F^H; B_k^H)$. Hence the two maps have the same set of Γ-degrees, i.e., the same d_H's, by inverting the relations of the Brouwer degrees (see below). That is,

$$[F]_\Gamma = [\tilde{F}]_\Gamma.$$

Furthermore, the preceding remark implies that $[F|_{B_k}]_{H_0} = [\tilde{F}|_{B_k}]_{H_0}$, thus

$$\deg(F|_{B_k}; B_k) = \deg(\tilde{F}|_{B_k}; B_k) = \deg(F^{\underline{H}}|_{B_k \cap V^{\underline{H}}}; B_k \cap V^{\underline{H}})\Big(\prod l_j\Big).$$

So,

$$\deg(F|_{B_k}; B_k) = \sum_{\underline{H} \le H \le H_0} \beta_{\underline{H}H} d_H |H_0/H|,$$

independently of F on $(V^{\underline{H}})^{\perp}$.

Corollary 3.1. *Ordering the subgroups H with $\underline{H} < H < H_0$, as in Definition* 3.1*, the relations of Theorem* 3.4 *may be expressed in the form*

$$\begin{pmatrix} \deg(F^{H_0}; B_k^{H_0}) \\ \vdots \\ \deg(F^{H_i}; B_k^{H_i}) \\ \vdots \\ \deg(F^{\underline{H}}; B_k^{\underline{H}}) \end{pmatrix} = \begin{pmatrix} 1 & & 0 \\ \vdots & & \vdots \\ \beta_{i1} & |H_0/H_j| & 0 \\ \vdots & \vdots & \vdots \\ \beta_{s1} & \beta_{sj}|H_0/H_j| & |H_0/\underline{H}| \end{pmatrix} \begin{pmatrix} d_0 \\ \vdots \\ d_j \\ \vdots \\ d_s \end{pmatrix}$$

with $\beta_{ij} = \beta_{H_i H_j}$, $d_j = d_{H_j}$.

This triangular matrix, since β_{ij} is non-zero if and only if $H_i < H_j$, in particular $i > j$, is invertible. Hence, the d_j's are completely determined by the Brouwer degrees on the left. One may use the Möbius inversion formula to get a compact expression for the inverse. Note that $\beta_{ij} = 1$, if $H_i < H_j$, for $V = \mathbb{R}^k \times W$.

Example 3.1. Consider the action of $\mathbb{Z}_2 \times \mathbb{Z}_2$ on $\mathbb{R} \times \mathbb{R}^3$ given by $(t, \gamma_1 x, \gamma_2 y, \gamma_1\gamma_2 z)$, with $\gamma_1^2 = \mathrm{Id}$, $\gamma_2^2 = \mathrm{Id}$. One has the following information.

Isotropy H	V^H	$\mathcal{C}_H$
$H_0 = \mathbb{Z}_2 \times \mathbb{Z}_2$	$(t, 0, 0, 0)$	$(t, 0, 0, 0)$
$H_1 = \mathbb{Z}_2 \times \{1\}$	$(t, 0, y, 0)$	$(t, 0, y > 0, 0)$
$H_2 = \{1\} \times \mathbb{Z}_2$	$(t, x, 0, 0)$	$(t, x > 0, 0, 0)$
$H_3 = \{(1, 1), (-1, -1)\}$	$(t, 0, 0, z)$	$(t, 0, 0, z > 0)$
$H_4 = \{(1, 1)\}$	(t, x, y, z)	$(t, x > 0, y > 0, z)$

$\lvert\Gamma/H\rvert$	$\tilde{F}_H$
1	$(2t - 1, x, y, z)$
2	$(2t + 1 - 2y^2, x, (y^2 - 1)y, z)$
2	$(2t + 1 - 2x^2, (x^2 - 1)x, y, z)$
2	$(2t + 1 - 2z^2, x, y, (z^2 - 1)z)$
4	$(2t + 1 - 2x^2y^2, (x^2 - 1)x, (y^2 - 1)y, z)$

Since $B_0 = I \times B_R$, any map in $\Pi^{\Gamma}_{S^V}(S^V)$ can be written as

$$[F] = \sum_0^4 d_j[\tilde{F}_j],$$

and, if i_j is the degree of F on B^{H_j}, one obtains

$$\begin{pmatrix} i_0 \\ i_1 \\ i_2 \\ i_3 \\ i_4 \end{pmatrix} = \begin{pmatrix} 1 & 0 & 0 & 0 & 0 \\ 1 & 2 & 0 & 0 & 0 \\ 1 & 0 & 2 & 0 & 0 \\ 1 & 0 & 0 & 2 & 0 \\ 1 & 2 & 2 & 2 & 4 \end{pmatrix} \begin{pmatrix} d_0 \\ d_1 \\ d_2 \\ d_3 \\ d_4 \end{pmatrix}$$

In particular, $\deg(F; I \times B_R) = \deg(F^{\Gamma}; I) + 2p$, a Borsuk–Ulam result. For instance, if $F(t, x, y, z) = (2t - 1, -x, -y, -z)$, one has $i_0 = d_0 = 1$, $i_j = -1 = d_j$ for $j = 1, 2, 3$, $i_4 = -1$ and $d_4 = +1$.

Example 3.2. Assume $V = \mathbb{R}^k \times W$, Then $\beta_{HK} = 1$ if $H \le K$ and 0 otherwise. Define the following Möbius function:

$$\mu_{HH} = 1$$

$$\mu_{HK} = \begin{cases} -\sum_{H \le L < K} \mu_{HL} = -\sum_{H < L \le K} \mu_{LK}, & \text{if } H < K \\ 0, & \text{otherwise.} \end{cases}$$

Thus, μ_{HK} is integer-valued and can be computed iteratively. Then, if

$$i_H = \sum_{H \le K \le H_0} |H_0/K| d_K,$$

one has

$$|H_0/H| d_H = \sum_{H \le K \le H_0} \mu_{HK} i_K.$$

In fact, if one writes

$$i_K = \sum_L \delta_{KL} |H_0/L| d_L,$$

with $\delta_{KL} = 1$ if $K \le L \le H_0$ and 0 otherwise, the substitution in the formula for $|H_0/H| d_H$ gives

$$\sum_{H \le K} \sum_{K \le L} \mu_{HK} \delta_{KL} |H_0/L| d_L = \sum_{H \le L} \Big(\sum_{H \le K \le L} \mu_{LK} \delta_{KL} \Big) |H_0/L| d_L,$$

where one has changed the order of the sums. Since $\sum_{H \le K \le L} \mu_{HK} = 0$, for H a strict subgroup of L, one obtains the result.

In the preceding example, the μ_{HK}'s give the matrix

$$\begin{pmatrix} 1 & 0 & 0 & 0 & 0 \\ -1 & 1 & 0 & 0 & 0 \\ -1 & 0 & 1 & 0 & 0 \\ -1 & 0 & 0 & 1 & 0 \\ 2 & -1 & -1 & -1 & 1 \end{pmatrix}$$

In many examples i_H corresponds to the index of an isolated solution, i.e., to the sign of the determinant of A^H, where A is an H_0-equivariant matrix: see Property 3.4. and Theorem 5.3 of Chapter 1. That is, $i_H = \pm 1$ for all H's. One has the following:

Proposition 3.1. *Assuming $V = \mathbb{R}^k \times W$ and $i_H = \pm 1$ for all H's, with $\underline{H} \le H \le H_0$, for a map satisfying the conditions of Theorem 3.4, then*

$d_H = 0$ if V^H has a coordinate where H_0 acts as $\mathbb{Z}_m$, $m \ge 3$,

$d_{H_0} = i_{H_0}$,

$d_{H_j} = (i_{H_j} - i_{H_0})/2$, for all maximal H_j's, with $H_0/H_j \cong \mathbb{Z}_2$,

d_H and i_H are completely determined by i_{H_j}, the above H_j's, for all H's not included in the above list.

Proof. Let $\tilde{V}$ be the subspace of V where H_0 acts trivially or as $\mathbb{Z}_2$, i.e., $\tilde{V}$ corresponds to the "real" representations of H_0. Let $\tilde{H}$ be the isotropy of $\tilde{V}$. Then any γ in H_0 is such that γ^2 belongs to $\tilde{H}$, since the action of H_0 on a "real" coordinate is by ± 1. Thus, $H_0/\tilde{H} \cong \mathbb{Z}_2 \times \cdots \times \mathbb{Z}_2$ and one cannot have a "complex" coordinate in $V^{\tilde{H}}$, i.e., with an action of H_0 as $\mathbb{Z}_m$, with $m \geq 3$: any generator γ of $\mathbb{Z}_m$ would need to have γ^2 in $\tilde{H}$, hence, if that coordinate would be in $V^{\tilde{H}}$, one has $m = 2$. That is, $V^{\tilde{H}} = \tilde{V}$.

Let H be such that V^H contains at least one complex coordinate. Then,

$$i_H = \sum_{\substack{\tilde{H} \leq K \leq H_0 \\ H \leq K}} |H_0/K| d_K + \sum_{H<K} |H_0/K| d_K + |H_0/H| d_H,$$

where the first sum is on the real coordinates of V^H, with isotropy $\hat{H}$, hence equal to $i_{\hat{H}}$. The second sum is over those isotropy subgroups K, different from H, with V^K containing at least one complex coordinate.

Assume, by induction, that $d_K = 0$ in the second sum, then

$$i_H = i_{\hat{H}} + |H_0/H| d_H.$$

Since $|H_0/H| > 2$, because of the complex coordinate, the only possibility is $d_H = 0$ and $i_H = i_{\hat{H}}$. Note that if $F(t, X) = AX$, with A an H_0-equivariant matrix, then, due to the block diagonal structure, $A|_{V^H \cap (V^{\hat{H}})^\perp}$ is complex, hence with determinant 1 and $i_H = i_{\hat{H}}$.

It remains to prove the last point of the proposition: let H correspond to some isotropy of real coordinates, then $|H_0/H| = 2^m$, for some $m > 1$. Then,

$$i_H = d_{H_0} + 2 \sum_{H<H_j} d_{H_j} + \sum_{\alpha=2}^{m-1} 2^\alpha \Big(\sum_{\substack{H<K \\ |H_0/K|=2^\alpha}} d_K \Big) + 2^m d_H,$$

where the first sum corresponds to the maximal H_j's containing H. Given d_{H_0}, d_{H_j}'s (hence i_{H_0} and i_{H_j}'s), assume by induction that the d_K's in the second sum are completely determined, and that one has the above identity for i_H and d_H and for i'_H and d'_H. Then the difference will give

$$i_H - i'_H = 2^m (d_H - d_{H'})$$

which is not possible, since $m > 1$, unless $i_H = i'_H$ and $d_H = d'_H$. □

This proposition implies that any change of the Γ-degree, at this stage, is detected by changes of d_{H_0} or on d_{H_j}, with $H_0/H_j \cong \mathbb{Z}_2$. This fact will lead to period doubling.

3.4 Borsuk–Ulam results

One of the first uses of symmetry, to give information on a map, is the Borsuk–Ulam theorem, which states that the Brouwer degree of an odd map, with respect to a ball centered at the origin, is an odd integer. There is a vast literature on extensions of this result to different situations. In this section, we shall indicate how the ideas of the three preceding sections may be used to give sharp results for the Brouwer degree of a Γ-map, when Γ is abelian. This section is not central to the book and is more of a topological interest.

In this section V and W are two arbitrary finite dimensional representations of Γ.

Our first result will yield a classification of Γ-maps in a context different from that of Theorem 2.3, i.e., where one may have no complementing maps but where the problematic isotropy subgroups have a finite Weyl group.

Theorem 4.1 (Hopf classification). *Assume* $\dim V^\Gamma \geq 1$ *and suppose that* ($\tilde{\mathrm{H}}$) *holds (i.e.,* $\mathrm{Sign}(\det \gamma)\,\mathrm{Sign}(\det \tilde{\gamma}) > 0$ *for all* γ *in* Γ*). Let* $\hat{J}$ *be the set of all elements* H *of* $\mathrm{Iso}(V)$ *with the property that for all* $K \leq H$ *one has*

$$\dim V^K \leq \dim W^K, \ \text{ if } \dim \Gamma/K = 0,$$

$$\dim V^K < \dim W^K + \dim \Gamma/K, \ \text{ if } \dim \Gamma/K > 0.$$

(Note that $J \subset \hat{J} \subset \tilde{J}$*, where* J *and* $\tilde{J}$ *are defined in Corollary* 1.2*).*

Then, if F *and* F_0 *are two equivariant maps which are* Γ*-homotopic on* $\bigcup_{H\in \hat{J}^c} S^{V^H}$*, one has integers* d_H *such that*

$$[F]_\Gamma = [F_0]_\Gamma + \sum_I d_H[\tilde{F}_H]_\Gamma,$$

where the sum is over the subset I *of* $\hat{J}$ *of* H*'s, with* $\dim V^H = \dim W^H$ *and* $\dim \Gamma/H = 0$*, and* $\tilde{F}_H$ *is the extension given in Example* 1.4*, of the map* F_H *with extension degree* 1 *in* $\Pi(H)$*. If* $\hat{J} = \mathrm{Iso}(V)$*, then* F_0 *is not present.*

Proof. Let $\hat{\Pi} = \{[F]_\Gamma : F : \bigcup_{H\in \hat{J}^c} S^{V^H} \to W\backslash\{0\}\}$, with $\hat{\Pi} = (1, 0)$ if $\hat{J} = \mathrm{Iso}(V)$. As in Sections 1 and 2, it is easy to see that $\hat{\Pi}$ is a group. Let R be the morphism, from $\Pi^\Gamma_{S^V}(S^W)$ into $\hat{\Pi}$, induced by restriction to the isotropy subgroups in $\hat{J}^c$.

From Corollary 1.2 (c), since $\hat{J} \subset \tilde{J}$, any element in $\hat{\Pi}$ extends to an element in $\Pi^\Gamma_{S^V}(S^W)$, that is R is onto. Furthermore, any $[F]$ in $\ker R$ is such that F has a non-zero Γ-extension to $\bigcup_{H\in \hat{J}^c} B^H$.

Let $[F_0]$ be in $\ker R$ and let H_1 be an element of $\hat{J}$ which is maximal in I. Thus, if $H > H_1$, then either H is in $\hat{J}^c$ or $\dim V^H < \dim W^H + \dim \Gamma/H$. In both cases, F_0 has a non-zero Γ-extension to B^H: use Theorem 1.1 in the second case. This implies that $F_0^{H_1}$ belongs to $\Pi(H_1)$, as defined in Definition 1.3, and its extendability to B^{H_1} is characterized by its extension degree, given by

$$\deg(F_0^H; B^{H_1}) = |\Gamma/H_1| \deg_E(F_0) = |\Gamma/H_1| d_{H_1},$$

from Theorem 1.2. From Example 1.4, $\deg_E(F_0)$ depends only on F_0 and there is a generator F_{H_1}, of $\Pi(H_1)$, which has an extension $\tilde{F}_{H_1}$ to $\Pi^\Gamma_{S^V}(S^W)$. $\tilde{F}_{H_1}$ is also in ker R, from the construction of F_{H_1}. Let

$$[F_1]_\Gamma \equiv [F_0]_\Gamma - d_{H_1}[\tilde{F}_{H_1}]_\Gamma.$$

Then, $[F_1]$ is in ker R and is extendable to B^{H_1}. Let $\hat{J}_1 = \hat{J}\backslash\{H \geq H_1\}$. Define as above $\hat{\Pi}_1$, over $\hat{J}_1^c$, and the projection R_1 onto $\hat{\Pi}_1$. It is clear that $[F_1]$ belongs to ker R_1 and that one may repeat the above construction with another maximal H_2. After a finite number of steps, one will arrive at

$$[F_0]_\Gamma - \sum_I d_H[F_H]_\Gamma = 0.$$

Finally, if F and F_0 are as in the statement of the theorem, then $[F] - [F_0]$ is in ker R and has an expression as a combination of the $[\tilde{F}_H]$. □

Example 4.1. Let us consider Example 6.1 of Chapter 1: one has the actions of $\mathbb{Z}p^2q$, with p and q relatively prime, on $V = \mathbb{C}^2$, as $(e^{2\pi ik/p^2}, e^{2\pi ik/pq})$ and on $W = \mathbb{C}^2$, as $(e^{2\pi ik/p}, e^{2\pi ik/p^2q})$. Then, on $I \times B$, with $B = \{(z_1, z_2) : |z_i| \leq 2\}$, one has the isotropy subgroups:

$$\begin{aligned}
&\Gamma \cong \mathbb{Z}_{p^2q}, && \text{with } V^\Gamma = \{(t, 0, 0)\}, && W^\Gamma = \mathbb{R},\\
&H \cong \mathbb{Z}_q, && \text{with } V^H = \{(t, z_1, 0)\}, && W^H = \mathbb{R} \times \{(\xi_1, 0)\},\\
&K \cong \mathbb{Z}_p, && \text{with } V^K = \{(t, 0, z_2\}, && W^K = \mathbb{R} \times \{(\xi_1, 0)\},\\
&L \cong \{e\}, && \text{with } V^L = I \times V, && W^L = \mathbb{R} \times W.
\end{aligned}$$

Thus, $\hat{J} = \mathrm{Iso}(V)$ and any equivariant map F from $I \times V$ into $\mathbb{R} \times W$, which is non-zero on $\partial(I \times B)$, may be written as

$$[F]_\Gamma = d_\Gamma[F_\Gamma] + d_H[F_H] + d_K[F_K] + d_L[F_L],$$

where, if $\alpha q + \beta p = 1$, the generators are the following:

$$\begin{aligned}
F_\Gamma &= (2t - 1, z_1^p + z_2^q, z_1^\alpha z_2^\beta)\\
F_H &= (2t + 1 - 2|z_1|^2, (z_1^{p^2} - 1)z_1^p, z_1^\alpha z_2^\beta)\\
F_K &= (2t + 1 - 2|z_2|^2, (z_2^{pq} - 1)z_2^q, z_2^\alpha, z_1^\alpha z_2^\beta)\\
F_L &= (2t + 1 - 2|z_1z_2|^2, (z_1^{p^2} - 1)z_1^p, (\bar{z}_1^p z_2^q - 1)z_1^\alpha z_2^\beta).
\end{aligned}$$

The zeros of $F_\Gamma - (0, \varepsilon, 0)$ are at $(1/2, 0, \varepsilon^{1/q}e^{2k\pi i/q})$ and $(1/2, \varepsilon^{1/p}e^{2k\pi i/p}, 0)$ with index α and β respectively. Hence, $\deg(F_\Gamma) = \alpha q + \beta p = 1$. Similarly, $\deg F_\Gamma^H = p$, $\deg F_\Gamma^K = q$. It is then not difficult to show that

$$\begin{pmatrix} \deg F^\Gamma \\ \deg F^H \\ \deg F^K \\ \deg F \end{pmatrix} = \begin{pmatrix} 1 & 0 & 0 & 0 \\ p & p^2 & 0 & 0 \\ q & 0 & pq & 0 \\ 1 & \beta p^2 & \alpha pq & p^2q \end{pmatrix} \begin{pmatrix} d_\Gamma \\ d_H \\ d_K \\ d_L \end{pmatrix}.$$

In particular, if $\tilde{F}$ is an equivariant map from V into W, then $F = (2t - 1, \tilde{F})$, has $d_\Gamma = 1$ and

$$\deg F = \deg \tilde{F} = 1 + mp$$

for some integer m.

Example 4.2. Let V and W be S^1-spaces with $\dim V = \dim W$. Then on a coordinate z_j of $(V^{S^1})^\perp$ one has the action $e^{im_j\varphi}$ and on a coordinate ξ_j in $(W^{S^1})^\perp$ the action is as $e^{in_j\varphi}$. One has, of course, that ($\tilde{\mathrm{H}}$) always holds. Recall that a negative m_j means conjugates. Then, the following statement holds

Proposition 4.1. *If* $\dim V = \dim W$ *and* F *is an* S^1*-map, from* $I \times V$ *into* $\mathbb{R} \times W$, *which is non-zero on* $\partial(I \times B)$, *then*

(a) *If* $\dim V^{S^1} \neq \dim W^{S^1}$, *one has* $\deg(F; I \times B) = 0$.

(b) *If* $\dim V^{S^1} = \dim W^{S^1}$, *then*

$$\deg(F; I \times B) = \beta \deg(F^{S^1}; I \times B^{S^1}),$$

where β *is the integer* $\left(\prod n_j\right)/\left(\prod m_j\right)$.

Proof. We shall use the following useful trick: Let $\tilde{V}$ be the S^1-space defined as $\{(X_0, Z_1, \ldots, Z_k)\}$, where X_0 is in V^{S^1}, the action of S^1 on Z_j is as $e^{i\varphi}$ and $k = \dim V - \dim V^{S^1}$, i.e., $\dim \tilde{V} = \dim W$ and S^1 acts semi-freely on $\tilde{V}$. Furthermore, the map

$$F_0(X_0, Z_1, \ldots, Z_k) = (X_0, Z_1^{m_1}, \ldots, Z_k^{m_k})$$

is an S^1-equivariant map from $\tilde{V}$ into V. Moreover, if F is an S^1-equivariant map from $I \times V$ into $\mathbb{R} \times W$, then

$$\tilde{F}(t, X_0, Z_1, \ldots, Z_k) = F(t, X_0, Z_1^{m_1}, \ldots, Z_k^{m_k})$$

is an S^1-equivariant map from $I \times \tilde{V}$ into $\mathbb{R} \times W$, which is non-zero on the sphere $\partial(I \times F_0^{-1}(B))$.

Since $\tilde{V}$ has only two isotropy subgroups, then, if $\dim V^{S^1} \le \dim W^{S^1}$, the set $\hat{J}$, of Theorem 4.1, is $\mathrm{Iso}(V)$ and

$$[\tilde{F}]_\Gamma = 0, \quad \text{if} \dim V^{S^1} < \dim W^{S^1},$$

or

$$[\tilde{F}]_\Gamma = d[\tilde{F}_\Gamma]_\Gamma, \quad \text{if} \dim V^{S^1} = \dim W^{S^1},$$

since, in the first case I is empty and, in the second, $I = \Gamma = S^1$. On the other hand

$$[\tilde{F}]_\Gamma = [\tilde{F}_0]_\Gamma, \ \text{if} \dim V^{S^1} > \dim W^{S^1},$$

where $\tilde{F}_0$ is any S^1-map, with $\tilde{F}_0^\Gamma$ homotopic to $\tilde{F}^\Gamma$: in this case, $\hat{J}$ reduces to $\{e\}$ and I is empty. One may choose

$$\tilde{F}_0 = (F^{S^1}, Z_1^{n_1}, \dots, Z_k^{n_k}, 0, \dots, 0),$$

since $\tilde{F}^{S^1} = F^{S^1}$ and $\dim(W^{S^1})^\perp > \dim(\tilde{V}^{S^1})^\perp$. If one replaces one 0 by ε in $\tilde{F}_0$ one obtains a non-zero map (of course non-equivariant), that is, $\deg(\tilde{F}_0; I \times B) = 0$.

Thus, if $\dim V^{S^1} \neq \dim W^{S^1}$, one has $\deg(\tilde{F}; I \times F_0^{-1}(B)) = 0$. In the remaining case, $d = \deg(F^{S^1}; I \times B^\Gamma)$, by definition of the extension degree. Furthermore, the map $\tilde{F}_0 = (F^{S^1}, Z_1^{n_1}, \dots, Z_k^{n_k})$ has the same invariant part and the same d: this implies that

$$[\tilde{F}]_\Gamma = [\tilde{F}_0]_\Gamma.$$

From the product theorem for the Brouwer degree, one gets

$$\deg(\tilde{F}; I \times F_0^{-1}(B)) = \Big(\prod n_j\Big) \deg(F^{S^1}; I \times B^{S^1}).$$

The proof of the proposition will be complete, once one uses the formula for the degree of a composition which yields

$$\deg(\tilde{F}; I \times F_0^{-1}(B)) = \Big(\prod m_j\Big) \deg(F; I \times B).$$

The fact that β is an integer follows from the next result. □

Corollary 4.1. *If V and W are S^1-spaces with* $\dim V^{S^1} = \dim W^{S^1}$, *then if $(2t - 1, X_0)$, X_0 in V^{S^1}, has a non-zero S^1-extension $\tilde{F}_{S^1}$ from $\partial(I \times B)$ into $\mathbb{R} \times W$, one has*

(a) $\dim V^H \leq \dim W^H$, *for all H in* $\mathrm{Iso}(V)$,

(b) $[F]_{S^1} = \deg(F^{S^1}; I \times B^{S^1})[\tilde{F}_{S^1}]$,

(c) $\Pi_{S^V}^{S^1}(S^W) \cong \mathbb{Z}$.

Proof. If $\Pi_{S^V}^{S^1}(S^W)$ has an element F with $\deg(F^{S^1}; I \times B^{S^1}) \neq 0$, for instance $\tilde{F}_{S^1}$, then, if for some H, one has $\dim V^H > \dim W^H$, consider F_0^H the restriction of F^H to a subspace V_0 of V^H with dimension equal to $\dim W^H$, i.e., with at least one coordinate z_0 equal to 0. From Proposition 4.1, one has that $\deg(F_0^H; I \times V_0) \neq 0$. But one may deform $(t, X_0, z_j$ in $V_0)$ to $(1/2, 0, \dots, 0, z_0 = R)$ and F_0^H to $F^H(1/2, 0, \dots, 0, R)$, a constant map with degree 0. This contradiction implies (a).

But then, from Theorem 4.1, one has that $\hat{J} = \mathrm{Iso}(V)$ and the only element of I is S^1. This implies (b) and that $(2t - 1, X_0)$ has the extension $\tilde{F}_{S^1}$ (hence to assume that there is a map F, with non-zero degree for F^{S^1}, is equivalent to assuming that $\tilde{F}_{S^1}$ exists). Then, any element in $\Pi_{S^V}^{S^1}(S^W)$ is classified by $\deg(F^{S^1}; I \times B^{S^1})$.

Finally, one has that $\beta = \deg(\tilde{F}_{S^1}; I \times B)$, hence an integer. Note that, if all maps F have $\deg(F^{S^1}; I \times B^{S^1}) = 0$, then the fact that β is an integer or not is irrelevant. □

Compare this result with Theorem 2.2 (b), where one has assumed, in this case, the existence of a complementing map $F_\perp$: there $[F] = [F^{S^1}, F_\perp]$.

The results above generalize to the case of an action of a torus T^n.

Proposition 4.2. *Let T^n act on V and W and take F in $\Pi_{S^V}^{T^n}(S^W)$, then*

(a) *If* $\dim V^{T^n} \neq \dim W^{T^n}$, *but* $\dim V = \dim W$, *one has* $\deg(F; I \times B) = 0$.

(b) *If* $\dim V^{T^n} = \dim W^{T^n}$ *and* $\dim V = \dim W$, *then*

$$\deg(F; I \times B) = \beta \deg(F^{T^n}; I \times B^{T^n}),$$

where β is a non-zero integer, independent of F.

(c) *If* $\dim V^{T^n} = \dim W^{T^n}$ *and $(2t-1, X_0)$, X_0 in V^{T^n}, has a non-zero extension $\tilde{F}_{T^n}$, from $\partial(I \times B)$ into $\mathbb{R} \times W$, then*

(α) $\dim V^H \leq \dim W^H$, *for all H in* $\mathrm{Iso}(V)$

(β) $[F]_{T^n} = \deg(F^{T^n}; I \times B^{T^n})[\tilde{F}_{T^n}]$

(γ) $\prod_{S^V}^{T^n}(S^W) \cong \mathbb{Z}$

(δ) $|\beta| = \Big(\prod_{l=1}^{k} a_l'\Big) / \Big(\prod_{l=1}^{k} a_l\Big)$, *if* $\dim V = \dim W$,

where a_l is the greatest common divisor of $(|n_1^l|, \dots, |n_n^l|)$ and the action of T^n on the coordinate z_l is given by $\exp i < N^l, \Phi >$, *with $N^l = (n_1^l, \dots, n_n^l)$ and $\Phi = (\varphi_1, \dots, \varphi_n)$, $l = 1, \dots, k$. The integer a_l' is given analogously by the action of T^n on W.*

Proof. From Lemma 2.5 of Chapter 1, one has an action of S^1 given by $\varphi_j = M_j\varphi$, such that $\langle N^l, M\rangle \not\equiv 0, [2\pi]$, unless $N^l = 0$ and $V^{S^1} = V^{T^n}$. This implies most of the proposition, since (a) and (b) are consequences of Proposition 4.1 (a) and (b), with

$$\beta = \Big(\prod_1^k \langle N'^l, M\rangle\Big) / \Big(\prod_1^k \langle N^l, M\rangle\Big).$$

Furthermore, if $\tilde{F}_{T^n}$ is a T^n-extension of $(2t-1, X_0)$, then under the above morphism, it is also an S^1-extension, and its restriction to any V^K, $K < T^n$, is a valid S^1-extension for maps from V^K into W^K (here we are using the fact that, since T^n is abelian, V^K and W^K are T^n-representations). From Corollary 4.1, with $H = \{e\} < S^1$, one has $\dim V^K \leq \dim W^K$, for any K strictly contained in T^n, hence $\dim T^n/K > 0$.

But then, the set $\hat{J}$ of Theorem 4.1 is $\mathrm{Iso}(V)$ and there is only one element in $I : T^n$. As in Corollary 4.1, this implies (β) and (γ).

It remains only to prove (δ): note first that β is independent of the chosen morphism from S^1 into T^n, provided $\langle N^l, M\rangle$ and $\langle N'^l, M\rangle$ are not multiples of 2π. Since the number k of terms in the quotient is fixed and the same in the numerator and the denominator one may take the components of M to be rational, provided the new $\langle N^l, M\rangle$ and $\langle N'^l, M\rangle$ are not congruent to 0 modulo 2π, and by denseness, for real M. Hence, β is the quotient of homogeneous polynomials of degree 1. This implies that for each l there is a q, such that $\langle N'^l, M\rangle = c_{lq}\langle N^q, M\rangle$. Thus, $N'^l = c_{lq}N^q$ and, if $n_j^q = a_q m_j^q, n_j'^l = a_l' m_j'^l$, one has that $c_{lq}a_q/a_l' = m_j'^l/m_j^q = m'/m$ for all $j = 1, \ldots, n$, where m' and m are relatively prime. Hence, m' divides all $m_j'^l$ and m divides all m_j^q, which is impossible, from the fact that a_q and a_l' are largest common divisors, unless $|m'| = |m| = 1$ and $c_{lq} = \pm a_l'/a_q$, that is, β is the expression of (δ). □

For a general abelian group, one has the following Borsuk–Ulam result.

Theorem 4.2 (Borsuk–Ulam result). *Let V and W be two arbitrary representations of Γ with* $\dim V = \dim W$ *and let $F : V\backslash\{0\} \to W\backslash\{0\}$ be an equivariant map. Then:*

(a) $\deg(F; B) = 0$ *if* ($\tilde{\mathrm{H}}$) *does not hold or if* $\dim V^{T^n} \neq \dim W^{T^n}$.

(b) *If* ($\tilde{\mathrm{H}}$) *holds and* $\dim V^{T^n} = \dim W^{T^n}$, *then*

$$\deg(F; B) = \beta \deg(F^{T^n}; B^{T^n}),$$

where β is the non-zero integer given in Proposition 4.2.

(c) *Let $\hat{J}' = \{H \in \mathrm{Iso}(V^{T^n}) : \forall K, T^n \leq K \leq H, \dim V^K \leq \dim W^K\}$ and $I' = \{H \in \hat{J}' : \dim V^H = \dim W^H\}$, then, if the hypothesis of* (b) *holds and $F_0 : V^{T^n}\backslash\{0\} \to W^{T^n}\backslash\{0\}$ is Γ-homotopy to F on $\bigcup_{H\notin\hat{J}'} S^{V^H}$, one has for any H in I'*

$$\deg(F^H; B^H) = \deg(F_0^H; B^H) + \sum_{I'} d_K \beta_{HK} |\Gamma/K|,$$

where $\beta_{HK} = 0$ if H is not a subgroup of K, $\beta_{KK} = 1$, β_{HK} are integers independent of F and F_0, while d_K are integers depending of F and F_0.

If $\hat{J}' = \mathrm{Iso}(V^{T^n})$, then F_0 is absent. Furthermore, if $W^\Gamma = \{0\}$, one has to add, on the right, a term $\beta_{H\Gamma}$.

Proof. If ($\tilde{\mathrm{H}}$) does not hold, then $\deg(F; B) = 0$ follows from Remark 1.1. If $V^\Gamma = \{0\}$, one may complement F by $2t - 1$ and obtain an element of $\Pi^\Gamma_{SV}(S^W)$. Thus, from Proposition 4.2, one obtains (a) and (b). Furthermore, from Theorem 4.1, applied to V^{T^n}, one has

$$[F^{T^n}] = [F_0^{T^n}] + \sum_{I'} d_K[\tilde{F}_K^{T^n}].$$

Thus, $\deg(\tilde{F}_K^{T^n}; B^H) = \beta_{HK}|\Gamma/K|$, from Example 1.4, where β_{HK} has the properties listed in the theorem.

Finally, if $W^\Gamma = \{0\}$, then $V^\Gamma = \{0\}$ since F^Γ maps the second space into the first. Hence, when supplementing by $2t - 1$, one has $\mathrm{Iso}(\mathbb{R} \times V) = \mathrm{Iso}(V) \cup \Gamma$, $\hat{J}'$ remains the same unless $\hat{J}' = \mathrm{Iso}(V^{T^n})$, since this is the only possibility for Γ to belong to the new $\hat{J}'$. In that case, I' has to be supplemented by Γ and

$$[2t - 1, F^{T^n}] = \sum_{I'} d_K[\tilde{F}_K^{T^n}] + [\tilde{F}_\Gamma^{T^n}],$$

where $\tilde{F}_\Gamma^\Gamma = 2t - 1$ and $\beta_{H\Gamma} = \deg[\tilde{F}_\Gamma^H; I \times B^H]$. □

In order to get congruence results, characteristic of Borsuk–Ulam theorems, it is interesting to know when one may construct F_0 such that $\deg(F_0^H; B^H) = 0$ for all H's in I', or at least for $H = T^n$. In that case $\deg(F; B)$ would be a multiple of the greatest common divisor of the $|\Gamma/K|$'s, for K in I'. Besides the case where $\hat{J}' = \mathrm{Iso}(V^{T^n})$, one has the following

Corollary 4.2. *Let $\mathcal{M}$ be the set of minimal elements K_j of $\hat{J}'^c$, i.e.,* $\dim V^{K_j} > \dim W^{K_j}$ *but* $\dim V^H \le \dim W^H$, *for any $H \ge T^n$, strict subgroup of K_j. Assume that the hypothesis of Theorem* 4.2 (c) *holds. Then we have the following.*

(a) *For any H in I', $H < K_j$ for some K_j in $\mathcal{M}$, one has*

$$\deg(F^H; B^H) = \sum_{K \in I_j} d_K^j \beta_{HK}^j |K_j/K|,$$

where $I_j = \{K \in I', K < K_j\}$, β_{HK}^j are integers independent of F and with $\beta_{HK}^j = 0$ if H is not a subgroup of K and $\beta_{HH}^j = 1$.

(b) *If for each K_j in $\mathcal{M}$, there is an equivariant map*

$$F_\perp^j : (V^{K_j})^\perp \backslash \{0\} \to (W^{K_j})^\perp \backslash \{0\},$$

then one may construct F_0 in Theorem 4.2 (c) *with*

$$\deg(F_0; B^{T^n}) = 0.$$

(c) *If $\mathcal{M}$ has a unique element K_0 and there is a complementing map $F_\perp^0$, then, for all H in I', one has*

$$\deg(F_0^H; B^H) = 0.$$

(d) *If $\mathcal{M} = \Gamma$, then the conclusion of* (c) *holds because $F_\perp^0$ exists.*

Proof. The proof of (a) will be a consequence of (d), hence we shall prove (b) first. Let $\hat{\Pi}$ be as in Theorem 4.1 and R be the morphism from $\Pi^\Gamma_{S^{V^{T^n}}}(S^{W^{T^n}})$ onto $\hat{\Pi}$. Let $[F_0] = R[F]$. Then, define, for some K_1 in $\mathcal{M}$

$$[F_1] = [F_0] - R[F_0^{K_1}, F_\perp^1],$$

where, from Borsuk equivariant extension theorem, one may take $F_1^{K_1} = (1, 0)$. For another element K_2 of $\mathcal{M}$, define

$$[F_2] = [F_1] - R[F_1^{K_2}, F_\perp^2],$$

with $F^{K_2} = (1, 0)$. Since $[F_1^{K_2}, F_\perp^2]^{K_1} = [F_1|_{V^{K_1} \cap V^{K_2}}, F_\perp^{2K_1}] = [(1, 0), F_\perp^{2K_1}]$ which is Γ-deformable to (1, 0), one may use the equivariant Borsuk theorem and assume that $F_2^{K_1} = (1, 0)$.

Continuing this process, one arrives at a final map F_s with $F_s = (1, 0)$ on $\bigcup S^{V^{K_j}}$, i.e., with $R[F_s] = 0$. Hence,

$$[F_0] = \sum_{p=1}^{s} R[F_{j-1}^{K_j}, F_\perp^j].$$

Since the maps on the right have obvious extensions to $S^{V^{T^n}}$, one may construct F_0 as $\sum[F_{j-1}^{K_j}, F_\perp^j]$.

Now, if H is in I' and $H < K_j$, then $\dim V^H \cap (V^{K_j})^\perp < \dim W^H \cap (W^{K_j})^\perp$ and $F_\perp^{jH}$ is deformable (non-equivariantly) to a non-zero constant map. This implies that

$$\deg((F_{j-1}^{K_j}, F_\perp^j)^H; B^H) = 0.$$

Since $T^n < K_j$ for all j's and T^n is in I', one obtains $\deg(F_0; B^{T^n}) = 0$, proving (b). The proof of (c) follows from the same argument, since one may take $[F_0] = [F^{K_0}, F_\perp^0]$, and any H in I' is a subgroup of K_0. If furthermore, $K_0 = \Gamma$, then $\dim(V^\Gamma)^\perp \cap V^H < \dim(W^\Gamma)^\perp \cap W^H$, for any strict subgroup of Γ (and containing T^n). Hence, from Corollary 1.2 (c), the complementing map $F_\perp^0$ exists.

Finally, for each K_j, consider F as a K_j-equivariant map. Then, the isotropy subgroups for K_j are those H in $\mathrm{Iso}(V)$ with $H < K_j$. Thus, the corresponding set of minimal elements reduces to K_j and I' reduces to I_j. One may apply (d) with a K_j-equivariant F_0, with degree equal to 0 on any B^H, $H < K_j$. Apply then Theorem 4.2. □

Corollary 4.3. *Assume that Γ/T^n is a p-group, i.e., $|\Gamma/T^n| = p^k$, for some prime number p. If V and W are two arbitrary representations of Γ with* $\dim V = \dim W$ *and $F : V\backslash\{0\} \to W\backslash\{0\}$ is an equivariant map, then* $\deg(F; B)$ *is a multiple of p, unless hypothesis* (H) *holds for V^{T^n}, in which case*

$$\deg(F^H; B^H) = \sum_{H\leq K} d_K\Big(\prod_{KH} l_i\Big)|\Gamma/K|,$$

for all H in $\text{Iso}(V^{T^n})$*, where the l_i's are given in Lemma* 6.2 *of Chapter* 1 *and correspond to the variables in $(V^K)^{\perp} \cap V^H$. Here $|\Gamma/K|$ is a multiple of p, except for $K = \Gamma$ and $d_\Gamma = \deg(F^\Gamma; B^\Gamma)$.*

Proof. If ($\tilde{\text{H}}$) does not hold or if $\dim V^{T^n} \neq \dim W^{T^n}$, then $\deg(F; B) = 0$. Otherwise if $\hat{J}'$ is not all of $\text{Iso}(V^{T^n})$, take any minimal element K_j of $\mathcal{M}$, then for any element K of I_j, $|K_j/K|$ is a positive power of p. Thus, from Corollary 4.2 (a), $\deg(F; B)$ is a multiple of p.

Hence, if this degree is not a multiple of p, then ($\tilde{\text{H}}$) holds, $\dim V^{T^n} = \dim W^{T^n}$ and (H) holds on V^{T^n}, in particular $\dim V^H \leq \dim W^H$, for all H with $T^n \leq H \leq \Gamma$. Now, if there is H such that $\dim V^H < \dim W^H$, then viewing F^{T^n} as a H-map, one should have

$$\deg(F^{T^n}; B^{T^n}) = \sum_{K<H} d_K^H \beta^H_{T^n K}|H/K|,$$

for K in I', and since H is not in I', $|H/K|$ is a positive power of p, and therefore $\deg(F^{T^n}; B^{T^n})$ would be a multiple of p, that is, for all H in $\text{Iso}(V^{T^n})$, one has $\dim V^H = \dim W^H$.

Finally, if K and H in $\text{Iso}(V^{T^n})$ are such that $\dim V^H \cap V^K$ and $\dim W^H \cap W^K$ are different, consider F^K, from V^K into W^K, as an H-equivariant map. The fixed point subspaces for the action of H on V^K and W^K are $V^H \cap V^K$ and $W^H \cap W^K$ respectively. Since H is also a p-group, from the arguments above, one gets that $\deg(F^K; B^K)$ is a multiple of p. Now, regarding F^{T^n} as a K-map, one has from Theorem 4.2, since $\hat{J}' \cap \{H \leq K\} = I' \cap \{H \leq K\}$ is the set $\text{Iso}(V^{T^n}) \cap \{H \leq K\}$, that

$$\deg(F^{T^n}; B^{T^n}) = a\deg(F^K; B^K) + bp,$$

hence, in this case a multiple of p. The contradiction with

$$\deg(F; B) = \beta\deg(F^{T^n}; B^{T^n}),$$

and not a multiple of p, implies that (H) holds for V^{T^n}.

In conclusion, one has, in this case,

$$[F^{T^n}]_\Gamma = \sum d_H[\tilde{F}_H]_\Gamma,$$

where each generator $\tilde{F}_H$ is of the form $(F_H, x_j^{l_j})$, with $\deg(F_H; B^H) = |\Gamma/H|$. □

Example 4.3. If $\Gamma = \mathbb{Z}_2$, then $\deg(F; B)$ is even unless $\dim V^\Gamma = \dim W^\Gamma$ in which case, $\deg(F; B) = \deg(F^\Gamma; B^\Gamma) + 2d$, where $\deg(F^\Gamma; B^\Gamma)$ is replaced by 1 if $V^\Gamma = \{0\}$, by adding $2t - 1$. In particular, the degree of an odd map is odd and the degree of an even map is even (in that case $V^\Gamma = \{0\}$, $W^\Gamma = W$ and, if V is odd dimensional, then ($\tilde{\mathrm{H}}$) does not hold and the degree is 0).

Example 4.4. Let $f : \mathbb{C}^n \to \mathbb{C}^n$, or $\mathbb{R}^n \to \mathbb{R}^n$, be such that $f(x) = P(x) + g(x)$, where each component P_j of P is a homogeneous polynomial of degree k_j. Assume that $P(x)$ has an isolated zero at the origin and that $g(x)$ is small with respect to $P(x)$ near the origin. Then,

$$\mathrm{Index}(f) = \mathrm{Index}(P) = \prod k_j$$

in the complex case and modulo 2 in the real case.

The first equality is clear. For the second, put the standard S^1-action on the first copy of $\mathbb{C}^n$ and the action given by $e^{ik_j\varphi}$ on the second copy (in the real case replace S^1 by $\mathbb{Z}_2$ and φ by $k\pi$). The map $P(x)$ is clearly equivariant. In the complex case, $\mathrm{Index}(P) = \beta$, independently of P, from Theorem 4.2. Taking $P_j(x) = x_j^{k_j}$, it is clear that β is $\prod k_j$. In the real case, either all k_j are odd and $\mathrm{Index}(P)$ is odd, or otherwise $V^\Gamma = \{0\}$ and $\dim W^\Gamma > 0$, hence from the preceding example, the degree is even.

Example 4.5. One may wonder if Corollary 4.2 (b) depends really on the existence of complementing maps. Here is an example to the contrary. Let $\mathbb{Z}_{12}$ act on two copies of $\mathbb{C}^6$ in the following way: on the first copy, as $e^{2\pi ik/4}$ on x_1, x_2, x_3, x_4 and as $e^{2\pi ik/6}$ on y_1 and y_2; on the second copy, as $e^{2\pi ik/2}$ on ξ_1, ξ_2, ξ_3 and as $e^{2\pi ik/12}$ on η_1, η_2, η_3. The elements of $\mathrm{Iso}(V)$ are

$\mathrm{Iso}(V)$	V^H	W^H
$K = \mathbb{Z}_3$ (for k multiple of 4)	$\{x_1, x_2, x_3, x_4\}$	$\{\xi_1, \xi_2, \xi_3\}$
$H = \mathbb{Z}_2$ (for k a multiple of 6)	$\{y_1, y_2\}$	$\{\xi_1, \xi_2, \xi_3\}$
$\{e\}$	V	W
Γ if one adds a dummy variable	t	$\mathbb{R}$

Hence $\hat{J} = \{H, \{e\}\}$, $I' = \{e\}$, $\mathcal{M} = \{K\}$. There is no equivariant map $F_\perp$ from $(V^K)^\perp \backslash \{0\}$ into $(W^K)^\perp \backslash \{0\}$, since any such map should map $(V^K)^\perp = V^H$ into $W^H = W^K$. If the conclusion of Corollary 4.2 (c) were true, one would have $\deg(F; B) = |\Gamma| d_e$, a multiple of 12.

However, the following map has degree 6:

$$F = (x_1^2 - \bar{x}_2^2 - \bar{y}_1^3, x_3^2 - \bar{x}_4^2 - \bar{y}_2^3, \mathrm{Re}\, x_1 x_2 + i\, \mathrm{Re}\, x_3 x_4 + y_1^2 y_2, \bar{x}_1 y_1^2, \bar{x}_3 y_2^2, \bar{x}_2 y_1^2 + \bar{x}_4 y_2^2).$$

The equivariance of F is clear. The fact that F has only one zero follows from the following considerations: subtract $\varepsilon > 0$ from the last equation. At a zero, one needs $y_1 y_2 = 0$, since if not one would have $x_1 = x_3 = 0$ and the 3rd component non-zero. Then, if $y_1 = 0$, one has $x_1 = \pm\bar{x}_2$ and $\operatorname{Re} x_1 x_2 = \pm|x_1|^2$, hence the 3rd equation implies $x_1 = x_2 = 0$. If y_2 is also 0, then $x_3 = \pm\bar{x}_4 = 0$, while if $y_2 \neq 0$ then $x_3 = 0$, $\bar{x}_4 y_2^2 = \varepsilon$ and $\bar{x}_4^2 + \bar{y}_2^3 = 0$, i.e., $-|y_2|^6 y_2 = \varepsilon^2$. In this case, the zero is $A = (0, 0, 0, \varepsilon^{3/7}, 0, -\varepsilon^{2/7})$.

On the other hand, if $y_1 \neq 0$, then $y_2 = 0$, $x_3 = \pm\bar{x}_4 = 0$, $x_1 = 0$, $\bar{x}_2 y_1^2 = \varepsilon$ and $\bar{x}_2^2 + \bar{y}_1^3 = 0$, and the zero is

$$B = (0, \varepsilon^{3/7}, 0, 0, -\varepsilon^{2/7}, 0).$$

In order to compute the degree of F it is enough to compute the index at A and B.

Near A one may deform linearly $\bar{x}_3 y_2^2$ to $\bar{x}_3 \varepsilon^{4/7}$ and to $\bar{x}_3$. Then x_3 can be deformed to 0 in the other equations. Then $y_1^2 y_2$ is deformed to y_1^2 and the term $\bar{x}_2 y_1^2$ to 0. One obtains the product of three maps:

$$\begin{aligned}
&\bar{x}_3 \text{ with index } -1\\
&(x_1^2 - \bar{x}_2^2 - \bar{y}_1^3, \operatorname{Re} x_1 x_2 + y_1^2, \bar{x}_1 y_1^2)\\
&(-\bar{x}_4^2 - \bar{y}_2^3, \bar{x}^4 y_2^2 - \varepsilon).
\end{aligned}$$

In order to compute the index of the second map at its only zero, the origin, perturb the second equation by $-i\varepsilon$. The zeros of the perturbed map are for $x_1 = 0$, $y_1^2 = i\varepsilon$. One may deform x_1 in the first two equations to 0 and y_1^2 to $i\varepsilon$ in the third. The degree will be

$$-\deg(-\bar{x}_2^2 - \bar{y}_1^3, y_1^2 - i\varepsilon).$$

Taking ε to 0 and $\bar{y}_1^3$ to 0, one obtains a degree which is $-(-2)(2) = 4$.

For the third map, with a unique zero, one may deform ε to 0 and consider the map

$$(\bar{x}_4^2 + \bar{y}_2^3 - \varepsilon, \bar{x}_4 y_2^2)$$

with 3 zeros of the form $(x_4 = 0, |y_2|^3 = \varepsilon)$, each of index $(-1)(-1) = 1$, and two zeros of the form $(|x_4|^2 = \varepsilon, y_2 = 0)$, each of index $(-1)(2) = -2$. Hence, the degree of the third map is -1, and the index of F at A is 4.

For B, one follows the same steps, except that the term $y_1^2 y_2$, which was deformed to y_1^2, is now deformed to y_2. Otherwise, one interchanges (x_1, x_2) with (x_3, x_4) and y_1 with y_2. The index of the second map is now 2 instead of 4, and the index of F at B is 2. Thus,

$$\deg(F; B) = 6.$$

By replacing the term $y_1^2 y_2$ by $y_1^{2+6n} y_2$, where a negative exponent means conjugation, the index at A is changed to $2(2 + 6n)$, while that of B in unchanged. Hence, any odd multiple of 6 is achieved as the degree of a Γ-map from V into W.

Furthermore, if two Γ-maps F and F_0 coincide on V^K, then

$$[2t - 1, F]_\Gamma = [2t - 1, F_0]_\Gamma + d[F_e]_\Gamma,$$

where

$$F_e = (2t+1-2|x_1y_1|^2, x_1^2(x_1^4-1), x_1^2(\bar{x}_1x_2-1), x_1^2(\bar{x}_1x_3-1),$$
$$\bar{x}_1y_1^2(\bar{x}_1^2y_1^3-1), \bar{x}_1y_1^2(\bar{y}_1y_2-1), \bar{x}_1y_1^2(\bar{x}_1x_4-1)).$$

It is easy to check that F_e is Γ-equivariant, with 12 zeros at $|x_i| = |y_i| = 1$, each of index 1, i.e., F_e is the generator for $\Pi(e)$. Then,

$$\deg(F; I \times B) = \deg(F_0; I \times B) + 12d.$$

By choosing F_0 the map of the example, one generates, for maps from $\mathbb{R} \times V$ into $\mathbb{R} \times W$, all odd multiples of 6, while if one replaces $(2t-1, F_0)$ by $(2t+1, 0)$, with degree equal to 0, one obtains all even multiples of 6 by varying d.

Hence, for maps from $\mathbb{R} \times V$ into $\mathbb{R} \times W$, all multiples of 6 are achieved.

The simplest case is when hypothesis (H) holds on V^{T^n}, i.e., $\dim V^H = \dim W^H$ for all H, with $T^n \le H \le \Gamma$, and there is an equivariant map $\{x_j^{l_j}\}$ from V^{T^n} into W^{T^n}.

Corollary 4.4. *If* $\dim V = \dim W$ *and* (H) *holds on* V^{T^n}*, then, if m is the greatest common divisor of* $\{(\prod l_j)|\Gamma/H|, \text{for } T^n \le H < \Gamma \text{ and } x_j \text{ in } (V^H)^\perp \cap V^{T^n}\}$*, one has*

$$\deg(F; B) = \beta \deg(F^{T^n}; B^{T^n})$$
$$\deg(F^{T^n}; B^{T^n}) = \left(\prod l_j\right) \deg(F^\Gamma; B^\Gamma) + dm$$

where any integer d is achieved. The term $\deg(F^\Gamma; B^\Gamma)$ *is replaced by* 1 *if* $V^\Gamma = \{0\}$.

Proof. Since (H) holds, one has $\hat{J}' = \mathrm{Iso}(V^{T^n}) = I'$ and $\beta_{T^nH} = \deg(F_\perp^H; (B^H)^\perp)$, where $F_\perp^H$ is the complementing map. Thus, $\beta_{T^nH} = (\prod l_j)$, for x_j in $(V^H)^\perp \cap V^{T^n}$. From Theorem 4.2, one has

$$\deg(F^{T^n}; B^{T^n}) = d_\Gamma\, \beta_{T^n\Gamma} + \sum_{H<\Gamma} d_H \beta_{T^nH}|\Gamma/H|$$
$$\deg(F^\Gamma; B^\Gamma) = d_\Gamma.$$

Hence $\deg(F^{T^n}; B^{T^n})$ has the form of the corollary. Moreover, if $m_j = (\prod l_i)|\Gamma/H_j|$, then from Darboux theorem, one has

$$m = \sum \alpha_j m_j,$$

where $(\alpha_1, \ldots, \alpha_r)$ are relatively prime. Let $[F^{T^n}] = d \sum \alpha_j [\tilde{F}_{H_j}]$, where F_{H_j} is the generator of $\Pi(H_j)$ and $\tilde{F}_{H_j} = (F_{H_j}, F_\perp^{H_j})$. □

This result may be refined by considering the greatest common divisor of $\{|\Gamma/H|, T^n \leq H < \Gamma\}$, see [I.V. 2, Proposition 4.3] and other references in the section on bibliographical remarks. One of its main applications is the following observation, which is used very often in order to prove the existence of non-trivial zeros.

We shall consider only one of the simplest cases: when Γ acts freely on V and $W^\Gamma = \{0\}$. Then $\Gamma \cong S^1$ or $\mathbb{Z}_m$ (see Definition 1.3 in Chapter 1). If $\Gamma \cong \mathbb{Z}_m$, then the action on a coordinate x_j of V is of the form $\exp(2\pi ikm_j/m)$, with m_j and m relatively prime, in particular there is an integer p_j such that $p_j m_j \equiv 1, [m]$. On a coordinate ξ_j of W the action is of the form $\exp(2\pi ikn_j/m)$, with $0 < n_j < m$. Recall that $x_j^{l_j}$, with $l_j = p_j n_j$, is an equivariant map.

Corollary 4.5. *If Γ acts freely on V and $W^\Gamma = \{0\}$, then*

(a) *If $\Gamma \cong S^1$ and* $\dim V > \dim W$, *then any equivariant map from ∂B_R into W must have a zero on ∂B_R.*

(b) *If $\Gamma \cong S^1$ and* $\dim V = \dim W$, *then any equivariant map $\partial B_R \to W\backslash\{0\}$ has a degree equal to $\pm(\prod n_j)$.*

(c) *If $\Gamma \cong \mathbb{Z}_m$ and* $\dim V > \dim W$, *then if $\prod n_j$ is not a multiple of m, for instance if m is a prime, any equivariant map from ∂B_R into W must have a zero on ∂B_R.*

(d) *If $\Gamma \cong \mathbb{Z}_m$ and* $\dim V = \dim W$, *then any equivariant map $\partial B_R \to W\backslash\{0\}$ has a degree equal to $\prod l_j + dm$, where any d is achieved.*

Proof. Adding the variable t and the component $2t - 1$, one may use the previous results. In particular, if $\Gamma \cong S^1$, then the action on z_j is by $e^{i\varphi}$ or $e^{-i\varphi}$ and on ξ_j by $e^{in_j\varphi}$. If $\dim V = \dim W$, then, from Proposition 4.1, one has

$$\deg((2t-1, F); I \times B_R) = \deg(F; B_R) = \Big(\prod n_j\Big)/\Big(\prod m_j\Big),$$

where $m_j = \pm 1$, for any Γ-map from ∂B_R into $W\backslash\{0\}$. This proves (b). Furthermore, if $\dim V > \dim W$ and there is a Γ-map from ∂B_R into $W\backslash\{0\}$, choose $\tilde{V}$ a Γ-subspace of V, with $\dim \tilde{V} = \dim W$. Then, if $x = \tilde{x} \oplus x_\perp$, $\tilde{x}$ in $\tilde{V}$ and $x_\perp$ in a Γ-complement, one has

$$\deg(F(\tilde{x}, 0); B_R \cap \tilde{V}) = \pm \prod n_j.$$

Let $x_\perp = (R, 0)$ then $F(\cos \tau\tilde{x},\ \sin \tau R, 0)$ is a valid deformation for $\|\tilde{x}\| = R$, that is $F(\tilde{x}, 0)$ is homotopic (not equivariantly) to the constant $F(0, R, 0)$, hence with degree 0. This contradiction implies (a).

If $\Gamma \cong \mathbb{Z}_m$ and $\dim V = \dim W$, then hypothesis (H) is satisfied and, from Corollary 4.4, one has

$$\deg(F; B) = \prod l_j + dm.$$

Furthermore, consider the equivariant map

$$F(x) = \Big(z_1^{l_1}(z_1^{md}\prod_{j\geq 2}|z_j|^{\alpha_j} - 1), z_2^{l_2}(\bar z_1^{p_1 m_2} z_2 - 1), \dots, z_s^{l_s}(\bar z_1^{p_1 m_s} z_s - 1)\Big),$$

where s is the number of variables in V, and $\alpha_2, \dots, \alpha_s$ are positive and chosen so that $m\,d - p_1 \sum_2^s \alpha_j m_j$ is non-zero. Recall that, if $m = 2$, then $m_j = n_j = 1 = l_j$.

The zeros of F are for $x = 0$, with an index equal to $\prod l_j$ (± 1 for $m = 2$) and $m\,d$ zeros of the form $(z_1^0, z_1^{0 p_1 m_2}, \dots, z_1^{0 p_1 m_s})$, with $z_1^{0\,m\,d} = 1$ (this is where the condition on the α_j's is used).

Near one of these zeros, one may deform $z_j^{l_j}$, via $((1-\tau)z_j + \tau z_1^{0 p_1 m_j})^{l_j}$ to a constant. The deformation, in $|z_j|$, via $(1-\tau)z_j + \tau/\bar z_1^{p_1 m_j}$ followed by a linear deformation of z_1 to z_1^0 in $\bar z_1^{-p_1 m_j}$ and, finally, another linear deformation of $|z_1|^{-p_1 \Sigma \alpha_j m_j}$ to 1, will leave the map, near the zero,

$$(z_1^{md} - 1, \bar z_1^{0 p_1 m_2} z_2 - 1, \dots, \bar z_1^{0 p_1 m_s} z_s - 1),$$

which has an index 1 at that zero. Hence,

$$\deg(F; B_R) = \prod l_j + dm.$$

If $\dim V > \dim W$, take any $\tilde V$ with $\dim \tilde V = \dim W$. As before, one has, if there is Γ-map from ∂B_R into $W\backslash\{0\}$:

$$\deg(F|_{\tilde V}; B_R \cap \tilde V) = \prod l_j + dm = 0,$$

which would lead to the desired contradiction if $\prod l_j$ is not a multiple of m. Since $l_j = p_j m_j$, one could think that l_j depends on the choice of $\tilde V$, through p_j. However $p_j m_j \equiv 1, [m]$, hence $n_j \equiv m_j l_j, [m]$ and $\big(\prod l_j\big)\big(\prod m_j\big) \equiv \prod n_j, [m]$. Thus, if $\prod l_j$ is a multiple of m, so is $\prod n_j$. Conversely, if $\prod n_j$ is a multiple of m, since m_j and m have no common factor, one needs to have $\big(\prod l_j\big) = km$. Thus, $\prod n_j \not\equiv 0, [m]$ if and only if $\prod l_j \not\equiv 0, [m]$, for any choice of $\tilde V$. □

Example 4.6. If $\dim V > \dim W$, then for any $\tilde V$, with $\dim \tilde V = \dim W$, one has $\deg(F|_{\tilde V}, B \cap \tilde V) = 0$, as a necessary condition for a non-zero map from ∂B into W. For instance, in Example 4.5, one has the action of $\mathbb{Z}_4$ on $V^K \cong \mathbb{C}^4$ and $W^K \cong \mathbb{C}^3$. On the other hand, a zero degree may often be used to construct non-zero equivariant maps from ∂B into W. For instance, let $\mathbb{Z}_{p^2}$ act freely on $\mathbb{C}^2$ (i.e., as $e^{2\pi i k/p^2}$) and as $e^{2\pi i k/p}$ on a second copy of $\mathbb{C}^2$. Consider the equivariant map from $\mathbb{C}^2$ into $\mathbb{C}^2$:

$$f(z_1, z_2) = (z_1^p(\bar z_1^{p^2}|z_2| - 1), z_2^p(\bar z_1 z_2 - 1)).$$

The zeros of f, are the origin, with index p^2 and the p^2 points (z_1^0, z_1^0), with $z_1^{p^2} = 1$, each of index -1. Thus, $\deg(f; B_R) = 0$ and there is a (non-equivariant) deformation $f_\tau(z_1, z_2)$, from ∂B_R into $\mathbb{C}^2$, with $f_1(z_1, z_2) = (1, 0)$.

Consider the fundamental cell on $\mathbb{C}^5$ given by

$$\mathcal{C} = \{0 \leq |z| \leq R,\ 0 \leq \operatorname{Arg} z = \varphi < 2\pi/p^2,\ |z_j| \leq R, j = 1, 2, 3, 4\}.$$

Consider the following map defined from $\mathcal{C}$ into $\mathbb{C}^4$:

$$F(z, z_1, z_2, z_3, z_4) = \begin{cases} (f_{2|z|\cos^2(\varphi p^2/2)/R}(z_1, z_2), (f_{2|z|\sin^2(\varphi p^2/2)/R}(z_3, z_4)) \\ \text{for } |z| \leq R,\ 0 \leq \varphi \leq \pi/p^2 \\ (e^{2\pi i/p} f_{2|z|\cos^2(\varphi p^2/2)/R})(e^{-2\pi i/p^2} z_1, e^{-2\pi i/p^2} z_2), f_{2|z|\sin^2(\varphi p^2/2)/R}(z_3, z_4) \\ \text{for } |z| \leq R,\ \pi/p^2 \leq \varphi \leq 2\pi/p^2, \end{cases}$$

where $f_\tau = (1, 0)$, for $\tau \geq 1$.

Since f_0 is equivariant, one has

$$F(e^{2\pi i/p^2}|z|, e^{2\pi i/p^2} z_1, \dots) = e^{2\pi i/p} F(|z|, z_1, \dots),$$

hence when using the action of $\mathbb{Z}_{p^2}$ to cover B_R by images of $\mathcal{C}$, one obtains an equivariant map from B_R into $\mathbb{C}^4$. Furthermore, if $|z_j| = R$ for some j, F is non-zero, since f_τ is non-zero, in that case. For $|z| = R$, then for any φ, one has either $2\cos^2(\varphi p^2/2) \geq 1$ or $2\sin^2(\varphi p^2/2) \geq 1$. Hence F is non-zero on ∂B_R.

Example 4.7. Another way of constructing equivariant maps from V into W with $\dim W > \dim V$, can be illustrated as follows: Let p and q be relatively prime, hence there are α and β such that $\alpha q + \beta p = 1$, and let $\mathbb{Z}_{pq}$ act on $\mathbb{C}^2$ as $\gamma^k = e^{2\pi i k/p}$ on z_1 and $\gamma^k = e^{2\pi i k/q}$ on z_2. Consider the map

$$f(z_1, z_2) = (z_1(\bar{z}_1^{\beta p} - 1), z_2(\bar{z}_2^{\alpha q} - 1)(\bar{z}_1^{\beta p} z_2^{\alpha q} + 1)).$$

Then $f(z_1, z_2)$ is equivariant, from $\mathbb{C}^2$ into itself, and its zeros are: $(0, 0)$ with index 1; $(0, \bar{z}_2^{\alpha q} - 1 = 0)$, that is $|\alpha| q$ zeros each of index -1 if $\alpha > 0$ and index 1 if $\alpha < 0$; $(\bar{z}_1^{\beta p} = 1, 0)$, that is $|\beta| p$ zeros each of index -1; $(\bar{z}_1^{\beta p} = 1, \bar{z}_2^{\alpha q} = 1)$, with $|\alpha\beta| pq$ zeros with index 1; and $(\bar{z}_1^{\beta p} = 1, z_2^{\alpha q} = -1)$, $|\alpha\beta| pq$ zeros of index -1. Then

$$\deg(f; B_R) = 0$$

for $R > 1$. Let $f_\tau(z_1, z_2)$ be an ordinary homotopy of f on ∂B_R to $(1, 0)$. Denote by $\hat{f}_\tau = f_\tau / \|f_\tau\|$ and $\hat{f} = f/\|f\|$.

Assume that $\mathbb{Z}_{pq}$ acts on z as $e^{2\pi i k/pq}$, and consider the fundamental cell, for the action on $\mathbb{C}^3$, given by

$$\mathcal{C} = \{0 \leq |z| \leq R, 0 \leq \operatorname{Arg} z < 2\pi/pq, |z_1|, |z_2| \leq R\}.$$

On $\partial\mathcal{C}$ define the following non-zero map:

$$F(z, z_1, z_2) = \begin{cases} f_{|z|/R}(R\tilde{f}(z_1, z_2)), & \text{if } \operatorname{Arg} z = 0 \\ \gamma f_{|z|/R}(R\gamma^{-1}\hat{f}(z_1, z_2)), & \text{if } \operatorname{Arg} z = 2\pi/pq \\ (e^{iq\varphi}, 0), & \text{if } |z| = R. \end{cases}$$

From the construction, one has $F(e^{2\pi i/pq}|z|, z_1, z_2) = \gamma F(|z|, \gamma^{-1}z_1, \gamma^{-1}z_2)$, hence F has the right symmetry and is well defined at $z = 0$. Replacing $\hat{f}$ by $\hat{f}_\tau$ and deforming next γ to Id and $e^{iq\varphi}$ to 1, one has that, on $\partial\mathcal{C}$, F is homotopic to $f_{|z|/R}(R, 0)$, which is a non-zero path in $\mathbb{C}^2$, from $(R(R^{\beta p} - 1), 0)$ to $(1, 0)$. Since $\mathbb{C}^2\backslash\{0\}$ is simply connected, one may deform this path to $(1, 0)$. Thus, F has a non-zero continuous extension to $\mathcal{C}$ and, using the action of the group, one may extend F to an equivariant map from $\mathbb{C}^3\backslash\{0\}$ into $\mathbb{C}^2\backslash\{0\}$.

Example 4.8. When one has more than one isotropy subgroup, then the situation may be very complicated. For instance, consider the action of $\mathbb{Z}_{p^2q}$ on $V = \mathbb{C}^{n+m}$ and on $W = \mathbb{C}^{r+s}$ in the following manner:

On $(x_1, \dots, x_n)$ as $e^{2\pi ik/p^2}$, with isotropy $K \cong \mathbb{Z}_q$,
On $(y_1, \dots, y_m)$ as $e^{2\pi ik/pq}$, with isotropy $H \cong \mathbb{Z}_p$,
On $(\xi_1, \dots, \xi_r)$ as $e^{2\pi ik/p}$, giving $W^H = W^K$,
On $(\eta_1, \dots, \eta_s)$ as $e^{2\pi ik/p^2q}$.

Assume p and q are relatively prime, hence one has $\alpha q + \beta p = 1$. Suppose $n + m = r + s$ and $n > r \geq 2$ (the existence of an equivariant map from V^K into W^K follows from Example 4.6). Note that Γ acts on V^H as $\mathbb{Z}_{pq}$, with a free action of Γ/H, that is, applying Corollary 4.5, with $\prod n_j = q^r$, which is not a multiple of pq, one obtains that $\dim V^H \leq \dim W^H$, that is $m \leq r$, if there is a non-zero Γ-map F on ∂B.

One has the following result.

Proposition 4.3. *For the above situation, one has*

(a) *If* $m = r$, *then* $\deg(F; B) = \alpha^{n-m+1}q + dpq \not\equiv 0, [pq]$.

(b) *If* $m = r - 1$, *then* $\deg(F; B) = \alpha^{n-m}pq + dp^2q \not\equiv 0, [p^2q]$.

(c) *If* $m < r - 1$, *then* $\deg(F; B) = dp^2q$.

Proof. We shall indicate only the proof of (a), since (b) and (c) have proofs which are tedious, and refer to [IV2, Proposition 4.1]. It is enough to say that the proof is based on a construction of F_0, extension of F^K, so that one may apply Theorem 4.2, with $\deg(F_0; B) \equiv \alpha^{n-m}pq, [p^2q]$, if $m = r - 1$ and 0 if $m < r - 1$, then the term dp^2q comes from Theorem 4.2. The extension of F involves terms of the form $x^p + y^q$ on W^H and $x^\alpha y^\beta$ on $(W^H)^\perp$.

If $m = r$, viewing F as a K-map, one may use Corollary 4.3 and one has that $\deg(F; B)$ is a multiple of q. If we view F as an H-map (hence if $m < r$ one has that $\deg(F; B)$ is a multiple of p), one has

$$\deg(F; B) = \deg((F^H, x_1^\alpha, \dots, x_n^\alpha); B) + dp = \alpha^n \deg(F^H; B^H) + \hat{d}p.$$

But viewing F^H as a Γ-map, one has

$$\deg(F^H; B^H) = \deg((2t-1, y_1^q, \dots, y_m^q) + \tilde{d}pq = q^m + \tilde{d}pq.$$

Thus,

$$\deg(F; B) = \alpha^n(q^m + \tilde{d}pq) + \hat{d}p = cq.$$

This implies that $\hat{d}$ is a multiple of q. Writing $\alpha^n q^m = \alpha^{n-m+1}q(\alpha q)^{m-1}$ and using $\alpha q = 1 - \beta p$, one obtains

$$\deg(F; B) = \alpha^{n-m+1}q + dpq,$$

in particular, this number is not 0, nor a multiple of pq, since α and β are relatively prime and $n > m$. □

3.5 The one parameter case

Let $V = \mathbb{R} \times U$ and assume that U and W satisfy condition (H), i.e., $\dim U^H = \dim W^H$, for all H in $\mathrm{Iso}(U)$, and there is a Γ-equivariant map $\{x_j^{l_j}\}$ from U into W. From Corollary 2.1 and Theorem 3.2, one has

$$\Pi_{S^V}^{\Gamma}(S^W) = \Pi_{S^{\tilde{V}}}^{\tilde{\Gamma}}(S^{\tilde{W}}) \times \mathbb{Z} \times \cdots \times \mathbb{Z},$$

with one $\mathbb{Z}$ for each isotropy subgroup H with $\dim \Gamma/H = 1$, and $\tilde{\Gamma} = \Gamma/T^n$, $\tilde{V} = V^{T^n}$, $\tilde{W} = W^{T^n}$.

Then, any element of $\Pi_{S^V}^{\Gamma}(S^W)$ may be written as

$$[F]_\Gamma = [\tilde{F}]_\Gamma + \sum d_H[\tilde{F}_H] = [F^{T^n}, x_j^{l_j}]_\Gamma + \sum d_H[\tilde{F}_H],$$

where $\{d_H\}$ are given, for the special case of well-defined Poincaré sections, in Corollary 3.1.

The purpose of this section is to compute $[\tilde{F}]$ which is the suspension by a complementing map on $(V^{T^n})^\perp$ of an element of $\Pi_{S^{\tilde{V}}}^{\tilde{\Gamma}}(S^{\tilde{W}})$, with $\tilde{\Gamma}$ a finite group. Thus, one may assume $\Gamma = \tilde{\Gamma}$ a finite group and one wishes to compute $\Pi_{S^{\mathbb{R}\times U}}^{\Gamma}(S^W)$. This is the case of a Hopf bifurcation when an autonomous equation is perturbed by a $2\pi/p$-periodic nonlinearity, breaking the action of S^1 to an action of $\mathbb{Z}_p$ (see Example 5.1 in Chapter 2).

From Theorem 2.3 one has, for Γ a finite group,

$$\Pi^{\Gamma}_{SV}(S^W) = \bigoplus \tilde{\Pi}(H),$$

for H in $\mathrm{Iso}(V)$, where $\tilde{\Pi}(H)$ is the suspension by the complementing map of $\Pi(H)$. Recall that, from Lemma 2.2, this particular suspension is one-to-one.

Any element in V^H is written as $(t, \mu, X_0, y_1, \dots, y_s, z_1, \dots, z_r)$, where (t, X_0) is in $U^\Gamma \cong \mathbb{R}^{n+1}$, μ is the parameter, Γ acts on y_j, in $\mathbb{R}$, with $\Gamma/H_j \cong \mathbb{Z}_2$ and on z_j in $\mathbb{C}$, with $\Gamma/H_j \cong \mathbb{Z}_{m_j}$. Define $B^H = \{0 \le t \le 1, |\mu|, \|X_0\|, |y_j|, |z_j| \le 2\}$ and set $\lambda = 2t - 1 + i\mu$.

Lemma 5.1. *If $H = \Gamma$, then $\Pi(\Gamma) = \Pi_{n+1}(S^n)$, i.e., 0 if $n \le 1$, $\mathbb{Z}$ if $n = 2$, $\mathbb{Z}_2$ if $n \ge 3$. The part of Γ-degree on $\Pi(H)$ is given by $[F^\Gamma]$.*

Proof. See Remark 8.1 in Chapter 1. Recall that the Hopf map η generates $\Pi_3(S^2)$ and its suspension $\tilde{\eta}$ generates $\Pi_{n+1}(S^n)$. □

Let $\tilde{F}$ be $(F^\Gamma, F^\perp_\Gamma)$, where $F^\perp_\Gamma$ is the complementing map, and let $[F_1] = [F] - [\tilde{F}]$. Then, one may assume that $F_1^\Gamma = (1, 0)$. The next isotropy subgroups are those corresponding to y_j, i.e., such that $\Gamma/H_j \cong \mathbb{Z}_2$.

Theorem 5.1. *If $\Gamma/H \cong \mathbb{Z}_2$, with* $\dim V^\Gamma = n + 2$, $\dim V^H - \dim V^\Gamma = s$, *then*

$$\Pi(H) \cong \begin{cases} \mathbb{Z}_2 \times \mathbb{Z}_2 & \text{if } s > 2 \\ \mathbb{Z} \times \mathbb{Z}_2 & \text{if } s = 2 \text{ and } n > 0 \\ \mathbb{Z} \times \mathbb{Z} & \text{if } s = 2 \text{ and } n = 0 \\ \Pi_{n+2}(S^{n+1}) & \text{if } s = 1. \end{cases}$$

Proof. (a) If $s = 1$, i.e., $V^H = \{t, \mu, X_0, y_1\}$, the fundamental cell $\mathcal{C}$ is $B^\Gamma \times \{y_1 \ge 0\}$, F_1 is given for $y_1 = R$ and it is $(1, 0)$ for $y_1 = 0$. Then, the obstruction for the extension to $\mathcal{C}$ is $[F_1|_{\partial\mathcal{C}}]$ in $\Pi_{n+2}(S^{n+1})$.

If $n > 0$, let $X_0 = (x_0, \tilde{X}_0)$ and let η_1 be the map, with $\lambda = 2t - 1 + i\mu$:

$$\big((1/4 - (y_1 - 1)^2 - x_0^2)(1/4 - (y_1 + 1)^2 - x_0^2), \tilde{X}_0,\\ \mathrm{Re}(\lambda(y_1^2 - 1 + ix_0)), y_1 \,\mathrm{Im}(\lambda(y_1^2 - 1 + ix_0))\big).$$

The map η_1 is equivariant, i.e., all but the last components are even in y_1 and the last is odd in y_1. The zeros of η_1 are for $\tilde{X}_0 = 0, \lambda = 0, x_0^2 + (y_1 \pm 1)^2 = 1/4$. On $\partial\mathcal{C}$, it is easy to see that η_1 is homotopic to the suspension $\tilde{\eta}$ of the Hopf map. Hence, there is a d_1, in $\Pi_{n+2}(S^{n+1})$, such that $[F_1|_{\partial\mathcal{C}}] - d_1[\eta_1|_{\partial\mathcal{C}}] = 0$. Thus, this difference has an extension to $\mathcal{C}$ and, by the action of Γ/H, to B^H. Thus,

$$[F_1^H] = d_1[\eta_1].$$

If $n = 0$, then $\Pi_2(S^1) = 0$ and $[F_1|_{\partial\mathcal{C}}] = 0$.

Note that $[\eta_1|_{\partial B^H}] = 2\tilde{\eta}$ and that changing λ to λ^{d_1} one realizes $d_1[\eta_1]$.

(b) If $s > 1$, the fundamental cell is $\mathcal{C} = B^H \cap \{y_1 \geq 0\}$, of dimension $n+s+2$. From Theorem 1.1, one has a Γ-extension to $\mathcal{C} \cap \{y_1 = y_2 = 0\}$ and an obstruction, an integer, to extension to the set $B^H \cap \{y_1 = 0\}$, which is the degree of the extension $\tilde{F}_1$ on $\partial(B^H \cap \{y_1 = 0, y_2 \geq 0\})$. Note that on the space $V^H \cap \{y_1 = 0\}$, hypothesis ($\tilde{\text{H}}$) does not hold, hence the obstruction may be not unique (except if $s = 2$), and the degree on $\partial(B^H \cap \{y_1 = 0\})$ is 0. Let

$$d\eta = (1 - y_1^2 - y_2^2, X_0, \lambda^d(y_1 + iy_2), y_j).$$

It is easy to see that $\deg(d\eta;\, B^H \cap \{y_1 = 0, y_2 \geq 0\}) = (-1)^{n+1}d$. Thus, for some d, $[\hat{F}_1] = [\tilde{F}_1] - d[\eta]$ has an equivariant extension to $B^H \cap \{y_1 = 0\}$.

Note that, from Theorem 8.3 in Chapter 1, $(\lambda^d(y_1 + iy_2), y_j)$ represents $A^d(\lambda)y$, with $A(\lambda) = \operatorname{diag}(\lambda, \mathrm{Id})$, in $\Pi_1(\mathrm{GL}(\mathbb{R}^s))$ and that, if $s > 2$, $A^d(\lambda)$ is deformable to $A^{d+2}(\lambda)$, thus, only the parity of d is important here. If $s = 2$, d may be any integer.

As before, the next obstruction will be the class of $\hat{F}_1$ in $\Pi_{n+s+1}(S^{n+s})$ given by $\hat{F}_1|_{\partial\mathcal{C}}$. Let $d_1\eta_1$ be the equivariant map

$$\big((4/3)^2(1/4 - (y_1-1)^2 - y_1^2y_2^2)(1/4 - (y_1+1)^2 - y_1^2y_2^2),\\ y_1^2X_0, \lambda^{d_1}(y_1(y_1^2-1) + iy_1^2y_2), y_1^2y_j\big).$$

Again, it is easy to see that $d_1\eta_1 = (1, 0)$ for $y_1 = 0$, that $d_1\eta_1|_{\partial\mathcal{C}} = d_1[\tilde{\eta}]$, and that $d_1\eta_1|_{\partial B^H} = 2d_1[\tilde{\eta}]$, where $\tilde{\eta}$ is the suspension of the Hopf map generating $\Pi_{n+s+1}(S^{n+s})$. Hence, there is a d_1 (in $\mathbb{Z}_2$ if $n+s > 2$, in $\mathbb{Z}$ if $n = 0, s = 2$) such that $[\hat{F}] - d_1[\eta_1] = 0$. Thus,

$$[F_1]_\Gamma = d[\eta]_\Gamma + d_1[\eta_1]_\Gamma.$$

By forgetting the action of Γ, one obtains on ∂B^H, that if $n+s > 2$, one has $[F_1] = d[\tilde{\eta}]$ in $\Pi_{n+s+1}(S^{n+s}) \cong \mathbb{Z}_2$, hence the parity of d is uniquely determined by F_1 and the first invariant d is unique (in $\mathbb{Z}_2$ if $n+s > 2$, in $\mathbb{Z}$ if $s = 2$). Therefore, from the above formula, d_1 (in $\mathbb{Z}$ if $s = 2$ and $n = 0$, in $\mathbb{Z}_2$ if $n+s > 2$) is also unique. □

Consider now the case of a general isotropy subgroup H, with fundamental cell $\mathcal{C}$. As in the proof of Theorem 1.1, we shall extend and modify a given element $[F]$ of $\Pi(H)$ to an equivariant map $\tilde{F}$ without zeros on $\partial\mathcal{C}$. There will be obstructions to modifications on each of the faces of $\partial\mathcal{C}$ (i.e., with just one $y_j = 0$ or one z_j with $\operatorname{Arg} z_j = 0$). As seen in the proof of Theorem 1.1, the value of F on an edge ($\operatorname{Arg} z_j = 0$, $\operatorname{Arg} z_i = 2\pi/k_i$) may be given by the value on a face $\operatorname{Arg} z_k = 0$ for some $k < i$. Thus, one has to start with the first face, modify F so that the new map will have a non-zero extension on that face and work the way up on the faces. For $\tilde{F}$ one will have a last obstruction, in $\mathbb{Z}_2$, for the extension to $\mathcal{C}$. This construction will be broken up in several lemmas.

In order to simplify the argument we shall *assume that if y_j is a real variable with isotropy H_j, then there is at least another coordinate y'_j with the same isotropy*. In that case, if $z_j = y'_j + iy_j$, the face $y_j = 0$ in $\mathcal{C}$, with $y_j \geq 0$ and $k_j = 2$, corresponds to $\operatorname{Arg} z_j = 0$.

Consider first the face $y_1 = 0$. From the presence of y'_1, the isotropy of the face is still H and one has a fundamental cell $\mathcal{C}'$ for that face where y_1 is not present and one has $y'_1 \geq 0$ but the other variables and k_j's are the same as for $\mathcal{C}$. From Theorem 1.1, one has a Γ-extension to $B^H \cap \{y_1 = y'_1 = 0\}$ and an integer as an obstruction to Γ-extension to $\mathcal{C}'$. Note that ($\tilde{\mathrm{H}}$) is not satisfied for Γ. However, since $\Gamma/H = (\Gamma/H_1) \times (H_1/H)$, $\mathcal{C}'$ is still the fundamental cell for H_1-maps on $B^H \cap \{y_1 = 0,\, y'_1 \geq 0\}$ and there ($\tilde{\mathrm{H}}$) holds for H_1.

Lemma 5.2. *Let $\deg(F^H; B^H \cap \{y_1 = 0,\, y'_1 > 0\}) = d_1|H_1/H|$, then there is a Γ-map η_1 such that $[F]_\Gamma - d_1[\eta_1]_\Gamma$ has a non-zero Γ-extension to $B^H \cap \{y_1 = 0\}$, in particular to the face $y_1 = 0$.*

Proof. Let η_1 be the Γ-map

$$\eta_1 = \Big(2t + 1 - 2(y_1^2 + y_1'^2)\prod |x_j|^2,\, X_0,\, \tilde{\lambda}(y'_1 + iy_1), \ldots, (Q_i - 1)y_i,\, (P_i + 1)x_i^{l_i}\Big),$$

where $\tilde{\lambda} = \mu + i(y_1^2 + y_1'^2 - 1)$, $Q_i = y_i^2$ if $k_i = 2$ and $Q_i = 2$ if $k_i = 1$, $P_i = P_i(y'_1 + iy_1, \ldots, x_i)$ is the monomial of Theorem 3.3 if $k_i > 1$ and $P_i = 0$ for $k_i = 1$. The product in the first component is over all the variables with $k_j > 1$. For $y_1 = 0$, $y'_1 = 1$, the set $\{Q_i - 1,\, P_i + 1\}$ has exactly $|H_1/H|$ zeros, with $|x_j| = 1$, and just one in $\mathcal{C}'$.

It is easy to see that $\deg(\eta_1|_{y_1=0}; \mathcal{C}') = (-1)^n$.. Since this is an orientation factor, due to the chosen order of the components, changing $\tilde{\lambda}$ to its conjugate, if necessary, we may assume that $\eta_1|_{y_1=0}$ is the generator for $\Pi_1(H)$, where this group stands for H_1-maps defined on $\{y_1 = 0,\, y'_1 \geq 0\}$.

Since F is in $\Pi(H)$, i.e., it has a Γ-extension to B^K for $K > H$, then $F|_{\{y_1=0,\, y'_1 \geq 0\}}$ belongs to $\Pi_1(H)$ and d_1 is its extension degree given in Theorem 1.2. Then $[F_1] = [F] - d_1[\eta_1]$, has a non-zero H_1-extension to $B^H \cap \{y_1 = 0,\, y'_1 \geq 0\}$. Since $F_1|_{y_1=y'_1=0}$ is a Γ-map, if $\tilde{F}_1$ is the H_1-extension for $y'_1 \geq 0$, define $\tilde{F}_1(-|y'_1|, x) = \tilde{\gamma}\tilde{F}_1(|y'_1|, \gamma^{-1}x)$, for any γ in Γ such that $\gamma y'_1 = -y'_1$. If γ_1 and γ_2 satisfy this relation then $\gamma_1\gamma_2^{-1}$ is in H_1, one may write $\gamma_2^{-1} = (\gamma_1\gamma_2^{-1})\gamma_1^{-1}$ and use the H_1-equivariance of $\tilde{F}_1$ to prove that the new map is well defined and a Γ-equivariant extension of $F_1|_{y_1=y'_1=0}$. □

For a face of the form $\operatorname{Arg} z_j = 0$, or for a pair of real variables with $z_j = y'_j + iy_j$ as above, if one considers $\mathcal{C} \cap \{z_j = 0\}$, then there is always, from Theorem 1.1, a non-zero Γ-extension to $B^H \cap \{z_j = 0\}$ (if the isotropy of the face is $K > H$, then the extension is given a priori).

Lemma 5.3. *Assume F has been modified to $[F_{j-1}]_\Gamma = [F]_\Gamma - \sum_{i<j} d_i[\eta_i]_\Gamma$, a map in $\Pi(H)$ without zeros on the faces* $\operatorname{Arg} z_i = 0, i < j$. *Then, if d_j is defined by*

$$\Big(\prod_{i \neq j} k_i\Big) d_j = \deg(F_{j-1}; B^H \cap \{\operatorname{Arg} z_j = 0\})$$

one has a Γ-map η_j such that

$$[F_j]_\Gamma = [F_{j-1}]_\Gamma - d_j[\eta_j]_\Gamma$$

belongs to $\Pi(H)$ and has no zeros on the faces $\operatorname{Arg} z_i = 0, i \leq j$.

Proof. Let H_j be the isotropy subgroup of z_j. In order to get an H_j-equivariant extension to the ball $B_j = B^H \cap \{\operatorname{Arg} z_j = 0\}$ one needs to consider the extension degree of F_{j-1} on the fundamental cell $\mathcal{C}_j$ for the action of H_j on B_j: since F_{j-1} is in $\Pi(H)$ it is also in $\Pi_j(H)$, the group for the action of H_j. Furthermore, from the dimension, F_{j-1} has a non-zero H_j-extension to $\partial\mathcal{C}_j$.

From Theorem 1.2, one has that this extension degree d_H is given by

$$\deg(F_{j-1}; B_j) = d_H|H_j/H|.$$

Now, the ball B_j is covered by $|H_j/H|$ disjoint replicae of $\mathcal{C}_j$ and F_{j-1} has degree d_H on each of them. Note that $k_i = |\tilde{H}_{i-1}/\tilde{H}_i|$ is the same, for $i > j$, for $\mathcal{C}$ and $\mathcal{C}_j$, hence F_{j-1} is non-zero on $\partial(\mathcal{C} \cap \operatorname{Arg} z_j = 0)$, by hypothesis if $\operatorname{Arg} z_i = 0, i < j$, and by the action of Γ for $\operatorname{Arg} z_i = 2\pi/k_i$ for $i < j$; for $i = j$, by the dimension for $z_j = 0$ and by definition for $z_j = R$; and for $i > j$, since F_{j-1} has a non-zero H_j-extension to $\partial\mathcal{C}_j$. Thus, F_{j-1} has a well-defined degree on $\mathcal{C} \cap \{\operatorname{Arg} z_j = 0\}$, and also on $\mathcal{C} \cap \{\operatorname{Arg} z_j = \varphi\}$ for any φ, such that all these degrees are equal, using φ as a deformation parameter.

Now, we know that B^H is covered by the $|\Gamma/H|$ disjoint replicae of $\mathcal{C}$. Thus, $B_j = B^H \cap \{\operatorname{Arg} z_j = 0\}$ is covered by the intersections of the sets $\gamma\mathcal{C}$ with $\operatorname{Arg} z_j = 0$. Recall that the action of Γ on z_j is as $e^{2\pi i k/m_j}$ and that $\tilde{H}_{j-1}$ acts on z_j as $e^{2\pi i k/k_j}$. Taking $k = m_j$ in the second expression, one should have a trivial action, since $\tilde{H}_{j-1}$ is a subgroup of Γ and $\gamma^{m_j}\big|_{z_j} = \operatorname{Id}$. Thus, k_j divides m_j. Notice that, since

$$|\Gamma/H| = |\Gamma/H_j||H_j/H| = m_j|H_j/H|, \text{ then } |H_j/H| = \Big(\prod_{i \neq j} k_i\Big)(k_j/m_j).$$

Now, if $\gamma\mathcal{C} \cap \{\operatorname{Arg} z_j = 0\}$ is not empty, then this set comes from the subset of $\mathcal{C}$ with $\operatorname{Arg} z_j = 2\pi k/m_j$, for some $k = 0, \ldots, m_j/k_j - 1$. If γ_j acts on z_j as $e^{2\pi i/m_j}$, one may write $\gamma = \gamma_j^{-k}\tilde{\gamma}$, with $\tilde{\gamma}$ in H_j. This implies that, for each such k, the number of γ's such that $\gamma\mathcal{C}$ intersects $\{\operatorname{Arg} z_j = 0\}$ is the same and is equal to $|H_j/H|$. Thus, one arrives at a total of $\prod_{i \neq j} k_i$ sets of the form $\gamma(\mathcal{C} \cap \operatorname{Arg} z_j = 2\pi k/m_j)$ covering

B_j and, on each of them, F_{j-1} has the same degree. This implies that $\deg(F_{j-1}; B_j)$ is a multiple of $\prod_{i \neq j} k_i$ and d_H is a multiple of m_j/k_j. Let η_j be

$$\eta_j = \Big(2t + 1 - 2\prod |x_i|^2, X_0, (Q_i - 1)y_i, \tilde{\lambda} z_j^{l_j}, \{(P_i + 1)x_i^{l_i}\}_{i \neq j}\Big),$$

where $\tilde{\lambda} = \mu + i(|z_j|^2 - 1)$ or its conjugate if $n = \dim X_0$ is odd, the product is over all the variables, except y_i with $k_i = 1$, and the set $\{Q_i - 1, P_i + 1\}$ has, for $z_j = 1, \prod_{i \neq j} k_i$ zeros, with $|x_i| = 1$ and just one of them on the face of $\mathcal{C}$ corresponding to $\operatorname{Arg} z_j = 0$ (one may have to change 1, in $P_i + 1$, to ε_i, with $|\varepsilon_i| = 1$, as in Theorem 3.3, in order to have this last property).

Then, the degree of η_j on that face is 1, η_j is trivial when restricted to the faces of $\mathcal{C}$ given by $\operatorname{Arg} z_i = 0$ for $i < j$ (since P_i is a monomial in $x_1, \dots, x_i$ and the zeros of $(P_1 + 1, \dots, P_i + 1)$ are not on these faces). Finally, $\deg(\eta_j; B_j) = \prod_{i \neq j} k_i$ and, if one replaces $\tilde{\lambda}$ by $\tilde{\lambda}^d$, one obtains a map Γ-homotopic to $d\eta_j$.

Hence, $d\eta_j$ generates all possible obstructions on the face $\mathcal{C} \cap \{\operatorname{Arg} z_j = 0\}$ and does not modify the previous construction. Choosing d as in the statement of the present lemma, one obtains that $[F_{j-1}] - d_j[\eta_j]$ has a non-zero extension on the faces of $\mathcal{C}$ with $\operatorname{Arg} z_i = 0, i \leq j$. This extension is then reproduced by the action of Γ on the other faces. □

Remark 5.1. Note first that d_j, in the above construction, depends only on the extension to $\mathcal{C} \cap \{z_j = 0\}$, from the H_j-extension argument and the formula for d_j. This dependence will be used to compute $\Pi(H)$ and see that one may have several values for d_j. Note also that, at each step, F is modified on the subsequent faces. Furthermore, in the formula for F_{j-1}, the sum stands for a Γ-homotopy on ∂B^H and for extensions to the faces $\mathcal{C} \bigcap \{\operatorname{Arg} z_i = 0\}_{i<j}$. However, the homotopy is not extended to these faces and, in particular, there is no relationship between the ordinary degrees of F on the face $\{\operatorname{Arg} z_j = 0\}$ and the sum of the degrees of $d_i \eta_i$ on that face and even less with respect to the degrees on B_j, except in particular cases, such as for the first face for which F has no extension, given in Lemma 5.2, and in the case of Theorem 5.3 below. This lack of relationship will be demonstrated in Example 5.1.

Lemma 5.4. *Any F in $\Pi(H)$ can be written as*

$$[F]_\Gamma = \sum d_j [\eta_j]_\Gamma + \tilde{d}[\tilde{\eta}]_\Gamma,$$

where d_j and η_j are given in Lemma 5.3, $\tilde{d}$ is 0 or 1 and $\tilde{\eta}$ is a Γ-map which is non-zero on $\partial\mathcal{C}$ and is deformable, on $\partial\mathcal{C}$, to the suspension of the Hopf map.

Proof. From Lemma 5.3, one may construct a step by step modification of F, such that the last one, say F_s, with $[F_s]_\Gamma = [F]_\Gamma - \sum d_j[\eta_j]_\Gamma$, is non-zero on $\partial\mathcal{C}$. In order to extend F_s to $\mathcal{C}$ one has a last obstruction, this time in $\mathbb{Z}_2$, if $\dim \mathcal{C} = \dim W^H + 1 > 4$, and in $\mathbb{Z}$ if $\dim \mathcal{C} = 4$. In that case, $V^H \cap (V^\Gamma)^\perp$ has only one complex variable, d_1

is uniquely determined by $\deg(F; B^H \cap \{\operatorname{Arg} z = 0\})/|\Gamma/H|$ and $\Pi(H)$ turns out to be $\mathbb{Z} \times \mathbb{Z}$. Let the generator $\tilde{\eta}$ be

$$\tilde{\eta} = \Big(\big|\prod_{i<n} |x_i| P_n - \varepsilon_n\big|^2 - \varepsilon^2, X_0, (Q_i - 1)y_i, (P_i - \varepsilon_i)x_i^{l_i}, \lambda(P_n - \varepsilon_n)x_n^{l_n}\Big),$$

where $\lambda = \mu + i(2t - 1)$, the constants ε_i, with $|\varepsilon_i| = 1$, are chosen such that the set $(Q_i - 1, P_i - \varepsilon_i)$ has $|\Gamma/H|$ zeros, with $|x_i| = 1$, and just one, X^0, in $\mathcal{C}$. The product is over all variables, except y_j with $k_j = 1$ and x_n. Note that k_n may be 1. The positive constant ε is chosen so small that the disc $\|X - X^0\| \leq \varepsilon$ is contained in $\mathcal{C}$. Hence, the only zeros of $\tilde{\eta}$ are for $x_i = x_i^0$, $\lambda = 0$, $|x_n - x_n^0|^2 = \varepsilon^2$. In fact, if $x_i = 0$, the first component reduces to $1 - \varepsilon^2$, since $|\varepsilon_n| = 1$. Furthermore, on $\partial\mathcal{C}$, one may deform $\tilde{\eta}$ to the suspension of $(|x_n - x_n^0|^2 - \varepsilon^2, \lambda(x_n - x_n^0))$, which is the Hopf map (deform $x_i^{l_i}$ to $x_i^{0l_i}$ by a linear path joining X to X^0 and the deformation is done at the same time in $\prod |x_i|$, and $P_i - \varepsilon_i$ to $x_i - x_i^0$).

Replacing λ by λ^d, one generates $d\tilde{\eta}$. If $d = 2$, then $2\tilde{\eta}$ has a non-zero continuous extension to $\mathcal{C}$ and, by the action of Γ, to B^H. Thus, $2[\tilde{\eta}]_\Gamma = 0$, if $\dim \mathcal{C} > 4$. □

Theorem 5.2. *Assume that whenever $k_j > 1$, $j = 1, \ldots, s$, the corresponding variable z_j has a double z_j' with the same isotropy if z_j is complex and is repeated twice to y_j, y_j', y_j'' if real. Then, $\Pi(H)$ is a finite group generated by $[\eta_j]_\Gamma$, $j = 1, \ldots, s$ and $[\tilde{\eta}]_\Gamma$, with the relations, with d_{ji} integers and $\tilde{d}_j = 0$ or 1:*

$$2[\tilde{\eta}]_\Gamma = 0$$

$$k_j[\eta_j]_\Gamma + \sum_{i>j} d_{ji}[\eta_i]_\Gamma + \tilde{d}_j[\tilde{\eta}]_\Gamma = 0,$$

in particular, one has

$$\begin{aligned} 2k_s[\eta_s]_\Gamma &= 0 \\ 2k_{s-1}k_s[\eta_{s-1}]_\Gamma &= 0 \\ &\vdots \\ 2\prod k_i[\eta_1]_\Gamma &= 0 \\ 2|\Gamma/H|[F]_\Gamma &= 0, \end{aligned}$$

for any F in $\Pi(H)$.

Proof. Let Π be the following morphism from $\mathbb{Z} \times \cdots \times \mathbb{Z} \times \mathbb{Z}_2$ into $\Pi(H)$:

$$\Pi(d_1, \ldots, d_s, \tilde{d}) = \sum d_j[\eta_j]_\Gamma + \tilde{d}[\tilde{\eta}]_\Gamma$$

It is easy to see that if $K > H$, the sum is non-zero on V^K and, from Lemma 5.4, Π is onto. Thus, one has to study $\ker \Pi$, which is the set of all possible d_j's and $\tilde{d}$'s corresponding to the trivial element $(1, 0)$.

The extension of $(1,0)$ will be studied by following the steps of Lemmas 5.2 and 5.3. Hence, if $k_1 = 2$ and corresponding to real variables y_1, y_1', y_1'', there is always a Γ-extension to the set $\{y_1 = y_1' = 0\}$ which has a fundamental cell of the form $\{y_1'' > 0\} \times \mathcal{C}''$, where in $\mathcal{C}''$ one has the same variables as in $\mathcal{C}$. By the dimension argument, on $\{y_1'' = 0\} \times \mathcal{C}''$ all Γ-extensions are homotopic, hence one may assume that this extension is still $(1,0)$. If one extends as $(1,0)$ to $y_1 = y_1'' = 0$, $y_1' > 0$, then $B^H \cap \{y_1 = 0, y_1' > 0\}$ can be divided into two pieces, according to the sign of y_1'', the map is $(1,0)$ on its boundary, except for $y_1 = y_1' = 0$, $-2 < y_1'' < 2$, where it is a Γ-map. Hence, from Theorem 1.2,

$$\deg(F^H; B^H \cap \{y_1 = 0, y_1' > 0\}) = 2\deg(F^H; B^H \cap \{y_1 = 0, y_1' > 0, y_1'' > 0\})$$

is a multiple of $|\Gamma/H|$. Thus, d_1 is even.

If $F_{j-1} \equiv (1,0)_\Gamma - \sum_{i<j} d_i[\eta_i]_\Gamma$ has been extended to $\{\operatorname{Arg} z_i = 0\}_{i<j}$, then $(1,0)$ has a Γ-extension F for $z_j = 0$, as in Lemma 5.3, by dimension. Furthermore, $B^H \cap \{z_j = 0\}$ is covered by $|\Gamma/H|$ replicae of the fundamental cell $\mathcal{C}'$ which has the same form as $\mathcal{C}$, except that z_j' replaces z_j. From the dimension, $\dim \partial\mathcal{C}' = \dim W^H - 1$, one may deform the map F on $\partial\mathcal{C}'$ to $(1,0)$ without changing the homotopy type of the map F on $\mathcal{C}'$, relative to its boundary.

Now, the set $B^H \cap \{\operatorname{Arg} z_j = 0\}$ is covered by the ball $\mathcal{C}' \times \{\operatorname{Arg} z_j = 0\}$ and its $|\Gamma/H|$-replicae, where z_j is considered as a parameter. The map F is $(1,0)$ on the boundary, except on $\mathcal{C}' \times \{z_j = 0\}$ where it is a Γ-map. Thus, as before, $\deg(F; B^H \cap \{\operatorname{Arg} z_j = 0\})$ depends only on the extension F and is a multiple of $|\Gamma/H|$.

Now, if $d_i = 0$, for $i < j$, or if $d_i\eta_i$ are trivial on $\operatorname{Arg} z_j = 0$, for $i < j$, then F_{j-1} is homotopic to F on $\partial(B^H \cap \operatorname{Arg} z_j = 0)$ and they have the same degree, that is, in this case d_j is a multiple of k_j (see Remark 5.1).

Consider the map, with $\tilde{\lambda} = \mu + i\varepsilon(|z_j|^2 - 1)$, $\varepsilon = (-1)^{\dim X_0}$,

$$F_j = \Big(2t + 1 - 2|z_j'| \prod_{i \neq j} |x_i|, X_0, (Q_1' - 1)y_i, \tilde{\lambda} z_j'^{l_j}, (P_j' - \varepsilon_j) z_j'^{l_j}, \{(P_i' - \varepsilon_i) z_i^{l_i}\}_{i \neq j}\Big),$$

where $\{P_i'\}$ is the usual set of invariant polynomials but with z_j replaced by z_j', such that the set $(Q_i' - 1, P_i' - \varepsilon_i)$ has $|\Gamma/H|$ zeros of the form γX^0, with $|x_i^0| = 1$, none of which is on the faces $\operatorname{Arg} z_i = 0, i \neq j$ or for $\operatorname{Arg} z_j' = 0$. The zeros of F_j are for $X_0 = 0$, $\mu = 0$, $X = \gamma X^0$, $|z_j| = 1$ and $t = 1/2$. For $\operatorname{Arg} z_j = 0$, the degree of this map is $\prod k_i = |\Gamma/H|$ and $F_j \neq 0$ for $\operatorname{Arg} z_i = 0, i \neq j$. Hence, in $[F_j]_\Gamma = \sum d_{ji}[\eta_i]_\Gamma + \tilde{d}_j[\tilde{\eta}]$, one has $d_{ji} = 0$ for $i < j$ and, from Lemma 5.3, $d_{jj} = k_j$, since one does not need to modify F_j.

Furthermore, on B^H, one may Γ-deform $\tilde{\lambda}$ to $\mu + i\varepsilon(|z_j|^2 + 2\tau - 1)$, since one does not have any more the restriction $0 \leq z_j \leq R$. But, for $\tau = 1$ the map has no zeros in B^H, that is $[F_j]_\Gamma = 0$, proving the relations.

Since $2[\tilde{\eta}] = 0$, one obtains $2k_s[\eta_s]_\Gamma = 0$ and, iteratively, $\big(2\prod_{i \geq j} k_i\big)[\eta_j] = 0$ and, from Lemma 5.4, $2|\Gamma/H|[F]_\Gamma = 0$ for any F in $\Pi(H)$, that is any element of $\Pi(H)$ has, at most, order $2|\Gamma/H|$.

Finally, if one has a representation of the trivial map,

$$0 = \sum d_j[\eta_j]_\Gamma + \tilde{d}[\tilde{\eta}]_\Gamma,$$

then we know that $d_1 = p_1 k_1$. Since $k_1[\eta_1] = -\big(\sum_2^s d_{1j}[\eta_j] + \tilde{d}_1[\tilde{\eta}]\big)$, upon substituting this in the above equality, one obtains

$$0 = \sum_2^s (d_j - p_1 d_{1j})[\eta_j] + (\tilde{d} - p_1\tilde{d}_1)[\tilde{\eta}].$$

From the argument at the beginning of the proof (absence of η_1, and the sum equal to 0) one has that

$$d_2 - p_1 d_{12} = p_2 k_2.$$

Substituting the equality for $k_2[\eta_2]$, one gets

$$0 = \sum_3^s (d_j - p_1 d_{1j} - p_2 d_{2j})[\eta_j] + (\tilde{d} - p_1\tilde{d}_1 - p_2\tilde{d}_2)[\tilde{\eta}].$$

Continuing this argument, one concludes

$$\begin{pmatrix} d_1 \\ d_2 \\ \vdots \\ d_s \end{pmatrix} = \begin{pmatrix} k_1 & 0 & 0 & \dots & 0 \\ d_{12} & k_2 & 0 & \dots & 0 \\ \vdots & & & & \\ d_{1s} & d_{2s} & d_{3s} & \dots & k_s \end{pmatrix} \begin{pmatrix} p_1 \\ p_2 \\ \vdots \\ p_s \end{pmatrix}$$

together with the relation

$$\tilde{d} = \sum p_i \tilde{d}_i \mod 2.$$

On the other hand, one may take the p_i's to be arbitrary integers and prove for them that $\sum d_j[\eta_j] + \tilde{d}[\tilde{\eta}] = 0$. That is, we have proved that $\ker \Pi$ is generated by the above relations. □

Remark 5.2. The computation of d_{ij} and $\tilde{d}_j$ is involved. A way of doing it is indicated in [I.V. 1, Theorem 8.2]. Here we shall only give it in the particular case where $V = \mathbb{R} \times W$, after studying two examples.

Example 5.1. Suppose $\Gamma \cong \mathbb{Z}_9$ acts on (z_1, z_2) as $(e^{2\pi ik/3}z_1, e^{2\pi ik/9}z_2)$ together with their twins (z_1', z_2'). Taking $\mathcal{C}$ to be $\{|z_j| \le 2, 0 \le \operatorname{Arg} z_j < 2\pi/3, j = 1, 2\}$, with $k_1 = k_2 = 3$, let

$$\begin{aligned} \eta_1 &= (2t + 1 - 2|z_1||z_2|, (\mu + i(|z_1| - 1))z_1, z_1', (z_1^2 z_2^3 + 1)z_2, z_2') \\ \eta_2 &= (2t + 1 - 2|z_1||z_2|, (z_1^3 + 1))z_1, z_1', (\mu + i(|z_2| - 1))z_2, z_2') \\ \tilde{\eta} &= (\varepsilon^2 - |z_1|^2|z_1^2 z_2^3 + 1|, (z_1^3 + 1)z_1, z_1', (\mu + i(2t - 1))(z_1^2 z_2^3 + 1)z_2, z_2'). \end{aligned}$$

Note that the generators given in Lemma 5.3 should have as first component $(2t+1-2|z_1z_1'z_2z_2'|)$ and $(\bar{z}_jz_j'+1)z_j'$ instead of z_j'. If these generators are denoted by η_1' and η_2', one has, from Lemma 5.4,

$$\begin{aligned}\eta_1 &= d_1\eta_1' + d_1'\eta_2' + \tilde{d}_1\tilde{\eta}\\ \eta_2 &= d_2'\eta_2' + \tilde{d}_2\tilde{\eta},\end{aligned}$$

since $\deg(\eta_1|_{\operatorname{Arg} z_1=0};\ B^H \cap \operatorname{Arg} z_1 = 0) = 3$, thus, $d_1 = 1$. Also, $\eta_2|_{\operatorname{Arg} z_1=0}$ is non-zero and $\deg(\eta_2;\ B^H \cap \operatorname{Arg} z_2 = 0) = 3$, hence $d_2' = 1$. One may then express η_1' and η_2' in terms of η_1 and η_2 (and $\tilde{\eta}$) and choose η_1 and η_2 as generators. One has the relations

$$\begin{aligned}3\eta_1 + d_2\eta_2 + \tilde{d}_1\tilde{\eta} &= 0\\ 3\eta_2 + \tilde{d}_2\tilde{\eta} &= 0.\end{aligned}$$

On the other hand, one may choose $\mathcal{C}' = \{|z_2| \le 2,\ 0 \le \operatorname{Arg} z_2 \le 2\pi/9\}$ with the generators

$$\begin{aligned}\eta' &= (2t+1-2|z_2|, z_1, z_1', (\mu + i(|z_2|-1))z_2, z_2'),\\ \tilde{\eta}' &= (\varepsilon^2 - |z_2^9+1|, z_1, z_1', (\mu + i(2t-1))(z_2^9+1)z_2, z_2'),\end{aligned}$$

since the same argument about the generator given in Lemma 5.3 is valid: in fact, the term $(P_j+1)z_j^{l_j}$, with $k_j = 1$, is useful only when $l_j > 1$; if $l_j = 1$, one replaces it by z_j.

One has the relation

$$9\eta' + \tilde{d}\tilde{\eta}' = 0,$$

and looking at $\deg(\eta_j;\ B^H \cap \operatorname{Arg} z_2 = 0)$, for $j = 1, 2$, one has

$$\begin{aligned}\eta_1 &= 2\eta' + d_1'\tilde{\eta}',\\ \eta_2 &= 3\eta' + d_2'\tilde{\eta}_1',\\ \tilde{\eta} &= \tilde{d}'\eta'.\end{aligned}$$

(The last relation comes from the fact that $\tilde{\eta}$ is not zero on $\operatorname{Arg} z_2 = 0$).

Now, if one forgets the action of Γ, one obtains maps from $\mathbb{R}^{10}$ into $\mathbb{R}^9$, hence elements of $\Pi_9(S^8)$, which is generated by the suspension of the Hopf map.

For η_2 and $\tilde{\eta}$ one may take small neighborhoods of the three zeros of z_1^3+1 and get that

$$\begin{aligned}\eta_2 &= 3[(2t+1-2|z_2|, z_1, z_1', (\mu + i(|z_2|-1))z_2, z_2')]\\ \tilde{\eta} &= 3[(\varepsilon^2 - |z_2^3+1|, z_1, z_1', (\mu + i(2t-1))(z_2^3+1)z_2, z_2')].\end{aligned}$$

In the first map one may replace $|z_2|-1$ by $(1-\tau)(|z_2|-1)+\tau(t-1/2)$ and then, in the first component, $2t+1$ by $(1-\tau)(2t+1)+2\tau$. On the other hand, near the three zeros of z_2^3+1, one obtains

$$\tilde{\eta} = 9[(\varepsilon^2 - |z_2 - z_2^0|, z_1, z_1', \lambda(z_2 - z_2^0), z_2')].$$

Thus, $\eta_2 = 3\eta$, $\tilde{\eta} = 9\eta$, where η is the suspension of the Hopf map.

The same argument yields $\eta' = \eta$, $\tilde{\eta}' = 9\eta$ in $\Pi_9(S^8)$. From here one obtains $\tilde{d}_2 = 1$, $\tilde{d} = 1$, $\tilde{d}' = 1$, $d_2' = 0$.

Substitution of these values in the relation for $3\eta_1$ yields

$$(6 + 3d_2)\eta' + (3d_1' + \tilde{d}_1)\tilde{\eta} = 0,$$

hence $6 + 3d_2 = 9k$ and $3d_1' + \tilde{d}_1 = k$. Thus, $d_2 = 3k - 2$ and two values of d_2 differ by 3, which, given the relation $3\eta_2 + \tilde{\eta} = 0$, changes only the second term. Hence, one may take $k = 1$ and get $d_1' + \tilde{d}_1 = 1$ and

$$\begin{aligned} 3\eta_1 + \eta_2 + \tilde{d}_1\tilde{\eta} &= 0, & 9\eta' + \tilde{\eta} &= 0, & \eta_1 &= 2\eta' + d_1'\tilde{\eta} \\ 3\eta_2 + \tilde{\eta} &= 0, & \tilde{\eta}' &= \tilde{\eta}, & \eta_2 &= 3\eta'. \end{aligned}$$

In order to compute the class of η_1 in $\Pi_9(S^8)$, one may perform the following sequence of deformations:

1. $$\begin{pmatrix} (1-\tau)\tilde{\lambda} & \tau \\ -\tau\tilde{\lambda}z_2 & (1-\tau)z_2 \end{pmatrix} \begin{pmatrix} z_1 \\ z_1^2 z_2^3 + 1 \end{pmatrix}$$
   where $\tilde{\lambda} = \mu + i(|z_1|^2 - 1)$. If $\tilde{\lambda}z_2 \neq 0$, then $z_1 = 0$ and η_1 has no zeros. If $\tilde{\lambda} = 0$, then $z_1^2 z_3^2 + 1 = 0$ and $|z_1| = |z_2| = 1$, $t = 1/2$, hence the deformation is valid.

2. In $\tilde{\lambda}$ change $|z_1|^2 - 1$ to $(1-\tau)|z_1|^2 + \tau(t + 1/2)^6 - 1$: on a zero, one has $|z_2| = |z_1|^{-2/3}$, $|z_1 z_2| = t + 1/2 = |z_1|^{1/3}$, hence $|z_1| = 1$ and the zeros are inside B^H.

3. If $x = t + 1/2$, then $x^6 - 1 = (x-1)(x^5 + x^4 + x^3 + x^2 + x + 1)$ and the second term may be deformed linearly to 6 and then to 1.

4. Deform linearly $2t + 1$ to 2 in the first component. One has obtained the map
   $$(1 - |z_1 z_2|, \lambda z_1 z_2, z_1^2 z_2^3 + 1).$$

5. Replace $z_1^2 z_2^3 + 1$ by $z_1^2 z_2^2 z_2 + 1 - \tau + \tau(z_1 z_2)^2(\bar{z}_1\bar{z}_2)^2$, where on a zero, with $|z_1 z_2| = 1$, one obtains $|z_2| = 1$.

6. The rotation
   $$\begin{pmatrix} 1-\tau & \tau z_2^2 \\ -\tau z_1^2 & (1-\tau)z_1^2 z_2^2 \end{pmatrix} \begin{pmatrix} \lambda z_1 z_2 \\ z_2 + (\bar{z}_1\bar{z}_2)^2 \end{pmatrix}$$
   gives the map
   $$(1 - |z_1 z_2|, z_2^3 + \bar{z}_1^2|z_2|^4, -\lambda z_1^3 z_2).$$

7. The deformation $z_2^3 + \bar{z}_1^2|z_2|^2((1-\tau)|z_2|^2 + \tau)$ is valid, since on a zero, with $|z_1 z_2| = 1$, one would obtain

$$|z_2|^3 = (1-\tau)|z_2|^2 + \tau, \quad \text{i.e.,} \quad (|z_2| - 1)(|z_2|^2 + \tau|z_2| + \tau) = 0,$$

which has a unique zero at $|z_2| = 1$.

8. Deform $1 - |z_1 z_2|$ to $1 - (1-\tau)|z_1 z_2| - \tau|z_2|^{3/2}$, which is valid since a zero of the second component gives $|z_2| = |z_1|^2$ or $z_2 = 0$.

9. Replace $z_2^3 + \bar{z}_1^2|z_2|^2$ by $z_2^3 + \bar{z}_1^2((1-\tau)|z_2|^2 + \tau)$.

10. Deform $1 - |z_2|^{3/2}$ to $\tau(|\lambda| - 1) + (1-\tau)(1 - |z_2|^{3/2})$: on a zero of the map one has either $z_1^3 z_2 = 0$ and then $z_1 = z_2 = 0$ from the second equation. On the boundary of B, one would have $|\lambda| = 2$ and the above expression is positive. The other possibility is $\lambda = 0$, then, if $|z_1| = 2$, one gets $|z_2| = |z_1|^{2/3} = 2^{2/3} > 1$ and, if $|z_2| = 2$, one has also $|z_2| > 1$ and the above deformation is negative, hence the deformation is valid. One has arrived at the map

$$(|\lambda| - 1, z_2^3 + \bar{z}_1^2, -\lambda z_1^3 z_2).$$

11. Replace z_1^3 by $z_1^3 - \tau$, obtaining, for $\tau = 1$, the following zeros:

    (a) $|\lambda| = 1$, $z_2 = 0$, $z_1 = 0$, where the map is locally deformable to $(|\lambda| - 1, \bar{z}_1^2, \lambda z_2)$, which is -2η, where η is the suspension of the Hopf map.

    (b) $|\lambda| = 1$, $z_1 = z_1^0$ with $z_1^0 = 1$ or $e^{2\pi ki/3}$, $z_2 = -e^{-4\pi ki/9}$, $k = 0, \dots, 8$, where the map is locally deformable to $(|\lambda| - 1, z_2 - z_2^0, -\lambda(z_1 - z_1^0))$, i.e., to η.

12. Since the additivity for the degree in $\Pi_9(S^8)$ is valid one obtains that the class of η_1 in $\Pi_9(S^8)$ is 7η. i.e., since this group is $\mathbb{Z}_2$, η_1 is η in this group and $\tilde{d}_1 = 0$, $d_1' = 1$. The relations are

$$\begin{aligned} 3\eta_1 + \eta_2 &= 0, \quad 9\eta' + \tilde{\eta} = 0, \quad \eta_1 = 2\eta' + \tilde{\eta} \\ 3\eta_2 + \tilde{\eta} &= 0, \quad \eta_2 = 3\eta'. \end{aligned}$$

    Thus, choosing η' as the generator, one obtains that $\Pi(H) \cong \mathbb{Z}_{18}$.

Note that the other isotropy subgroup is $K = \mathbb{Z}_3$, with $V^K = V^\Gamma \times \{(z_1, z_1')\}$. For $\Pi(K)$, one has the generators

$$\begin{aligned} \eta_0 &= (2t + 1 - 2|z_1|^2, (\mu + i(|z_1|^2 - 1))z_1, z_1', z_2, z_2') \\ \tilde{\eta}_0 &= (\varepsilon^2 - |z_1^3 + 1|, \lambda(z_1^3 + 1)z_1, z_1', z_2, z_2') \end{aligned}$$

with the relations

$$3\eta_0 + d\tilde{\eta}_0 = 0, \quad 2\tilde{\eta}_0 = 0.$$

By following the preceding deformations, it is easy to show that, in $\Pi_9(S^8)$, one has $\tilde{\eta}_0 = 3\eta$ and $\eta_0 = \eta$ and then $d = 1$.

Thus, $\Pi(K) \cong \mathbb{Z}_6$.

If one adds X_0, with dim $X_0 \geq 3$, one gets

$$\Pi^{\Gamma}_{S^V}(S^W) \cong \mathbb{Z}_2 \times \mathbb{Z}_6 \times \mathbb{Z}_{18}.$$

Note that $\deg(\eta_1; B^H \cap \{\operatorname{Arg} z_2 = 0\}) = -2$.

In fact, the zeros of η_1 on this set are for $\mu = 0, t = 1/2, z_1 = \pm i, z_2 = 1$ and one may perform the following deformations:

1. $(z_1^2 z_2^3 + 1)((1-\tau)z_2 + \tau)$, since $z_2 \geq 0$.

2. $\mu + i((1-\tau)|z_1| + \tau(t+1/2)^3 - 1)$, since, on a zero, one has $|z_2| = |z_1|^{-2/3}$ and $|z_1 z_2| = |z_1|^{1/3} = t + 1/2$. Deform next $(t+1/2)^3 - 1$ to $t - 1/2$ and $2t + 1 - 2|z_1 z_2|$ to $2(1 - |z_1 z_2|)$ and next to $1 - |z_1 z_2|$. One has obtained the map
$$(1 - |z_1 z_2|, \lambda z_1, z_1^2 z_2^3 + 1).$$

3. The deformation $(1-\tau)(1 - |z_1 z_2|) + \tau(|z_2|^{1/2} - 1)$ is valid (again, on a zero, one has $|z_1| = |z_2|^{-3/2}$ and $|z_1 z_2| = |z_2|^{-1/2}$).

4. Multiplying the first component by $|z_2|^{1/2} + 1$ and deforming z_2^3 to 1, one has the map
$$(z_2 - 1, \lambda z_1, z_1^2 + 1)$$
which has degree -2.

   However, one has that $\deg(3\eta_1 + \eta_2; B^H \cap \{\operatorname{Arg} z_2 = 0\}) = -6 + 3 = -3$, i.e., this justifies the second part of Remark 5.1, that the relation $3\eta_1 + \eta_2 = 0$ is valid in $\Pi^{\Gamma}_{S^V}(S^W)$ but not in $B^H \cap \{\operatorname{Arg} z_2 = 0\}$.

Nevertheless, there is one important case where one may compute the coefficients of the relations from the degrees of restrictions. That is

Theorem 5.3. *Assume that $V = \mathbb{R} \times W$ and $k_j = m_j = |\Gamma/H_j|$ for all j's with $k_j > 1$. Then $[F]_\Gamma = \sum d_j[\eta_j]_\Gamma + \tilde{d}[\tilde{\eta}]_\Gamma$, where*

$$d_j = \deg(F; B^H \cap \{\operatorname{Arg} z_j = 0\}) \Big/ \Big(\prod_{i \neq j} m_i\Big)$$

and the relations for $\Pi(H)$ are

$$\begin{aligned} m_j([\eta_j]_\Gamma + [\tilde{\eta}]_\Gamma) &= 0 \\ 2[\tilde{\eta}]_\Gamma &= 0. \end{aligned}$$

In particular, $[F]_\Gamma = 0$ if and only if $d_j = a_j m_j$ and $\tilde{d}$ has the parity of $\sum d_j$.

Proof. The condition $k_j = m_j$ means that one has s coordinates with the same k_j's regardless of the order in the construction of $\mathcal{C}$, the other variables are just a suspension. Hence, one may take $P_j = x_j^{m_j}$ and the generators η_j are the same, independently of the order. Thus, if one chooses x_j as the first coordinate in $\mathcal{C}$, then d_j will have the form stated after using Lemma 5.2.

On the other hand, taking x_j as the last coordinate in $\mathcal{C}$, one has the relation

$$m_j[\eta_j] + \tilde{d}_j[\tilde{\eta}_j] = 0$$

where

$$\begin{aligned}\eta_j &= \Big(2t + 1 - 2|z_j| \prod |x_i|, X_0, (Q_i - 1)y_i, (x_i^{m_i} + 1)x_i, \tilde{\lambda} z_j, z_j'\Big)\\ \tilde{\eta}_j &= \Big(\varepsilon^2 - \prod_{i \neq j} |x_i||x_j^{m_j} + 1|^2, X_0, (Q_i - 1)y_i, (x_i^{m_i} + 1)x_i, \lambda(z_j^{m_j} + 1)z_j, z_j'\Big).\end{aligned}$$

Note that, by construction, $\tilde{\eta}_j$ is non-zero on $\partial\mathcal{C}$ and has the class of η in $\mathcal{C}$, this implies that all $\tilde{\eta}_j$'s are homotopic on $\partial\mathcal{C}$ and Γ-homotopic on ∂B^H to a single map $\tilde{\eta}$.

Note also that, in η_j, one may perform the deformations, in the first component and in λ, given by

$$(1 - \tau)\Big(2t - 1\Big) + 2(1 - |z_j| \prod |x_i|), \mu + i((1 - \tau)(|z_j| - 1) + \tau(2t - 1)\Big):$$

on a zero of the map one has $|x_i| = 1$ and the above components can be written as

$$\begin{pmatrix} 1 - \tau & -2 \\ \tau & (1 - \tau) \end{pmatrix} \begin{pmatrix} 2t - 1 \\ |z_j| - 1 \end{pmatrix}$$

which gives an admissible deformation. Hence, with $\lambda = \mu + i\varepsilon(2t - 1)$

$$\eta_j = \Big(1 - |z_j| \prod |x_i|, X_0, (Q_i - 1)y_i, (x_i^{m_i} + 1)x_i, \lambda z_j, z_j'\Big).$$

Denote by $\mathcal{A}$ the vector $(X_0, (Q_i - 1)y_i, \{(x_i^{m_i} + 1)x_i\}_{i \neq j})$ and by $A = \prod |x_i|$. For a lighter notation we shall drop the index j in z_j, z_j' and m_j. Consider then the map

$$F = (1 - A(|z| + |z'|), \mathcal{A}, \lambda z', \lambda^{m-1} z^{(m-1)^2})$$

which is equivariant since $z^{m^2 - 2m}$ is invariant. Take the equivariant deformation

$$((1 - \tau)\lambda z' - \tau \bar{z}^{m-1}, \tau \bar{z}'^{\,m-1} + (1 - \tau)\lambda^{m-1} z^{(m-1)^2}).$$

On a zero, conjugate the first equation and take its $(m - 1)$-power. One obtains the system

$$\begin{pmatrix} (1 - \tau)^{m-1} \bar{\lambda}^{m-1} & -\tau^{m-1} \\ \tau & (1 - \tau)\lambda^{m-1} \end{pmatrix} \begin{pmatrix} \bar{z}'^{m-1} \\ z^{(m-1)^2} \end{pmatrix} = 0.$$

The only zero of the deformed map is for $\lambda = 0$, $\tau = 0$, $|x_i| = 1$ and $|z| + |z'| = 1$. Furthermore, for $\tau = 1$, the map has no zeros, that is F is trivial.

One may also perform the equivariant deformation

$$((1-\tau)\lambda z' + \tau\lambda^m z^{(m-1)^2}, -\tau z' + (1-\tau)\lambda^{m-1} z^{(m-1)^2})$$

which deforms F to

$$F' = (1 - A(|z| + |z'|), \mathcal{A}, \lambda^m z^{(m-1)^2}, -z').$$

This map is non-zero on the faces of $\mathcal{C}$, except for $\operatorname{Arg} z = 0$, on which it has degree m. Hence,

$$0 = [F] = m_j[\eta_j] + \tilde{d}_j[\tilde{\eta}].$$

One may also rotate λ^m and obtain

$$(1 - A(|z| + |z'|), \mathcal{A}, \lambda^m z', z^{(m-1)^2}).$$

Replace $z^{(m-1)^2}$ by $(z^m + \tau 2^{-m})^{m-2} z$. For $\tau = 1$, one has a map with zeros at $|x_i| = 1$, $i \neq j$, $\lambda = 0$ and either $z = 0$, $|z'| = 1$, or $|z| = 1/2 = |z'|$.

Divide B^H into two invariant sets: $B_1 = B^H \cap \{|z| < 1/4\}$ and its complement B_2. One may compute the Γ-degree on each one and, from Remark 2.3 of Chapter 2, one has

$$\deg_\Gamma(F; B^H) = \deg_\Gamma(F; B_1) + \deg_\Gamma(F; B_2).$$

Now, $\deg_\Gamma(F; B_1) = \Sigma_0[F'|_{B_1}]_\Gamma$. But, on B_1 one may deform z^m to 0, rotate back λ^m and obtain the map

$$(1 - A(|z| + |z'|), \mathcal{A}, \lambda^m z, -z') = m[\eta_j].$$

Furthermore, from the form of the generators, Σ_0 is an isomorphism.

On ∂B_2, one may deform linearly the first component to $(1/2 - A(|z^m + 2^{-m}| + |z'|)$ and deform $z^m + 2^{-m}$ to $z^m + 1$. Rotate back λ^m to get the map

$$\tilde{F} = (1/2 - A(|z^m + 1| + |z'|), \mathcal{A}, \lambda^m (z^m + 1)^{m-2} z, -z')$$

once one has noticed that any disk with center at a point with $|z| = 1$ and with $1/2 - |z^m + 1| = 0$ does not intersect ∂B_2.

The map $\tilde{F}$ has no zeros on the faces of $\mathcal{C}$, hence its class is a multiple of $[\tilde{\eta}]_\Gamma$, which is given by its ordinary class with respect to $\mathcal{C}$, where the set $(Q_i - 1, z_i^{m_i} + 1)$ has just one zero. It is easy to see that this class is $m(m-2)$-times the Hopf map in $\mathbb{Z}_2$. Thus, $[\tilde{F}]_\Gamma = m[\tilde{\eta}]_\Gamma$, proving the theorem, since for the last point, one has that $\ker \Pi$, in the proof of Theorem 5.2, is given by multiples of m_j, with the stated congruence for d (one has $d_{ij} = 0$, for $i \neq j$, and $\tilde{d}_i = m_i$). □

Although Theorem 5.3 seems to be a very special case, it will enable us to give another description of $\Pi(H)$.

Assume $V = \mathbb{R} \times W$ and $\Gamma/H \cong \mathbb{Z}_{p_1} \times \cdots \times \mathbb{Z}_{p_m}$, generated by $\gamma_1, \ldots, \gamma_m$. Let $X = \{Z_1, Z_1', \ldots, Z_m, Z_m'\}$ be a new Γ-space with the following action: $\gamma_j Z_j = e^{2\pi i/p_j} Z_j$, $\gamma_j Z_i = Z_i$, $i \neq j$, and Z_j' is the duplicate of Z_j.

For any F in $\Pi(H)$ one has the suspension $\Sigma^X F = (F, X)$ from $V^H \times X^H$ into $W^H \times X^H$. If one takes $\mathcal{C} \times B_X^H$ as fundamental cell, then any Γ-map G from $V^H \times X^H$ into $W^H \times X^H$ which is non-zero on the ball in $(V \times X)^K$, for all $K > H$, is classified by the formula

$$[G]_\Gamma = \sum d_j [\Sigma^X \eta_j]_\Gamma + \tilde{d} [\Sigma^X \tilde{d}\eta]_\Gamma,$$

since the suspensions $\Sigma^X \eta_j$ and $\Sigma^X \tilde{\eta}$ are clearly the generators for the group $\Pi(H)$ corresponding to $V \times X$. This formula proves also that $\Sigma^X \Pi(H) \cong \Pi(H)$.

But one may choose the cell $\mathcal{C}'$ given by $\{0 \leq \operatorname{Arg} Z_j < 2\pi/p_j\}$, with the generators

$$\begin{aligned} \Sigma^V \eta_j' &= \Big(1 - \prod |Z_i|, X_0, \{x_i\}, (Z_i^{p_i} + 1) Z_i, Z_i', \lambda Z_j, Z_j'\Big) \\ \Sigma^V \tilde{\eta}' &= \Big(\varepsilon^2 - \prod_{i<m} |Z_i| |Z_m^{p_m} + 1|, X_0, \{x_i\}, \{(Z_i^{p_i} + 1) Z_i\}_{i<m}, \lambda (Z_m^{p_m} + 1) Z_m\Big). \end{aligned}$$

Then, $[G]_\Gamma = \sum d_j' [\Sigma^V \eta_j']_\Gamma + \tilde{d}' \, [\Sigma^V \tilde{\eta}']$ as in Theorem 5.3. Furthermore, the relations of that theorem hold and Σ^V is an isomorphism. We have proved the following

Theorem 5.4. *If $V = \mathbb{R} \times W$ and $\Gamma/H \cong \mathbb{Z}_{p_1} \times \cdots \times \mathbb{Z}_{p_m}$, then any F in $\Pi(H)$ is given by*

$$\Sigma^X [F]_\Gamma = \sum d_j' [\Sigma^V \eta_j']_\Gamma + \tilde{d}' [\Sigma^V \tilde{\eta}']_\Gamma,$$

with the relations

$$\begin{aligned} 2[\tilde{\eta}']_\Gamma &= 0 \\ p_j([\eta_j']_\Gamma + [\tilde{\eta}']_\Gamma) &= 0. \end{aligned}$$

For instance, one may have $p_j = k_j$ coming from the fundamental cell.

Note that in order to compute d_j' with the formula of Theorem 5.3, one has to perturb $\Sigma^X F$ so that it has no zeros on the edges of $\mathcal{C}'$, that is for $Z_j = 0$. However, $[F]_\Gamma = 0$ if and only if $d_j' = a_j p_j$ and $\sum d_j' + \tilde{d}'$ is even, since Σ^X and Σ^V are isomorphisms.

One may give a better presentation of the above relations. For example, let

$$\begin{aligned} [\eta_j]_\Gamma &= [\eta_j']_\Gamma + [\tilde{\eta}']_\Gamma, \quad j = 1, \ldots, m \\ [\eta_0]_\Gamma &= [\tilde{\eta}']_\Gamma. \end{aligned}$$

Then $\Pi(H)$ is presented by $[\eta_j]_\Gamma$, $j = 0, \ldots, m$, with the relations

$$p_j [\eta_j]_\Gamma = 0, \quad p_0 = 2.$$

Theorem 5.5. *If* $V = \mathbb{R} \times W$, *then* $\Pi(H) \cong Z_2 \times \Gamma/H$.

Proof. Any F in $\Pi(H)$ is given by $\Sigma^X[F]_\Gamma = \sum d_j'[\Sigma^V \eta_j']_\Gamma + \tilde{d}'[\Sigma^V \tilde{\eta}']_\Gamma = \sum d_j'[\Sigma^V \eta_j]_\Gamma + (\tilde{d}' - \sum d_j')[\Sigma^V \eta_0]_\Gamma$. In particular $[F]_\Gamma = 0$ if and only if d_j' is a multiple of p_j, for $j = 0, \dots, m$. Thus, each η_j generates a cyclic group of order p_j.

Note however that the generators η_j are more difficult to write down explicitly. □

Another presentation of $\Pi(H)$ is the following

Theorem 5.6. *If* $V = \mathbb{R} \times W$ *and* $\Gamma/H \cong \mathbb{Z}_{p_1} \times \cdots \times \mathbb{Z}_{p_m}$, *then* $\Pi(H) \cong \mathbb{Z}_{q_0} \times \cdots \times \mathbb{Z}_{q_m}$, *with* $q_0 = (2 : p_1 : \dots p_m)$ *the largest common divisor,* q_m *is the least common multiple of* 2, $p_1, \dots, p_m$. *Furthermore,* $q_j = h_{j+1}/h_j$ *for* $j = 0, \dots, m$, *where* $h_0 = 1$, h_j *is the largest common divisor of all possible products of* j *among the numbers* $p_0, \dots, p_m$.

Proof. Given p_i, p_j, let $p = (p_i : p_j)$, then there are k_i, k_j such that $p_i k_i + p_j k_j = p$. Let

$$\begin{aligned} \xi_i &= (p_i/p)\eta_i - (p_j/p)\eta_j \\ \xi_j &= k_j \eta_i + k_i \eta_j. \end{aligned}$$

Then, $(p_i/p)\xi_j = \eta_j + k_j \xi_i$, $(p_j/p)\xi_j = \eta_i - k_i \xi_i$. Furthermore,

$$p\xi_i = p_i \eta_i - p_j \eta_j = 0, \quad (p_i p_j/p)\xi_j = (p_j k_j/p) p_i \eta_i + (p_i k_i/p) p_j \eta_j = 0.$$

Without taking into account the relations, one may express, on the basis ξ_i, ξ_j, these equations in the form

$$\begin{pmatrix} p & 0 \\ 0 & p_i p_j/p \end{pmatrix} = \begin{pmatrix} 1 & -1 \\ p_j k_j/p & p_i k_i/p \end{pmatrix} \begin{pmatrix} p_i & 0 \\ 0 & p_j \end{pmatrix} \begin{pmatrix} k_i & p_j/p \\ -k_j & p_i/p \end{pmatrix},$$

where the non-diagonal matrices have determinant equal to 1, i.e., they are invertible over $\mathbb{Z}$. Thus, one may replace η_i, η_j by ξ_i, ξ_j and (p_i, p_j) by $p = (p_i : p_j)$ and the least common multiple of p_i and p_j. Note that p may be 1 and that if $p = \min(p_i, p_j)$, say $p = p_i$, then one may take $k_i = 1$, $k_j = 0$ and the change of variables does not change the relations.

Continuing this process, it is easy to see that one arrives at a new set of generators $\zeta_0, \dots, \zeta_m$ and relations $q_j \zeta_j = 0$, where $q_0 = (p_0 : p_1 : \cdots : p_m)$, $q_m = l.c.m.(p_0, p_1, \dots, p_m)$, q_j divides q_{j+1} and one has matrices M and N, invertible over $\mathbb{Z}$, such that

$$Q = MPN,$$

where

$$Q = \operatorname{diag}(q_0, q_1, \dots, q_m) \quad \text{and} \quad P = \operatorname{diag}(p_0, p_1, \dots, p_m).$$

This is the content of the Fundamental Theorem for abelian groups [Jo, p. 57]. The integers q_j are called the invariant factors of P.

If $h_i(A)$ is the greatest common divisor of the principal $(i \times i)$-minors of a matrix A with integers entries, then one may prove that, if $Q = MPN$, with M and N invertible over $\mathbb{Z}$, then $h_i(P) = h_i(Q)$.

Furthermore, since q_j divides q_{j+1}, we have that $h_i(Q) = \prod_{j<i} q_j$ and $h_i(P)$ is the greatest common divisor of all possible products of i of the p_j's. Hence, $q_i = h_{i+1}(P)/h_i(P)$. Since the above results are true for arbitrary P and Q, they will also hold for the triangular matrix T of Theorem 5.2, where $h_i(T) = h_i(\mathrm{diag}(2, k_1, \dots, k_s))$, since principal minors of triangular matrices reduce to products of terms on the diagonal. □

Corollary 5.1. *If $V = \mathbb{R} \times W$ and $\Gamma/H \cong \mathbb{Z}_n$, then*

$$\Pi(H) \cong \begin{cases} \mathbb{Z}_2 \times \mathbb{Z}_n, & \text{if } n \text{ is even} \\ \mathbb{Z}_{2n}, & \text{if } n \text{ is odd.} \end{cases}$$

Corollary 5.2. *If $V = \mathbb{R} \times W$ and $\Gamma/H \cong \mathbb{Z}_{p_1} \times \cdots \times \mathbb{Z}_{p_m}$, where any two p_i, p_j are relatively prime and odd, then*

$$\Pi(H) \cong \mathbb{Z}_{2|\Gamma/H|}.$$

Note that if one applies the same presentation to Γ/H in this last corollary, one has that $\Gamma/H \cong \mathbb{Z}_{|\Gamma/H|}$. In fact, from a purely algebraic point of view, one may reformulate Theorem 5.5, i.e., $\Pi(H) \cong \mathbb{Z}_2 \times \Gamma/H$, in the following form (losing the starting point of the action on W and the construction of the fundamental cell).

Theorem 5.7. *If $V = \mathbb{R} \times W$ and $\Gamma/H \cong \mathbb{Z}_{p_1} \times \cdots \times \mathbb{Z}_{p_m}$, where p_j divides p_{j+1} and $p_0 = 1$, then*

$$\Pi(H) \cong \mathbb{Z}_{p_1} \times \cdots \times \mathbb{Z}_{p_{j_0-1}} \times \mathbb{Z}_{2p_{j_0}} \times \mathbb{Z}_{p_{j_0+1}} \times \cdots \times \mathbb{Z}_{p_m},$$

where j_0 is the largest index j with p_j odd.

Proof. When taking products of j terms among $2, p_1, \dots, p_m$, one gets as largest common divisor $2p_1 \dots p_{j-1}$ or $p_1 \dots p_j$, if one takes 2 in the first case, since p_i divides p_{i+1}. Hence

$$h_j = \begin{cases} p_1 \dots p_{j-1}, & \text{if } p_j \text{ is odd} \\ 2p_1 \dots p_{j-1}, & \text{if } p_j \text{ is even.} \end{cases}$$

Since $q_j = h_{j+1}/h_j$, with $h_0 = 1$, $p_0 = 1$, one gets $q_j = h_j$ if p_{j+1} is odd (hence p_j is also odd), $q_{j_0} = 2p_{j_0}$ since p_{j_0+1} is even and p_{j_0} is odd, and $q_j = p_j$ for $j > j_0$ since then p_{j+1} and p_j are even. □

Example 5.2. The relations of the last theorems are a good source of problems where there is no bifurcation, i.e., with no non-trivial solutions. For instance, if $\mathbb{Z}_m$ acts on $\mathbb{C}^2$, then the equivariant map

$$\begin{pmatrix} \lambda z_1 \\ \lambda^{m-1} z_2^{(m-1)^2} \end{pmatrix} + \tau \begin{pmatrix} -\bar{z}_2^{m-1} \\ \bar{z}_1^{m-1} \end{pmatrix} = 0,$$

where $\lambda = \mu + i\nu$ and $\tau = (|z_1|^2 + |z_2|^2)$, has no solution but $z_1 = z_2 = 0$: use the argument of Theorem 5.3.

The action of Γ/H on the auxiliary space is, in a certain sense, arbitrary. In studying Hopf bifurcation for non-autonomous problems we shall encounter the following situation: Let V be a $\mathbb{Z}_p$-space of the form (μ, z_1, z_m), with action of $\mathbb{Z}_p$ as $\exp(2\pi ik/p)$ on z_1 and as $\exp(2\pi imk/p)$ on z_m, with m and p relatively prime. One has then the following generators, with $\lambda = \mu + i(2t - 1)$

$$\eta_1 = (1 - |z_1|, \lambda z_1, z_m), \tilde{\eta} = (\varepsilon - |z_1^p + 1|, \lambda(z_1^p + 1)z_1, z_m)$$

with the relations

$$p(\eta_1 + \tilde{\eta}) = 0, \quad 2\tilde{\eta} = 0.$$

One could have taken instead η_m and $\tilde{\eta}_m$ which are defined as η_1 and $\tilde{\eta}$ but with z_1 and z_m interchanged.

Proposition 5.1. *There is an integer n such that $nm \equiv 1$, modulo p, and $|n|$ is odd, with the property*

$$\eta_m = n\eta_1, \quad \tilde{\eta}_m = \tilde{\eta}.$$

Proof. Since m and p are relatively prime, there is n such that $nm \equiv 1$, modulo p, with $n > 0$, or else $nm + \alpha p = 1$. If p is even, then n and m are odd, while if p is odd and n is even, replace n by $n - p$ (and α by $\alpha + m$), with $|n - p|$ odd.

Recall, from Theorem 5.3, that any $\mathbb{Z}_p$-map f such that $f|_{z_1=0}$ is non-zero, can be written as

$$[f]_\Gamma = d[\eta_1]_\Gamma + \tilde{d}[\tilde{\eta}]_\Gamma,$$

where $d = \deg(f; B \cap \{\operatorname{Arg} z_1 = 0\})$.

Take $f = (\varepsilon - |z_m^p + 1|, z_1 - z_m^n, \lambda(z_m^p + 1)z_m)$ which is $\mathbb{Z}_p$-deformable to $\tilde{\eta}_m$. Since, on a zero, z_m is close to a p-th root of $e^{\pi i}$, hence the argument of z_m^n will be close to $\pi n(1 + 2k)$, which is not close to $2\pi l$, since n is odd. Thus, f is non-zero for $\operatorname{Arg} z_1 = 0$ and one has $[\tilde{\eta}_m]_\Gamma = \tilde{d}[\tilde{\eta}]_\Gamma$. Since one may interchange z_1 and z_m (and z_m^n by z_1^m), one has $[\tilde{\eta}]_\Gamma = \tilde{d}_m[\tilde{\eta}_m]_\Gamma$, hence $\tilde{d}\tilde{d}_m = 1$ and one has $\tilde{\eta} = \tilde{\eta}_m$.

Consider now the maps

$$f_1 = (1 - |z_1|^2 - |z_m|^2, z_1^{mn}, \lambda z_m)$$
$$f_2 = (1 - |z_1|^2 - |z_m|^2, \lambda^n z_1^{mn}, z_m).$$

From the above rules, one has $[f_1] = mn[\eta_m] + d_1[\tilde{\eta}_m]$ and $[f_2] = n[\eta_1] + d_2[\tilde{\eta}]$.

Now, on $\partial B^2 \times \partial B$, one has the $\mathbb{Z}_p$-deformations

$$((1-\tau)z_1^{mn} + \tau\lambda^n z_m^n, (1-\tau)\lambda z_m - \tau z_1^m)$$

(which, by taking the expression $(1-\tau)\lambda z_m = \tau z_1^m$ to the n-th power, is zero only if $z_1 = \lambda z_m = 0$), hence f_1 is $\mathbb{Z}_p$-homotopic to

$$(1 - |z_1|^2 - |z_m|^2, \lambda^n z_m^n, -z_1^m).$$

Then, the rotation $((1-\tau)z_m^n + \tau z_1^{nm}, -(1-\tau)z_1^m + \tau z_m)$ gives that f_1 is $\mathbb{Z}_p$-homotopic to f_2. Hence,

$$\begin{aligned} m_n\eta_m + d_1\tilde{\eta} &= (1+kp)\eta_m + d_1\tilde{\eta} = \eta_m + \tilde{d}_1\tilde{\eta} = n\eta_1 + d_2\tilde{\eta} \\ \text{and } \eta_m &= n\eta_1 + d\tilde{\eta}. \end{aligned}$$

If p is odd, then from $p\eta_m = -p\tilde{\eta}$, $p\eta_1 = -p\tilde{\eta}$, one gets $pd \equiv p(n-1)$, [2]. But, since n is odd, one gets $d \equiv 0$, [2].

If p is even, then m and n are odd. Let $mn = 1 + kp$ and on $\partial\{(\lambda, z_1, z_1', z_m) : |\lambda| \le 1, |Z| \le 2\}$, consider the following maps, which are $\mathbb{Z}_p$-homotopic to a suspension of f_1:

$$\begin{aligned} f_3 &= (1 - |Z|^2, z_1', z_1(z_1^{kp} + \tau 2^{-kp}), \lambda z_m) \\ f_4 &= (1 - |Z|^2, \lambda^n z_1', z_1(z_1^{kp} + \tau 2^{-kp}), z_m) \end{aligned}$$

(the second map comes from a rotation of z_1^{mn} and z_1' in f_2).

Decompose the set $\{Z : |z_1| \le 2\}$ in $B_1 = \{z_1 : |z_1| \le 1/4\}$ and its complement B_2. The $\mathbb{Z}_p$-degree of the above maps is the sum of the degrees on B_1 and B_2.

On B_1 one may deform z_1^{kp} to 0 and obtain η_m for f_3 and $n\eta_1$ for f_4. On B_2, one may use the homotopy

$$(1-\tau)(1 - |Z|^2) + \tau(\varepsilon^{2kp} - |z_1^{kp} + 2^{-kp}|^2 - |z_1'|^2 - |z_m|^2),$$

where ε is so small that any disk, with center at a point with $z_1' = 0$, $z_m = 0$, $|z_1| = 1/2$, and of radius ε does not intersect ∂B_2, hence, for $\tau = 1$, the degree is the same on B_2 and on the full set $\{Z : |Z| \le 2\}$. Now, for $\varepsilon < 1/2$, the maps are non-zero if $\operatorname{Arg} z_1 = 0$. Hence, the classes of the maps are multiples of $\tilde{\eta}$, which may be computed on the boundary of the fundamental cell $\{0 \le \operatorname{Arg} z_1 < 2\pi/p\}$.

There, one may deform z_1 to 1 and obtain k zeros of $z_1^{kp} + 2^{-kp}$ in the cell. It is then easy to see that, for the first map, one obtains $k\tilde{\eta}$ and $kn\tilde{\eta}$ for the second. Hence,

$$[f_3] = \eta_m + k\tilde{\eta}, \quad [f_4] = n\eta_1 + kn\tilde{\eta}.$$

From the fact that $[f_3] = [f_4]$ and $\eta_m = n\eta_1 + d\tilde{\eta}$ one gets

$$d\tilde{\eta} = k(n-1)\tilde{\eta}.$$

Since we have chosen n to be odd, we have $d \equiv 0$, [2], and the proposition is proved. □

3.6 Orthogonal maps

Recall that if the abelian group $\Gamma = T^n \times \mathbb{Z}_{m_1} \times \cdots \times \mathbb{Z}_{m_s}$ acts on the finite dimensional space V, then a Γ-orthogonal map $F(x)$ from V into itself is an equivariant map with the property that

$$F(x) \cdot A_j x = 0\ ,\ \ j = 1, \ldots, n,$$

where A_j is an infinitesimal generator for the action of T^n. (see §7 of Chapter 1). Gradients of Γ-invariant functionals are Γ-orthogonal maps. In §4 of Chapter 2, we have defined an orthogonal degree for such maps, as elements of $\Pi^{\Gamma}_{\perp S^V}(S^V)$, the abelian group of all orthogonal Γ-homotopy classes of S^V into itself. One has the following important result

Theorem 6.1. 1. $\Pi^{\Gamma}_{\perp S^V}(S^V) \cong \mathbb{Z} \times \cdots \times \mathbb{Z}$, *with one* $\mathbb{Z}$ *for each isotropy subgroup of* Γ.

2. $[F]_\perp = \sum d_H [F_H]_\perp$, *with explicit generators* F_H. *If* $d_H \neq 0$, *then* F *has a zero in* V^H.

Proof. Let F be an orthogonal Γ-map, from B into V, which is non-zero on ∂B (recall that we are including the variable t in V). We shall decompose $[F]_\perp$ by modifying it on the different isotropy subspaces.

Step 1. If $\tilde{V} = V^{T^n}$, then $[F^{T^n}]$ is an element of $\Pi^{\Gamma}_{S^{\tilde{V}}}(S^{\tilde{V}})$ and as such, one has from Corollary 2.1 and Theorem 3.2

$$[F^{T^n}] = \sum_{T^n \leq H} d_H [F_H].$$

Since $A_j x = 0$ on $\tilde{V}$, then $[F_1]_\perp \equiv [F]_\perp - [F^{T^n}, Z]_\perp$ has a non-zero orthogonal Γ-extension to B^{T^n}. Thus, $F_1(X, Z)$ may be written as $(F_1^{T^n}(X, Z), F_\perp(X, Z))$, with $F_1(X, 0) \neq 0$ and $F_\perp(X, Z)$ orthogonal to $A_j Z$.

Step 2. Recall that the action of T^n on the k-th coordinate of Z is of the form $\exp i\langle N^k, \Phi\rangle$, where $N^k = (n_1^k, \ldots, n_n^k)$. Assume, without loss of generality, that $n_1^1 \neq 0$ and define $\lambda_j = n_1^j / n_1^1$ for $j = 2, \ldots n$. Let

$$V_1 = V^{T^n} \times \{z_k : n_1^k \neq 0 \text{ and } n_j^k = \lambda_j n_1^k,\ j \geq 2\}.$$

(Of course, by removing the condition $n_1^k \neq 0$, one includes directly V^{T^n} in V_1). Then, on V_1, one has $A_j x = \lambda_j A_1 x$ and $V_1 = V^{T_1}$, where T_1 is the $(n-1)$-torus $(-\sum_2^n \lambda_j \varphi_j, \varphi_2, \ldots, \varphi_n)$. Let B_1 be the ball B^{V_1}, then the map $F_1(x) + \lambda A_1 x$ is non-zero on $\partial(I \times B_1)$, where λ is in $I = [-1, 1]$: in fact, $F_1(X, 0) \neq 0$ and, from the fact that F_1 is orthogonal to $A_1 x$ a zero of the above map is such that $F_1(x) = 0$ and $\lambda A_1 X = 0$. That is, if $Z \neq 0$, then $\lambda = 0$ and $F_1(x) + \lambda A_1 x$ defines an element of $\Pi^{\Gamma}_{S^{\mathbb{R}\times V_1}}(S^{V_1}) \cong \Pi^{\Gamma}_{S^{\mathbb{R}\times \tilde{V}}}(S^{\tilde{V}}) \times \mathbb{Z} \times \cdots \times \mathbb{Z}$, with one $\mathbb{Z}$ for each isotropy subgroup H with $\dim \Gamma/H = 1$ (see Corollary 2.1). Since $F_1^{T^n} \neq 0$, one has that

$$[F_1(x) + \lambda Ax] = 0 + \sum_{T_1 \le H < T^n} d_H[\tilde{F}_H],$$

where $\tilde{F}_H$ is given in Theorem 3.2:

$$\tilde{F}_H(\lambda, x) = \Big(2t + 1 - |z_1| \prod |x_i|, X_0, (\varepsilon(|z_1| - 1) + i\lambda)z_1, (Q_i - 1)y_i, (P_i + 1)z_i\Big),$$

where Q_i, P_i are the familiar monomials, $\varepsilon = \pm 1$ is such that the degree of $\tilde{F}_H$ on the fundamental cell is 1. Let

$$F_H(x) = \tilde{F}_H(0, x) - (\tilde{F}_H(0, x), \tilde{A}_1(x))\tilde{A}_1(x),$$

where $\tilde{A}_1(x) = A_1x/\|A_1x\|$, is as in Theorem 7.1 of Chapter 1. By construction, $F_H(x)$ is an orthogonal Γ-map. Its z_1-component is $(\varepsilon(|z_1| - 1) - i\alpha(x)n_1^1)z_1$, where $\alpha(x) = (\tilde{F}_H(0, x), \tilde{A}_1(x))/\|A_1x\|$. Furthermore, the first component of F_H is $\big(2t+1-|z_1| \prod |x_i|\big)$. Thus, the zeros of F_H are those of $\tilde{F}_H(0, x)$ and F_H defines an element of $\Pi^{\Gamma}_{\perp S^V}(S^V)$. Moreover, $\tilde{F}_H(\lambda, x)$ is Γ-homotopic to $F_H(x) + \lambda\varepsilon_1 A_1 x$, where $\varepsilon_1 = \operatorname{Sign} n_1^1$: in fact, the z_k-component of this last map is $(P_k + 1 + i(\lambda\varepsilon_1 - \alpha(x))n_1^k)z_k$, while the z_1-component has the form $(\varepsilon(|z_1| - 1) + i(\lambda\varepsilon_1 - \alpha(x))n_1^1)z_1$. Since zeros must be with $z_1 \neq 0$ (first component), one may deform $\lambda\varepsilon_1 - \alpha(x)$ to 0 in the z_k-component, $\alpha(x)$ to 0 in the z_1-component, and then $\varepsilon_1 n_1^1$ to 1 and arrive at $\tilde{F}_H(\lambda, x)$. Note that the zeros of $F_H(x) + \lambda\varepsilon_1 A_1 x$ are only at $\lambda = 0$ and with $F_H(x) = 0$, since F_H is orthogonal to $A_j x$. Hence, $F_H(x) + \lambda\varepsilon_1 A_1 x$ can be taken as a generator for $\Pi^{\Gamma}_{S^{\mathbb{R}\times V_1}}(S^{V_1})$.

Complementing F_H by the identity on $V_1^{\perp}$, one has that

$$[F_2]_{\perp} \equiv [F_1]_{\perp} - \sum_{T_1 \le H < T^n} d_H[F_H]_{\perp}$$

is orthogonal to $A_j x$ and $F_2(x) + \lambda A_1 x$, is Γ-extendable on $\partial(I \times B_1) \cup B^{T^n}$ to a non-zero Γ-map $F(\lambda, x)$ on $I \times B_1$.

Claim 6.1. *$F_2|_{V_1}$ has a non-zero orthogonal extension to B_1, i.e., $[F_2^{V_1}]_{\perp} = 0$.*

Proof. The proof follows the lines of Theorem 1.1 by working on V_1^H, for H in decreasing order. Thus, if H is maximal (hence any $K > H$ must contain T^n), one may extend $[F_2']_{\perp} = [F_1]_{\perp} - d_H[F_H]_{\perp}$ in such a way that the resulting orthogonal map is non-zero on $\partial\mathcal{C}_H$: in fact, this is true on V^K, for $K > H$, since there F_1^K is non-zero and by a dimension argument, since $\dim \partial\mathcal{C}_H = \dim V^H - 2$, as in Corollary 1.2. Thus, one may assume that $F_2'(x) + \lambda A_1 x$ is non-zero on $\partial(I \times \mathcal{C}_H)$ and has a zero extension degree, i.e., the degree with respect to $I \times \mathcal{C}_H$.

Now, on $\mathcal{C}_H$ one has that z_1 is in $\mathbb{R}^+$ and, since $F_2'(x) \neq 0$ for $z_1 = 0$, one may compute this obstruction degree on the ball $A \equiv I \times \mathcal{C}_H \cap \{z_1 > \varepsilon\}$, for some small ε. If F_2' is written as $(f_1, f_2, F_\perp)$, where $f_1 + if_2$ corresponds to the z_1-component, one may perform on ∂A the homotopy

$$F_2'(x) + \lambda(\tau A_1 x + (1-\tau)A_1 z_1).$$

In fact, taking the scalar product with $F_2'(x)$ one has $|F_2'|^2 + \lambda(1-\tau)(F_2', A_1 z_1) = 0$, on a zero of the homotopy, that is, from the orthogonality,

$$|F_2'|^2 - \lambda(1-\tau)(F_\perp, A_1 y) = |F_2'|^2 + \lambda^2(1-\tau)|A_1 y|^2 = 0$$

since on a zero of the homotopy one has $F_\perp = -\lambda\tau A_1 y$. Hence, $F_2'(x) = 0$ and $\lambda A_1 z_1 = 0$, but, since $z_1 \geq \varepsilon$, this means $\lambda = 0$, that is, the zeros of the homotopy are inside A. The resulting map $(f_1, f_2 + \lambda n_1^1 z_1, F_\perp)$ is linearly deformable on ∂A to $(f_1, \lambda, F_\perp)$, since from the orthogonality one has, for z_1 real: $f_2 z_1 = -(F_\perp, A_1 y)$, assuming $n_1^1 > 0$.

From the product theorem, one obtains that $\deg(f_1, F_\perp; \mathcal{C}_H \cap \{z_1 > \varepsilon\}) = 0$, i.e., $(f_1, F_\perp)$ has a non-zero extension, $(\tilde{f}_1, \tilde{F}_\perp)$, to $\mathcal{C}_H \cap \{z_1 \geq \varepsilon\}$. Defining, on this set, $\tilde{f}_2 = (-\tilde{F}_\perp, A_1 y)/z_1$, one obtains a non-zero orthogonal extension $\tilde{F}_2'(x)$ of $F_2'(x)$ first on $\mathcal{C}_H$ (since for $0 \leq z_1 \leq \varepsilon$, one has the given map F_2') and then, by the action of the group Γ, on V_1^H.

For a general H, one assumes by induction that

$$[F_2']_\perp = [F_1]_\perp - \sum_{H \leq K} d_K [F_K]_\perp$$

has been extended, as a non-zero orthogonal map to all V_1^K, for $H < K$. That is, together with a dimension argument, one has a non-zero map on $\partial\mathcal{C}_H$, in particular for the corresponding $z_1 = 0$. Then, one repeats the above argument in order to obtain a non-zero orthogonal extension F_2' on V_1^H. □

Step 3. On $V_1^\perp$ consider the first coordinate z_k with $n_1^k \neq 0$ and repeat the above construction in order to get $\tilde{V}_1 = V^{\tilde{T}_1}$. Clearly, $\tilde{V}_1 \cap V_1 = V^{T^n}$ and one obtains a non-zero orthogonal extension on $\tilde{V}_1$ of F^{T^n}. Since the generators for F_2 are trivial on $V_1^\perp$, one obtains a compatible extension. One repeats this construction until all coordinates with $n_1^k \neq 0$ are exhausted and then, with

$$V_2 = V^{T^n} \times \{z_k : n_1^k = 0,\ n_2^k \neq 0,\ n_j^k = \lambda_j n_2^k,\ j > 2,\ \text{where } \lambda_j = n_j^{k_0}/n_2^{k_0}\},$$

and so on.

Hence, if H is such that $\dim \Gamma/H = 1$, one has one z_1 with $\dim \Gamma/H_1 = 1$ and $|H_1/H| < \infty$, and one has an extension

$$[F_2]_\perp = [F_1]_\perp - \sum_{\dim \Gamma/H = 1} d_H [F_H]_\perp,$$

which is orthogonal and non-zero on $\bigcup_{\dim \Gamma/H=1} V^H$.

Step 4. The next stage is for two-dimensional Weyl groups. Assume

$$\det \begin{pmatrix} n_1^1 & n_2^1 \\ n_1^2 & n_2^2 \end{pmatrix} = \det A \neq 0$$

and define, for $j \geq 3$, λ_j^1 and λ_j^2 by

$$\begin{pmatrix} n_j^1 \\ n_j^2 \end{pmatrix} = A \begin{pmatrix} \lambda_j^1 \\ \lambda_j^2 \end{pmatrix}.$$

Let $V_2 = \{z_k : n_j^k = \lambda_j^1 n_1^k + \lambda_j^2 n_2^k, j \geq 2\}$.

Then, on V_2, one has $A_j x = \lambda_j^1 A_1 x + \lambda_j^2 A_2 x$ for $j \geq 3$ and $V_2 = V^{T_2}$, where T_2 is the $(n-2)$-torus $(-\sum \lambda_j^1 \varphi_j, -\sum \lambda_j^2 \varphi_j, \varphi_3, \dots, \varphi_n)$. In particular, any isotropy subgroup H for V_2 has $\dim \Gamma/H \leq 2$ and the action of T^n on z_k in V_2 is $\exp i(n_1^k \psi_1 + n_2^k \psi_2)$, where $\psi_1 = \varphi_1 + \sum \lambda_j^1 \varphi_j$, $\psi_2 = \varphi_2 + \sum \lambda_j^2 \varphi_j$: see Lemmas 2.4, 7.1, and Remark 2.1 of Chapter 1.

Consider the map $F_2(x) + \lambda_1 A_1 x + \lambda_2 A_2 x$, λ_1, λ_2 in $I = [-1, 1]$, where $F_2(x) \neq 0$ if $\dim \Gamma/\Gamma_x \leq 1$ and F_2 is an orthogonal Γ-extension of $F(x)$. Hence, a zero of this map will give a zero of F_2 and with $\lambda_1 = \lambda_2 = 0$: in fact, since $A_j x$ is tangent to the orbit Γx, here at most two-dimensional, and that $F_2(x) \neq 0$ if Γx is one-dimensional. Hence, on zeros of F_2, $A_1 x$ and $A_2 x$ are linearly independent. (We are assuming here that $\det A > 0$; if not, change λ_1 to $-\lambda_1$).

Thus, $[F_2(x) + \lambda_1 A_1 x + \lambda_2 A_2 x]$ is an element of $\Pi^{\Gamma}_{S^{\mathbb{R}^2 \times V_2}}(S^{V_2})$, the group of all Γ-homotopy classes of maps from $\partial(I^2 \times B_2)$ into $V_2 \backslash \{0\}$, where B_2 is the ball B^{V_2}. Now, this group is $A \times \mathbb{Z} \times \cdots \times \mathbb{Z}$, with A corresponding to isotropy subgroups H on V_2 with $\dim \Gamma/H \leq 1$ and there is one $\mathbb{Z}$ for each H with $\dim \Gamma/H = 2$: see Theorem 3.2. Then,

$$[F_2(x) + \lambda_1 A_1 x + \lambda_2 A_2 x]_\Gamma = 0 + \sum_{T_2 \leq H} d_H [\tilde{F}_H]_\Gamma,$$

with $\dim \Gamma/H = 2$. Here $\tilde{F}_H$ is the following map

$$\begin{aligned} \tilde{F}_H(\lambda, x) = \Big(& 2t + 1 - 2|z_1 z_2| \prod |x_i|, X_0, (Q_i - 1)y_i, (P_j + 1)z_j, \\ & (|z_1|^2 - 1 + i(n_1^1 \lambda_1 + n_2^1 \lambda_2))z_1, \\ & (\varepsilon(|z_2|^2 - 1) + i(n_1^2 \lambda_1 + n_2^2 \lambda_2))z_2\Big), \end{aligned}$$

where x_j, X_0, z_j, P_j are as in the first step. The zeros of $\tilde{F}_H$ are at $|z_1| = 1$, $|z_2| = 1$, $|x_i| = 1$, $t = 1/2$, $\lambda_1 = \lambda_2 = 0$. For z_1 and z_2 real and positive the index of each zero is equal to $\pm\varepsilon$ Sign det A, and ε is chosen such that this index is 1. Thus, $\tilde{F}_H$ may be taken as a generator. Let

$$F_H(x) = \tilde{F}_H(0, x) - (\tilde{F}_H(0, x), \tilde{A}_1(x))\tilde{A}_1(x) - (\tilde{F}_H(0, x), \tilde{A}_2(x))\tilde{A}_2(x),$$

as in Theorem 7.1 of Chapter 1. By construction $F_H(x)$ is an orthogonal map. Writing $F_H(x) = \tilde{F}_H(0, x) - \alpha(x)A_1x - \beta(x)A_2x$, one easily sees that the zeros of F_H are those of $\tilde{F}_H(0, x)$, looking at the (z_1, z_2)-components, and that for them one has $\alpha(x) = \beta(x) = 0$. Furthermore, as a Γ-map, $F_H(x) + \lambda_1 A_1 x + \lambda_2 A_2 x$ is linearly deformable to $\tilde{F}_H(0, x) + \lambda_1 A_1 x + \lambda_2 A_2 x$ (the zeros of the deformation are for $\lambda_1 = \tau\alpha$, $\lambda_2 = \tau\beta$ and $\tilde{F}_H(0, x) = 0$, for which $\alpha = \beta = 0$). Then, this last map is deformable to $\tilde{F}_H(\lambda, x) = \tilde{F}_H(0, x) + \lambda_1 A_1 Z + \lambda_2 A_2 Z$, with $Z^T = (z_1, z_2)$. This means that one may take $F_H(x) + \lambda_1 A_1 x + \lambda_2 A_2 x$ as a generator in $\Pi^{\Gamma}_{S^{\mathbb{R}^2 \times V_2}}(S^{V_2})$. Let then

$$[F_3]_\perp = [F_2]_\perp - \sum_{\substack{T_2 \leq H \\ \dim \Gamma/H = 2}} d_H [F_H]_\perp.$$

Hence, F_3 is an orthogonal Γ-map and $F_3(x) + \lambda_1 A_1 x + \lambda_2 A_2 x$ is Γ-extendable on $\partial(I^2 \times B_2) \bigcup_{\dim \Gamma/H \leq 1} V^H$ to a non-zero map $F(\lambda, x)$ on $I^2 \times B_2$.

Claim 6.2. *In fact,* $[F_3]_\perp = 0$.

Proof. As before, one proceeds on isotropy subspaces of increasing dimension by considering on each fundamental cell $\mathcal{C}_H$ an orthogonal map F_3' which, by induction and dimension arguments, is non-zero on $\partial\mathcal{C}_H$. In particular, $F_3'(x) \neq 0$ for $0 \leq z_1 \leq \varepsilon$ or $0 \leq z_2 \leq \varepsilon$ and the obstruction degree d_H is the degree of $F_3'(x) + \lambda_1 A_1 x + \lambda_2 A_2 x$ on the ball $\mathcal{A} = I^2 \times \mathcal{C}_H \cap \{z_1, z_2 \geq \varepsilon\}$. If $F_3'(x) = (f_1 + if_2, g_1 + ig_2, F_2) = (F, F_\perp)$, then one may deform linearly $F_3'(x) + \lambda_1 A_1 x + \lambda_2 A_2 x$ to $F_3'(x) + \lambda_1 A_1 Z + \lambda_2 A_2 Z$ with $Z^T = (z_1, z_2)$: by taking the scalar product with $F_3'(x)$ one obtains, on a zero of the homotopy,

$$|F_3'(x)|^2 + (1 - \tau)(\lambda_1(F, A_1 Z) + \lambda_2(F, A_2 Z)) = 0.$$

But, by the orthogonality, $(F, A_i Z) = -(F_\perp, A_i Y)$ and, on a zero, $F_\perp = -\tau(\lambda_1 A_1 Y + \lambda_2 A_2 Y)$, hence

$$|F_3'|^2 + \tau(1 - \tau)(\lambda_1^2 |A_1 Y|^2 + 2\lambda_1\lambda_2(A_1 Y, A_2 Y) + \lambda_2^2 |A_2 Y|^2) = 0.$$

Since the quadratic form is non-negative, this implies that $F_3'(x) = 0$ and $\lambda_1 A_1 Z + \lambda_2 A_2 Z = 0$, that is $\lambda_1 = \lambda_2 = 0$, since, on $\mathcal{A}$, the vectors $A_1 Z$ and $A_2 Z$ are linearly independent. Hence, the zeros of the deformation are inside $\mathcal{A}$. The resulting map

$$\left(f_1, g_1, \begin{pmatrix} z_1 & 0 \\ 0 & z_2 \end{pmatrix} A \begin{pmatrix} \lambda_1 \\ \lambda_2 \end{pmatrix} + \begin{pmatrix} f_2 \\ g_2 \end{pmatrix}, F_\perp\right)$$

is linearly deformable to

$$\left(f_1, g_1, A\begin{pmatrix}\lambda_1\\ \lambda_2\end{pmatrix}, F_\perp\right),$$

since, from the orthogonality,

$$A^T\begin{pmatrix}z_1 f_2\\ z_2 g_2\end{pmatrix} = -\begin{pmatrix}(F_\perp, A_1 Y)\\ (F_\perp, A_2 Y)\end{pmatrix}$$

and a zero of $F_\perp$ on $\mathcal{A}$ will give $f_2 = g_2 = 0$ and then z_1 and z_2 may be deformed to 1.

This last map is a product and since the extension degree is 0 one has that $(f_1, g_1, F_\perp)$ has a degree equal to 0 on $\mathcal{C}_H \cap \{z_1, z_2 \geq \varepsilon\}$ and, therefore, a non-zero extension $(\tilde{f}_1, \tilde{g}_1, \tilde{F}_\perp)$ to this set. Defining $\tilde{f}_2$ and $\tilde{g}_2$ on this set via

$$A^T\begin{pmatrix}z_1 \tilde{f}_2\\ z_2 \tilde{g}_2\end{pmatrix} = -\begin{pmatrix}(\tilde{F}_\perp, A_1 Y)\\ (\tilde{F}_\perp, A_2 Y)\end{pmatrix},$$

one obtains a non-zero orthogonal extension $\tilde{F}_3(x)$ of $F_3'(x)$ first on $\mathcal{C}_H$ and then, by the action of the group Γ, on V_2^H. □

The rest of the proof of (1) in Theorem 6.1 is then clear: exhaust all isotropy subgroups H with $\dim \Gamma/H = 2$ and then go on to higher dimensional Weyl groups.

Now, if $[F]_\perp = \sum d_H [F_H]_\perp$, then $[F^K]_\perp = \sum d_H [F_H^K]_\perp$ and, in fact, the sum reduces to those $H \geq K$, since $F_H^K \neq 0$ if K is not a subgroup of H, in which case $V^H \cap V^K$ is a strict subspace of V^H, hence there is at least one $x_j = 0$ and the first component of F_H^K is non-zero. For $K \leq H$, it is easy to see that F_H^K is the generator for the group $\Pi^\Gamma_{\perp S^K}(S^K)$. Hence, if $F^K \neq 0$, one has $d_H = 0$ for all $K \leq H$, proving (2). □

The last results of this section concern the computation of an orthogonal class by approximations by normal maps or by reduction to Poincaré sections as in §3.

Lemma 6.1. *For any fixed H, any map F in $\Pi^\Gamma_{\perp S^V}(S^V)$ is orthogonally Γ-homotopic to a map $\tilde{F}(x) = (F^H(x), \tilde{F}_\perp(x))$, such that $\tilde{F}_\perp(x) = x_\perp$ if $|x_\perp| \leq \varepsilon$ and $x = x_H \oplus x_\perp$, with x_H in V^H. In case of a gradient, the homotopy is also a gradient.*

Proof. This lemma is parallel to Lemma 3.1: let $\psi : (V^H)^\perp \to [0,1]$ be such that $\psi(x_\perp)$ is 1 if $|x_\perp| < \varepsilon$ and 0 if $|x_\perp| > 2\varepsilon$. If $F(x)$ is written as

$$(F^H(x_H, x_\perp), F_\perp(x_H, x_\perp)),$$

then $F(x)$ is Γ-orthogonally homotopic to the map

$$(F^H(x_H, (1-\psi)x_\perp), (1-\psi)F_\perp(x_H, (1-\psi)x_\perp) + \psi x_\perp),$$

since $A_j x$ is orthogonal to $x_\perp$ and to $F(x)$. Since $F_\perp(x_H, 0) = 0$ and $F^H(x_H, 0) \neq 0$ on ∂B^H, one may choose ε so small that $F^H(x_H, x_\perp) \neq 0$ for $|x_\perp| < 2\varepsilon$, justifying the homotopy.

In the case of a gradient, if $F(x) = \nabla\Phi(x)$, let

$$\Phi(x) = \psi(x_\perp)(\Phi(x_H) + |x_\perp|^2/2) + (1 - \psi(x_\perp))\Phi(x_H, x_\perp).$$

Then,

$$\begin{aligned}\nabla\tilde{\Phi}(x) = (F^H(x) + \psi(F^H(x_H) - F^H(x)), (1 - \psi)F_\perp(x) \\ + \psi x_\perp + (\tilde{\Phi}(x_H) - \Phi(x) + |x_\perp|^2/2)\nabla\psi).\end{aligned}$$

If $|x_\perp| > 2\varepsilon$, then $\nabla\tilde{\Phi}(x) = F(x)$, while if $|x_\perp| < \varepsilon$, one has $\nabla\tilde{\Phi}(x) = (F^H(x_H), x_\perp)$. If on ∂B^H one has that $|F^H(x_H)| > \eta$, one chooses ε so small that on $\partial B^H \times \{x_\perp : |x_\perp| \leq 2\varepsilon\}$, one has $|F^H(x) - F^H(x_H)| < \eta/2$. Thus, $\nabla\tilde{\Phi}$ is Γ-homotopic to $\nabla\Phi$. □

Lemma 6.2. *Any F in $\Pi^\Gamma_{\perp S^V}(S^V)$ is orthogonally Γ-homotopic to a normal orthogonal map F_N.*

Proof. As in Lemma 3.2, working in stages, one gets that F is orthogonally Γ-homotopic to F_N, where $F_N(x_H, x_\perp) = (F^H_N(x_H, x_\perp), x_\perp)$, for any H, provided $|x_\perp| < \varepsilon$, i.e., a normal map. Similarly, for the case of gradients, $\nabla\Phi$ is Γ-homotopic to $\nabla\Phi_N$. □

Finally, as in Theorem 3.4, we shall study the following situation: let H be an isotropy subgroup such that $\dim \Gamma/H = k$. Then, there are complex coordinates $z_1, \dots, z_k$ with isotropy $H_0 > H$ and $|H_0/H| < \infty$.

Assume that the orthogonal map F, from B into V, is non-zero on ∂B and on each set given by $z_j = 0$ for each $j = 1, \dots, k$. Let $\underline{H}$ be the torus part of H_0 (see Lemma 2.6 of Chapter 1). If $A^{\underline{H}}$ is the $N \times n$ matrix with $A^{\underline{H}}_{ij} = n^i_j$, $i = 1, \dots, N = \dim V^{\underline{H}}$, $j = 1, \dots, n$, then $A^{\underline{H}}$ has rank k and has an invertible submatrix A, for instance n^i_j, for $i, j = 1, \dots, k$, corresponding to $z_1, \dots, z_k$ and $\varphi_1, \dots, \varphi_k$. If for $j > k$ one defines λ^i_j from

$$\begin{pmatrix}\lambda^1_j \\ \vdots \\ \lambda^k_j\end{pmatrix} = A^{-1}\begin{pmatrix}n^1_j \\ \vdots \\ n^k_j\end{pmatrix},$$

then, the subspace $V^{\underline{H}}$ is given by those coordinates z_l for which $n^l_j = \sum_1^k \lambda^s_j n^l_s$ for $j > k$ (if for some j and l one does not have equality then $A^{\underline{H}}$ would have rank bigger than k). See Remark 2.1 of Chapter 1.

Note that $A_j x = \sum_1^k \lambda^s_j A_s x$ for $j > k$ and x in $V^{\underline{H}}$ and $A_1 x, \dots, A_k x$ are linearly independent if x has its coordinates $z_1, \dots, z_k$ non-zero.

Proposition 6.1. *Let F be as above, then $[F]_\perp = \sum_{H_j < H_0} d_j [F_j]_\perp$. If $B_k^j = B^{H_j} \cap \{z_1, \dots, z_k \in \mathbb{R}^+\}$, then for $H_i > \underline{H}$, the corresponding d_i are given by the formulae*

$$\deg\left(\left(F + \sum_1^k \lambda_l A_l x\right)^{H_i}; B_k^i\right) = \sum_{H_i < H_j < H_0} d_j |H_0/H_j|.$$

Proof. If K is not a subgroup of H_0, then for some j, $j = 2, \dots, k$, one has $z_j = 0$ in V^K. Hence, from Theorem 6.1 (2), the corresponding d_K is 0. Also, one has that $[F^{\underline{H}}]_\perp = \sum d_j [F_j^{\underline{H}}]_\perp$, where the sum is over those j with $\underline{H} < H_j < H_0$ (for the others $[F_j^{\underline{H}}]_\perp = 0$). This equality means that there is a Γ-orthogonal homotopy $F(\tau, x)$ between both sides. It is clear that $F(\tau, x) + \sum_1^k \lambda_l A_l x$ provides the Γ-homotopy to prove that $[(F + \sum \lambda_l A_l x)^{\underline{H}}]_\Gamma = \sum d_j [(F_j + \sum \lambda_l A_l x)^{\underline{H}}]_\Gamma$. From the construction of Theorem 6.1 and Theorem 3.4, one has the above formula.

Note that these formulae can be arranged as a lower triangular invertible matrix, as in Corollary 3.1, which will yield d_j for $\underline{H} < H_j < H_0$. The other components d_j, with $\dim \Gamma/H_j \neq k$, have to be computed in special cases as for that of an isolated orbit in next chapter. □

Remark 6.1. For the correct application of Proposition 6.1, it is important to take the generators $[F_j]_\perp$ such that $(F_j + \sum \lambda_l A_l x)^{\underline{H}}$ has index 1 on the fundamental cell, that is, $(F_j + \sum \lambda_l A_l x)^H$ is Γ-homotopic to the generator $F_j(\lambda, x)$ of Theorem 3.3 and given in the proof of Theorem 6.1. Now, due to Theorem 1.3, $[F_j(\lambda, x)]_\Gamma$ is unique up to conjugations. However, $[F_j]_\perp$ is *not unique* since it depends on the choice of the set of k linearly independent $A_j x$'s. For instance, if T^2 acts on $\mathbb{C}$ as $e^{i(\varphi_1 - \varphi_2)}$, then the Γ-orthogonal map $F(t, z) = (2t + 1 - 2|z|^2, (1 - |z|^2)z)$ is such that $F(t, z) + \lambda A_1 z$ has extension degree equal to 1 and is Γ-homotopic to $(2t + 1 - 2|z|^2, (|z|^2 - 1 - i\lambda)z) = F'(t, z) + \lambda A_2 z$, via the rotation $e^{\pi i \tau}$. However, $F + \lambda A_2 z$ has a degree (on the set $z \in \mathbb{R}^+$) equal to -1, as a map from (t, λ, z) into $\mathbb{R}^3$. Thus, from Proposition 6.1, one has $[F]_\perp = -[F']_\perp$. In fact, one has

Proposition 6.2. *Let F_H^* be the generators obtained by orthogonalization of $F_H^*(0, x)$, where $F_H^*(\lambda, x)$ is the generator of $\Pi(H)$ given in Theorem 3.3. Then, if F_H is the generator obtained from $F_H + \sum_1^k \lambda_l A_l x$, one has*

$$[F_H]_\perp = \operatorname{Sign} \det A [F_H^*]_\perp,$$

where $A_{ij} = n_j^i$ is the $k \times k$ matrix of $A_1 x, \dots, A_k x$ for $x = (z_1, \dots, z_k)$, with $z_j = 1$. The generators F_H^* will be called the ***normalized*** *generators.*

Proof. Note first that F_H^* was constructed in Step 2 of the proof of Theorem 6.1, while

F_H was constructed in Step 4. Thus, if

$$F_H^*(\lambda, x) = \Big(2t + 1 - 2\prod |x_j|, \ldots, (1 - |z_1|^2 + i\lambda_1)z_1, \\ \ldots, (\varepsilon^*(1 - |z_1|^2) + i\lambda_k)z_k, \ldots\Big),$$

where ε^* is such that the index on $\mathcal{C}_H$ is 1, then

$$\begin{aligned} F_H^*(x) &= F_H^*(0, x) - \sum_1^k (F_H^*(0, x), \tilde{A}_l(x))\tilde{A}_l(x) \\ &= F_H^*(0, x) - \sum \alpha_l(x)A_l x. \end{aligned}$$

Then, as in the proof of Theorem 6.1, $F_H^*(x) + \sum \lambda_l A_l x$ is Γ-homotopic to the map

$$\Big(2t+1-\prod |x_j|, \ldots, (1-|z_1|^2+i\sum \lambda_l n_l^1)z_1, \ldots, (\varepsilon^*(1-|z_k|^2)+i\sum \lambda_l n_l^k)z_k, \ldots\Big)$$

which has index, on $\mathcal{C}_H$, equal to Sign det A, since, for F_H, ε^* is replaced by ε^* Sign det A. □

Example 6.1. Suppose S^1 acts on (z_1, z_2) as $(e^{i\varphi}z_1, e^{-i\varphi}z_2)$, that is the representation on z_2 is conjugate to that on z_1, but as real representations they are the same. Then one may take as generators of Π^Γ the map

$$\begin{aligned} F(t, \lambda, z_1, z_2) &= (2t + 1 - 2|z_1 z_2|, (1 - |z_1|^2 + i\lambda)z_1, (z_1 z_2 - 1)z_2) \\ F'(t, \lambda, z_1, z_2) &= (2t + 1 - 2|z_1 z_2|, (z_2 z_1 - 1)z_1, (1 - |z_2|^2 + i\lambda)z_2), \end{aligned}$$

since, on $\mathcal{C} = \{z_1 \in \mathbb{R}^+\}$, the first map has degree 1 while the second has degree 1 on $\mathcal{C}' = \{z_2 \in \mathbb{R}^+\}$. However, on $\mathcal{C}'$ the first map has degree -1: for $z_2 > 0$ the only zero is for z_1 real and positive, i.e., for $z_1 = 1$. If $z_1 = x + iy$, the map is locally deformable to $(2t - 1, 1 - x^2 - y^2, \lambda, xz_2 - 1, yz_2)$ and then to $(2t - 1, 1 - x, \lambda, z_2 - 1, y)$.

Thus, one has

$$[F]_\Gamma = -[F']_\Gamma,$$

as expected, since conjugation changes the sign of the degree: see Remark 3.1. Here $F^*(t, z_1, z_2) = F(t, 0, z_1, z_2) - \alpha(z)Az$, where $\alpha(z) = i|z_2|^2(z_1 z_2 - \bar{z}_1\bar{z}_2)/2$, while $F'^*(t, z_1, z_2) = F'(t, 0, z_1, z_2) - \alpha'(z)Az$, where $\alpha'(z) = -i|z_1|^2(z_1 z_2 - \bar{z}_1\bar{z}_2)/2$. One has $[F^* + \lambda Az]_\Gamma = [F(t, 0, z) + \lambda Az]_\Gamma = [F]_\Gamma$, by following the steps of the proof of Theorem 6.1. On the other hand, $[F'^* + \lambda Az]_\Gamma = [F'(t, 0, z) + \lambda Az]_\Gamma = -[F']_\Gamma = [F]_\Gamma$. Thus, $[F^*]_\perp = [F'^*]_\perp$.

Now, consider the orthogonal maps

$$\begin{aligned} F_0(t, z_1, z_2) &= (2t + 1 - 2|z_1|, (1 - |z_1|^2)z_1, z_2) \\ F_0'(t, z_1, z_2) &= (2t + 1 - 2|z_2|, z_1, (1 - |z_2|^2)z_2). \end{aligned}$$

Clearly, $[F_0 + \lambda Az]_\Gamma = [F]_\Gamma$, since the degree of $F_0 + \lambda Az$ on $\mathcal{C}$ is 1. On the other hand, $[F_0' + \lambda Az]_\Gamma = -[F']_\Gamma$. Thus, $[F_0]_\perp = [F_0']_\perp$.

Note that, if one changes z_2 to $\bar{z}_2 = z_2'$, then F has to be modified in its last component to $(\bar{z}_1 z_2' - 1)z_2'$, which gives a degree 1 for F on $\mathcal{C}$ and on $\mathcal{C}'$. (Here F' has to be modified to $(\bar{z}_2' z_1 - 1)z_1$); and $[F_0]_\perp = [F_0']_\perp$. In this case the map $z_2 \to z_2'$ is equivariant (with action $e^{-i\varphi}$ on z_2 and $e^{i\varphi}$ on z_2') with a complementing map $\bar{z}_2$, i.e., with $l = -1$.

In general, one has the following.

Proposition 6.3. *The normalized generators are independent of conjugations.*

Proof. If, as in Remark 3.1, one has two fundamental cells, $\mathcal{C}$ and $\mathcal{C}'$, such that one has coordinates $z_1, \dots, z_s$ with $k_j = \infty$ (hence $s \leq k$) in $\mathcal{C}$ and $z_1', \dots, z_s'$ in $\mathcal{C}'$, with action on z_j' conjugate from that on z_j, then one has $[F]_\Gamma = (-1)^s[F']_\Gamma$, where F and F' are the generators of Theorem 3.3.

Now, if F^* and F'^* are the orthogonalizations of $F(0,x)$ and $F'(0,x)$ one has $[F^* + \sum_1^k \lambda_l A_l x]_\Gamma = [F(0,x) + \sum \lambda_l A_l x]_\Gamma = \operatorname{Sign}\det A[F]_\Gamma$, by following the proof of Theorem 6.1, where A corresponds to the matrix of $A_1 z, \dots, A_k z$, on the coordinates $z_1, \dots, z_k$ of $\mathcal{C}$ (which include $z_1, \dots, z_s$). Similarly, $[F'^* + \sum_1^k \lambda_l A_l x]_\Gamma = \operatorname{Sign}\det A'[F']_\Gamma$. But, since A' has s lines which are the opposite of those of A, one has $\operatorname{Sign}\det A' = (-1)^s \operatorname{Sign}\det A$ and $[F']_\Gamma = (-1)^s[F]_\Gamma$. Thus,

$$\left[F^* + \sum \lambda_l A_l x\right]_\Gamma = \left[F'^* + \sum \lambda_l A_l x\right]_\Gamma \quad \text{and} \quad [F^*]_\perp = [F'^*]_\perp \qquad \square$$

3.7 Operations

In the last section of this chapter, we shall examine how the Γ-homotopy groups of spheres behave under different operations: suspension, reduction of the group, products and composition, for the case of parameters and that of orthogonal maps. These operations will enable us to acquire a certain number of tools for applications. This section is the continuation of § 6 of Chapter 2.

3.7.1 Suspension

We have seen in § 8 of Chapter 1 that the suspension operation enables us to go to the infinite dimensional setting, when a map may be approximated with maps of finite dimensional range.

The setting will be that of Theorem 3.2, i.e., that $V = \mathbb{R}^k \times U$, where U and W satisfy (H), that is $\dim U^H \cap U^K = \dim W^H \cap W^K$ for all H, K in $\operatorname{Iso}(U)$. Let then V_0 *be an irreducible* representation of Γ, generated by a real or complex variable x with isotropy subgroup H_0 (hence Γ/H_0 is trivial or $\mathbb{Z}_2$ in the first case, $\mathbb{Z}_m$, $m \geq 3$

or S^1 in the second case). From Theorem 2.3, one has that

$$\Pi^{\Gamma}_{S^V}(S^W) \cong \bigoplus_H \tilde{\Pi}(H) \quad \text{and} \quad \Pi^{\Gamma}_{S^{V\times V_0}}(S^{W\times V_0}) \cong \bigoplus_{H'} \tilde{\Pi}(H'),$$

where H is in $\text{Iso}(V)$ and H' in $\text{Iso}(V \times V_0)$. The group H' will be of the form H with H in $\text{Iso}(V)$ or $H \cap H_0$. Then, if H_0 is not an isotropy subgroup for V, there will be more isotropy types for $V \times V_0$ (at least H_0) and the equivariant group for $V \times V_0$ will have more components (unless trivial). In order to make clearer our statements, we shall use also the notation $\Pi_V(H)$, respectively $\Pi_{V\times V_0}(H')$, for the subgroup H in $\text{Iso}(V)$, respectively H' in $\text{Iso}(V \times V_0)$.

Theorem 7.1. *For any H in $\text{Iso}(V)$, one has*

(a) Σ^{V_0} *maps* $\Pi_V(H)$ *into* $\Pi_{V\times V_0}(H)$.

(b) Σ^{V_0} *is an isomorphism if H is not a subgroup of H_0.*

(c) *If H is a subgroup of H_0 and for all K in $\text{Iso}(V)$, with $H < K$ and $K \cap H_0 = H$, one has* $\dim W^H - \dim W^K \geq k+1-\dim \Gamma/H$, *then Σ^{V_0} is onto. If for these K's, one has* $\dim W^H - \dim W^K \geq k+2-\dim \Gamma/H$, *then Σ^{V_0} is also one-to-one.*

(d) *If there are no K's as above, then Σ^{V_0} is onto if* $\dim W^H \geq k+1-\dim \Gamma/H$, *and Σ^{V_0} is an isomorphism from $\Pi_V(H)$ onto $\Pi_{V\times V_0}(H)$ if* $\dim W^H \geq k+2-\dim \Gamma/H$. *(Note that t is not taken to be part of V).*

Proof. Let us consider first the case where H' is not in $\text{Iso}(V)$, i.e., $H' = H \cap H_0$. Let $\bar{H}$ be the isotropy subgroup of $W^{H'}$ (see Definition 2.1 of Chapter 1), then $W^{H'} = W^{\bar{H}}$ and $H' < \bar{H}$. Furthermore, $H' < H$ implies $W^H \subset W^{H'} = W^{\bar{H}}$, hence $\bar{H} < H$. Thus, $H' = H \cap H_0 < \bar{H} \cap H_0 < H \cap H_0$, that is $H' = \bar{H} \cap H_0$ and $W^{H'} = W^{\bar{H}}$. This implies that if F belongs to $\Pi(H')$, then F maps $(B^{V\times V_0})^{K'}$ into $(W \times V_0)^{K'} \backslash \{0\}$ for all $K' > H'$, in particular for $K' = \bar{H}$ which is not a subgroup of H_0. Then, $(V \times V_0)^{\bar{H}} = V^{\bar{H}}$ and $F^{\bar{H}} = F|_{x=0} \neq 0$. That is, F cannot come from the suspension of a non-trivial element.

On the other hand, if $H' = H$, then $\Pi(H')$ consists of maps from $(V \times V_0)^H$ into $(W \times V_0)^H$ which map $(V \times V_0)^{K'}$ into $(W \times V_0)^{K'} \backslash \{0\}$ for all $K' > H$, with $K' = K$ or $K \cap H_0$, i.e., for $K > H$. Thus, if H is not a subgroup of H_0, then $(V \times V_0)^H = V^H$ and $(V \times V_0)^{K'} = V^K$ (there are no K' of the form $K \cap H_0 > H$ in this case) and any element of $\Pi(H')$ is in $\Pi(H)$. If $H < H_0$, then $(V \times V_0)^{K'} = V^{K'} \times V_0 = V^{\bar{K}} \times V_0$ if $H < K' < H_0$, $\bar{K} \cap H_0 = K'$ with $V^{\bar{K}} = V^{K'}$ if K' is not in $\text{Iso}(V)$, while $(V \times V_0)^K = V^K$ if K is not a subgroup of H_0. Thus, if F belongs to $\Pi(H)$, $(F, x)^{K'}$ will be $(F^{\bar{K}}, x)$ in the first case (or (F^K, x) if $H < K' = K \cap H_0$), or F^K if K is not a subgroup of H_0, that is, in all cases, different from 0. That is, $\Sigma^{V_0} F$ belongs to $\Pi(H')$ if $H' = H$.

Hence, if $[F]_\Gamma$ belongs to $\Pi^\Gamma_{S^V}(S^W)$, then $[F]_\Gamma = \sum [F_H]_\Gamma$, where F_H belongs to $\Pi(H)$ and $\Sigma^{V_0} F_H$ belongs to $\Pi(H')$ with $H' = H$. Thus, Σ^{V_0} is an isomorphism if H is not a subgroup of H_0. While if $H < H_0$, the elements F_H will be given by a sum of obstruction classes for extensions of a map to edges and to faces of the fundamental cell, respecting the symmetries and ending with a map from $\partial \mathcal{C}_H$ into $W^H \backslash \{0\}$ (it is enough to follow the argument given for the case $k = 1$). Thus, the obstruction will come from elements of $\Pi_{S^n}(S^{W^H})$, for $n \leq \dim \mathcal{C} = \dim V^H - \dim \Gamma/H$, while the obstruction classes for $\Pi(H')$ will be in $\Pi_{S^n \times V_0}(S^{W \times V_0})$. In particular, if x remains as a dummy variable at each stage of the extension, then for (F_H, x) the obstruction classes would be the suspension of the classes for F_H. From the ordinary suspension theorem, one would have an isomorphism if $n \leq 2 \dim W^H - 2$, for any $n \leq \dim V^H - \dim \Gamma/H$. In this case Σ^{V_0} would be one-to-one from $\Pi(H)$ into $\Pi(H')$ and for any element G in $\Pi(H')$, the obstruction classes would be the suspensions by x, that is $[G]_\Gamma = \sum [F_K, x] = [\sum F_K, x]$, therefore Σ^{V_0} would be onto.

There is however a delicate point here: the new variable x has really to remain a dummy variable in this process. In fact, if F is an element of $\Pi_V(H)$, then F^K is non-zero for any $K > H$ and, of course, (F^K, x) is non-zero in $(V \times V_0)^{K'}$, where $K' = K \cap H_0$. If K' is strictly larger than H, then x is still a dummy variable, however, if $K' = H$, one may have a first obstruction, for an element of $\Pi_{V \times V_0}(H)$, for a non-zero Γ-extension to $V^K \times V_0$ of a map $F(x_K, x)$, with $F(x_K, 0) \neq 0$. Clearly, if $F(x_K, x) = (F^K(x_K), x)$, there is always the non-zero extension given by (F^K, x). But, for a general map $F(x_K, x)$, one needs to look at the fundamental cell $\mathcal{C}_{K'} = \mathcal{C}_K \times \{x : 0 \leq \operatorname{Arg} x < k_0\}$, where $k_0 = |K/H|$ if finite, or $k_0 = 0$ if $\dim K/H = 1 = \dim \Gamma/H_0$.

From Theorem 1.1, one will have a non-zero Γ-extension to $V^K \times V_0$ if

$$\dim(V^K \times V_0) - \dim \Gamma/H < \dim(W \times V_0)^H.$$

Since $\dim V^K = k + \dim W^K$ and $\dim(W \times V_0)^H = \dim W^H + \dim V_0$, because $H < H_0$, one has that, if $\dim W^H - \dim W^K > k$, there is no additional obstruction for $\Pi_{V \times V_0}(H)$, coming from this K.

Subsequent obstructions, by adding new variables to V^K, will be obtained for $\Pi_V(H)$ and $\Pi_{V \times V_0}(H)$, for the edges of $\mathcal{C}_H$, in such a way that the ordinary suspension is onto, due to the fact that $\dim W^H > k + \dim W^K > k$.

On the other hand, if for all K's with $K > H$, one has that $K \cap H_0$ is larger than H, then the condition $\dim W^H > k$ will suffice to give ontoness. This is the case if $H_0 = \Gamma$.

In order to prove injectivity, assume that F in $\Pi_V(H)$ is such that $\Sigma^{V_0} F$ is trivial, that is $(F(x_H), x)$ has a non-zero equivariant extension $\tilde{F}(x_H, x)$ to $B^{V \times V_0}$. Let $K > H$, with $K \cap H_0 = H$, and consider on $\partial(B^{V^K \times V_0} \times I)$ the Γ-map $\hat{F}(x_K, x, \tau)$

defined as

$$\hat{F}(x_K, x, \tau) = \begin{cases} (F^K(x_K), x) & \text{on } \partial(B^{V^K \times V_0}) \times I \\ (F^K(x_K), x) & \text{on } B^{V^K \times V_0} \times \{0\} \\ \tilde{F}(x_K, x) & \text{on } B^{V^K \times V_0} \times \{1\}. \end{cases}$$

Then, from Theorem 1.1, $\hat{F}$ has a non-zero Γ-extension to $B^{V^K \times V_0} \times I$ if

$$\dim(V^K \times V_0 \times I) - \dim \Gamma/H < \dim(W^H \times V_0),$$

that is, if $\dim W^H - \dim W^K > k+1-\dim \Gamma/H$. This non-zero extension provides a Γ-homotopy of $(F^K(x_K), x)$ on $B^{V^K \times V_0}$ to $\tilde{F}(x_K, x)$ fixing the value on $\partial(B^{V^K \times V_0})$. Thus, one may assume that $\tilde{F}(x_K, x)$ is of the form $(F^K(x_K), x)$. If $K \cap H_0$ is larger than H, then we know, from Hypothesis (H), that the obstructions are independent of the extensions to $V^{K \cap H_0}$.

At this point, one has started an induction argument: if one assumes that, on the boundary of some face, $\tilde{F}(x_H, x)$ is a suspension, then, from the fact that the ordinary suspension is one-to-one and the fact that $\tilde{F}(x_H, x)$ has an extension to the face, one has that F has a non-zero extension to that face. One arrives finally at the result that F is trivial. □

Remark 7.1. Given the explicit generators for the subgroups $\Pi(H)$, if $\dim \Gamma/H = k$ or if $k = 0$ or 1, it is apparent that Σ^{V_0} is one-to-one for any H_0 provided $\dim W^\Gamma \geq k+2$, if $H_0 = \Gamma$, and $\dim W^H - \dim W^K \geq 2$ (always true if W^H contains a complex variable which is not in W^K) for $\Pi(H)$ if $k = \dim \Gamma/H$. In this last case, the suspension is always onto, from $\Pi_V(H) \cong \mathbb{Z}$ onto $\Pi_{V \times V_0}(H) \cong \mathbb{Z}$, hence it has to be one-to-one. For the case $k = 1$, the condition $\dim W^H - \dim W^K \geq 3$ is consistent with the results of Section 5 and explains why we have asked for repetition of variables: see Theorems 5.1 and 5.3.

The properties of Σ^{V_0}, as a map from $\Pi^\Gamma_{S^V}(S^W)$ into $\Pi^\Gamma_{S^{V \times V_0}}(S^{W \times V_0})$, will follow from the study of the behavior of the suspension on $\Pi_V(H)$, for all H in $\mathrm{Iso}(V)$ with $H < H_0$. However, it is not necessary to check the dimension conditions for all of these H's.

Corollary 7.1. (a) *If H_0 is an isotropy subgroup for V, then the suspension map*

$$\Sigma^{V_0} : \Pi^\Gamma_{S^V}(S^W) \to \Pi^\Gamma_{S^{V \times V_0}}(S^{W \times V_0})$$

is one-to-one provided

α) $\dim W^\Gamma \geq k+2$, *if* $H_0 = \Gamma$, or

β) $\dim W^{H_0} - \dim W^K \geq k + 2 - \dim \Gamma/H_0$

for all $K > H_0$.

The map will be onto if $k+2$ is replaced by $k+1$ in the above inequalities.

(b) *If H_0 is not an isotropy subgroup for V, then Σ^{V_0} will not be, in general, onto unless $k = 0$ and* $\dim \Gamma/H_0 = 1$. *The suspension will be one-to-one if it satisfies the dimension conditions of Theorem* 7.1.

(c) *In all cases, Σ^{V_0} is one-to-one if the number of real coordinates with the same isotropy is at least $k+2$ and the number of complex coordinates with the same action of Γ is at least $k/2+1$, if this action is finite, or at least $k/2$, if the action of Γ is as S^1.*

Proof. If H_0 is an isotropy subgroup for V, then either $H_0 = \Gamma$ and one has condition (α) or for any K, strictly larger than H_0, one has $K \cap H_0 = H_0$ and condition (β) is given in Theorem 7.1. Let then $H < H_0$ be in Iso(V). One has $\dim W^H \geq \dim W^{H_0} \geq k+2-\dim \Gamma/H_0 \geq k+2-\dim \Gamma/H$. Furthermore, if $K > H$ is such that $K \cap H_0 = H$, let K_0 be the isotropy of $W^K \cap W^{H_0}$. Then, K_0 contains K and H_0, thus, $K_0 \cap H_0 = H_0$ and, if $K_0 = H_0$, one would have $K < H_0$ and $K \cap H_0 = K > H$, contradicting the equality $K \cap H_0 = H$. Hence, from (β),

$$\dim W^{H_0} - \dim W^{K_0} \geq k+2-\dim \Gamma/H_0.$$

But, $W^K \cap W^{H_0} \subset W^{K_0}$ which implies

$$\begin{aligned}\dim W^H - \dim W^K &= \dim W^{H_0} - \dim W^K \cap W^{H_0} \\ &\quad + \dim (W^{H_0})^{\perp} - \dim W^K \cap (W^{H_0})^{\perp} \\ &\geq k+2-\dim \Gamma/H_0 \geq k+2-\dim \Gamma/H,\end{aligned}$$

and the condition is verified for H. Replacing $k+2$ by $k+1$ one has the surjectivity result.

In order to prove (b), one has to show that $\Pi(H') = 0$ for all $H' = H \cap H_0$ which are not in Iso(V). Now, for $H' = H \cap H_0 = \bar{H} \cap H_0$, with $V^{\bar{H}} = V^{H'}$, the group $\Pi(H')$ will vanish, from Theorem 1.1, provided $\dim(V \times V_0)^{H'} - \dim \Gamma/H' < \dim(W \times V_0)^{H'}$, i.e., $\dim W^{\bar{H}} + k + \dim V_0 - \dim \Gamma/H' < \dim W^{\bar{H}} + \dim V_0$, or else if $k < \dim \Gamma/H'$. This inequality has to be true in particular for $H' = H_0$, where $\dim \Gamma/H_0$ is 0 or 1. Hence $k = 0$ and $\Gamma/H_0 \cong S^1$, then $\dim \Gamma/H' > 0$ for any $H' = H \cap H_0$, and the equivariant group for $V \times V_0$ has no new components. Furthermore, if $H < H_0$ one has $\dim \Gamma/H \geq 1$ and the other conditions for ontoness of Theorem 7.1 are trivially met. On the other hand, if $k \geq \dim \Gamma/H_0$, then the first obstruction for extension, in $\Pi_{V \times V_0}(H_0)$, will be in the group $\Pi_{k+n-\dim \Gamma/H_0}(S^n)$, where $n = \dim W^{\overline{H}} + \dim V_0$ and $W^{\overline{H}} = W^{H_0}$. Since this group is, in general, non-trivial, this explains the wording of Corollary 7.1.

Finally, under the condition of (c), one has $\dim W^H \geq \dim W^{\Gamma} \geq k+2$, and $\dim W^H - \dim W^K \geq k+2-\dim \Gamma/H$, for any pair $H < K$ in Iso(V), noting that if, on some coordinate of $(W^K)^{\perp} \cap W^H$ the group Γ acts as S^1, then $\dim \Gamma/H \geq 1$. □

Let us turn now to Theorem 8.2 in Chapter 1, under the following formulation.

Corollary 7.2. *Let V_0 be a not necessarily irreducible representation of Γ, with coordinates $x_1, \ldots, x_n$ and H_j the isotropy of x_j. Then:*

(a) *The suspension Σ^{V_0} is one-to-one, if whenever H_0 is an isotropy subgroup for V_0 one has that for all H and K in* Iso(V)*, with $H < H_0$, $H < K$ (strictly) and $K \cap H_0 = H$, the following inequality holds*

$$\dim W^H - \dim W^K \geq k + 2 - \dim \Gamma/H.$$

If there are no K's as above, then the inequality

$$\dim W^H \geq k + 2 - \dim \Gamma/H,$$

will suffice. If H_0 is in Iso(V)*, then $H = H_0$ is allowable and is enough for other H's.*

(b) *If* Iso$(V_0) \subset$ Iso(V)*, then Σ^{V_0} will be onto if*

$$\dim W^\Gamma \geq k + 1$$
$$\dim W^{H_j} - \dim W^K \geq k + 1 - \dim \Gamma/H_j,$$

for all K in Iso(V)*, with $K > H_j$ and all j's.*

(c) *If some H_j is not an isotropy subgroup for V, then Σ^{V_0} will not be, in general, onto unless $k = 0$ and* $\dim \Gamma/H_j = 1$ *for all such H_j's.*

(d) *If for all K, H in* Iso(V)*, with $H < K$, one has*

$$\dim W^\Gamma \geq k + 2$$
$$\dim W^H - \dim W^K \geq k + 2 - \dim \Gamma/H,$$

then any suspension will be one-to-one. This will be the case if one has the repetition of coordinates of Corollary 7.1 (c).

Proof. It is enough to note that Σ^{V_0} is the composition

$$\Pi_V(H) \xrightarrow{\Sigma^{V_1}} \Pi_{V \times V_1}(H) \xrightarrow{\Sigma^{V_2}} \Pi_{V \times V_1 \times V_2}(H) \to \cdots \to \Pi_{V \times V_0}(H),$$

where Σ^{V_j} is the suspension by the coordinate x_j, and likewise for the full equivariant groups. Note that the order of the suspensions is irrelevant. Now, for

$$\Sigma^{V_j} : \Pi_{V \times V_1 \times \cdots \times V_{j-1}}(H) \to \Pi_{V \times V_1 \times \cdots \times V_j}(H),$$

one has an isomorphism if H is not a subgroup of H_j, while if $H < H_j$ one needs $\dim(W \times V_1 \times \cdots \times V_{j-1})^H - \dim(W \times V_1 \times \cdots \times V_{j-1})^K \geq k + 2 - \dim \Gamma/H$, for any K in Iso$(V \times V_1 \times \cdots \times V_{j-1})$, with $H < K$ and $K \cap H_j = H$. But, if

$K < H_{i_1}, \ldots, H_{i_l}$ for $i_1, \ldots, i_l$ between 1 and $j-1$, then $K = \bar{K} \cap H_{i_1} \cdots \cap H_{i_l}$, where $\bar{K}$ is the isotropy of $W^K = W^{\bar{K}}$. Then, if $H_0 = H_{i_1} \cap \cdots \cap H_{i_l} \cap H_j$, one has $\bar{K} \cap H_0 = H$, hence $H < H_0$ and $H < K < \bar{K}$. Since

$$\dim(W \times V_1 \times \cdots \times V_{j-1})^H \geq \dim W^H + \sum_{s=1}^{l} \dim V_{i_s}$$

$$\dim(W \times V_1 \times \cdots \times V_{j-1})^K = \dim W^{\bar{K}} + \sum_{s=1}^{l} \dim V_{is},$$

the above inequality is true under the condition of (a). Of course, if there are no K's then the second condition of (a) is stronger than the one needed here. If H_0 is in $\mathrm{Iso}(V)$ and $H < H_0$, one may repeat the argument of Corollary 7.1, in order to show that one does not need to check the inequalities for H.

Part (b) is then clear since Σ^{V_j} is onto at each stage, since $\dim(V_1 \times \cdots \times V_{j-1})^{H_j} \geq \dim(V_1 \times \cdots \times V_{j-1})^K$ and using Corollary 7.1. On the other hand, if H_j is not an isotropy subgroup for V, one may start the above sequence by

$$\Sigma^{V_j} : \Pi^{\Gamma}_{S^V}(S^W) \to \Pi^{\Gamma}_{S^{V \times V_j}}(S^{W \times V_j}),$$

which will be, in general, not onto unless $k = 0$ and $\dim \Gamma/H_j = 1$. If $k = 0$ and $\dim \Gamma/H_j = 1$ for all such j's, then the inequalities of (b) hold for any H_j in $\mathrm{Iso}(V)$ and Σ^{V_0} is onto. The word "in general" has to be taken in this context. Finally the conditions of (d) cover all possible suspensions. □

Remark 7.2. In the case of orthogonal maps, without parameters, the explicit construction of Theorem 6.1 implies that Σ^{V_0} is one-to-one. This implies that the approximation by finite dimensional orthogonal maps is valid and that one may take the direct limit of these groups to give an alternative definition of the orthogonal degree in the infinite dimensional case, as in § 3 of Chapter 2.

3.7.2 Symmetry breaking

Let Γ_0 be a subgroup of Γ. If a map is Γ-equivariant it is also Γ_0-equivariant and, in case it is Γ-orthogonal it will be Γ_0-orthogonal, since the torus part of Γ_0 is a subgroup of the torus part of Γ. One has then two morphisms

$$P_* : \Pi^{\Gamma}_{S^V}(S^W) \to \Pi^{\Gamma_0}_{S^V}(S^W)$$
$$P_\perp : \Pi^{\Gamma}_{\perp S^V}(S^W) \to \Pi^{\Gamma_0}_{\perp S^V}(S^W)$$

corresponding to the reduction of the group from Γ to Γ_0.

Under hypothesis ($\tilde{\mathrm{K}}$) we have seen, in Theorem 2.3, that $\Pi^{\Gamma}_{S^V}(S^W)$ is of the form $\bigoplus \tilde{\Pi}(H)$. It is thus important to determine first the relation between the isotropy subgroups for Γ and for Γ_0 and then the relation between the subgroups $\Pi(H)$, for Γ, and $\Pi_0(H_0)$, for Γ_0.

Lemma 7.1. (a) *Any isotropy subgroup H_0 for Γ_0 is of the form $H \cap \Gamma_0$, with H an isotropy subgroup for Γ. For a given H_0, there may be several H's, but there is a minimal one $\underline{H}$, for which $V^{\underline{H}} = V^{H_0}$.*

(b) *For all H with $H_0 = H \cap \Gamma_0$, one has* $\dim \Gamma_0/H_0 \leq \dim \Gamma/H$. *In case of equality, if $\tilde{H}_0$ is the isotropy of the k variables with $k_j = \infty$ ($k = \dim \Gamma/H$) and $\tilde{H}_0^0 = \tilde{H}_0 \cap \Gamma_0$, then $|\tilde{H}_0^0/H_0|$ divides $|\tilde{H}_0/H|$. Moreover, $\tilde{H}_0$ is Γ if $k = 0$.*

(c) *If* $\dim \Gamma_0/H_0 = \dim \Gamma/\underline{H}$ *and* Sign det γ Sign det $\tilde{\gamma} > 0$, *for all γ in Γ, then P_* maps $\Pi(\underline{H})$ into $\Pi_0(H_0)$ and if* $\dim W^{\underline{H}} = \dim V^{\underline{H}} - \dim \Gamma/\underline{H}$, *then*

$$P_*[F^{\underline{H}}]_\Gamma = \begin{cases} |\tilde{H}_0/\underline{H}|/|\tilde{H}_0^0/H_0|[F^{H_0}]_{\Gamma_0}, & \text{if } W^{\underline{H}} = W^{H_0} \\ 0, & \text{otherwise}, \end{cases}$$

where $F^{\underline{H}}$ generates $\Pi(\underline{H})$ and F^{H_0} generates $\Pi_0(H_0)$.

Proof. If $H_0 = \Gamma_{0x} = \{\gamma \in \Gamma_0 : \gamma x = x\}$, then clearly $H_0 = \Gamma_x \cap \Gamma_0$. Hence, $\underline{H}$ is the intersection of all such H's and the isotropy subgroup for V^{H_0} (see Definition 2.1 of Chapter 1). If z_i is a coordinate in this space with the subgroups $\tilde{H}_{i-1} = H_1 \cap \ldots H_{i-1}$ and $\tilde{H}_i = \tilde{H}_{i-1} \cap H_i$, then the corresponding subgroups for Γ_0 will be $\tilde{H}_i^0 = \tilde{H}_i \cap \Gamma_0$. Furthermore, if $k_i = |\tilde{H}_{i-1}/\tilde{H}_i|$ is finite, then any γ in $\tilde{H}_{i-1}$ may be written as $\gamma = \gamma_i^{\alpha_i}\tilde{\gamma}$, with $0 \leq \alpha_i < k_i$, $\gamma_i^{k_i}$ and $\tilde{\gamma}$ in $\tilde{H}_i$. Thus, if γ is also in Γ_0, then γ^{k_i} is in $\Gamma_0 \cap \tilde{H}_i$, that is $k_i^0 = |\tilde{H}_{i-1}^0/\tilde{H}_i|$ is finite and divides k_i. Hence, if x_l is the last coordinate in V^H, then $\tilde{H}_l = H$. Thus, $\tilde{H}_l^0 = H_0$ and $k_i^0 = 1$ for $i > l$. Since there are at most $k = \dim \Gamma/H$ coordinates with k_i^0 infinite, one has that $\dim \Gamma/H_0 \leq \dim \Gamma/H$ and the rest of (b) is then clear, since $|\tilde{H}_0^0/H_0| = \prod k_i^0$ divides $|\tilde{H}_0/H| = \prod k_i$.

For (c) one has that the fundamental cell $\mathcal{C}_0$ for H_0 is made of $\prod k_i/\prod k_i^0$ copies of the fundamental cell $\mathcal{C}$ for $\underline{H}$. If F belongs to $\Pi(\underline{H})$ and $K_0 > H_0$, then $K_0 = \underline{K} \cap \Gamma_0$, where $\underline{K}$ is minimal and the isotropy subgroup for $V^{K_0} \subset V^{H_0}$. Thus, $\underline{K} > \underline{H}$ and $F^{\underline{K}} \neq 0$, by definition of $\Pi(\underline{H})$, i.e., $F|_{V^{K_0}} \neq 0$ and F is in $\Pi_0(H_0)$. If $\dim V^{\underline{H}} = \dim W^{\underline{H}} + \dim \Gamma/\underline{H}$, $\dim \Gamma_0/H_0 = \dim \Gamma/\underline{H}$, since $H_0 = \underline{H} \cap \Gamma_0 < \underline{H}$, then $W^{\underline{H}} \subset W^{H_0}$. Hence, if $W^{\underline{H}} = W^{H_0}$, one obtains that $\dim V^{H_0} = \dim W^{H_0} + \dim \Gamma/H_0$ and F in $\Pi_0(H_0)$ is characterized by its extension degree $\deg_{\Gamma_0}(F)$, such that, from Theorem 1.2,

$$\deg(F; B_k) = |\tilde{H}_0/\underline{H}| \deg_\Gamma(F) = |\tilde{H}_0^0/H_0| \deg_{\Gamma_0}(F).$$

Since $\deg_\Gamma(F^H) = 1$, $\deg_{\Gamma_0}(F^{H_0}) = 1$, by definition, one obtains the equality of (c), since $P_*[F^{\underline{H}}]_\Gamma = \deg_{\Gamma_0}(F^{\underline{H}})[F^{H_0}]_{\Gamma_0}$. Finally, if $W^{\underline{H}}$ is strictly contained in W^{H_0}, then $\dim V^{H_0} < \dim W^{H_0} + \dim \Gamma_0/H_0$ and, from Theorem 1.1, one has $\Pi_0(H_0) = 0$. □

Lemma 7.2. (a) *If for all H's there is a complementing map $F_H^\perp$, then this is also true for all H_0's. In this case, P_* maps $\tilde{\Pi}(H)$ into $\tilde{\Pi}_0(H_0)$.*

(b) *If $V = \mathbb{R}^k \times U$, such that U and W satisfy hypothesis* (H) *for Γ (i.e.,* $\dim U^H = \dim W^H$ *for all H in* $\operatorname{Iso}(U)$ *and one has the Γ-equivariant map $x_j^{l_j}$) and if $W^{\underline{H}} = W^{H_0}$ for all H_0, then U and W satisfy hypothesis* (H) *for Γ_0. This is the case if $V = \mathbb{R}^k \times W$.*

Proof. Since $V^{H_0} = V^{\underline{H}}$, any complementing map for $\underline{H}$ will also work for H_0. Now, if $H_0 = H \cap \Gamma_0$ and $\underline{H}$ is a strict subgroup of H, let F^H be in $\Pi(H)$ and consider $(F^H, F_\perp^H)^{\underline{H}}$. Take $K_0 > H_0$, hence as above, $K_0 = \underline{K} \cap \Gamma_0$ with $\underline{K} > \underline{H}$. If $(F^H, F_\perp^H)^{K_0}(x) = 0$, then x is in V^H, since $F_\perp^H$ is zero only at the origin, and $F^H(x) = 0$. But x is in $V^{K_0} = V^{\underline{K}}$, thus x is fixed by H and $\underline{K}$. But H cannot be a strict subgroup of Γ_x, since one would have $F^H(x) \neq 0$ for F^H in $\Pi(H)$. That is, $\Gamma_x = H$ and $\underline{K} \leq H$. But the relation $\underline{H} < \underline{K}$ would imply $H_0 = K_0$, which is a contradiction. Thus, $(F^H, F_\perp^H)^{K_0} \neq 0$ if $K_0 > H_0$ and the pair $(F^H, F_\perp^H)$ belongs to $\tilde{\Pi}_0(H_0)$.

(b) is clear, since $U^{H_0} = U^{\underline{H}}$ and $U^{K_0} = U^{\underline{K}}$ and the Γ-equivariant map $x_j^{l_j}$ is also Γ_0-equivariant. □

Proposition 7.1. *If $V = \mathbb{R}^k \times U$, where U and W satisfy hypothesis* (H) *and $W^{\underline{H}} = W^{H_0}$, for all H in* $\operatorname{Iso}(U)$*, then, for any H such that* $\dim \Gamma_0/H_0 = \dim \Gamma/H = k$*, one has*

$$P_*[F^H, F_\perp^H]_\Gamma = \deg((F_\perp^H)^{\underline{H}}) \frac{|\tilde{H}_0/H|}{|\tilde{H}_0^0/H_0|} [F_0^{H_0}, F_\perp^{\underline{H}}]_{\Gamma_0},$$

where F^H generates $\Pi(H)$ and $F_0^{H_0}$ generates $\Pi_0(H_0)$.

Proof. Since F^H is in $\Pi(H)$, one has that $(F^H, F_\perp^H)$ is non-zero on ∂B_k and, from Theorem 3.4, for any $H_i \leq \tilde{H}_0$, with $\dim \Gamma/H_i = k$, one has

$$\deg(F^{H_i}; B_k^{H_i}) = \sum_{H_i \leq H_j \leq \tilde{H}_0} \beta_{ij} d_j |\tilde{H}_0/H_j|,$$

where $\beta_{ij} = \deg((F_\perp^{H_j})^{H_i})$.

Then, for $F = (F^H, F_\perp^H)$, the degree on the left is a product and the degree of $(F^H)^{H_i}$ corresponds to $V^H \cap V^{H_i}$, which has isotropy larger than H, i.e., there $F^K \neq 0$, unless $H_i < H$, in which case the degree is $\beta_{H_i H} |\tilde{H}_0/H|$. On the right-hand side, one has $d_j = 0$, except for $d_H = 1$. Hence, $\deg(F^H; B_k^H) = |\tilde{H}_0/H|$.

From the product theorem, one has

$$\deg(F^K, F_\perp^H; B_k) = |\tilde{H}_0/H| \deg(F_\perp^H).$$

Now, as a Γ_0-map, one has, from Lemma 7.2, that

$$P_*[F^H, F_\perp^H]_\Gamma = a[F_0^{H_0}, F_\perp^{\underline{H}}]_{\Gamma_0},$$

where a is an integer, recalling that $F_{\perp}^{H}$ is a complementing map for H_0. Since $\deg(F_0^{H_0}, B_k^{H_0}) = |\tilde{H}_0^0/H_0|$, one gets

$$a = |\tilde{H}_0/H|/|\tilde{H}_0^0/H_0| \deg(F_{\perp}^{H})/\deg(F_{\perp}^{H})$$

and the result follows. □

Corollary 7.3. *If* $V = \mathbb{R}^k \times W$ *and*

$$[F]_\Gamma = [\tilde{F}]_\Gamma + \sum_{\dim \Gamma/H=k} d_H[\tilde{F}_H],$$

where $\tilde{F}$ *corresponds to isotropy subgroups* K *with* $\dim \Gamma/K < k$*, then*

$$P_*[F]_\Gamma = P_*[\tilde{F}]_\Gamma + \sum_{\substack{H_0=H\cap\Gamma_0 \\ \dim \Gamma_0/H_0=k}} \frac{|\tilde{H}_0/H|}{|\tilde{H}_0^0/H_0|} d_H[\tilde{F}_{H_0}]_{\Gamma_0}.$$

Proof. Since $\dim \Gamma_0/H_0 \leq \dim \Gamma/H$, one has that $P_*[\tilde{F}]$ corresponds to subgroups with Weyl group of dimension less than k. Similarly, if $\dim \Gamma/K > k$, then the component of F on that $\Pi(K)$ is 0 and so it does not appear in $\Pi_0(K_0)$, even if $\dim \Gamma_0/K_0 = k$. Here, $\beta_{\underline{H}H} = 1$. □

Example 7.1. If $k = 0$ and $\Gamma_0 = \{e\}$, with $V = W$, then for

$$[F]_\Gamma = \sum_{\dim \Gamma/H=0} d_H[\tilde{F}_H]_\Gamma,$$

one obtains $P_*[F]_\Gamma = \deg(F; B)[\mathrm{Id}] = \left(\sum d_H|\Gamma/H|\right)[\mathrm{Id}]$. (Compare with Corollary 3.1)

For instance, if $\mathbb{Z}_n$ acts on $\mathbb{C}^2$, via $(x, e^{2\pi ik/n}z)$, $k = 0, \ldots, n-1$, then the map

$$F(t, x, z) = (2t + 1 - 2(|x| + |z|), \bar{x}(x^n - 1), z(z^n - 1))$$

has zeros at $(x = 0, z^n = 1)$, with index -1 and at $(x^n = 1, z = 0)$, with index 1, and $t = 1/2$. One has $\deg(F^\Gamma; B^\Gamma) = n$, $\deg(F, B) = 0$. From Corollary 3.1,

$$\begin{pmatrix} n \\ 0 \end{pmatrix} = \begin{pmatrix} 1 & 0 \\ 1 & n \end{pmatrix} \begin{pmatrix} d_0 \\ d_1 \end{pmatrix}.$$

Hence, $d_0 = n$ and $d_1 = -1$. As a $\mathbb{Z}_n$-map, one has

$$[F]_\Gamma = n[F_0]_\Gamma - [F_1]_\Gamma,$$

with $F_0 = (2t - 1/2, x_0, z)$ and $F_1 = (2t + 1 - 2|z|, x, z(z^n - 1))$, while $[F]_{\{e\}} = 0$.

There are two other cases where we have explicit generators: the case $k = 1$ and for orthogonal maps. Let us consider first the case $k = 1$.

Let $V = \mathbb{R} \times W$, then $\Pi^{\Gamma}_{S^V}(S^W)$ is generated, in its free part, by $[F^H]_\Gamma$ as above for $\dim \Gamma/H = 1$, and for H with $\Gamma/H = A = \mathbb{Z}_{p_1} \times \cdots \times \mathbb{Z}_{p_m}$, by η'_j and $\tilde{\eta}'$, $j = 1, \dots, m$, given in terms of the auxiliary space $X = (Z_1, \dots, Z_m)$ with action of Γ/H on Z_j given by $\exp(2\pi i/p_j)$. Then η'_j and $\tilde{\eta}'$ are given in Theorem 5.4 and one has the relations

$$p_j(\eta'_j + \tilde{\eta}') = 0, \quad 2\tilde{\eta}' = 0.$$

Similar definitions hold for $\Gamma_0/H_0 = A_0$.

Proposition 7.2. (a) *If* $\dim \Gamma/H = \dim \Gamma_0/H_0 = 1$, *then*

$$P_*[F^H]_\Gamma = \frac{|\tilde{H}_0/H|}{|\tilde{H}^0_0/H_0|}[F^{H_0}]_{\Gamma_0}.$$

(b) *If* $\dim \Gamma/H = \dim \Gamma_0/H_0 = 0$, *then for* $j = 1, \dots, m$

$$P_*[\eta'_j]_\Gamma = \frac{|A|}{|A_0|}\frac{p_{0j}}{p_j}[\eta'_{0j}]_{\Gamma_0} + \tilde{d}_j[\tilde{\eta}'_0]_{\Gamma_0},$$

where $\tilde{d}_j$ *is* 0 *or* 1 *and* $\tilde{d}_j = 0$ *if* $|A_0|$ *or* p_j *is odd. Moreover,*

$$P_*[\tilde{\eta}']_\Gamma = \frac{|A|}{|A_0|}[\tilde{\eta}'_0]_{\Gamma_0}.$$

(c) *If* $\dim \Gamma/H = 1$ *and* $\dim \Gamma_0/H_0 = 0$, *then*

$$P_*[F'^H]_\Gamma = \frac{|\tilde{H}_0/H|}{|A_0|}p_{01}[\eta'_{01}]_{\Gamma_0} + \tilde{d}[\tilde{\eta}'_0]_{\Gamma_0},$$

where $\tilde{d} = 0$ *if* $|A_0|$ *is odd, and* F'^H *is given below.*

Proof. (a) was already proved in the previous proposition. For (b), notice that if Γ acts as $\exp(2\pi i/p_j)$ on Z_j, then Γ_0 has to act as $\exp(2\pi i/p_{0j})$, where p_{0j} divides p_j. Hence, $|A_0|$ divides $|A|$.

From Theorem 5.4 and 5.7, the components of $P_*[\eta'_j]_\Gamma$ on η'_{0i} can be computed via $\deg(\eta'_j; B^{H_0} \cap \{\operatorname{Arg} Z_i = 0\})/\prod_{i \neq k} p_{0k}$. Since

$$\eta'_j = \Big(1 - \prod |Z_j|, X_0, \{x_i\}, (Z_i^{p_i} + 1)Z_i, Z'_i, \lambda Z_j, Z'_j\Big),$$

it is clear that this degree is 0 if $i \neq j$ and $\prod_{j \neq k} p_k / \prod_{j \neq k} p_{0k}$, if $i = j$.

Now, if one computes the ordinary class of both sides in $\Pi_{n+1}(S^n)$, one obtains that $[P_*\eta'_j] = \big(\prod_{j \neq k} p_k\big)\eta$, where η is the suspension of the Hopf map, while, on the right-hand side, one has $|A|/p_j\eta + \tilde{d}_j|A_0|\eta$. Thus, $\tilde{d}_j|A_0|\eta = 0$, in particular, $\tilde{d}_j = 0$ if $|A_0|$ is odd.

Furthermore, since $\tilde{\eta}'$ is the Hopf map based on the fundamental cell for Γ/H and the fundamental cell for Γ_0/H_0 is generated by $|A|/|A_0|$ copies of the first one, with a suspension on the variables on $X^{\underline{H}} \cap (X^H)^\perp$, one has

$$P_*[\tilde{\eta}']_\Gamma = |A|/|A_0|[\tilde{\eta}'_0]_{\Gamma_0}.$$

Then, from the relations $p_j(\eta'_j + \tilde{\eta}') = 0$, one has

$$(|A|/|A_0|)(p_{0j}\eta'_{0j} + p_j\tilde{\eta}'_0) + p_j\tilde{d}_j\tilde{\eta}'_0 = 0,$$

or else

$$p_j\tilde{d}_j + (|A|/|A_0|)(p_j - p_{0j}) \text{ is even.}$$

Hence, if p_j is odd, p_{0j}, which divides p_j, is also odd and one has $\tilde{d}_j = 0$.

For (c), one has $\Gamma/H \cong S^1 \times \mathbb{Z}_{p_2} \times \cdots \times \mathbb{Z}_{p_m}$ and, using the auxiliary space X, one may take the action of Γ on Z_1, as $e^{i\varphi}$ and on Z_j as $e^{2\pi i/p_j}$, while Γ_0 acts as $e^{2\pi i/p_{0j}}$. One may take

$$F'^H = (2t + 1 - 2\Pi|Z_j|, X_0, x_i, \tilde{\lambda}Z_1, (Z_j^{p_j} + 1)Z_j),$$

where $\tilde{\lambda} = \mu + i(|Z_1| - 1)$, see Theorem 3.3.

Again, the components of $P_*[F^H]_\Gamma$ on η'_{0j} are given by

$$\deg(F^H; B^{H_0} \cap \{\operatorname{Arg} Z_j = 0\}) / \prod_{k\neq j} p_{0k}.$$

Hence, these components are 0 if $j > 1$ (since $Z_j^{p_j} + 1 \neq 0$, for Z_j real and positive) and $\left(\prod_{k>1} p_k\right)/\left(\prod_{k>1} p_{0k}\right)$ for $j = 1$.

The fact that $\tilde{d}$ is 0 if $|A_0|$ is odd is proved as above. □

Example 7.2. Consider the action of S^1 on Fourier series, i.e., as $\exp(im\varphi)$ on z_m, for $m \geq 0$, which is broken to a $\mathbb{Z}_p$-action for $\Gamma_0 = \{\varphi = 2k\pi/p, k = 0, \dots, p-1\}$. Here $V = \mathbb{R} \times W$ and, according to Theorem 3.2, one has

$$\Pi^{S^1}_{S^{\mathbb{R}\times W}}(S^W) \cong \mathbb{Z}_2 \times \mathbb{Z} \times \cdots \times \mathbb{Z},$$

where $\mathbb{Z}_2$ corresponds to the invariant part and is generated by the suspension of the Hopf map $(1-|x_0|^2-|x_1|^2, \lambda(x_0+ix_1))$, where x_0, x_1 are in W^Γ and $\lambda = \mu+i(2t-1)$. The isotropy subgroups are of the form $H_m = \{\varphi = 2k\pi/m, k = 0, \dots, m-1\}$, with $W^{H_m} = \{z_n, n \text{ multiple of } m\}$ and the corresponding generator is the suspension of

$$\eta_m = (1 - |z_m|, \lambda z_m).$$

Note that Theorem 3.3 gives the generator $(2t+1-2|z_m|, \tilde{\lambda} z_m) = \eta'_m$ with $\tilde{\lambda} = \mu + i(|z_m|-1)$. The homotopy

$$((1-\tau)(2t-1)+2-2|z_m|, (\mu + i((1-\tau)(|z_m|-1)+\tau(2t-1)))z_m)$$

is valid, since if $z_m = 0$, one has $(2-\tau)(2t-1)+2 \geq 1$, and otherwise one has a rotation between $2t-1$ and $1-|z_m|$. Thus, $\eta_m = \eta'_m$. Hence, any S^1-map will be of the form

$$[F]_{S^1} = \sum_{m \geq 0} d_m [\eta_m]_{S^1}.$$

On the other hand, the isotropy subgroups of Γ_0 are of the form $H_0 = \{\varphi = 2kp'\pi/p,\ k = 0, 1, \dots, p/p'-1\}$, where p' divides p, that is $H_0 \cong \mathbb{Z}_{p/p'}$ and $\Gamma_0/H_0 \cong \mathbb{Z}_{p'}$. Furthermore, z_m belongs to W^{H_0} if m is a multiple of p/p' and z_m will have exactly the isotropy H_0 if $m = kp/p'$, with k and p' relatively prime (if $k/p' = k'/p''$ then z_m belongs to $W^{H'_0}$, with $H'_0 = \mathbb{Z}_{p/p''} > H_0$). One may write $k = m' + k'p'$, with $1 \leq m' < p'$ such that m' and p' are relatively prime. Hence, $m = m'p/p' + k'p$.

Now, any $\mathbb{Z}_p$-map can be written as

$$[G]_{\mathbb{Z}_p} = \sum_{p'} (d'_{p/p'}[\eta'_{p/p'}]_{\mathbb{Z}_p} + \tilde{d}_{p/p'}[\tilde{\eta}_{p/p'}]_{\mathbb{Z}_p}),$$

where for any divisor p' of p, one has that $\eta'_{p/p'}$ is the suspension of

$$(1-|z_{p/p'}|, \lambda z_{p/p'}).$$

In this case, according to Proposition 7.2 (c), one has that

$$P_*[\eta_{p/p'}] = [\eta'_{p'}]$$

since in this case $|A_0| = p' = p_{01}$ and $|\tilde{H}_0/H| = 1$.

Now, one could have taken a different generator for $\Pi_0(H_0)$, for instance $(1-|z_m|, \lambda z_m)$, with $m = m'p/p'+k'p$. From Proposition 5.1, we have that $[\eta_m]_{\mathbb{Z}_p} = n'[\eta'_{p/p'}]_{\mathbb{Z}_p}$, where $|n'|$ is odd and $n'm' \equiv 1$, modulo p' (in fact $(m'+k'p')n' \equiv 1$, modulo p').

Proposition 7.3. *Under the above hypothesis, if*

$$[F]_{S^1} = \sum_{m \geq 0} d_m [\eta_m]_{S^1},$$

then

$$P_*[F] = \sum_{p'|p} d'_{p/p'}[\eta'_{p/p'}]_{\mathbb{Z}_p},$$

where

$$d'_{p/p'} = \sum_j n_j \Big(\sum_{k \geq 0} d_{m_j p/p' + kp} \Big)$$

with $|n_j|$ *odd,* $n_j m_j \equiv 1$*, modulo* p'*, and* $1 \leq m_j < p'$*, with* m_j *and* p' *relatively prime. The number* $d'_{p/p'}$ *is in* $\mathbb{Z}_{p'}$ *if* p' *is even and in* $\mathbb{Z}_{2p'}$ *if* p' *is odd. The number* d'_p *is in* $\mathbb{Z}_2$*, corresponds to* $H_0 \equiv \mathbb{Z}_p$ *and is*

$$d'_{\Gamma_0} = d'_p = \sum_{k \geq 0} d_{kp}.$$

For instance, if $p = 2$, then one has

$$d'_{\Gamma_0} = \sum d_{2k} \bmod 2, \qquad d'_{\{e\}} = \sum d_{2k+1} \bmod 2.$$

For $p = 3$, one has

$$d'_{\Gamma_0} = \sum d_{3k} \bmod 2, \qquad d'_{\{e\}} = \sum (d_{3k+1} - d_{3k+2}) \bmod 6.$$

For $p = 4$, one has

$$\begin{aligned} d'_{\Gamma_0} &= \sum d_{4k} \bmod 2, \qquad d'_{\mathbb{Z}_2} = \sum d_{4k+2} \bmod 2, \\ d'_{\{e\}} &= \sum (d_{4k+1} - d_{4k+3}) \bmod 4. \end{aligned}$$

For $p = 5$ one has

$$d'_{\Gamma_0} = \sum d_{5k} \bmod 2,$$

$$d'_{\{e\}} = \sum (d_{5k+1} - d_{5k+4}) + 3 \sum (d_{5k+2} - d_{5k+3}) \bmod 10.$$

For $p = 6$, one has

$$\begin{aligned} d'_{\Gamma_0} &= \sum d_{6k} \bmod 2, d'_3 = \sum d_{6k+3} \bmod 2, \text{ for } p' = 2, \\ d'_2 &= \sum (d_{6k+2} - d_{6k+4}) \bmod 6, \text{ for } p' = 3, \\ d'_{\{e\}} &= \sum (d_{6k+1} - d_{6k+5}) \bmod 6. \end{aligned}$$

For $p = 7$, one has

$$\begin{aligned} d'_{\Gamma_0} &= \sum d_{7k} \bmod 2, \\ d'_{\{e\}} &= \sum (d_{7k+1} - d_{7k+6}) - 3 \sum (d_{7k+2} - d_{7k+5}) + 5 \sum (d_{7k+3} - d_{7k+4}) \bmod 14. \end{aligned}$$

In order to illustrate this sort of result, consider the following system in $\mathbb{R}^2 \times \mathbb{C}^2$, with action on z_j as $e^{ij\varphi}$, for $j = 1$ and $p - 1$:

$$f = (1 - |z_1|^2 - |z_{p-1}|^2, \lambda z_1, \lambda z_{p-1}).$$

This map is S^1-equivariant and non-zero on $\partial\{(\lambda, z_1, z_{p-1}) : |\lambda| \leq 1, |z_j| \leq 2, j = 1, p-1\}$. One may perturb λz_{p-1} to $(\lambda - \varepsilon)z_{p-1}$, for $|\varepsilon| < 1$, obtaining the zeros $(\lambda = 0, |z_1| = 1, z_{p-1} = 0)$ and $(\lambda = \varepsilon, z_1 = 0, |z_{p-1}| = 1)$. The S^1-degree is then the sum of two S^1-degrees, the first one, near $\lambda = 0$, is $[\eta_1]$, while the second one, near $\lambda = \varepsilon$, is $[\eta_{p-1}]$. Hence,

$$[f]_{S^1} = [\eta_1]_{S^1} + [\eta_{p-1}]_{S^1}.$$

(This result may also be obtained by using Whitehead's homomorphism: this method will be used, in next chapter, when discussing bifurcation).

One may perturb f to get a $\mathbb{Z}_p$-map:

$$f_\varepsilon = (1 - |z_1|^2 - |z_{p-1}|^2, \lambda z_1 + \varepsilon \bar{z}_{p-1}, \lambda z_{p-1} - \varepsilon \bar{z}_1).$$

However, conjugating the third equation, one has

$$\begin{pmatrix} \lambda & \varepsilon \\ -\varepsilon & \bar{\lambda} \end{pmatrix} \begin{pmatrix} z_1 \\ \bar{z}_{p-1} \end{pmatrix}$$

whose only zero, for $\varepsilon \neq 0$, is $z_1 = z_{p-1} = 0$, i.e., f_ε is never zero. Here $n_1 = 1$, $n_{p-1} = -1$ and $[f_\varepsilon]_{\mathbb{Z}_p} = 0$.

Let us conclude this subsection by considering the case of orthogonal maps, i.e., the morphism

$$P_\perp : \Pi^\Gamma_{\perp S^V}(S^V) \to \Pi^{\Gamma_0}_{\perp S^V}(S^V),$$

where $\Gamma_0 = T^{n_0} \times \ldots$ is a subgroup of $\Gamma = T^n \times \ldots$ It is clear that one may choose T^{n_0} to correspond to $\varphi_1, \ldots, \varphi_{n_0}$.

We have seen, in Theorem 6.1, that $\Pi^\Gamma_{\perp S^V}(S^V)$. has explicit generators F_H, which are orthogonal maps, for each isotropy subgroup H. Furthermore, if $\dim \Gamma/H = k$, with $A_1 x, \ldots, A_k x$ linearly independent in V^H, then,

$$F(\lambda_1, \ldots, \lambda_k, x) = F_H(x) + \sum_1^k \lambda_j A_j x$$

may be taken as the generator of $\Pi(H)$ in $\Pi^\Gamma_{S^{\mathbb{R}^k \times V}}(S^V)$.

From Proposition 7.1, we know that if $\dim \Gamma_0/H_0 = \dim \Gamma/H = k$, then

$$P_*[F(\lambda_1, \ldots, \lambda_k, x)]_\Gamma = \frac{|\tilde{H}_0/H|}{|\tilde{H}_0^0/H_0|}[F_0(\lambda_1, \ldots, \lambda_k, x)]_{\Gamma_0},$$

where $F_0(\lambda_1, \ldots, \lambda_k, x) = F_{H_0}(x) + \sum_1^k \lambda_j A_j^0 x$ is the generator for $\Pi_0(H_0)$, and A_j^0 are the infinitesimal generators for the action of Γ_0. We shall prove the following

Proposition 7.4.

$$P_\perp\Big(\sum_{H<\Gamma} d_H[F_H]_\perp\Big) = \sum_{H_0<\Gamma_0}\Big(\sum\nolimits_1 d_H \frac{|\tilde{H}_0/H|}{|\tilde{H}_0^0/H_0|}\Big)[F_{H_0}]_\perp,$$

where the sum $\sum_1$ *is over all* H *with* $H_0 = H \cap \Gamma_0$ *and* $\dim \Gamma/H = \dim \Gamma_0/H_0$. *In particular,* $P_\perp[F_H]_\perp = 0$ *if* $\dim \Gamma_0/H_0 < \dim \Gamma/H$.

Proof. From the proof of Theorem 6.1, it is clear that one may take the generators for the parametrized problem as $F_H(\lambda_1, \dots, \lambda_k, x)$. If $k = k_0$, then $A_1x, \dots, A_kx$ are linearly independent for x with $\Gamma_x = H$ and $\Gamma_{0x} = H_0$. Proposition 7.1 will give part of the answer.

On the other hand, if $k_0 < k$ for some H, then, since $\underline{H} < H$, where $\underline{H}$ is the minimal isotropy subgroup such that $H_0 = H \cap \Gamma_0$, one has $\dim \Gamma/\underline{H} \geq \dim \Gamma/H$. But, since $V^{\underline{H}} = V^{H_0}$, the only possibility is that $n_0 < n$ and the action of T^{n_0} on $V^{\underline{H}}$ reduces the number of linearly independent A_jx from k to k_0. Assume then that $A_1x, \dots, A_{k_0}x$ correspond to Γ_0 and are linearly independent if $\Gamma_{0x} = H_0$, while $A_1x, \dots, A_kx$ correspond to Γ and are linearly independent if $\Gamma_x = H$ (and a fortiori if $\Gamma_x = \underline{H}$).

Consider the map $F_H(x) + \tilde{A}_{k_0+1}(x)$, where $\tilde{A}_{k_0+1}(x)$ is the vector constructed from the Gram–Schmidt process and orthogonal to $A_1x, \dots, A_{k_0}x$ (see § 7 of Chapter 1), hence it is a Γ_0-orthogonal map in V^{H_0}. Now, the zeros of this map are such that $F_H(x) = (F_H^H(x_H), Z) = 0$ and $A_{k_0+1}x$ is a linear combination of $A_1x, \dots, A_{k_0}x$ (since $F_H(x)$ is Γ-orthogonal to all A_jx). But then, $Z = 0$, x_H which has isotropy H is such that $A_1x_H, \dots, A_kx_H$ are linearly independent. This means that this map has no zeros. But $P_\perp[F_H]_\perp = [F_H + \tilde{A}_{k_0+1}(x)]_\perp = 0$ (since as Γ_0-orthogonal map, F_H and $F_H + \tilde{A}_{k_0+1}(x)$ are Γ_0-homotopic). □

3.7.3 Products

We have considered, in § 6 of Chapter 2, a product of maps $(f_1(x_1), f_2(x_2))$ defined on a product $\Omega = \Omega_1 \times \Omega_2$ from $V_1 \times V_2$ into $W_1 \times W_2$, where f_1 and f_2 are Γ-equivariant, and Ω_i are Γ-invariant, open and bounded. The associated maps, which define the Γ-degree, are $F_i(t_i, x_i) = (2t_i + 2\varphi_i(x_i) - 1, f_i(x_i))$.

As shown in Lemma 6.1 of Chapter 2, $[F_1, F_2] = \Sigma_0 \deg_\Gamma((f_1, f_2); \Omega_1 \times \Omega_2)$, where Σ_0 is the suspension by $2t_2 - 1$.

Note that if $[F_i]$ belongs to $\Pi^\Gamma_{S^{V_i}}(S^{W_i})$, then $[F_1, F_2]$ is in $\Pi^\Gamma_{S^{V_1\times\mathbb{R}\times V_2}}(S^{W_1\times\mathbb{R}\times W_2})$ and one has a morphism of groups, i.e.,

$$\begin{aligned}
[F_1 + G_1, F_2] &= [F_1, F_2] + [G_1, F_2]\\
[F_1, F_2 + G_2] &= [F_1, F_2] + [F_1, G_2],
\end{aligned}$$

where, for this last operation, with the sum defined on t_2, one has to translate this sum on t_1. This is done as in any text on homotopy and is left to the reader. Hence, if $[F_1]$ and $[F_2]$ are expressed as sums, as we have already seen in several examples, one may expand $[F_1, F_2]$ in terms of elementary products of the generators. Let $V = V_1 \times \mathbb{R} \times V_2$ and $W = W_1 \times \mathbb{R} \times W_2$. ($t_2$ will be absorbed in V_2).

Lemma 7.3. (a) *Any isotropy subgroup H for V is of the form $H_1 \cap H_2$, with H_i in* Iso(V_i). *There are minimal isotropy subgroups $\underline{H}_i$, with $H = \underline{H}_1 \cap \underline{H}_2$, $V^{\underline{H}_i} = V_i^H$ and*

$$\dim \Gamma/\underline{H}_i \leq \dim \Gamma/H \leq \dim \Gamma/\underline{H}_1 + \dim \Gamma/\underline{H}_2.$$

(b) *If $[F_i]$ is in $\Pi(\underline{H}_i)$, then $[F_1, F_2]$ is in $\Pi(H)$. If for any H_i there are complementing maps $F_\perp^i$, then, if $[F_i, F_\perp^i]$ is in $\Pi(H_i)$, we have that $[F_1, F_\perp^1, F_2, F_\perp^2]$ is in $\tilde{\Pi}(H)$.*

(c) *If $V_i = \mathbb{R}^{k_i} \times U_i$ and hypothesis* (H) *holds for U_i and W_i and furthermore $W^{\underline{H}_i} = W_i^H$, then $V = \mathbb{R}^{k_1+k_2} \times \mathbb{R} \times U$ and hypothesis* (H) *holds for U and W. This is the case if $V_i = \mathbb{R}^{k_i} \times W_i$.*

Proof. If $H = \Gamma_{(x_1,x_2)}$, then $H = \Gamma_{x_1} \cap \Gamma_{x_2} = H_1 \cap H_2$, by recalling that $\Gamma_x = \bigcap H_j$ over the isotropy subgroups of the non-zero variables in x. Then $V^H = V_1^H \times \mathbb{R} \times V_2^H$. Now, if $\underline{H}_i$ is the isotropy of V_i^H (see Definition 2.1 of Chapter 1), then $H < \underline{H}_i < H_i$ and $V_i^{\underline{H}_i} = V_i^H$. Since $H = H_1 \cap H_2$, one has $\dim \Gamma/H_i \leq \dim \Gamma/\underline{H}_i \leq \dim \Gamma/H$.

In the decomposition of Γ/H over the isotropy subgroups of the coordinates of V, one obtains the groups $\tilde{H}'_{i-1}/\tilde{H}'_i$ for the first coordinates, corresponding to $V_1^{\underline{H}_1}$, with order k_i^1, and then $H_1 \cap \tilde{H}^2_{i-1}/H_1 \cap \tilde{H}^2_i$, for the coordinates of $V_2^{\underline{H}_2}$, with order $\tilde{k}_i^2$. We shall denote by k_i^2 the order of $\tilde{H}^2_{i-1}/\tilde{H}^2_i$, corresponding to the coordinates of $V_2^{\underline{H}_2}$. If k_i^2 is finite, then any γ in $\tilde{H}^2_{i-1}$ can be written as $\gamma_i^\alpha \tilde{\gamma}$, where $0 \leq \alpha < k_i^2$, $\gamma_i^{k_i^2}$ and $\tilde{\gamma}$ are in $\tilde{H}_i^2$. In particular, for γ in $H_1 \cap \tilde{H}^2_{i-1}$, one has that $\gamma^{k_i^2}$ is in $H_1 \cap \tilde{H}_i^2$, then $\tilde{k}_i^2$ divides k_i^2.

Thus, the number of k_i's infinite for V^H is the sum of the number for those of $V_1^{\underline{H}_1}$ and a quantity less or equal to the number of those for $V_2^{\underline{H}_2}$. Note that when $H_1 \cap \tilde{H}^2_{i-1} = H$, then $\tilde{k}_j^2 = 1$ for $j \geq i$.

For (b), if $K = \underline{K}_1 \cap \underline{K}_2 > H_1 \cap H_2$, then $V^K = V_1^{\underline{H}_1} \times \mathbb{R} \times V_2^{\underline{K}_2}$ is strictly contained in $V^H = V_1^{\underline{H}_1} \times \mathbb{R} \times V_2^{\underline{H}_2}$. Then, either $\underline{K}_1 > \underline{H}_1$, or $\underline{K}_2 > \underline{H}_2$ and the corresponding $F_i^{\underline{K}_i} \neq 0$, i.e., $[F_1, F_2]$ is in $\Pi(H)$.

Also, if $(F_1, F_\perp^1, F_2, F_\perp^2)$ has a zero at (x_1, x_2) in V^K for $K > H$, then, since $F_\perp^i$ is zero only at the origin, (x_1, x_2) must be in $V_1^{H_1} \times V_2^{H_2}$, with $\Gamma_{(x_1,x_2)} \leq H_1 \cap H_2 = H$, leading to a contradiction. Thus, the above map is in $\tilde{\Pi}(H)$.

Finally, if (H) holds for $V_i = \mathbb{R}^{k_i} \times U_i$ and W_i, let $K = \underline{K}_1 \cap \underline{K}_2$ and $H = \underline{H}_1 \cap \underline{H}_2$. It is then clear that $\dim U^H \cap U^K = \dim W^H \cap W^K$, since $U^H = U_1^{\underline{H}_1} \times U_2^{\underline{H}_2}$ and likewise for K and one has $W_i^{\underline{H}_i} = W_i^H$. Note that in general $W_i^{\underline{H}_i} \subset W_i^H$. □

Proposition 7.5. (a) *If* $\dim V_i^{\underline{H}_i} = \dim W_i^{\underline{H}_i} + \dim \Gamma/\underline{H}_i$, $i = 1, 2$ *and* $\dim \Gamma/H = \dim \Gamma/\underline{H}_1 + \dim \Gamma/\underline{H}_2$, *then, for* $[F_i]$ *in* $\Pi(\underline{H}_i)$, *one has*

$$\deg_E(F_1, F_2) = \deg_E(F_1)\deg_E(F_2)\prod(k_i^2/\tilde{k}_i^2)$$

if $W_i^{\underline{H}_i} = W_i^H$ *and* 0 *otherwise.*

(b) *If* $V_i = \mathbb{R}^{k_i} \times U_i$ *and let* U_i *and* W_i *satisfy hypothesis* (H) *and* $W^{\underline{H}_i} = W_i^H$. *Assume* $\dim \Gamma/H_i = k_i$ *and* $\dim \Gamma/H = k_1 + k_2$, *then, for* $[F_i, F_\perp^i]$ *in* $\tilde{\Pi}(H_i)$, *one has* $[F_1, F_\perp^1, F_2, F_\perp^2] = d_H[F_H]$, *where* F_H *is the generator for* $\tilde{\Pi}(H_1 \cap H_2)$ *and*

$$d_H = \beta_{\underline{H}_1 H_1}\beta_{\underline{H}_2 H_2}\frac{|\tilde{H}_1^0/H_1| \cdot |\tilde{H}_2^0/H_2|}{|\tilde{H}_1^0 \cap \tilde{H}_2^0/H_1 \cap H_2|}.$$

Here $\tilde{H}_i^0$ *is the maximal isotropy subgroup containing* H_i, *with* $\dim \Gamma/\tilde{H}_i^0 = k_i$ *and* $\beta_{\underline{H}_i H_i} = \deg((F_\perp^i)^{\underline{H}_i})$.

(c) *Furthermore, if* $[F_i]_\Gamma = \sum d_j^i[F_{H_j^i}]_\Gamma + [\tilde{F}_i]_\Gamma$ *with* $\dim \Gamma/H_j^i = k_i$ *and* $\tilde{F}_i$ *in* $\tilde{\Pi}_{k_i-1}$, *then*

$$[F_1, F_2]_\Gamma = \sum d_j^1 d_k^2 d_{H_j \cap H_k}[F_{H_j \cap H_k}]_\Gamma + [\tilde{F}]_\Gamma,$$

where the sum is over all (j, k)*'s such that* $\dim \Gamma/H_j \cap H_k = k_1 + k_2$, $d_{H_j \cap H_k}$ *is as above and* $[\tilde{F}]_\Gamma$ *belongs to* $\tilde{\Pi}_{k_1+k_2-1}$, *as defined in Theorem* 3.2.

Proof. It is clear that the fundamental cell for $\underline{H}_1 \cap \underline{H}_2$ is the product of the fundamental cell for $\underline{H}_1$ by the fundamental cell for $\underline{H}_1 \cap \underline{H}_2$ on V_2. The dimension conditions imply that $\tilde{k}_2^j$ is infinity exactly when $k_2^j = \infty$. From Theorem 1.2, one has

$$\deg_E(F_1, F_2) = \deg((F_1, F_2); B_{k_1} \times B_{k_2})\Big/\Big(\prod k_j^1\Big)\Big(\prod \tilde{k}_j^2\Big)$$

if $W^H = W^{\underline{H}_1} \times \mathbb{R} \times W^{\underline{H}_2}$ and 0 otherwise. From the degree of a product, one obtains the result.

For (b), from Lemma 7.4 (b) and (c), one sees that it is enough to compute d_H. Now, as in Proposition 7.1, the map $[F_1, F_\perp^1, F_2, F_\perp^2]$ is non-zero if $z_j = 0$ for any j with k_j^1 or k_j^2 (hence $\tilde{k}_j^2$) infinite. That is, one may apply Theorem 3.4 (on global Poincaré sections). Thus

$$\begin{aligned}\beta_{H_1}\beta_{H_2}\deg\left(F_1^{H_1}|_{B_{k_1}}, F_2^{H_2}|_{B_{k_2}}\right) &= \beta_{H_1}\beta_{H_2}\deg(F_1^{H_1}; B_{k_1})\deg(F_2^{H_2}; B_{k_2})\\ &= \beta_{H_1}\beta_{H_2}|\tilde{H}_1^0/H_1||\tilde{H}_2^0/H_2|\\ &= \beta_H d_H|\tilde{H}_1^0 \cap \tilde{H}_2^0/H_1 \cap H_2|,\end{aligned}$$

since clearly $\tilde{H}_1^0 \cap \tilde{H}_2^0$ is the maximal isotropy subgroup for $H_1 \cap H_2$ (recalling the dimension hypothesis of (b)). Here, $\beta_{H_i} = \deg(F_\perp^i) = \deg(F_\perp^i; V_i^{\underline{H}_i})\beta_{\underline{H}_i}$. Since one may complement F_H by $(F^1_\perp, F^2_\perp)\,|_{(V^H)^\perp}$ with degree equal to $\beta_{\underline{H}_1}\beta_{\underline{H}_2}$, one obtains the result.

Note that, we have $\tilde{H}^0_1 \cap \tilde{H}^0_2/H_1\cap H_2 = (\tilde{H}^0_1\cap\tilde{H}^0_2/H_1\cap\tilde{H}^0_2)\,(H_1\cap\tilde{H}^0_2/H_1\cap H_2)$. The first quotient has order $\prod k^1_j$ since the coordinates coming from $\tilde{H}^0_2$ have $k^2_j=\infty$, and the second has order $\prod \tilde{k}^2_j$. Hence (a) and (b) give the same result for $H_i = \underline{H}_i$.

For (c) it is enough to note that if $[\tilde{F}_1]_\Gamma$ belongs to $\tilde{\Pi}_{k_1-1}$, i.e., to subgroups with $\dim \Gamma/K < k_1$, then, from Lemma 7.4 (a), $[\tilde{F}_1, F_2]$ is in $\tilde{\Pi}_{k_1+k_2-1}$. Then one applies the bilinearity of the product. □

Example 7.3. If $V_1 = \mathbb{R}\times W_1$ and $V_2 = W_2$, then the only relevant isotropy subgroups for the product are those H_1 with $\dim\Gamma/H_1 \le 1$ and those H_2 with $\dim\Gamma/H_2 = 0$. Assume $\dim\Gamma/H_1 = 1$ with $\Pi(H_1)$ generated by η_1 and $\Pi(H_2)$ generated by η_2. Then, from Lemma 7.4, one has $\dim\Gamma/H = 1$ and $\Pi(H)$ generated by η. From Proposition 7.5 (b) one has

$$[\eta_1,\eta_2]_\Gamma = \frac{|\tilde{H}^0_1/H_1|\cdot|\Gamma/H_2|}{|\tilde{H}^0_1/H_1\cap H_2|}[\eta]_\Gamma.$$

Suppose now that $V_1 = \mathbb{R}\times W_1$, $V_2 = W_2$ and that $\dim\Gamma/H_1 = \dim\Gamma/H_2 = 0$. Then, from Lemma 7.4, one has $\dim\Gamma/H = 0$. We shall consider the presentations of Γ/H_1 and Γ/H given by the fundamental cell decomposition. That is, $\Gamma/H_1 = (\Gamma/\tilde{H}^1_1)(\tilde{H}^1_1/\tilde{H}^1_2)\dots(\tilde{H}^1_s/H_1)$, with $k^1_j = |\tilde{H}^1_{j-1}/\tilde{H}^1_j|$, as in Lemma 7.4. Similarly, Γ/H_2 will have the decomposition in $\prod(\tilde{H}^2_{j-1}/\tilde{H}^2_j)$, with order k^2_j and Γ/H with subgroups of order k^1_j for the coordinates of $V_1^{H_1}$ and of order $\tilde{k}^2_j = |H_1\cap\tilde{H}^2_{j-1}/H_1\cap\tilde{H}^2_j|$ for the coordinates of $V_2^{H_2}$, with $\tilde{k}^2_j$ dividing k^2_j and the coordinates of $V_i^{\underline{H}_i}\cap(V_i^{H_i})^\perp$ staying as suspensions.

As in §5, we shall use auxiliary spaces with a special action of Γ: namely the spaces X_1, X_2 and X, with

$$\begin{aligned} X_1 &= (Z_1, Z'_1,\dots,Z_{s_1},Z'_{s_1})\\ X_2 &= (Y_1, Y'_1,\dots,Y_{s_2},Y'_{s_2})\\ X &= X_1\times X_2 \end{aligned}$$

where s_i is the number of k^i_j which are larger than 1, the action on Z_j and Z'_j is by γ_j in $\tilde{H}^1_{j-1}/\tilde{H}^1_j$ and as a cyclic group of order k^1_j, while γ_j acts trivially on the other coordinates. If $k_j = 2$ and corresponds to a real variable of V_1, then Z_j is complex. The action of Γ on X_2 is similar but with k^2_j, while the action on X coincides for X_1 but, on X_2, it is as cyclic groups of order $\tilde{k}^2_j$. If $\tilde{k}^2_i = 1$, then the action is trivial. Then, on $X_1\times V_1$, one has the following generators for $\Pi(H_1)$:

$$\Sigma^{W_1}\eta^1_j = \Big(1-\prod|Z_i|, w_1, (Z_i^{k^1_i}+1)Z_i, Z'_i, \lambda_j Z_j, Z'_j\Big)$$

$$\Sigma^{W_1}\tilde{\eta}^1 = \Big(\varepsilon^2 - \prod_{i<s_1} |Z_i||Z_{s_1}^{k_{s_1}^1} + 1|, w_1, (Z_i^{k_i^1} + 1)Z_i, Z_i', \lambda(Z_{s_1}^{k_{s_1}^1} + 1)Z_{s_1}\Big),$$

where w_1 is in W_1 and $\lambda = \mu + i(2t_1 - 1)$.

The generator of $\Pi(H_2)$ on $X_2 \times V_2$, will be

$$\Sigma^{W_2}\eta_2 = \Big(2t_2 + 1 - 2\prod |Y_j|, w_2, (Y_j^{k_j^2} + 1)Y_j, Y_j'\Big),$$

with w_2 in W_2 and j going from 1 to s_2.

Finally, the generators of $\Pi(H)$ on $X \times V$ will be

$$\Sigma^W \eta_j = \Big(1 - \prod |X_i|, w, (X_i^{k_i} + 1)X_i, X_i', \lambda X_j, X_j'\Big),$$

with w in W and k_j being k_j^1 for $j = 1, \ldots s_1$ and $\tilde{k}_j^2$ afterward. The other generator $\Sigma^W \tilde{\eta}$ is constructed similarly.

For other presentations of $\Pi(H_i)$ we refer to Proposition 6.4 in [IV2]: the proof is much longer than the one for the present special case. Note that, from Theorem 7.1, all these Γ-suspensions are isomorphisms.

Proposition 7.6. *Under the above hypothesis one has*

$$\begin{aligned}
[\Sigma^{W_1}\eta_j^1, \Sigma^{W_2}\eta_2]_\Gamma &= \frac{|\Gamma/H_1| \cdot |\Gamma/H_2|}{|\Gamma/H_1 \cap H_2|}[\Sigma^W \eta_j]_\Gamma + \tilde{d}_j[\Sigma^W \tilde{\eta}]_\Gamma \\
[\Sigma^{W_1}\tilde{\eta}^1, \Sigma^{W_2}\eta_2]_\Gamma &= \frac{|\Gamma/H_1| \cdot |\Gamma/H_2|}{|\Gamma/H_2 \cap H_2|}[\Sigma^W \tilde{\eta}]_\Gamma
\end{aligned}$$

where $k_j^1 \tilde{d}_j$ is even.

Proof. Note first that $(\Sigma^{W_1}\eta_j^1, \Sigma^{W_2}\eta_2)$ is non-zero if $X_i = 0$, and that the action of Γ on $X \times V$ is such that the hypothesis of Theorem 5.3 may be applied, i.e.,

$$[F]_\Gamma \equiv [\Sigma^{W_1}\eta_j^1, \Sigma^{W_2}\eta_2]_\Gamma = \sum d_i[\Sigma^W \eta_i]_\Gamma + \tilde{d}_j[\tilde{\eta}],$$

where

$$d_i = \deg(F; B^H \cap \operatorname{Arg} X_i = 0) / \prod_{i \neq j} k_j.$$

Here, for $i \neq j$, $(Z^{k_i^1} + 1)Z_i$ or $(Y_i^{k_i^2} + 1)Y_i$ is 0 for $\operatorname{Arg} X_i = 0$, only if $X_i = 0$, in which case the first equation for η_j^1 or η_2 is non-zero. Hence $d_i = 0$ for $i \neq j$. On the other hand, it is easy to compute d_j as

$$d_j = \Big(\prod_{i \neq j} k_i^1\Big)\Big(\prod k_i^2\Big) / \prod_{i \neq j} k_i,$$

that is $d_j = (|\Gamma/H_1|/k_j^1)|\Gamma/H_2|/(|\Gamma/H|/k_j)$.

Since $k_j = k_j^1$ one has the first result.

For the same reasons as above, one has that $(\Sigma^{W_1}\tilde{\eta}_1, \Sigma^{W_2}\eta_2)$ is non-zero on the boundary of the fundamental cell for $X \times V$ (i.e., if $\operatorname{Arg} X_i = 0$ or $2\pi/k_i$), this implies that the class of this map is a multiple of $\Sigma^W\tilde{\eta}$. Counting the zeros of the map in the fundamental cell, one gets that

$$[\Sigma^{W_1}\tilde{\eta}^1, \Sigma^{W_2}\eta_2]_\Gamma = \prod (k_j^2/\tilde{k}_j^2)[\Sigma^W\tilde{\eta}]_\Gamma,$$

giving the second formula, since one obtains the suspension of the Hopf map.

Finally, since

$$k_j^1(\Sigma^{W_1}\eta_j^1 + \Sigma^{W_1}\tilde{\eta}^1) = 0 \quad \text{and} \quad k_j^1(\Sigma^W\eta_j + \Sigma^W\tilde{\eta}) = 0,$$

one obtains that $k_j^1\tilde{d}_j$ is even. □

Example 7.4. When studying Hopf bifurcation, one will need to compute the class of $[\eta^1, -y]_\Gamma$, where Γ acts on y as $\mathbb{Z}_2$ and $\eta^1 = (1 - |z|^2, \lambda z)$ with Γ acting on z as S^1 or $\mathbb{Z}_n$. Consider the map

$$F_2 = (2t_2 - 1, -y, Y),$$

where y and Y have isotropy H_2, with $\Gamma/H_2 \cong \mathbb{Z}_2$. Now, one may look at the map $\eta_0 = (2t_2 - 1, y, Y)$, which generates $\Pi(\Gamma)$ for $V_2 = W_2$. But η_0 may be deformed to $(2t_2 - 1, y^3, Y)$ and then to $(2t_2 - 1, y(y^2 - 1), Y)$, whose Γ-degree is decomposed on the set $|y| < 1/2$, giving F_2 and on the set $|y| > 1/2$, where it is linearly deformable to $\eta_2 = (2t_2 + 1 - 2y^2, y(y^2 - 1), Y)$. Hence,

$$[F_2]_\Gamma = [\eta_0]_\Gamma - [\eta_2]_\Gamma.$$

Since F_2 is the suspension of $-y$, one may compute as well $[\eta^1, F_2]_\Gamma$. For $[\eta^1, \eta_0]$, η_0 is just a suspension, hence this part is $\Sigma^{W_2}\eta^1 \equiv \eta_1$, which generates $\Pi(H_1)$ for V.

For $[\eta^1, \eta_2]$, assume first that $\dim \Gamma/H_1 = 1$, i.e., Γ acts as S^1 on z. Then, in Example 7.3, one has $\tilde{H}_1^0 = H_1$, $|\Gamma/H_2| = 2$ and $|H_1/H| = 2$, since H_2 is maximal and H_1 cannot be a subgroup of H_2: in fact the elements of H_1 are of the form (Φ, K) such that $\langle N, \Phi\rangle + 2\pi\langle K, L/M\rangle$ is an integer. Hence, for any K in $\mathbb{Z}_{m_1} \times \cdots \times \mathbb{Z}_{m_s}$, there is a $\Phi(K)$ such that $(\Phi(K), K)$ is in H_1. On the other hand, $H_2 = T^n \times A_2$, with $Z_{m_1} \times \cdots \times \mathbb{Z}_{m_s}/A_2 \cong \mathbb{Z}_2$. If H_1 is a subgroup of H_2, then one would have $\mathbb{Z}_{m_1} \times \cdots \times \mathbb{Z}_{m_s} = A_2$, which is not true.

In this case, $[\eta^1, \eta_2]_\Gamma = [F_{H_1 \cap H_2}]$ and one has, if $\dim \Gamma/H_1 = 1$,

$$[1-|z|^2, \lambda z, -y]_\Gamma = [1-|z|^2, \lambda z, y]_\Gamma - [1-|z|\cdot|y|, \lambda z, (y^2-1)y]_\Gamma = [\eta_1]_\Gamma - [\eta_{12}]_\Gamma.$$

On the other hand, if $\Gamma/H_1 \cong \mathbb{Z}_n$, then H_1 may be a subgroup of H_2 if n is even, since then $\Gamma/H_2 < \Gamma/H_1$, hence one has a γ such that $\gamma y = -y$ and $\gamma z = e^{2\pi i/n}$.

If H_1 is not a subgroup of H_2, then $|\Gamma/H_1 \cap H_2| = 2n$, since $k^2 = 2$ and $\tilde{k}^2 = 2$. Then, from Proposition 7.6 one has

$$[1-|z|^2, \lambda z, -y]_\Gamma = [\eta_1]_\Gamma - [\eta_{12}]_\Gamma - d[\tilde{\eta}],$$

where $\tilde{\eta} = (\varepsilon - |y||z^n+1|, \lambda(z^n+1)z, (y^2-1)y)$. We shall see, in the next chapter, Lemma 4.2, that in fact $d = 0$. Recall that $n(\eta_{12}+\tilde{\eta}) = 0$ and $2\tilde{\eta} = 0$.

If H_1 is a subgroup of H_2, then $k^2 = 2$, $\tilde{k}^2 = 1$ and

$$[\eta^1, \eta_2]_\Gamma = 2[\eta_1] + d_1[\tilde{\eta}_1],$$

where $\tilde{\eta}_1 = (\varepsilon - |z^n+1|, \lambda(z^n+1)z, y)$, in which case one has

$$[1-|z|^2, \lambda z, -y]_\Gamma = -[\eta_1]_\Gamma - d_1[\tilde{\eta}_1]_\Gamma.$$

We shall see, in the next chapter (Theorem 4.1) that $d_1 = 1$ if $n = 2m$ with m odd. Since n is even one has $n\eta_1 = 0$ and $2\tilde{\eta}_1 = 0$ in this case. In particular, if $n = 2$, then $[\eta^1, \eta_2] = [\tilde{\eta}_1]$.

As we have done with the previous operations, we shall end this subsection by looking at products of orthogonal maps. Clearly, Lemma 7.3 is still valid, with the orthogonal degree on the right hand side.

Proposition 7.7. *Let $V_i = W_i$ and for any isotropy subgroup H_j, with* $\dim \Gamma/H_j = k$, *let $\tilde{H}_j^0$ be the isotropy of the k coordinates with $k_j = \infty$. Let F_i, in $\Pi^\Gamma_{\perp S^{V_i}}(S^{V_i})$, be written, for $i = 1, 2$, as*

$$[F_i]_\perp = \sum d^i_H [F^i_H]_\perp,$$

then

$$[F_1, F_2]_\perp = \sum d^1_{H_1} d^2_{H_2} \frac{|\tilde{H}_1^0/H_1||\tilde{H}_2^0/H_2|}{|\tilde{H}_1^0 \cap \tilde{H}_2^0/H_1 \cap H_2|} [F_{H_1\cap H_2}]_\perp,$$

where the sum is over all H_1 in $\mathrm{Iso}(V_1)$, *H_2 in* $\mathrm{Iso}(V_2)$, *with* $\dim \Gamma/H_1 + \dim \Gamma/H_2 = \dim \Gamma/(H_1 \cap H_2)$.

Proof. It is clearly enough to compute the class $[F^1_{H_1}, F^2_{H_2}]_\perp$ for the generators. Writing V^H as $(V_1^{H_1} \times V_2^{H_2}) \times (V_1^{H_1})^\perp \times (V_2^{H_2})^\perp$, one has for the action of

$$\Gamma/H = (\Gamma/H_1) \times (H_1/H_1 \cap H_2)$$

k_1 coordinates of $V_1^{H_1}$, $z_1, \ldots, z_{k_1}$, giving $A_1x_1, \ldots, A_{k_1}x_1$ linearly independent, and $k - k_1$ coordinates of $V_2^{H_2}$, $\tilde{z}_1, \ldots, \tilde{z}_{k-k_1}$ for the action of H_1 on that space. Here, $k_i = \dim \Gamma/H_i$ and $k = \dim \Gamma/H$. Note that, given the order chosen in V^H, the coordinates of $(V_1^{H_1})^\perp$ and of $(V_2^{H_2})^\perp$ do not contribute, in a non-trivial way, to the fundamental cell.

Now, as in Lemma 7.1 of Chapter 1, one may write the action of T^n on $V_1^{H_1}$ as $C(\psi_1, \dots, \psi_{k_1})^T$, that is under a reparametrization of T^n, one gets $A_j x_1 = 0$ for $j > k_1$ and x_1 in $V_1^{H_1}$. Assume that $\psi_{k_1+1}, \dots, \psi_{k-k_1}$ give $A_j x_2$ linearly independent for the action of H_1 on $V_2^{H_2}$, then, one may suppose, changing the parametrization, that $A_j x_2 = 0$ for $j > k$ and that $A_j x_2$ are linearly independent for $k_1 < j \le k - k_1$. There are also $k_1 + k_2 - k$ linearly independent vectors $A_j x_2$ for $j \le k_1$.

Now, if $k = k_1 + k_2$, then $\left[F_{H_1}^1 + \sum_1^{k_1} \lambda_j A_j x_1, F_{H_2}^2 + \sum_{k_1+1}^{k} \lambda_j A_j x_2\right]$ has been computed in Proposition 7.5 and gives $\alpha\left[F_{H_1 \cap H_2} + \sum_1^k \lambda_j A_j x\right]$, where α is the integer of the proposition: recall, from Theorem 6.1, that $F_H + \sum \lambda_j A_j x$ may be taken as generator of $\Pi(H)$, whenever F_H generates $\Pi_\perp(H)$.

On the other hand, if $k < k_1 + k_2$, one has to add to $F_{H_2}^2 + \sum_{k_1+1}^{k} \lambda_j A_j x_2$ the sum $\sum \lambda_j A_j x_2$ for j in a subset J of $k_1 + k_2 - k$ elements of $\{1, \dots, k_1\}$ in order to get the generator of $\Pi(H_2)$ in $\Pi^{\Gamma}_{S^{\mathbb{R}^{k_2}\times V_2}}(S^{V_2})$. But for this second sum one may deform λ_j to 0 and then to $\varepsilon_j \neq 0$ fixed, without affecting the class of

$$\left[F_{H_1}^1 + \sum\nolimits_1^{k_1} \lambda_j A_j x_1, F_{H_2}^2 + \sum\nolimits_J \lambda_j A_j x_2 + \sum\nolimits_{k_1+1}^{k} \lambda_j A_j x_2\right];$$

a zero of the pair implies $F_{H_i}^i(x_i) = 0$, either $x_1 = 0$ or $\lambda_j = 0$ for $j = 1, \dots, k_1$, but the zeros of $F_{H_1}^1(x_1)$ have isotropy H_1, with $\dim \Gamma/H_1 = k_1$, hence $x_1 \neq 0$. But then, $\sum_J \lambda_j A_j x_2$ is 0.

The last map is never 0, since $F_{H_2}^2(x_2) = 0$ implies $\lambda_j = 0$ for j in J and for $j = k_1 + 1, \dots, k$. In particular, $\varepsilon_j A_j x_2$ implies that the map is never 0. Thus, $[F_{H_1}^1, F_{H_2}^2]_\perp = 0$, using Proposition 6.1, since this pair is $\alpha\ [F_{H_1 \cap H_2}]_\perp$.

Note that one may use Proposition 6.1 to prove this result: in fact $(F_{H_1}^1, F_{H_2}^2)$ is non-zero on ∂B_k and thus,

$$[F_{H_1}^1, F_{H_2}^2]_\perp = \sum_{H_j < \tilde{H}_1^0 \cap \tilde{H}_2^0} d_j [F_j]_\perp$$

with, for any $H_i > \underline{H}$, the torus part of $H_1 \cap H_2$, one has, with $H_0 = \tilde{H}_1^0 \cap \tilde{H}_2^0$:

$$\deg\left(\left(F_{H_1}^1 + \sum\nolimits_1^k \lambda_l A_l x_1, F_{H_2}^2 + \sum\nolimits_1^k \lambda_l A_l x_2,\right)^{H_i}; B_k^i\right) = \sum\nolimits_{H_i < H_j < H_0} d_j |H_0/H_j|.$$

Now, a zero of the pair gives (x_1, x_2) with $\Gamma_{x_1} = H_1, \Gamma_{x_2} = H_2$ and $\lambda_l = 0$ for $l = 1, \dots, k$. Thus, the degree on the left hand side is 0 if H_i is not a subgroup of $H_1 \cap H_2$. Furthermore, $F_{H_1}^1 = (F_{H_1}^{H_1}, x_{\perp H_1})$, hence for H_i a strict subgroup of $H_1 \cap H_2$, the degree is the degree for $H_1 \cap H_2$. From this, we deduce that $d_j = 0$, except for $H_j = H_1 \cap H_2$, in which case

$$[F_{H_1}^1, F_{H_2}^2]_\perp = d[F_{H_1 \cap H_2}]_\perp$$

with $|H_0/H_1 \cap H_2| d = \deg(F_{H_1}^1 + \sum_1^k \lambda_l A_l x_1, F_{H_2}^2 + \sum_1^k \lambda_l A_l x_2)^{H_1 \cap H_2}; B_k^{H_1 \cap H_2})$.

If $k < k_1 + k_2$, we have already seen that this degree is 0. While, if $k = k_1 + k_2$, then $B_k^{H_1 \cap H_2} = B_{k_1}^{H_1} \times B_{k_2}^{H_2}$, one may deform $\lambda_l A_l x_2$ to 0 for $l \leq k_1$, and one obtains a product:

$$|H_0/H_1 \cap H_2| d$$
$$= \deg\left(\left(F_{H_1}^1 + \sum_1^{k_1} \lambda_l A_l x_1\right)^{\underline{H_1}}; B_{k_1}^{\underline{H_1}}\right) \deg\left(\left(F_{H_2}^2 + \sum_{k_1+1}^{k} \lambda_l A_l x_2\right)^{\underline{H_2}}; B_{k_2}^{\underline{H_2}}\right).$$

From the fact that on $V^{\underline{H_i}}$ one has a suspension of $V_i^{H_i}$, one has

$$\deg\left(F_{H_1}^1 + \sum_1^{k_1} \lambda_l A_l x_1; B_{k_1}^{H_1}\right) = |\tilde{H}_1^0/H_1|$$
$$\deg\left(F_{H_2}^2 + \sum_{k_1+1}^{k} \lambda_l A_l x_2, B_{k_2}^{H_2}\right) = |\tilde{H}_2^0/H_2|$$

by repeating the application of Proposition 6.1 or from the construction of Theorem 6.1. This gives the result. □

3.7.4 Composition

The last operation which we shall consider is that of composition of maps. Consider three representations V, W and U of the group Γ and assume $f : V \to W$ and $g : W \to U$ are equivariant maps. Then $g \circ f$ is also equivariant. Let Ω be a bounded open invariant subset of V.

We have seen, under the hypothesis of Lemma 6.2 of Chapter 2, that

$$\deg_\Gamma(g \circ f; \Omega) = [G \circ F]_\Gamma,$$

where $[F]_\Gamma = \deg_\Gamma(f; \Omega)$ and $[G]_\Gamma = \deg_\Gamma(f; f(\Omega))$.

Furthermore, we have also seen in Lemma 6.3 of Chapter 2, that under certain hypothesis, one has that $[G \circ F]_\Gamma = [G \circ \hat{F}]_\Gamma$, where $\hat{F}(s, x) = F(s, x)/\|F(s, x)\|$, a fact which will enable us to use the algebraic properties of the Γ-homotopy groups of spheres.

In general, if $F(s, x) = (\varphi(s, x), f(s, x))$, is defined on $[-1, 1] \times \{x : \|x\| \leq 1\}$, and non-zero on the boundary of this cylinder, then $\hat{F} = F/\|F\|$ will belong to a cylinder with similar characteristics and one may take the composition with a Γ-map G, i.e., one obtains a pairing

$$\Pi_{S^V}^\Gamma(S^W) \times \Pi_{S^W}^\Gamma(S^U) \to \Pi_{S^V}^\Gamma(S^U)$$
$$([F]_\Gamma, [G]_\Gamma) \to [G \circ F]_\Gamma,$$

which is well defined on homotopy classes. Furthermore, since one may take $F(s, x) = (1, 0)$ if $s = \pm 1$ (Lemma 8.1 of Chapter 1), with $2t - 1 = s$, one has, for $\|x\| = 1$,

$$(F_1 \oplus F_2)(s, x) = \begin{cases} F_1(2s + 1, x), & \text{if } -1 \leq s \leq 0 \\ F_2(2s - 1, x), & \text{if } 0 \leq s \leq 1, \end{cases}$$

since, on each s-interval, the first argument of F_i must go form -1 to 1.

Lemma 7.4. (a) $[G \circ (F_1 \oplus F_2)]_\Gamma = [G \circ F_1]_\Gamma + [G \circ F_2]_\Gamma$

(b) $[(G_1 \oplus G_2) \circ \Sigma_0 f]_\Gamma = [G_1 \circ \Sigma_0 f]_\Gamma + [G_2 \circ \Sigma_0 f]_\Gamma$, *where* $(\Sigma_0 f)(s_1, x)$ *is the suspension by* s_1 *of* $f(x)$, *i.e.*, $(s_1, \|x\| f(x/\|x\|))$ *with* $\|f(x)\| = 1$ *whenever* $\|x\| = 1$.

Proof. The proof of (a) follows from the definition, while for (b) one has, for $\|x\| = 1$,

$$(G_1 \oplus G_2) \circ (\Sigma_0\ f) = \begin{cases} G_1(2s_1 + 1, f(x)), & \text{if } -1 \le s_1 \le 0 \\ G_2(2s_1 - 1, f(x)), & \text{if } 0 \le s_1 \le 1 \end{cases}$$

which corresponds to the second sum. As usual, one may perform the sum on s or on s_1 and here we may always assume that F is a suspension. □

Thus, if $[F]_\Gamma = \sum d_i [\tilde{F}_i]_\Gamma$ and $[G]_\Gamma = \sum e_j [\tilde{G}_j]_\Gamma$, as an application of Theorem 2.3, then

$$[G \circ \hat{F}]_\Gamma = \sum d_i e_j [\tilde{G}_j \circ \tilde{F}_i]_\Gamma.$$

Note that if $F^K|_{S^K}$ has an extension to V^K, then $F^K|_{S^K}$ is Γ-deformable to $(1, 0)$ and then $(G \circ F)^K$ is also Γ-deformable to $G(1, 0) = (1, 0)$. Similarly, if G^K has a non-zero extension to W^K, then this will be also true for $(G \circ F)^K$. It is thus important to study the composition for the generators.

Lemma 7.5. (a) *If* $V = \mathbb{R}^{k_1+k_2} \times V'$, $W = \mathbb{R}^{k_2} \times W'$ *and hypothesis* (H) *holds for* V' *and* W' *and for* W' *and* U, *and furthermore* $\dim V'^H = \dim U^H$ *for all* H *in* $\mathrm{Iso}(V)$, *then hypothesis* (H) *holds for* V' *and* U.

(b) *If, under the same hypothesis,* $\{x_i^{l_i}\}$ *is a complementing map from* $(V^H)^\perp$ *onto* $(W^H)^\perp$ *and* $\{z_j^{q_j}\}$ *is a complementing map from* $(W^H)^\perp$ *onto* $(U^H)^\perp$, *then* $\{x_i^{l_i q_i}\}$ *will be a complementing map from* $(V^H)^\perp$ *onto* $(U^H)^\perp$.

Proof. Let H and K be in $\mathrm{Iso}(V)$. Then $\dim(V'^H \cap V'^K) = \dim(W'^H \cap W'^K)$. Let $\tilde{H}$ be the isotropy of W'^H, then $H < \tilde{H}$ and $W'^{\tilde{H}} = W'^H$. One has $\dim(W'^{\tilde{H}} \cap W'^{\tilde{K}}) = \dim(U^{\tilde{H}} \cap U^{\tilde{K}})$. Now, $U^{\tilde{H}} \subset U^H$. From hypothesis (H), one has $\dim V'^H = \dim W'^H = \dim U^{\tilde{H}}$, hence the extra hypothesis implies that $U^H = U^{\tilde{H}}$, proving (a).

Now, the spaces $(V^H)^\perp, (W^H)^\perp = (W^{\tilde{H}})^\perp, (U^H)^\perp = (U^{\tilde{H}})^\perp$ have the same dimension and one has equivariant monomials between them, the composition will be a complementing map. □

Note that the extra dimension condition will be met if $\mathrm{Iso}(V) \subset \mathrm{Iso}(W)$, since then $U^{\tilde{H}} = U^H$, because H is in $\mathrm{Iso}(W)$. On the other hand, if $\tilde{H}$ is in $\mathrm{Iso}(W)$, then, if H is the isotropy of $V^{\tilde{H}}$, one has $\tilde{H} < H$, $V^H = V^{\tilde{H}}$ and $W^H \subset W^{\tilde{H}}$. In order to compare the Γ-degrees of $\tilde{F}_i$ and $\tilde{G}_j$, *we shall assume that* $\mathrm{Iso}(V) = \mathrm{Iso}(W)$. This is the case if $V = \mathbb{R}^{k_1} \times W$ and $W = \mathbb{R}^{k_2} \times U$.

Lemma 7.6. *Under the hypothesis of the preceding lemma and assuming that* $\text{Iso}(V) = \text{Iso}(W)$*, let* $F^{H_1} : V^{H_1} \to W^{H_1}$ *be in* $\Pi(H_1)$ *and* $G^{H_2} : W^{H_2} \to U^{H_2}$ *be in* $\Pi(H_2)$*. Define* $\tilde{F} = (F^{H_1}, x_i^{l_i})$ *and* $\tilde{G} = (G^{H_2}, z_j^{q_j})$ *and* $H = H_1 \cap H_2$*. Then:*

(a) $\dim \Gamma/H_i \le \dim \Gamma/H \le \dim \Gamma/H_1 + \dim \Gamma/H_2$.

(b) $(\tilde{G} \circ \tilde{F})^H$ *is in* $\Pi(H)$.

Proof. Since $H < H_i$, the first inequality is clear. Now, since H_2 is also in $\text{Iso}(V)$, then H is the isotropy subgroup for the space V_1 generated by V^{H_1} and V^{H_2}. Then, $V^H = V^{H_1} \times (V^{H_1})^\perp \cap V^{H_2} \times (V_1^\perp \cap V^H)$, hence, as in the proof of Lemma 7.4, one has $k_j = k_j^1$ for x_j in V^{H_1} and $k_j = \tilde{k}_j^2$, which divides k_j^2, in the second space, while $k_j = 1$ in the third. This proves the second inequality.

Note that $\tilde{G} \circ \tilde{F} = \{x_i^{l_i q_i}\}$ on $V_1^\perp$ and that if $H_1 < H_2$, then for any $K > H_1$, F^K is Γ-deformable to $(1, 0)$ and $(G \circ F)^{H_1}$ is in $\Pi(H_1) = \Pi(H)$. A similar result holds if $H_2 < H_1$. In general,

$$V = \mathbb{R}^{k_1} \times \mathbb{R}^{k_2} \times (V'^{H_1} \cap V'^{H_2}) \times (V'^{H_1} \cap V'^{H_2 \perp}) \times (V'^{H_1 \perp} \times V'^{H_2}) \times V_1^\perp$$

and any X in V is of the form $X = (\lambda_1, \lambda_2, X_0, X_1, X_2, X_\perp)$.

Similarly,

$$W = \mathbb{R}^{k_2} \times (W'^{H_1} \cap W'^{H_2}) \times (W'^{H_1} \cap W'^{H_2 \perp}) \times (W'^{H_1 \perp} \cap W'^{H_2}) \times W_1^\perp$$

and any Y in W is of the form $Y = (\lambda_2, Y_0, Y_1, Y_2, Y_\perp)$.

From the hypothesis on V' and W', these subspaces have the same dimension. A similar decomposition holds for U, and any element Z of U is of the form $Z = (Z_0, Z_1, Z_2, Z_\perp)$. One has

$$\tilde{F}(X) = ((F_\lambda, F_0, F_1)(\lambda_1, \lambda_2, X_0, X_1), X_2^l, X_\perp^l),$$

with $F_1|_{X_1=0} = 0$ and $(F_\lambda, F_0)|_{X_1=0} \ne 0$, since the isotropy of $V^{H_1} \cap V^{H_2}$ is strictly larger than H_1 and F^{H_1} is in $\Pi(H_1)$. Here $(X_2^l, X_\perp^l)$ stands for $\{x_i^{l_i}\}$ and one should normalize $\tilde{F}$ as $\tilde{F}/\|\tilde{F}\|$.

Similarly, one has

$$\tilde{G} \circ \tilde{F}(X) = (G_0(F_\lambda, F_0, X_2^l), F_1^q(\lambda_1, \lambda_2, X_0, X_1), G_2(F_\lambda, F_0, X_2^l), X_\perp^{lq}),$$

where $G_2(\lambda_2, Y_0, 0) = 0$ and $G_0(\lambda_2, Y_0, 0) \ne 0$ on $W^{H_1} \cap W^{H_2}$, since G^{H_2} is in $\Pi(H_2)$. Thus, $(\tilde{G} \circ \tilde{F})^{H_1}$, with $X_2 = X_\perp = 0$, has G_0 deformable to $(1, 0)$. Similarly, $(\tilde{G} \circ \tilde{F})^{H_2}$, with $X_1 = X_\perp = 0$, has $F_1 = 0$ and (F_λ, F_0) independent of X_2 and Γ-deformable to $(1, 0)$. Hence $(\tilde{G} \circ \tilde{F})^{H_2}$ is Γ-deformable to $(G_0(1, 0, X_2^l), 0, G_2(1, 0, X_2^l), 0)$ and then to $(1, 0)$. Thus, if H is a strict subgroup of H_i, $i = 1, 2$, then $\tilde{G} \circ \tilde{F}$ is trivial on $V^{H_1} \cup V^{H_2}$.

Let now $K > H$ and decompose V^K as above. One has a non-zero Γ-extension of $\tilde{G} \circ \tilde{F}$ on $V^K \cap (V^{H_1} \cup V^{H_2})$, i.e., for $X_2 = 0$ or $X_1 = 0$. If $V^K \cap V^{H_1}$ is strictly contained in V^{H_1}, then X_1 has some components $x_i = 0$ and the remaining variables, in X_1, have isotropy $\tilde{H}_1$ containing strictly H_1 (if not $V^K \cap V^{H_1} = V^{\tilde{H}_1}$ would be V^{H_1}). Hence, on $V^K \cap V^{H_1}$, one may extend $F^{\tilde{H}_1}$ to a map of norm 1. Then, for X, in the unit ball of V^K, one has either $\|X_2\| = 1$ and $(G_0, G_2) \neq 0$, or $\|X_2\| < 1$, in which case, from $\|F^{\tilde{H}_1}\| = 1$, either $\|F_1\| = 1$ and $\tilde{G} \circ \tilde{F} \neq 0$ or $\|F_1\| < 1$ and $\|(F_\lambda, F_0)\| = 1$ with $(G_0,\ G_2) \neq 0$. Hence, in this case one has a non-zero Γ-extension to V^K.

On the other hand, if $V^K \cap V^{H_1} = V^{H_1}$, then $V^K \cap V^{H_2}$ is strictly contained in V^{H_2} and (G_0, G_2) has a non-trivial Γ-extension to $W^K \cap W^{H_2}$. But (F_λ, F_0, F_1) has a Γ-extension to $V^{H_1} = V^K \cap V^{H_1}$ with norm one. If $F_1 \neq 0$, then $(\tilde{G} \circ \tilde{F})^K \neq 0$, while if $F_1 = 0$, then (F_λ, F_0) is in $V^K \cap V^{H_2}$ and (G_0, G_2) has the non-trivial Γ-extension. Thus, $(\tilde{G} \circ \tilde{F})^K$ has a non-trivial Γ-extension for all $K > H$, i.e., $(\tilde{G} \circ \tilde{F})^H$ is in $\Pi(H)$. □

Proposition 7.8. *Let* $V = \mathbb{R}^{k_1+k_2} \times V'$, $W = \mathbb{R}^{k_2} \times W'$, $\mathrm{Iso}(V) = \mathrm{Iso}(W)$ *and assume hypothesis* (H) *holds for* V' *and* W' *and for* W' *and* U. *If* $\dim \Gamma/H_i = k_i$ *and* $k = \dim \Gamma/H = k_1 + k_2$, *let* $\tilde{F}$ *and* $\tilde{G}$ *be the generators of* $\tilde{\Pi}(H_i)$. *Then,* $[\tilde{G} \circ \tilde{F}]_\Gamma = d[\tilde{F}_H]_\Gamma$, *where* $\tilde{F}_H$ *generates* $\tilde{\Pi}(H)$, $\hat{F} = \tilde{F}/\|\tilde{F}\|$ *and*

$$d = \beta_{HH_1} \tilde{\beta}_{HH_2} \frac{|\tilde{H}_1^0/H_1| \cdot |\tilde{H}_2^0/H_2|}{|\tilde{H}_1^0 \cap H_2^0/H_1 \cap H_2|},$$

where $\beta_{HH_1} = \prod l_i$ *for* x_i *in* $V^H \cap (V^{H_1})^\perp \cap (V^{\tilde{H}_2^0})^\perp$, $\tilde{\beta}_{HH_2} = \prod q_j$ *for* y_j *in* $W^H \cap (W^{H_2})^\perp$. *Here* $\tilde{H}_i^0$ *is the isotropy of the* k_i *coordinates with* $k_j = \infty$.

More generally, if $F^{H_1}|_{\partial B_{k_1}} \neq 0$ *and* $G^{H_2}|_{\partial B_{k_2}} \neq 0$, *with* F^{H_1} *in* $\Pi(H_1)$ *and* G^{H_2} *in* $\Pi(H_2)$, *then* $\tilde{G} \circ \tilde{F}$ *is in* $\Pi(H)$ *and has a non-zero extension* $\tilde{G} \circ \hat{F}$ *to* $\partial B_{k_1+k_2}$, *where* $\hat{F}(x) = \alpha \tilde{F}(x)$, *with*

$$\alpha^{-1}(x) = \min_{\partial B_{k_1}}(\|\tilde{F}\|) \max(\|\tilde{F}(x)\|/\min_{\partial B_{k_1}} \|\tilde{F}\|, 1 - \|x\|)$$

and one has, with the number d *above,*

$$\deg_E((\tilde{G} \circ \tilde{F})^H) = d \deg_E(F^{H_1}) \deg_E(G^{H_2}).$$

Proof. Let $z_1, \dots, z_{k_1}$ be the variables in V^{H_1} with $k_j^1 = \infty$ and $z_{k_1+1}, \dots, z_k$, be the variables in V^{H_2} with $k_j^2 = \infty$. From the fact that $k = k_1 + k_2$ one has that none of these variables are in $V^{H_1} \cap V^{H_2}$.

From Theorem 3.3 and rescaling the variables so that one works in a unit ball, one has, with $s = 2t - 1$, that $\tilde{F}$ is, up to normalization

$$\tilde{F} = \Big(s + 2 - 2\prod |2x_j|, X_0^0, \lambda_2, (\lambda_1^1 + i(|2z_1|^2 - 1))z_1^{l_1}, \dots,$$
$$(\lambda_1^{k_1} + i(|2z_{k_1}|^2 - 1))z_{k_1}^{l_{k_1}}, (P_j(2X_1, 2X_0) + 1)x_j^{l_j}, (Q_j(2y_j) - 1)y_j, X_2^l, X_\perp^l\Big),$$

where X_0^0 is in V^Γ, x_j is in V^{H_1}, with x_j in the first component standing for those coordinates with $k_j^1 > 1$ (including z_j and y_j), and P_j, Q_j have the usual meaning. By starting the fundamental cell with the components of X_1, P_j will be a monomial in the coordinates of X_1, for x_j in X_1. The zeros of $\tilde{F}$ in $B_{k_1+k_2}$ are for $s = 0$, $\lambda = 0$, $2z_j = 1$, $|2x_i| = 1$ and there are $|\tilde{H}_1^0/H_1|$ of them.

One has a similar expression for $\tilde{G}$, before normalization

$$\tilde{G} = \Big(s + 2 - 2\prod |2\tilde{x}_j|^2, X_0^0, Y_1^q, (\lambda_2^1 + i(|2z_{k_1+1}|^2 - 1))z_{k_1+1}^{q_{k_1+1}}, \dots,$$
$$(\lambda_2^{k_2} + i(|2z_k|^2 - 1))z_k^{q_k}, (\tilde{P}_j(2Y_2, 2Y_0) + 1)\tilde{x}_j^{q_j}, (\tilde{Q}_j(2\tilde{y}_j) - 1)\tilde{y}_j, Y_\perp^q\Big),$$

where $\tilde{x}_j$ are in W^{H_2}. By starting the fundamental cell with the coordinates of Y_2, $\tilde{P}_j$ will depend on these coordinates for $\tilde{x}_j$ in Y_2.

We leave to the reader the task of giving expressions for $\tilde{G}(\alpha\tilde{F})$ and to compute its degree on $B_{k_1+k_2}$. In fact, in general, if $F^{H_1}(X_0, X_1) = (F_0, F_1)$, with F_0 in $W^{H_1} \cap W^{H_2}$ and F_1 in $W^{H_1} \cap (W^{H_2})^\perp$, then $\tilde{F}(X_0, X_1, X_2, X_\perp) = (F_0, F_1, X_2^l, X_\perp^l)$, while, if $G^{H_2}(Y_0, Y_2) = (G_0, G_2)$, with G_0 in $U^{H_1} \cap U^{H_2}$ and G_2 in $U^{H_2} \cap (U^{H_1})^\perp$, then

$$\tilde{G}(Y_0, Y_1, Y_2, Y_\perp) = (G_0, Y_1^q, G_2, Y_\perp^q).$$

Then, $\tilde{G}(\alpha\tilde{F}) = (G_0(\alpha F_0, \alpha X_2^l), \alpha^q F_1^q(X_0, X_1), G_2(\alpha F_0, \alpha X_2^l), \alpha^q X_\perp^{ql})$. Note that $\alpha^{-1}(x) = \|\tilde{F}(x)\|$ if $\|x\| = 1$ (on ∂B) and on ∂B_{k_1} (there $\|\tilde{F}(x)\|/\min_{\partial B_{k_1}} \|\tilde{F}\| \geq 1 \geq 1 - \|x\|$). In general, if $\alpha^{-1}(x) = \|\tilde{F}(x)\|$, then $\tilde{G}(\alpha\tilde{F}(x))$ is non-zero (since $\tilde{G}$ is non-zero on the unit sphere), while if $\alpha^{-1}(x) = \min_{\partial B_{k_1}} \|\tilde{F}\|(1 - \|x\|) \geq \|\tilde{F}(x)\|$, then $\|\alpha\tilde{F}(x)\| \leq 1$.

For $k_1 < j \leq k$, z_j in B_{k_2} appears as $z_j^{l_j}$ in $\tilde{F}$, thus $\alpha\tilde{F}$ maps $B_{k_1+k_2}$ into B_{k_2} and $\partial B_{k_1+k_2}$ into ∂B_{k_2}, where $\tilde{G}$ is non-zero.

From Proposition 6.1 of Chapter 2, one has

$$\deg(\tilde{G}(\alpha\tilde{F})^H; B_{k_1+k_2}^H) = \deg(\tilde{G}; B_{k_2}^H)\deg(\alpha\tilde{F} - p; B_{k_1+k_2}^H),$$

where p is in B_{k_2}. The left hand side is

$$|\tilde{H}_1^0 \cap \tilde{H}_2^0/H| \deg_E(\tilde{G}(\tilde{F})),$$

while

$$\deg(\tilde{G}; B_{k_2}^H) = |\tilde{H}_2^0/H_2| \deg_E(G)\prod q_j,$$

for q_j corresponding to Y_1 and $Y_\perp$, hence $\prod q_j = \tilde\beta_{HH_2}$. On the other hand, one may choose all the components of p to be 0 except those corresponding to z_j, $j = k_1+1,\dots,k$, which may be taken to be $1/2$. One may deform α to 1 and use the product theorem, where $x_j^{l_j}$ will contribute l_j except for $z_j^{l_j} - 1/2$, for z_j real and positive, which contributes 1, i.e., a total degree equal to

$$\deg(F^{H_0}; B_{k_1})\beta_{HH_1} = |\tilde H_1^0/H_1| \deg_E(F^{H_1})\beta_{HH_1}. \qquad \square$$

Corollary 7.4. *Under the hypothesis of Proposition* 7.8, *if*

$$[F]_\Gamma = \sum d_i[\tilde F_i]_\Gamma + [\tilde F]_\Gamma$$
$$[G]_\Gamma = \sum e_j[\tilde G_j]_\Gamma + [\tilde G]_\Gamma,$$

with $\dim \Gamma/H_i = k_1$, $\dim\Gamma/H_j = k_2$, $[\tilde F]_\Gamma$ *in* Π_{k_1-1}, $[\tilde G]$ *in* Π_{k_2-1}, *then*

$$[G\circ F]_\Gamma = \sum f_l[\tilde K_l]_\Gamma + [\tilde K]_\Gamma \quad \text{with } f_l = \sum d_i e_j d_{ij},$$

where $[\tilde K]_\Gamma$ *is in* Π_{k-1} *and the second sum is over all* (i, j) *such that* $H_i \cap H_j = H_l$, *with* $\dim\Gamma/H_l = k_1 + k_2 = k$, *and* d_{ij} *is given in Proposition* 7.8.

Proof. From Lemma 7.4, one has $[G\circ F]_\Gamma = \sum d_i e_j[\tilde F_i \circ \tilde G_j]_\Gamma + [\tilde K]_\Gamma$ with $[\tilde K]_\Gamma$ in Π_{k-1} and $[\tilde F_i\circ\tilde G_j]_\Gamma = d_{ij}[\tilde K_{ij}]_\Gamma$, for $H_l = H_i\cap H_j$ with $\dim\Gamma/H_l = k$ and $d_{ij} = \beta_{HH_i}\beta_{HH_j}|\tilde H_i^0/H_i||\tilde H_j^0/H_j|/|\tilde H_i^0\cap\tilde H_j^0/H|$. $\square$

Example 7.5. Let $V = W$, hence $k_1 = 0$, and $V = \mathbb{R}^k \times U$. Then, $\beta_{HH_j} = \tilde\beta_{HH_j} = 1$, and from Lemma 7.6, one has $\dim\Gamma/H_l = \dim\Gamma/H_j$ for any H_i, with $\dim\Gamma/H_i = 0$. In this case $\tilde F = 0$, $\tilde H_i^0 = \Gamma$. For instance, assume that F consists in changing one real variable y, where Γ acts as $-\mathrm{Id}$, into $-y$, leaving the other coordinates unchanged. Then, from Example 7.4, one has

$$[F] = [F_\Gamma] - [F_1],$$

where $F_1(s, y, X) = (s + 2 - 8y^2, (4y^2-1)y, X)$ and $[F_\Gamma] = [s, y, X]$.

Then, if $[G]_\Gamma = \sum e_j[\tilde F_j]_\Gamma + [\tilde G]_\Gamma$, one has

$$[G\circ F]_\Gamma = [G]_\Gamma - [G\circ F_1]_\Gamma.$$

If $\dim\Gamma/H_j = k$, and H_1 is the isotropy of y, then, either $H_1\cap H_j = H_j$, i.e., $H_j < H_1$, and $d_{1j} = |\Gamma/H_1| = 2$, or $H_1\cap H_j$ is a strict subgroup of H_j, with $|\tilde H_j^0/H_1\cap H_j| = |\tilde H_j^0/H_j||H_j/H_1\cap H_j| = 2|\tilde H_j^0/H_j|$, since any γ in Γ, in particular in H_j, is such that γ^2 is in H_1, in which case $d_{1j} = 1$. Thus,

$$[G\circ F]_\Gamma = -\sum_{H_j<H_1} e_j[\tilde F_j]_\Gamma + \sum_{H_j\not< H_1} e_j([\tilde F_j]_\Gamma - [\tilde F_{H_1\cap H_j}]_\Gamma) + [\tilde K]_\Gamma,$$

for H_j with $\dim \Gamma/H_j = k$ and $[\tilde{K}]$ in Π_{k-1}.

The last result in this section will concern the case $k_1 = 1$, $k_2 = 0$, $V = \mathbb{R} \times W$, $W = U$. The case $\dim \Gamma/H_1 = \dim \Gamma/H = 1$, $\dim \Gamma/H_2 = 0$ was treated in the preceding proposition. There remains only the case $\dim \Gamma/H = \dim \Gamma/H_i = 0$.

Let $\{k_j^1\}$ corresponding to the fundamental cell decomposition for H_1 and $\{k_j^2\}$ for H_2. Then, for $H = H_1 \cap H_2$, one has the fundamental cell with $k_j = k_j^1$ for the variables in V^{H_1} and $\tilde{k}_j^2$, dividing k_j^2, for the variables in $V^{H_2} \cap (V^{H_1})^\perp$. Let $X_1 = (Z_1, Z_1' \dots, Z_{s_1}, Z_{s_1}')$, $X_2 = (Y_1, Y_1', \dots, Y_{s_2}, Y_{s_2}')$ and $X = X_1 \times X_2$ be as in Proposition 7.6, with the special action on Z_j and Y_j.

Then, on $(X \times V)^{H_1}$, one has the generators for $\Pi(H_1)$

$$\Sigma^W \eta_j^1 = \left(1 - \prod |2Z_i|, w, ((2Z_i)^{k_i^1} + 1)Z_i, Z_i', \dots, \lambda Z_j, Z_j', Y_i, Y_i'\right)$$

with $\lambda = \mu + i\, s$, $(s = 2t - 1)$, and

$$\begin{aligned}\Sigma^W \tilde{\eta}^1 = &\left(\varepsilon^2 - \prod_{i<s_1} |2Z_i||(2Z_{s_1})^{k_{s_1}^1} + 1|, w,\right.\\ &\left.((2Z_i)^{k_i^1} + 1)Z_i,\ Z_i', \dots, \lambda((2Z_{s_1})^{k_{s_1}^1} + 1)Z_{s_1}, Z_{s_1}'\right).\end{aligned}$$

On the other hand, on $(X \times V)^{H_2}$, one has the generator for $\Pi(H_2)$

$$\Sigma^W \eta_2 = \left(s + 2 - 2\prod |2Y_i|, w, ((2Y_i)^{k_i^2} + 1)Y_i, Y_i', Z_i, Z_i'\right).$$

Finally, the generators, on $(X \times V)^H$, for $\Pi(H)$, are similar to η_j^1 and $\tilde{\eta}^1$ but of the form

$$\Sigma^W \eta_j = \left(1 - \prod |2X_i|, w, ((2X_i)^{k_i} + 1)X_i, X_i', \dots, \lambda Z_j, Z_j'\right)$$

with $k_i = k_i^1$ for $X_i = Z_i$ and $1 \le i \le s_1$ and $k_i = \tilde{k}_i^2$ for Y_i.

As in Propositions 7.6 and 7.8, one has

$$[F]_\Gamma = [\Sigma^W \eta_2(\alpha \Sigma^W \eta_j^1)]_\Gamma = \sum d_i [\Sigma^W \eta_i]_\Gamma + \tilde{d}_j [\tilde{\eta}]$$

where α is the normalization of Proposition 7.8,

$$d_i = \deg(F; B^H \cap \operatorname{Arg} X_i = 0) / \prod_{i \neq l} k_l.$$

Since

$$\begin{aligned}\Sigma^W \eta_2(\alpha \Sigma^W \eta_j^1) = &\left(\alpha\left(1 - \prod |2Z_i|\right) + 2 - 2\prod |2Y_i|, \alpha w,\right.\\ &\left.\alpha((2Z_i)^{\alpha k_i^1} + 1)Z_i, \alpha Z_i', \alpha\lambda Z_j, \alpha Z_j', ((2Y_i)^{k_i^2} + 1)Y_i, Y_i'\right),\end{aligned}$$

it is clear that, if $Z_i \geq 0, i \neq j$, or if $Y_i \geq 0$, this map has no zeros.

Hence, $d_i = 0, i \neq j$. While

$$d_j = \Big(\prod_{i \neq j} k_i^1\Big)\Big(\prod k_i^2\Big) / \prod_{i \neq j} k_i.$$

We have proved the first part of

Proposition 7.9. *Under the above hypothesis, one has*

$$\begin{aligned}
[\Sigma^W \eta_2(\Sigma^W \eta_j^1)]_\Gamma &= \frac{|\Gamma/H_1| \cdot |\Gamma/H_2|}{|\Gamma/H_1 \cap H_2|}[\Sigma^W \eta_j]_\Gamma + \tilde{d}_j[\Sigma^W \tilde{\eta}]_\Gamma \\
[\Sigma^W \eta_2(\Sigma^W \tilde{\eta}^1)]_\Gamma &= \frac{|\Gamma/H_1| \cdot |\Gamma/H_2|}{|\Gamma/H_1 \cap H_2|}[\Sigma^W \tilde{\eta}]_\Gamma
\end{aligned}$$

where $k_j^1 \tilde{d}_j$ is even.

Proof. For the second equality, one has that $\Sigma^W \eta_2(\Sigma^W \tilde{\eta}^1)$ is non-zero on the fundamental cell for $X \times V$, hence its class is a multiple of $\Sigma^W \tilde{\eta}$. Counting the zeros of the map in the fundamental cell, one obtains $\prod(k_j^2/\tilde{k}_j^2)$ of them, which gives the equality. The fact that $k_j^1 \tilde{d}_j$ is even is proved as in Proposition 7.6. □

3.8 Bibliographical remarks

The problem of classification of equivariant homotopy classes of maps on spheres has been partially studied from the point of view of algebraic topology, essentially for finite groups and self-maps, but not necessarily linear actions. The obstruction approach has been used in the books by T. t. Dieck and Bredon. Study of the first obstruction has been given in the paper by Kosniowski. A complete result (with a proof corrected by Dancer) for self-maps and linear actions was given by Rubinstein.

This obstruction idea was used in [I0] for the group S^1 and a semi-free action.

The results on the extension problem and the first 3 sections of this chapter are taken from [I.V. 1–3]. The case of non-abelian actions is treated in the book of Kushkuley and Balanov, with an important contribution to the general Borsuk–Ulam problem.

This last subject, the ordinary degree of equivariant maps, has been extensively studied: see the survey papers by Steinlein, Zabrejko and interesting results by Nirenberg, Wang, Rabier and the book by Bartsch. As seen in Section 4, a complete answer is still lacking, even for abelian actions.

The one parameter case, in particular the problem of secondary obstructions, is taken from [IV2]. The recent papers by Balanov and Krawcewicz give results for non-abelian actions.

Orthogonal maps are classified in [IV3]. The results on operations are taken from [IV2] and [IV3]. Products of maps are also considered in the book by Krawcewicz and Wu.

The suspension Theorem 7.1 was given incompletely in [IV1].

Chapter 4

Equivariant Degree and Applications

In this last chapter we shall see how to apply the results for the equivariant homotopy groups, given in the previous chapter, to the computation of the equivariant degree of a map, in particular coming from differential equations. We shall first prove that any class in these homotopy groups is achieved as the Γ-degree of a map on a reasonable set Ω. Then, we shall compute the Γ-index of an isolated orbit, with several applications to bifurcation. The next section will concern the orthogonal index of an isolated orbit and an application to two mechanical systems. The last section regards the Γ-degree of a loop of orbits and its applications to Hopf bifurcation, systems with first integrals and similar problems.

4.1 Range of the equivariant degree

Recall that if V and W are two Γ-representations and if Ω is a bounded open and invariant subset of V, then for $f(x)$, a Γ-equivariant map from $\bar{\Omega}$ into W and non-zero on $\partial\Omega$, one defines the Γ-degree of f with respect to Ω as

$$\deg_\Gamma(f;\Omega) = [F]_\Gamma = [2t + 2\varphi(x) - 1, \tilde{f}(x)]_\Gamma,$$

where $\tilde{f}$ is a Γ-extension of f to a ball B_R containing Ω and $\varphi(x)$ is an invariant Uryson function with value 0 in $\bar{\Omega}$ and value 1 outside a neighborhood of $\bar{\Omega}$. Then, $[F]_\Gamma$ is an element of $\Pi_{S^V}^\Gamma(S^W)$. Hence, the first question is the following: given $[F]_\Gamma$ in the above group, does there exist a Γ-map f, from $\bar{\Omega}$ into W, such that $\deg_\Gamma(f;\Omega) = [F]_\Gamma$? In this section we shall give a partial, but explicit, answer to this question, that is, in all the cases studied in Chapter 3, where one had concrete generators for the above group, or at least its "free part". We shall also answer this question for the case of Γ-orthogonal maps.

Assume then that $V = \mathbb{R}^k \times U$ and U and W satisfy hypothesis (H), that is, for an abelian group Γ:

(a) $\dim U^H = \dim W^H$, for all H in $\mathrm{Iso}(V)$

(b) There is a Γ-equivariant map $\{x_i\} \to \{x_i^{l_i}\}$ from U into W.

Then we have seen in Theorem 3.2 of Chapter 3 that

$$\Pi_{S^V}^\Gamma(S^W) = \Pi_{k-1} \times \mathbb{Z} \times \cdots \times \mathbb{Z},$$

with one $\mathbb{Z} \cong \Pi(H)$ for each H with $\dim \Gamma/H = k$, and $\Pi(H)$ is generated by the maps F_H given in Theorem 3.3 of Chapter 3.

Furthermore, if $k = 1$, then

$$\Pi_0 = \bigoplus_{\dim \Gamma/H=0} \Pi(H) = \Pi^{\tilde{\Gamma}}_{S^{\tilde{V}}}(S^{\tilde{W}}),$$

where $\tilde{\Gamma} = \Gamma/T^n$, $\tilde{V} = V^{T^n}$, $\tilde{W} = W^{T^n}$, and $\Pi(H)$ is a finite group generated by η_j^H, $j = 1, \ldots, s_H$, and $\tilde{\eta}^H$, as given in Theorem 5.2 of Chapter 3 (here s_H is the number of k_j's which are larger than 1 and one has repetition of the variables).

Thus, any element $[F]_\Gamma$ in $\Pi^{\Gamma}_{S^V}(S^W)$ is written as

$$[F]_\Gamma = [\tilde{F}]_\Gamma + \sum d_H[\tilde{F}_H]_\Gamma,$$

where $[\tilde{F}]_\Gamma$ is in Π_{k-1} and d_H is an integer. The sum is over all H's with $\dim \Gamma/H = k$.

If $k = 1$, then one may write $[\tilde{F}]_\Gamma$ in the form

$$[\tilde{F}]_\Gamma = \sum_H \left(\sum d_{jH}[\eta_j^H]_\Gamma + \tilde{d}_H[\tilde{\eta}^H]_\Gamma\right),$$

where the sum is over all H's with $\dim \Gamma/H = 0$. If $U = W$, then one may use the presentation of $\Pi(H)$ given in Theorem 5.5 of Chapter 3.

Theorem 1.1. (a) *If $V = \mathbb{R}^k \times U$, where U and W satisfy* (H), *then, given any sequence $\{d_H\}$ of integers, there is a Γ-map f from $\bar{\Omega}$ into W, non-zero on $\partial\Omega$, such that*

$$\deg_\Gamma(f; \Omega) = [\tilde{F}]_\Gamma + \sum d_H[\tilde{F}_H]_\Gamma,$$

provided one takes $d_H = 0$ if $\Omega^H = \phi$ and $|\Gamma/H||d_H|$ at most equal to the number of components of Ω^H if $\dim V^H = 1$, hence $|\Gamma/H| \leq 2$.

(b) *If $k = 1$, then any $[F]_\Gamma$ in $\Pi^{\Gamma}_{S^V}(S^W)$ is the Γ-degree of a Γ-map f defined on Ω, provided the corresponding invariants $d_H, d_{jH}, \tilde{d}_H$ are taken to be 0, if $\Omega^H = \phi$ (the repetition of variables, of Theorem 5.2 in Chapter 3, is assumed here), and $d_\Gamma = 0$ if $\dim W^\Gamma \leq 2$.*

Proof. Note first that if there is an H with Ω^H empty then $\Omega^K = \phi$ for any $K > H$, in particular for $K = \Gamma$. On the other hand, if $\Omega^H \neq \phi$, then, since Ω^H is open in V^H, there is a $X^0 = (\lambda^0, X_0^0, y_i^0, z_j^0)$, with y_i^0 and z_j^0 different from 0, in Ω^H. Here, any point in $V = \mathbb{R}^k \times U$ is written as (λ, X), with λ in $\mathbb{R}^k$, X in U of the form $X = (X_0, y_i, z_j)$, where X_0 is in U^Γ, the group Γ acts as $\mathbb{Z}_2$ on y_i and as $\mathbb{Z}_m$ or S^1 on the complex coordinate z_j. By changing variables, we shall assume that $\lambda^0 = X_0^0 = 0$.

Our next step will be to show that any of the explicit generators given in the previous chapter may be taken as the Γ-degree of a map $f : \Omega \to W$, $\partial\Omega \to W\backslash\{0\}$.

(a) If $\dim \Gamma/H = k > 0$ and $\dim W^\Gamma \geq 1$ (hence $\dim U^\Gamma \geq 1$), let $X_0 = (x_0, \tilde{X}_0)$ be a decomposition of U^Γ. Define $x'_j = x_j/|x_j^0|$, for $x_j = y_j$ or z_j, and $x'_0 = x_0/R$ where R is the radius of a large ball containing Ω. Let

$$f(\lambda, X) = \Big(x'_0 - 2\big(\prod |x'_j|^2 - 1\big), \tilde{X}_0, \{(i\lambda_j - (|z'_j|^2 - 1))z_j^{l_j}\}_{j=1,\dots,k}, \\ (-Q_j + 1)y_j, (P_j + 1)z_j^{l_j}, x_s^{l_s}\Big),$$

where the product is over all x'_j in $U^H \cap (U^\Gamma)^\perp$, $z'_1, \dots, z'_k$ are the variables with $k_j = \infty$ and an isotropy subgroup $H_0 = \Gamma_{z'_1} \cap \dots \cap \Gamma_{z'_k}$ with $\dim \Gamma/H_0 = k$. The invariant polynomial Q_j is $y_j'^2$ if $k_j = 2$ and $P_j(y'_1, \dots, y'_j)$ if $k_j = 1$, where P_j is based on the real coordinates. For z_j, one takes $P_j = P_j(x'_1, \dots, x'_j)$ as the invariant polynomial of Lemma 6.3 in Chapter 1. Finally, x_s are the coordinates of $(V^H)^\perp$.

For any integer d, one may replace $\lambda_k + i(|z'_k|^2 - 1) = A$ by A^d, where A^d means $\bar{A}^{|d|}$ if d is negative.

Since $|x'_0| \leq 1$ in Ω^H, the zeros of f in Ω^H have $x_j \neq 0$ for all j's and for $|x'_j| = 1$, as in Theorem 3.2 of Chapter 3. For z_j in $\mathbb{R}^+$, for $j = 1, \dots, k$, there are $\prod k_j = |H_0/H|$ zeros, equal to γX^0, for some γ in Γ, and only one of these zeros is in $\mathcal{C}_H$, the fundamental cell for H. For the map $(2t - 1 + 2\varphi(\lambda, X), f(\lambda, X))$ one may deform φ to 0 on $\partial(I \times B^H)$, since the zeros of $f(\lambda, X)$ in $I \times B^H$ are the orbit of X^0, i.e., in Ω. Furthermore, one may rotate $2t - 1$ and x'_0 to obtain the map

$$\Big(-x'_0, 2t+1-2\prod |x'_j|^2, \tilde{X}_0, (i\lambda_j - (|z'_j|^2 - 1))z_j^{l_j}, (-Q_j+1)y_j, (P_j+1)z_j^{l_j}, x_s^{l_s}\Big).$$

After a rotation of the first two components, one obtains a Γ-map which is similar to the generator F_H of Theorem 3.3 in Chapter 3: it differs from the fact that here one has all the components x'_j and by the new definition of Q_j, while in F_H one had considered only y_j with $k_j = 2$. In any case, this map is in $\Pi(H)$, since, if $K > H$, one needs that one of the x_j to be 0. Furthermore, this map has an extension degree 1, up to an orientation factor which may be fixed by choosing $d = -1$. Thus, one may take this new map as a generator of $\Pi(H)$ and, by letting d to be arbitrary, have the complete $\Pi(H) \cong \mathbb{Z}$.

(b) If $k > 0$ and $\dim W^\Gamma = \{0\}$, then $f(\lambda, 0) = 0$ and one needs $\bar{\Omega}^\Gamma = \phi$ in order to define the Γ-degree of f. As before, let $(0, x_j^0)$ be a point of Ω^H with $x_j^0 \neq 0$ for all j's and define $\lambda'_k = \lambda_k/R$. Let

$$f(\lambda, x) = \Big(\{(i\lambda_j - (|z'_{j+1}|^2 - 1))z_j^{l_j}\}_{j=1,\dots,k-1}, \\ i\big(\lambda'_k + 2\sum(|x'_j|^2 - 1)^2 + i(|z'_1|^2 - 1)\big)z_k^{l_k}, \\ (-|z'_k|^2 Q_j + 1)y_j, (|z'_k|^2 P_j + 1)z_j^{l_j}, x_s^{l_s}\Big),$$

where $Q_j = Q_j(y'_1, \dots, y'_j)$ and $P_j = P_j(x'_1, \dots, x'_j)$ as before, and the sum in the z_k-component is over all j's. The factor A of $z_k^{l_k}$ may be replaced by A^d.

If $f(\lambda, X) = 0$ and $z_k = 0$, then $x_j = 0$ for all j's and $X = 0$, that is $(\lambda, 0)$ belongs to $\Omega^\Gamma = \phi$. Hence, $z_k \neq 0$, $|z'_j| = 1$ for $j = 1, \dots, k$, and if $x'_j = 0$ for $j > k$, one has $\lambda'_k + 2\sum(|x'_j|^2 - 1)^2 \geq \lambda'_k + 2 > 0$. Thus, the zeros in Ω^H are for $|x'_j| = 1, \lambda = 0, X = \gamma X^0$. For the map

$$F(t, \lambda, X) = (2t - 1 + 2\varphi(\lambda, X), f(\lambda, X)),$$

one may deform $\varphi(\lambda, X)$ to 0, obtaining an element of $\Pi(H)$. The extension degree of F on the fundamental cell $\mathcal{C}_H$ is 1, up to an orientation factor, and one may take $F(t, \lambda, X)$ as the generator of $\Pi(H)$.

(c) If $k = 0$ and one has at least one complex z_1 in V^H, then, if dim W^Γ is positive, the map

$$f(X) = \left(x'_0 - 2\left(\prod |x'_j|^2 - 1\right), \tilde{X}_0, -(Q_j - 1)y_j, (P_j^{d_j} + 1)z_j^{l_j}, x_s^{l_s}\right)$$

gives an element $F(t, X) = (2t - 1 + 2\varphi(X), f(X))$ which is in $\Pi(H)$ with extension degree equal to $\prod d_j$. While, if dim $W^\Gamma = 0$, then one defines

$$f(X) = ((-aQ_j + 1)y_j, (aP_j^{d_j} + 1)z_j^{l_j}, x_s^{l_s}),$$

where $a = \prod |x'_j|$. It is clear that one cannot have $a = 0$ in a zero, unless $X = 0$ which does not belong to Ω^H. Since $P_1(z'_1) = z_1'^{k_1}$, on a zero in Ω^H, one has $|x'_j| = a^{\alpha_j}$ and one may modify aP_1 to $a^p P_1$ in such a way that $\sum \alpha_j \neq 1$. Hence, on a zero, one has $a = 1$, $|x'_j| = 1$, i.e., $X = \gamma X^0$ and $\deg_\Gamma(f; \Omega) = \left(\prod d_j\right)[\tilde{F}_H]$, since for the zero in $\mathcal{C}_H$, one may deform $x_j^{l_j}$ to $x_j^{0l_j}$ and a to 1.

(d) If all coordinates in V^H are real, $k = 0$ and dim $W^\Gamma > 0$, then if $|\Gamma/H| > 2$, take two y's, say y_1 and y_2, with $k_1 = k_2 = 2$ and consider the map

$$\begin{aligned} f(X) = \Big(x'_0 - 2\left(\prod y_j'^2 - 1\right), \tilde{X}_0, &-(\mathrm{Re}(y_1'^2 - 1 + i(y_2'^2 - 1))^d y_1, \\ &-(\mathrm{Im}(y_1'^2 - 1 + i(y_2'^2 - 1))^d)y_2, -(Q_j - 1)y_j, x_s^{ls}\Big), \end{aligned}$$

where $Q_j = y_j'^2$ if $k_j = 2$ and $P_j(y'_1, \dots, y'_j)$ if $k_j = 1$. Again, the zeros of $f(X)$ are for $X_0 = 0$, $|y'_j| = 1$, with an extension degree equal to d (up to an orientation factor). While, if $|\Gamma/H| = 2$ and $k_1 = 2$, with $k_j = 1$ for $j > 1$, consider the above map but with $y_1'^2 - 1 + i(y_2'^2 - 1)$ replaced by $y_1'^2 - 1 + i(y'_1 y'_2 - 1)$, if dim $V^H \cap (V^\Gamma)^\perp > 1$. If this dimension is one, take the map

$$\begin{aligned} f(X) = (\mathrm{Im}(x'_0 - 2(y'^2 - 1) + iy'^2(y'^2 - 1))^d, \\ y\,\mathrm{Re}(x'_0 - 2(y'^2 - 1) + iy'^2(y'^2 - 1))^d, x_s^{ls}), \end{aligned}$$

which has only two zeros in Ω^H : $y' = \pm 1, x'_0 = 0$. Then, one proves that $(2t - 1 + 2\varphi(x), f(x))$ has an extension degree equal to d: near $(x'_0 = 0, y' = 1)$ deform y to 1, use $((1-\tau)y'^2 + \tau)(y'^2 - 1)$ and deform $(x'_0 - 2(y'^2 - 1) + i(y'^2 - 1))^d$ to $(x'_0 + i(y' - 1))^d$, with degree equal to d. If $H = \Gamma$ and $\dim W^\Gamma \geq 2$, the map $((x_0 + ix_1)^d, \tilde{X}_0, x_i^{l_i})$ gives, on W^Γ, a degree equal to d.

If $\dim W^\Gamma = 1$, then Ω^Γ is the union of disjoint intervals and with $\pm(x_0 - x_j)$, x_j a fixed point in the j'th interval, one achieves at most $\pm$ (the number of components of Ω^Γ).

(e) If all coordinates in V^H are real, $k = 0$ and $\dim W^\Gamma = 0$, then, if $\dim V^H > 1$, take $a = \prod y_j'^2$ and consider the map

$$f(X) = (-\operatorname{Re}(aQ_1 - 1 + ib(aQ_2 - 1))^d y_1,\\ -\operatorname{Im}(aQ_1 - 1 + ib(aQ_2 - 1))^d y_2, (aQ_j - 1)y_j),$$

where b is a positive number, depending on d, such that $(1 + ib)^d$ is neither real nor pure imaginary. Thus, $a = 0$ leads to $y_j = 0$ for all j, i.e., a zero $X = 0$ in $\Omega^\Gamma = \phi$. Hence, $a \neq 0$ and the zeros of $f(X)$ in Ω are such that $|y_j| = a^{\alpha_j}$. Modifying a Q_s to $a^2 Q_s$ if $\sum \alpha_j = 1$, one gets that $a = 1$ and $|y_j| = 1$. In particular, there is only one zero in $\mathcal{C}_H$, with $y_j = 1$ for all j's. Near that zero, one may deform $f(X)$ to $((aQ_1 - 1 + i(aQ_2 - 1))^d, aQ_j - 1)$, with index d.

If $\dim V^H = 1$, then Ω^H is the union of disjoint intervals, one has $\deg(2t + 2\varphi(y) - 1, \tilde{f}(y); B_0) = \deg(f(y); B_0 \cap \Omega^H)$ and, on each interval of $B_0 \cap \Omega^H = \Omega^H \cap \{y > 0\}$, the degree of f may be 0 or ± 1. It is then easy to construct an odd map with local index equal to ± 1 on each such interval.

(f) If $k = 1$ and $|\Gamma/H| < \infty$ with $\dim W^\Gamma \geq 1$, let

$$f_j(\mu, X) = \Big(x'_0 - 2\big(\prod |x'_i|^2 - 1\big), \tilde{X}_0, (Q_i - 1)y_i,\\ i(\mu + i(|z'_j|^2 - 1))^d z_j^{l_j}, \{(P_i + 1)x_i^{l_i}\}_{i \neq j}, x_s^{l_s}\Big),$$

where Q_i and P_i are functions of $x'_1, \dots, x'_i$. Here the repetition of variables of Theorem 5.2 of Chapter 3 is also assumed. Thus, if z_j corresponds to a couple of real variables with a $\mathbb{Z}_2$-action, then $z_j = y_1 + iy_2$. It is clear that $\deg(2t - 1 + 2\varphi(\mu, X), f_j; \mathcal{C} \cap \{\operatorname{Arg} z_j = 0\}) = d$ (up to an orientation factor), i.e., that f_j may replace η_j, when $d = 1$, in Lemma 5.4 of Chapter 3.

Similarly, choose $\varepsilon_1, \dots, \varepsilon_n$, with $|\varepsilon_i| = 1$, such that $\{(Q_i - 1, P_i - \varepsilon_i)\}$ has $|\Gamma/H|$ zeros, with $|x'_i| = 1$, and only one zero X^0 in $\mathcal{C}_H$. Take ε small enough and with $A = -\varepsilon_n^{-1} \prod_{i<n} |x'_i| P_n + 1$, define

$$\tilde{f}(\mu, X) = (|A|^2 - \varepsilon^2, \tilde{X}_0, (-\varepsilon^{-1}|A|Q_i + 1)y_i,\\ \{(-(\varepsilon\varepsilon_i)^{-1}|A|P_i + 1)x_i^{l_i}\}_{i<n}, i(\mu + ix_0)Ax_n^{l_n}, x_s^{l_s}),$$

recalling that $n > 1$ since at least x_n is repeated. Since $x_i = 0$ or $x_n = 0$ implies $|A| = 1$ and $A = 0$ implies, on a zero, $y_i = x_i = 0$, a contradiction, the zeros of $\tilde{f}(\mu, X)$ are such that $\mu = x_0 = \tilde{X}_0 = 0$, $X_i = \gamma X_i^0$, $|x_n - x_n^0| = \varepsilon$. Furthermore, on $\partial \mathcal{C}_H$, the map $(2t + 2\varphi(\mu, X) - 1, \tilde{f}(\mu, X))$ is homotopic to the generator $\tilde{\eta}$ of Lemma 5.4 in Chapter 3.

If $H = \Gamma$ and $n = \dim W^\Gamma$, then $\deg(F^\Gamma; \Omega^\Gamma)$, as defined in Chapter 2, belongs to $\Pi_{n+1}(S^n)$, due to the presence of t and μ. Then, if $n \geq 3$, the map $(x_1^2 + x_2^2 - \varepsilon^2, i(\mu + ix_0)(x_1 + ix_2), \dots)$ is the Hopf map, while if $n = 2$ and Ω is the ball $\{\mu^2 + x_0^2 + x_1^2 < 2\}$, then any F^Γ is homotopic, on the boundary of the ball, to a constant map (since $\Pi_2(S^1) = 0$), hence d_Γ must be 0 in this case. If $n < 2$, then $d_H = 0$.

(g) If $k = 1$ and $|\Gamma/H| < \infty$, with $W^\Gamma = \{0\}$, take the map

$$f_j(\mu, X) = \Big((-|z_j'|Q_i + 1)y_i, (|z_j'|P_i + 1)x_i^{l_i},\\ i\big(\mu' + 2\sum_1^n(|x_i'|^2 - 1)^2 + i(|z_j'|^2 - 1)\big)^d z_j^{l_j}, x_s^{l_s}\Big),$$

where Q_i and P_i are functions of $x_1', \dots, x_i'$ and $\mu' = \mu/R$. The zeros of f_j in Ω^H are for $|z_j'| = 1$, $|x_i'| = 1$ and $\mu = 0$ (since $\Omega^\Gamma = \phi$). As in Theorem 5.2 of Chapter 3, it is easy to compute $\deg(f_j; \mathcal{C}_H \cap \{\operatorname{Arg} z_j = 0\}) = d$ and to see that there are no zeros on the previous faces of $\mathcal{C}_H$. Hence, $\deg_\Gamma(f_j; \Omega) = d\eta_j$.

In order to get a map with Γ-degree equal to $\tilde{\eta}$, consider

$$\tilde{f}(\mu, X) = \Big(\{1 - \varepsilon^{-1}\varepsilon_i^{-1}|x_n'||A|P_i')x_i^{l_i}\}_{i<n},\\ i\big(\mu' + 2\sum_1^n(|x_i'|^2 - 1)^2 + i(\varepsilon^2 - |A|^2|x_n'|^2)\big)Ax_n^{l_n}, x_s^{l_s}\Big),$$

where P_i' corresponds to Q_i or P_i, functions of $x_1', \dots, x_i'$, the factor $A = 1 - \varepsilon_n^{-1}\big(\prod_{i<n}|x_i'|\big)P_n$, and the phases ε_j, with $|\varepsilon_j| = 1$, are chosen as above. Hence, if $x_n = 0$, then $x_i = 0$ for all i, i.e., a point in $\Omega^\Gamma = \phi$. While $A = 0$ leads to $x_i = 0$ for $i < n$, a contradiction. Thus, a zero of $\tilde{f}$, in Ω, gives $|x_i'| = 1$, $P_i' = \varepsilon_i$, for $i < n$, that is $x_i = \gamma x_i^0$ (since $x_i' = 0$ would not give a zero of $\mu' + 2\sum_1^n(|x_i'|^2 - 1)^2$), with $|x_n'||P_n - \varepsilon_n| = \varepsilon$ and $\mu' + 2(|x_n'|^2 - 1)^2 = 0$. Since $|\mu'| \leq 1$ in Ω, the last equality implies that $|x_n'|^2 \geq 1 - (1/2)^{1/2}$, hence, for ε small enough, x_n' is close to the unique zero x_n^0/R in $\mathcal{C}_H$. In order to get the Γ-degree of $\tilde{f}$ one has to compute the class of $(2t - 1 + 2\varphi(\mu, X), \tilde{f}(\mu, X))$ on $\partial \mathcal{C}_H$. In particular, since $\Omega^\Gamma = \phi$, one may assume that $\varphi(\mu, 0) = 1$. Replace, for $j < n$, $\{x_j\}$ by $x_j^\tau = (1 - \tau)x_j + \tau x_j^0$ in the terms $x_j^{l_j}$, $|x_j'|$ and in φ. A zero of the deformed map may have $x_n = 0$, but then $x_i^\tau = 0$ for all i, that is the path goes through the origin, but there φ has value 1. Furthermore, if the deformed A is 0, one gets $x_i^\tau = 0$ for $i < n$, again a contradiction. Thus, a zero of the deformed map will lead to $P_i' = \varepsilon_i$, $|x_n'||P_n - \varepsilon_n| = \varepsilon$ and $\mu' + 2(|x_n'|^2 - 1)^2 = 0$, with $x_i^\tau = x_i^0$ for $i < n$. As before, x_n should be close to x_n^0, hence a point inside $\mathcal{C}_H$. It is then easy to deform $x_n^{l_n}$ and $|x_n'|$ to $x_n^{0l_n}$ and 1, arriving at the map

$$(2t - 1 + 2\varphi, -\varepsilon^{-1}\varepsilon_i^{-1}|P_n - \varepsilon_n|P_i + 1, (\mu + i(\varepsilon^2 - |P_n - \varepsilon_n|^2))(P_n - \varepsilon_n)).$$

One may deform linearly the first component to $2t - 1 + (|P_n - \varepsilon_n|^2 - \varepsilon^2)$. Replacing $2t - 1$ by $(1 - \tau)(2t - 1)$ in this component and $i(\varepsilon^2 - |P_n - \varepsilon_n|^2)$ by $i(1 - \tau)(\varepsilon^2 - |P_n - \varepsilon_n|^2) + i\tau(2t - 1)$, one arrives at

$$(|P_n - \varepsilon_n|^2 - \varepsilon^2, -\varepsilon^{-1}\varepsilon_i^{-1}|P_n - \varepsilon_n|P_i + 1, (\mu + i(2t - 1))(P_n - \varepsilon_n)).$$

One may replace $\varepsilon^{-1}|P_n - \varepsilon_n|$ by 1 and get the map $\tilde{\eta}$ of Lemma 5.4 in Chapter 3.

Thus, up to here we have seen that all the known generators and their multiples are realized by the Γ-degree of some map defined on Ω. It remains to show that any sequence $\{d_H\}$ may be realized by the Γ-degree of a map.

Let then $\{d_H\}$ be any admissible sequence of integers (i.e., $d_H = 0$ if $\Omega^H = \phi$ and d_H limited by the number of components of Ω^H if $\dim V^H = 1$). We shall give two constructions, according to the case $k > 0$ or $k = 0$, leaving to the reader the task to extend each one to the other case.

(α) If $k > 0$, choose N values of $\lambda_k \equiv \mu$, labelled $\mu_1, \dots, \mu_N$, with $N = \sum |d_H|$ and $\mu_{j+1} - \mu_j \geq 4\varepsilon_1$, for some small ε_1 such that, for each j, there is an isotropy group H and a point $(\lambda_1^0, \dots \lambda_{k-1}^0, \mu_j, X_0^0, X_H^0)$ in Ω^H, with all the components of X_H^0 non zero. This is possible because Ω^H is open (and non-empty) in V^H and there are only a finite number of d_H's different from 0.

For each j, corresponding to a certain H and a possible face of $\mathcal{C}_H$, let f_j be one of the above generators with the following modifications:

1. Replace (λ, X_0) by $(\lambda - \lambda_0, X_0 - X_0^0)$, where $\lambda = (\lambda_1, \dots, \lambda_{k-1}, \mu)$ and $\lambda_0 = (\lambda_1^0, \dots, \lambda_{k-1}^0, \mu_j)$.

2. Let φ_j be a Uryson function depending only on μ, with value 1, if $|\mu - \mu_j| < \varepsilon_1$ and value 0, if $|\mu - \mu_j| > 2\varepsilon_1$. Then, in case (b), replace the factor of $z_k^{l_k}$ by

$$i\Big(\mu' + 4\sum_{j=1}^{k}(1 - \varphi|z_j|^2)^2 + 4\sum(1 - \varphi|z_k'|Q_j)^2 + 4\sum|1 + \varphi|z_k'|P_j|^2 + i(|z_1'|)^2 - 1\Big).$$

A similar modification is made for the first maps f_j in case (g).

For the map $\tilde{f}$ in (g) replace the factor of $x_n^{l_n}$ by

$$i\Big(\mu' + 2\sum_{1}^{n-1}|\varphi|x_n'||A|P_i\varepsilon^{-1} - \varepsilon_i|^2 + 2|\varphi^2|x_n'|^2 - 1|^2 - i(1 - \varphi^2|x_n'|^2|A|^2\varepsilon^{-2})\Big)A.$$

Define then (with a slight change, given below, for the maps $\tilde{f}$ of cases (f) and (g))

$$f(x) = \begin{cases} \varphi_j f_j(x) + (1-\varphi_j)(1, 0, \dots, x_i^{l_i}, \dots), & \text{if } |\mu - \mu_j| < 2\varepsilon_1 \\ (1, 0, \dots x_i^{l_i}, \dots), & \text{on the complement,} \end{cases}$$

where $(1, 0)$ has the usual meaning on W^Γ and is not present if $W^\Gamma = \{0\}$. We shall see below that $f(x)$ is non-zero on $\partial\Omega$, hence, if

$$\Omega_j = \Omega \cap \{\mu : |\mu - \mu_j| < 2\varepsilon_1\},$$

one has that, up to one suspension (which is an isomorphism for the $\Pi(H)$ of the theorem),

$$\deg_\Gamma(f; \Omega) = \sum \deg_\Gamma(f; \Omega_j).$$

Since $\Omega^\Gamma = \phi$ if $W^\Gamma = \{0\}$, the map $(1, 0, \dots, x_i^{l_i}, \dots)$ is never 0 in Ω. Furthermore, if $\phi f_j(x) + (1-\phi)(1, 0, \dots, x_i^{l_i}, \dots) = 0$ in Ω_j, then in case (a), the first component is

$$\varphi\Big(x_0' + 2\Big(1 - \prod |x_j'|^2\Big)\Big) + (1-\varphi) = 0$$

and the component of $z_k^{l_k}$ will give, by translating μ_j to 0,

$$(i\varphi\mu + 1 - \varphi|z_k'|^2)z_k^{l_k} = 0.$$

If $z_k = 0$, the first component would be positive (recall that $|x_0'| \leq 1$), hence $\varphi\mu = 0$. But $\varphi = 0$ gives a non-zero map, hence $\mu = 0$ and $\varphi = 1$, giving the original map f_j, generator of $\Pi(H)$. A change to $-\mu$ in the z_k component, will give the inverse of the generator.

In case (b), a zero of $f(x)$ with $z_k = 0$ leads to $y_j = 0$, $z_j = 0$, i.e., a point of the form $(\lambda, 0)$ in $\Omega^\Gamma = \phi$. Hence, on a zero, one has $z_k \neq 0$ and, since $|\mu'| \leq 1$ by construction and $Q_j = 0$ if $y_j = 0$, or $P_j = 0$, if $z_j = 0$, none of these variables may be 0. This implies that if $f(x) = 0$ in Ω_j, one has $\varphi|z_j|^2 = 1$, for $j = 1, \dots, k$, $\varphi|z_k'|Q_j = 1$, and $\varphi|z_k'|P_j + 1 = 0$, reducing the factor of $z_k^{l_k}$ to $i\mu'\varphi - \varphi|z_1'|^2 + 1 = 0$, that is $\mu'\varphi = 0$. Since $\varphi = 0$ cannot happen on a zero, one has $\mu' = 0$ and $\varphi(\mu) = 1$ and one gets the generator of (b).

For the maps for the faces of $\mathcal{C}_H$ in case (f), the argument is parallel to case (a) and for the maps for the faces of $\mathcal{C}_H$ in case (g), one follows the steps of case (b). Thus, the only remaining cases are for the Hopf map $\tilde{f}$ of cases (f) and (g). For these cases, one will modify the construction of $f(x)$ by defining it as

$$\varphi\tilde{f} + (1-\varphi)(1, 0, x_1^{l_1}, \dots, (\tilde{\varphi}Ab + (1-\tilde{\varphi}))x_n^{l_n}, \dots),$$

for $|\mu - \mu_j| \leq 4\varepsilon_1$, where $\tilde{\varphi}(\mu) = \varphi((\mu - \mu_j)/2)$, hence $\tilde{\varphi}(\mu) = 1$ if $|\mu - \mu_j| \leq 2\varepsilon_1$ and $\tilde{\varphi}(\mu) = 0$ if $|\mu - \mu_j| \geq 4\varepsilon_1$. The factor b is 1 in case (f) and $1 - \varepsilon^{-2}|A|^2|x_n'|^2\varphi^2$ in case (g).

Taking $\mu_j = 0$, the map, for case (f), has, for $|\mu| \le 2\varepsilon_1$, a factor of $x_n^{l_n}$ of the form $(i\mu\varphi - x_0 + (1-\varphi))A$ and a first component $\varphi(|A|^2 - \varepsilon^2) + (1-\varphi)$. Hence, if $x_n = 0$, then $A = 1$ and the first component is positive, while if $A = 0$, the other components of the map reduce to y_i of $x_i^{l_i}$ which are never 0 if $A = 0$. Hence, on a zero of the map, one has $\mu\varphi = 0$; but $\varphi = 0$ is not possible and then $\mu = 0, \varphi = 1$ and one is back to the original map. On the other hand, if $|\mu| > 2\varepsilon_1$, then $\varphi = 0$ and the map is not 0. (The case $H = \Gamma$ is covered by taking $A = (x_1^2 + x_2^2)^{1/2}$).

For the map $\tilde{f}$ of case (g), if $|\mu| > 2\varepsilon_1$, then a zero of the map implies that $x_j = 0$ for $j < n$ and $A = 1$, hence $x_n = 0$, which gives a point which is not in Ω, since $\Omega^\Gamma = \phi$. On the other hand, if $|\mu| \le 2\varepsilon_1$, then $\tilde{\varphi} = 1$ and if $x_n = 0$, thus, on a zero, one has that $x_j = 0$ for all j, hence not a point in Ω. If $A = 0$, then $x_j = 0$ for $j < n$, which contradicts the definition of A. Hence, a zero of the map will have all x_j's different from 0 and $A \neq 0$ (if $x_j = 0$ then the coefficient of $Ax_n^{l_n}$ has an imaginary part which is positive, since $|\mu'| \le 1$). On a zero, this coefficient is $i\varphi\mu' + 2i\varphi(\varphi^2|x_n'|^2 - 1)^2 + 1 - \varepsilon^{-2}|A|^2|x_n'|^2\varphi^2$. Hence, on a zero, one has $P_i = \varepsilon_i$ with solution x_i with $|x_i| = 1$ for $i < n$. Thus, $A = 1 - \varepsilon_n^{-1}P_n$, $\varphi|A||x_n'| = \varepsilon$, $\mu' + 2(\varphi^2|x_n'|^2 - 1)^2 = 0$. Since $|\mu'| \le 1$, one has, as before, that $\varphi|x_n'| \ge 1 - 1/\sqrt{2}$ hence $|A| \le C\varepsilon$ and x_n cannot be close to 0: in fact, $|x_n'|$ has to be close to 1 (for $|A|$ to be small, i.e., for $|P_n|$ close to 1), φ has to be close to 1, i.e., μ is close to 0 and the map is essentially the one given in (g) and, in fact, deformable to it on Ω_j.

(β) For the remaining cases, i.e., with $k = 0$, one will use the following construction:

Let H be any isotropy subgroup and write any point X in U as $X = X_0 \oplus X_H \oplus X_H^\perp$, where X_0 is in U^Γ, X_H in $U^H \cap (U^\Gamma)^\perp$ and $X_H^\perp$ in $(U^H)^\perp$. For some small ε, let the open set

$$\Omega_H = \left\{X \in \Omega,\ \|X_H^\perp\| < \varepsilon,\ |x_i| > 2\varepsilon \text{ for all } x_i \text{ components of } X_H\right\}.$$

Take an even function $\varphi(x)$, non increasing for $x > 0$, with value 1 if $|x| \le \varepsilon$, and value 0 if $|x| \ge 2\varepsilon$ and define

$$\varphi_H(X) = \prod_{x_i \in U^H \cap (U^\Gamma)^\perp} (1 - \varphi(x_i)) \prod_{x_i \in (U^H)^\perp} \varphi(x_i).$$

In particular, $\varphi_H(X) = 1$ if X belongs to Ω_H. Now, if K and H are two different isotropy subgroups then either $U^H \cap (U^K)^\perp$ or $U^K \cap (U^H)^\perp$ do not reduce to 0: in fact, $U^H \cap (U^K)^\perp = \{0\}$ if and only if $U^H \subset U^K$, hence, if both intersections are $\{0\}$, one has $U^H = U^K$ and, H and K being isotropy subgroups, one gets $H = K$. Thus, if x_i is a common component to U^H and $(U^K)^\perp$, one has that $\Omega_H \cap \Omega_K = \phi$ and $\varphi_K(X) = 0$ on Ω_H.

For each H and d_H, consider the maps given in (c)–(e), denoted as f_H and modified in the following way: if the coefficient of y_j or of $z_j^{l_j}$ is denoted by a_j then multiply a_j by $e^{i\psi}$ in such a way that a_j has a positive real part if y_j or z_j is 0. The angle ψ

will depend only on d_H. The maps of (c)–(e) have been set up in such a way that this condition is met when $d_H = 1$.

Since P_j or Q_j are 0 when z_j or y_j are 0, the maps in (c) do not need any adjustment, while, for those of (d) and (e), it is easy to figure out the rotation needed in the first two components. Define

$$f(X) = \sum_H \varphi_H(X) f_H(X) + \prod_H (1 - \varphi_H)(1, 0, x_i^{l_i}).$$

Then, $f(X) = f_H(X)$ if X is in Ω_H. Furthermore, if X belongs to $\Omega \backslash \bigcup_H \Omega_H$, then X has a non-zero component x_i in $(U^\Gamma)^\perp$ with $|x_i| \leq 2\varepsilon$ (if all such components are with norm $|x_i| > 2\varepsilon$, then X would be in Ω_{H_0}, with $H_0 = \Gamma_X$). But then the i-th component of $f(X)$ will be

$$\Big(\sum a_{H_i} \varphi_H + \prod (1 - \varphi_H)\Big) x_i^{l_i}.$$

One may choose ε small enough, since Ω is bounded, such that $\mathrm{Re}(a_{H_i}) > 0$ for $|x_i| \leq 2\varepsilon$. Then, if all φ_H's are 0, the product is 1.

Hence, $f(X)$ is non-zero on the complement of $\bigcup_H \Omega_H$ and

$$\deg_\Gamma(f; \Omega) = \sum \deg_\Gamma(f_H; \Omega_H) = \sum d_H [F_H],$$

since the suspension is an isomorphism. □

A similar result holds for orthogonal maps: recall that in this case, the abelian group Γ acts on the finite dimensional space V and one considers Γ-maps $F(x)$ from V into itself, such that

$$F(x) \cdot A_j x = 0, \quad j = 1, \dots, n = \dim \Gamma,$$

where A_j is an infinitesimal generator for the torus part of Γ. When considering the abelian group $\Pi^\Gamma_{\perp S^V}(S^V)$, of all orthogonal Γ-homotopy classes of S^V into itself, we have proved, in Theorem 6.1 of Chapter 3, that

$$\Pi^\Gamma_{\perp S^V}(S^V) \cong \mathbb{Z} \times \cdots \times \mathbb{Z},$$

with one $\mathbb{Z}$ for each isotropy subgroup of Γ, and that any $[F]_\perp$ in $\Pi^\Gamma_{\perp S^V}(S^V)$ can be written as

$$[F]_\perp = \sum d_H [F_H]_\perp,$$

with explicit generators F_H. Also, in §4 of Chapter 2, we have defined the orthogonal degree of a Γ-orthogonal map $f(x)$, defined on a Γ-invariant open bounded set Ω in V and non-zero on $\partial\Omega$, with the usual construction, as

$$\deg_\perp(f; \Omega) = [2t + 2\varphi(x) - 1, \tilde{f}(x)]_\perp = [F(t, x)]_\perp.$$

Theorem 1.2. *Any sequence of d_H's is the orthogonal degree of some orthogonal Γ-map defined on Ω, provided d_H is taken to be 0 if Ω^H is empty.*

Proof. As in the preceding theorem, we shall first construct orthogonal maps which have an orthogonal degree equal to $d[F_H]_\perp$, for each isotropy subgroup H, such that $\Omega^H \neq \phi$. If $\dim \Gamma/H = 0$, then the generators were already constructed in the preceding theorem, with $k = 0$, since then $V^H \subset V^{T^n}$ and any map, on V^H, is orthogonal.

Assume then that $\dim \Gamma/H = k > 0$ and that one has the components $z_1, \dots, z_k$, with action of T^n given on z_l by $\exp i\langle N^l, \Phi\rangle$, where $N^l = (n_1^l, \dots, n_n^l)$, and isotropy $H_0 > H$ such that $|H_0/H| < \infty$. Let N be the dimension of V^H and let A^H be the $N \times n$ matrix with $A_{ij}^H = n_j^i$, $i = 1, \dots, N$, $j = 1, \dots, n$. Then A^H has rank k and has an invertible submatrix A, for instance n_j^i, for $i, j = 1, \dots, k$, corresponding to $z_1, \dots, z_k$ and $\varphi_1, \dots, \varphi_k$. Then, if for $j > k$, one defines λ_j^i by

$$\begin{pmatrix} \lambda_j^1 \\ \vdots \\ \lambda_j^k \end{pmatrix} = A^{-1} \begin{pmatrix} n_j^1 \\ \vdots \\ n_j^k \end{pmatrix},$$

one has, for any coordinate z_l in V^H and $j > k$, the relation

$$n_j^l = \sum_{s=1}^{k} \lambda_j^s n_s^l.$$

See § 6 of Chapter 3 and Lemma 7.1 of Chapter 1. Furthermore, for X in V^H and $j > k$, one has

$$A_j X = \sum_1^k \lambda_j^s A_s X,$$

and $A_1 X, \dots, A_k X$ are linearly independent if X has its first coordinates, $z_1, \dots, z_k$, non-zero.

(a) If $\dim V^\Gamma \geq 1$, let the point $X_0 = (x_0^0, \tilde{X}_0^0, y_j^0, u_j^0, z_j^0)$ be in Ω^H, where $(x_0^0, \tilde{X}_0^0)$ is in V^Γ (by translation we shall assume it to be $(0, 0)$) and (y_j, u_j) is in V^{T^n}, with Γ acting as $\mathbb{Z}_2$ on y_j and as $\mathbb{Z}_m$ on the complex variable u_j. By perturbing a little, one may assume that y_j^0, u_j^0, z_j^0 are non zero, provided they are components of V^H. Let $x_j' = x_j/|x_j^0|$ for these components and $x_0' = x_0/R$, where $\Omega \subset B_R$. Consider the generator $f(\lambda, X)$, given in (a) of the preceding theorem,

$$\begin{aligned} f(\lambda, X) = \Big(x_0' - 2\big(\prod |x_j'|^2 - 1\big), \tilde{X}_0, \{(i\lambda_j - (|z_j'|^2 - 1))z_j\}_{j=1,\dots,k}, \\ (-Q_j + 1)y_j, (P_j + 1)z_j, x_s\Big) \end{aligned}$$

where some P_j may be changed to P_j^d if one wants an equivariant degree equal to $d[F_H]$. If there are no P_j, let $\varphi(x)$ be a smooth function for $x \geq 0$, with $\varphi(0) = 0$ and with exactly d solutions of $\varphi(x) = 1$, at $x_j = 1 + j\varepsilon$, for $j = 0, \dots, d-1$ and with $\varphi'(x_j) \neq 0$. Replace then the coefficient of z_k by $a_k = i\varphi'(|z_k'|)\lambda_k + 1 - \varphi(|z_k'|)$. For z_k real and positive, it is easy to see that the map $(\lambda_k, z_k) \to a_k$ has index 1 at $(0, z_k' = x_j)$ and degree d. Let φ_H be the Uryson map with value 0 if some $|x_j'| \leq \varepsilon$ and value 1 if all $|x_j'| \geq 2\varepsilon$, for x_j' a coordinate in $V^H \cap (V^\Gamma)^\perp$.

Define

$$f_0(\lambda, X) = \varphi_H f(\lambda, X) + (1 - \varphi_H)(1, 0).$$

The linear deformation $\tau f_0 + (1-\tau)f = (\tau\varphi_H + (1-\tau))f + \tau(1-\varphi_H)(1,0)$ has all its zeros fixed at $\lambda_j = 0$, and the orbit of X_0: in fact, if $x_j' = 0$, then the first component reduces to $(1-\tau)(x_0' + 2) + \tau \geq 1$, since then $\varphi_H = 0$. Thus, $f_0(\lambda, X)$ can be taken as generator for this part of $\Pi^\Gamma_{S^{\mathbb{R}^k \times V}}(S^V)$.

Now, since A is invertible, let $A(\tau)$ be a path of invertible matrices joining A, for $\tau = 1$, to I, if $\det A > 0$, or to $\begin{pmatrix} -1 & 0 \\ 0 & I \end{pmatrix}$ if $\det A < 0$, for $\tau = 0$. Replace in $f(\lambda, X)$, the vector $(\lambda_1, \dots, \lambda_k)^T$ by $A(\tau)(\lambda_1, \dots, \lambda_k)^T$. Then, one obtains again a Γ-homotopy to the generator (if $\det A < 0$, one may choose the map with $-\lambda_1$ as the generator). This implies that $f_0(\lambda, X)$ is Γ-homotopic to

$$f_1(\lambda, X) = f_0(0, X) + \varphi_H \sum_{j=1}^{k} \lambda_j A_j X;$$

the imaginary parts of the factor of z_j, $j = 1, \dots, k$, give $A(\lambda_1, \dots, \lambda_k)^T \equiv A\lambda$, with its only zero at $\lambda = 0$, since for $z_j = 0$ one has $\varphi_H = 0$.

Let $\mathcal{A}(X)$ be the $k \times k$ matrix with entries $a_{ij}(X) = (A_i X, A_j X)$. If $\mathcal{A}(X)\lambda = 0$, one has $\left(A_i X, \sum \lambda_j A_j X\right) = 0$, hence $\lambda = 0$ whenever the $A_j X$'s are linearly independent, in particular if $\varphi_H(X) > 0$. For such an X let $b(X)$ be the vector with i-th component $b_i(X)$ equal to $(A_i X, f(0, X))$ and define

$$\lambda(X) = -\mathcal{A}^{-1}(X) b(X).$$

Then, $(A_i(X), f_1(\lambda(X), X)) = \varphi_H(b_i(X) + (\mathcal{A}(X)\lambda(X))_i) = 0$. Thus, if $\tilde{\lambda}(X) = \varphi_H(X)\lambda(X)$, if $\varphi_H(X) > 0$ and 0 otherwise, one has a continuous vector and

$$f_H(X) = f_0(0, X) + \sum_{j=1}^{k} \tilde{\lambda}_j(X) A_j X$$

is a Γ-orthogonal map, recalling that on V^H, $A_j X$ is a linear combination of $A_1 X, \dots, A_k X$, for $j > k$, and that $f_0(0, X) = (1, 0)$ if $\varphi_H = 0$. Furthermore, the zeros of $f_H(X)$, in Ω, are ΓX_0 with $\tilde{\lambda}(X_0) = 0$.

If one considers the Γ-function

$$f_2(\lambda, X) = f_H(X) + \sum_{j=1}^{k} \lambda_j A_j X,$$

then

$$f_2(\lambda, X) = f_1(\lambda, X) + \sum_{j=1}^{k} (\varphi_H \lambda_j(X) + (1 - \varphi_H)\lambda_j) A_j X.$$

When $\varphi_H = 0$, one has $f(X) = (1, 0)$, hence $f_2(\lambda, X)$ is non-zero on this set. If $\varphi_H > 0$, one may deform linearly f_2 to f_1, since on a zero one would have $A(\tau\tilde{\lambda} + \lambda(\varphi_H + (1 - \varphi_H)\tau)) = 0$ and $f_0(0, X) = 0$, hence $\lambda = 0$, $X = \Gamma X_0$.

Then, either by recalling the proof of Theorem 6.1 in Chapter 3, or by using Proposition 6.1 and Corollary 3.1 of Chapter 3, one has that, if $[f_H(X)]_\perp = \sum d_j [F_j]_\perp$,

$$\deg(f_2(\lambda, X)^{H_i}; B_k^i) = \sum_{H_i < H_j < H_0} d_j |H_0/H_j|$$

and the same relations, with $[f_2(\lambda, X)]_\Gamma = \sum \tilde{d}_j [\tilde{F}_j]_\Gamma$. But, f_2 has all $\tilde{d}_j = 0$ except for H, where $\tilde{d}_H = 1$, or d if one has taken P_l^d, for some l. Hence, $d_j = \tilde{d}_j$ and $[f_H(X)]_\perp = [F_H]_\perp$.

Note that one may also compute directly the set of degrees for $f_2(\lambda, X)^{H_i}$, noticing first that if $V^H \cap (V^{H_i})^\perp \neq \{0\}$, then $f_2^{H_i}$ has a component x_j in V^H which is 0, that is $\varphi_H = 0$ for X in V^{H_i} and $f_H(X) = (1, 0)$: in this case the above degree is 0. Thus, one has to compute these degrees only for $H_i < H$. However, if H_i is a strict subgroup of H, then for some component x_s of $V^{H_i} \cap (V^H)^\perp$ the map will be $(\varphi_H + i \sum \lambda_j n_j^s) x_s$ which can be deformed to x_s, that is $f_2(\lambda, X)^{H_i}$ is a suspension of $f_2(\lambda, X)^H$, with the same degree. Since we have computed many times $\deg(f_2(\lambda, X)^H; B_k^H) = |H_0/H|$, the relations give

$$|H_0/H| = \sum_{H_i < H_j < H} d_j |H_0/H_j|$$

and $d_j = 0$ if H_j is not a subgroup of H. From here it is easy to see that $d_j = 0$ if $H_j \neq H$ and $d_H = 1$.

Then, if $f(X)$ is defined as

$$f(X) = \sum \varphi_H(X) f_H(X) + \prod (1 - \varphi_H(X))(1, 0, x_i),$$

one obtains an orthogonal map which reduces to $f_H(X)$ on Ω_H, as defined in the proof of the preceding theorem, and which is non-zero on $\Omega \backslash \bigcup \Omega_H$, giving that the orthogonal degree of f is

$$\deg_\perp(f; \Omega) = \sum d_H [F_H]_\perp.$$

(b) If $\dim V^\Gamma = 0$, then $\{0\}$ does not belong to Ω. As before, for each H with $\dim \Gamma/H = k$, let $\{x_j^0\}$ be a point of Ω^H with $x_j^0 \neq 0$ for all j's. If $k = 0$, one may use, in V^{T^n}, the generators of the preceding theorem. Thus, assume $k > 0$ and let $z_1, \dots, z_k$ be the coordinates with isotropy H_0 such that $|H/H_0| < \infty$ and the submatrix A is invertible. Let $f(\lambda, X)$ be the generator of (b) in the proof of the preceding theorem

$$\begin{aligned} f(\lambda, X') = \Big(&\{(i\lambda_j - (|z'_{j+1}|^2 - 1)z_j\}_{j=1,\dots,k-1}),\\ &i\big(\lambda'_k + 2\sum(|x'_j|^2 - 1)^2 + i(|z'_1|^2 - 1)\big)z_k,\\ &(-|z'_k|^2 Q_j + 1)y_j, (|z'_k|^2 P_j + 1)z_j, x_s\Big), \end{aligned}$$

and let φ_H be as before. Define

$$f_0(\lambda, X) = f(\lambda, \varphi_H X') + (1 - \varphi_H)X,$$

where, in $f(\lambda, \varphi_H X')$, one replaces the factors $|z'_1|^2, \dots, |z'_k|^2$ by $\varphi_H|z'_1|^2, \dots, \varphi_H|z'_k|^2$. On a zero of f_0 one has $\varphi_H > 0$ and, since $x_j \neq 0$, one gets $\lambda_1 = \cdots = \lambda_{k-1} = 0$; $1 - \varphi_H^2|z'_j|^2 = 0$, for $j = 1, \dots, k$; $Q_j(\varphi_H X') = P_j(\varphi_H X') = 1$. The last equalities imply that $\varphi_H|x'_j| = 1$, in particular, $|x'_j| \geq 1$ for all j's. But then $\varphi_H = 1$ and one has the orbit of X_0. Replacing P_j by P_j^d or by repeating the construction given in (a), one has that

$$\deg_\Gamma(f_0(\lambda, X); \Omega) = \deg_\Gamma(f(\lambda, X); \Omega_H) = d[F_H]_\Gamma.$$

Observe that, due to the multiplication by φ_H, one may deform the term $\sum(|x'_i|^2\varphi_H^2 - 1)^2$ to 0.

Define then $f_1(\lambda, X)$ as before, giving the same degree. Since X is orthogonal to $A_j X$, one may define $\lambda(X)$ as above (the term φ_H factors out) and with $\tilde{\lambda}(X) = \varphi_H(X)\lambda(X)$, the map

$$f_H(X) = f_0(0, X) + \sum_{j=1}^{k} \tilde{\lambda}_j(X) A_j X$$

is Γ-orthogonal and $f_2(\lambda, X)$, defined as before, has Γ-degree equal to $d[F_H]_\Gamma$, since the imaginary parts of the factors of $z_1, \dots, z_k$ are not affected by $(1 - \varphi_H)X$, that is the argument is the same as before. This proves that $[f_H]_\perp = d[F_H]_\perp$.

Defining the orthogonal map

$$f(X) = \sum_H \varphi_H(X) f_H(X) + \prod(1 - \varphi_H)X,$$

one has that the factor of x_j has real part equal to

$$\sum \varphi_H(1 - \varphi_H^2|z'_{j+1}|^2) + \prod(1 - \varphi_H)$$

for $j = 1, \ldots, k$ (with z'_{k+1} to be replaced by z'_1), or

$$\sum \varphi_H(1 + \varphi_H^2 |z'_k|^2 \operatorname{Re} P_j) + \prod (1 - \varphi_H),$$

with P_j replaced by $-Q_j$ for y_j. Hence, this real part is strictly positive if $|x'_j| \leq 2\varepsilon$, for ε small enough (for $j = 1, \ldots, k$, one has to consider the factor of z_{j-1} for $j = 1, \ldots, k$ and that of z_k for $j = 1$). Thus, $f(X)$ is non-zero on $\Omega \backslash \bigcup \Omega_H$ and its orthogonal degree is

$$\sum_H d_H [F_H]_\perp.$$

□

4.2 Γ-degree of an isolated orbit

One of the basic results in classical degree theory is that the index of Ax at 0, where A is an invertible matrix, is Sign det A. This fact is the building block for the analytic definition of the degree: if $f(x)$ is a continuous function defined from $\bar{\Omega}$ into $\mathbb{R}^n$, where Ω is an open and bounded subset of $\mathbb{R}^n$, and $f(x)$ is non-zero on $\partial\Omega$, then one approximates f, on $\partial\Omega$, by a smooth function $\tilde{f}(x)$ which, due to Sard's lemma, has 0 as a regular value. In particular, $\tilde{f}^{-1}(0)$ consists of a finite number of points (due to the compactness of $\bar{\Omega}$) with non-zero Jacobian. Then, the degree of $\tilde{f}$ with respect to Ω is the sum of the degrees of $\tilde{f}$ with respect to small neighborhoods of these points, so small that on each of them $\tilde{f}(x)$ is deformable to $D\tilde{f}(x_0)(x - x_0)$. Hence, one obtains

$$\deg(f(x); \Omega) = \deg(\tilde{f}(x); \Omega) = \sum_{x_0 \in \tilde{f}^{-1}(0),} \operatorname{Sign} \det D\tilde{f}(x_0).$$

Thus, one of the first questions, in case of equivariant maps, is what is the Γ-index of an isolated orbit? I.e., if x_0 is such that $f(x_0) = 0$, hence $f(\Gamma x_0) = 0$, and there is an invariant neighborhood Ω of the orbit Γx_0, what is $\deg_\Gamma(f(x); \Omega)$?

Definition 2.1. The Γ-*index* of an isolated orbit, Γx_0, will be denoted by $i_\Gamma(f; x_0)$ and is equal to $\deg_\Gamma(f(x); \Omega)$ for any small invariant neighborhood Ω of the orbit Γx_0.

It is clear that, since one has orbits of solutions, the answer to this question will be much more involved than in the non-equivariant case and will depend on the orbit type of x_0.

However, let us begin with the case of a linear map. As seen in §5 in Chapter 1, if there is a Γ-equivariant linear map A, between two representations of Γ, which is invertible, this implies that the representations are equivalent. Hence assume A is an invertible equivariant matrix on the finite dimensional space V. Then, from Theorem 5.3 in Chapter 1, we know that A has a diagonal structure

$$A = \operatorname{diag}(A^\Gamma, A_j^{\mathbb{R}}, A_l^{\mathbb{C}})$$

where A^Γ is the restriction of A to V^Γ, $A_j^{\mathbb{R}}$ are real matrices on each subspace of equivalent irreducible representations, where Γ acts as $\mathbb{Z}_2$, and $A_l^{\mathbb{C}}$ are complex matrices, corresponding to an action of Γ as $\mathbb{Z}_n$, $n \geq 3$, or S^1.

From Theorem 8.3 in Chapter 1, each of the real matrices A^Γ or $A_j^{\mathbb{R}}$ is Γ-deformable to

$$\begin{pmatrix} \text{Sign det } A^\Gamma & 0 \\ 0 & I \end{pmatrix} \quad \text{or} \quad \begin{pmatrix} \text{Sign det } A_j^{\mathbb{R}} & 0 \\ 0 & I \end{pmatrix},$$

while the complex matrices are Γ-deformable to the identity. Hence, if Ω is an invariant neighborhood of 0, one has

$$\deg_\Gamma(Ax;\Omega) = \deg_\Gamma(\varepsilon_0 x_0, \dots \varepsilon_j x_j, \dots; \Omega_0)$$

where $\varepsilon_0 = \text{Sign det } A^\Gamma$, $\varepsilon_j = \text{Sign det } A_j^{\mathbb{R}}$ and Ω_0 is a neighborhood of 0 in the space $\{(x_0, \dots, x_j, \dots, \}$, after using the suspension on the other variables. One has the following result

Proposition 2.1. *If A is a Γ-equivariant invertible matrix, then*

$$i_\Gamma(Ax;0) = \varepsilon_0\Big([F_0]_\Gamma + \sum(\varepsilon_j - 1)/2[F_j]_\Gamma + \sum d_H[F_H]_\Gamma\Big)$$

where $\varepsilon_0 = \text{Sign det } A^\Gamma$, $\varepsilon_0\varepsilon_j = \text{Sign det } A^{H_j}$, where $\Gamma/H_j \cong \mathbb{Z}_2$, and d_H are completely determined by ε_0 and $\{\varepsilon_j\}$'s, for H's which are intersections of more than one of the H_j's.

Proof. This is a direct consequence of Proposition 3.1 in Chapter 3. □

As a simple application of the above result, consider the bifurcation problem, for the Γ-equivariant function

$$f(\lambda, u) = (A - T(\lambda))u - g(\lambda, u),$$

from $\mathbb{R} \times E$ into the Γ-space E, where A is a Γ-compact perturbation of the identity, $\|T(\lambda)\| \to 0$ as λ goes to 0 and $g(\lambda, u) = o(\|u\|)$.

As seen in §9 of Chapter 1, the equation $f(\lambda, u) = 0$ is equivalent, near $(0, 0)$ to the bifurcation equation

$$B(\lambda)x + G(\lambda, x) = 0,$$

where x is in $\ker A$, $B(0) = 0$, $B(\lambda)$ is an equivariant matrix and $G(\lambda, x) = o(\|x\|)$.

Proposition 2.2. *Assume $B(\lambda)$ is invertible for $\lambda \neq 0$ and let*

$$\varepsilon_0(\lambda) = \text{Sign det } B(\lambda)^\Gamma, \quad \varepsilon_0(\lambda)\varepsilon_j(\lambda) = \text{Sign det } B(\lambda)^{H_j},$$

with $\Gamma/H_j \cong \mathbb{Z}_2$.

(a) *If $\varepsilon_0(\lambda)$ changes at $\lambda = 0$, one has global bifurcation in E^Γ.*

(b) *If* $\varepsilon_j(\lambda)$ *changes at* $\lambda = 0$, *one has global bifurcation in* E^{H_j}.

(c) *If all* $\varepsilon_j(\lambda)$ *remain constant, then there is an equivariant nonlinearity* $g(\lambda, u)$ *such that the only solution of* $f(\lambda, u) = 0$ *is* $u = 0$.

Proof. (1) and (2) follow directly from the above proposition and Theorem 5.2, Corollary 5.2 of Chapter 2. Part (3), the "necessary condition for linearized bifurcation", follows from Proposition 6.3 in [I], where the construction of $g(\lambda, u)$ is given. □

Remark 2.1. In this case there is also an orientation factor due to the invertible part of $A - T(\lambda)$: in fact if one writes, as in §9 of Chapter 1,

$$(A - T(\lambda))u = (A - QT(\lambda))H(\lambda, x, x_2) \oplus B(\lambda)x - (I - Q)T(\lambda)H(\lambda, x, x_2)$$

with $H(\lambda, x, x_2) = x_2 - (I - KQT(\lambda))^{-1}KQT(\lambda)x$, one may perform an equivariant deformation, using the fact that $T(0) = 0$, to

$$Ax_2 \oplus B(\lambda)x.$$

Since $A = I - T_0$, where T_0 is a compact operator, one may decompose equivariantly

$$E = \ker A^\alpha \oplus \text{Range } A^\alpha,$$

with α the ascent of A and $m = \dim\ker A^\alpha$, the algebraic multiplicity of 1 as eigenvalue of T_0. In this case, if one has chosen bases on $\ker A^\alpha$ so that the nihilpotent A is in Jordan form with d blocks of size $m_1, \dots, m_d$, with $\sum m_j = m$ and x having coordinates on the i-th block $(x_{i1}, x_{i2}, \dots, x_{im_i})$, with $(x_{i1}, 0, \dots, 0)$ corresponding to a generator of $\ker A$ and $(0, \dots, 0, 1)$ to a generator of coker A, then on $\ker A^\alpha$ the map $Ax_2 \oplus B(\lambda)x$ has the form

$$\begin{aligned}&(x_{11}, x_{12}, \dots, x_{1m_1}, x_{21}, x_{22}, \dots)\\ &\qquad \to (x_{12}, x_{13}, \dots, x_{1m_1}, b_{11}(\lambda)x_{11} + b_{12}x_{21} + \dots, x_{22}, \dots),\end{aligned}$$

which has a degree equal to Sign det $B(\lambda)(-1)^{m-d}$: in fact, the factor Sign det $B(\lambda)$ comes from the composition and, on the other hand, the second factor is the degree of the map

$$(x_{11}, x_{12}, \dots, x_{1m_1}, x_{21}, \dots) \to (x_{12}, x_{13}, \dots, x_{1m_1}, x_{11}, x_{21}, \dots),$$

that is $(-1)^{m_1-1}(-1)^{m_2-1}\dots(-1)^{m_d-1} = (-1)^{m-d}$, due to the necessary permutations.

If $T(\lambda) = \lambda T_0$, then we have seen in §9 of Chapter 1 that

$$B(\lambda) = \text{diag}(-\lambda^{m_1}/(1+\lambda)^{m_1-1}, \dots, -\lambda^{m_d}/(1+\lambda)^{m_d-1})$$

with Sign det $B(\lambda) = (-1)^d$ Sign λ^m, hence the contribution to the index is Sign$(-\lambda)^m$.

On the other hand, the contribution to the index for A restricted to $\text{Range}(A)^\alpha$ is $(-1)^{\sum m_j}$ where m_j is the algebraic multiplicity of λ_j as characteristic value of T_0, i.e., such that $I - \lambda_j T_0$ is not invertible, for $0 < \lambda_j < 1$: this is well known but a proof of this fact will be given in Theorem 2.4 below. Hence,

$$i_\Gamma(A - T(\lambda); 0) = \varepsilon_0([F_0]_\Gamma + \sum(\varepsilon_j - 1)/2[F_j]_\Gamma + \sum d_H[F_H]),$$

where

$$\begin{aligned}\varepsilon_0 &= (-1)^{\sum m_i^\Gamma} \operatorname{Sign}\det(-B(\lambda)^\Gamma)\\ \varepsilon_j\varepsilon_0 &= (-1)^{\sum m_i^{H_j}} \operatorname{Sign}\det(-B(\lambda)^{H_j}),\end{aligned}$$

with m_i^Γ the algebraic multiplicity of $(I-\lambda_i T_0)^\Gamma$ for $0 < \lambda_i \leq 1$ and $m_i^{H_j}$ the algebraic multiplicity of $(I - \lambda_i T_0)^{H_j}$ for $0 < \lambda_i \leq 1$.

In case A is a Fredholm operator of index 0 with an isolated eigenvalue at 0, then we have seen in §9 of Chapter 1 that

$$B(\lambda) = (\lambda^{m_1}, \dots, \lambda^{m_d}).$$

Hence, whenever defined one has

$$\begin{aligned}\varepsilon_0 &= (-1)^{m^\Gamma - d^\Gamma} \operatorname{Sign}\lambda^{m^\Gamma} \operatorname{Index}(A^\Gamma;\ \operatorname{Range}(A^\Gamma)^\alpha)\\ \varepsilon_j\varepsilon_0 &= (-1)^{m^{H_j} - d^{H_j}} \operatorname{Sign}\ \lambda^{m^{H_j}} \operatorname{Index}(A^{H_j}; \operatorname{Range}(A^{H_j})^\alpha).\end{aligned}$$

Example 2.1. If $\Gamma = \mathbb{Z}_2 \times \mathbb{Z}_2$ acts on $\mathbb{R}^3$ via $(x, \gamma_1 y, \gamma_2 z)$, then any linear map has the form $(\varepsilon_0 x, \varepsilon_1 y, \varepsilon_2 z)$ and one has

$$\begin{pmatrix}\varepsilon_0\\ \varepsilon_0\varepsilon_1\\ \varepsilon_0\varepsilon_2\\ \varepsilon_0\varepsilon_1\varepsilon_2\end{pmatrix} = \begin{pmatrix}1&0&0&0\\1&2&0&0\\1&0&2&0\\1&2&2&4\end{pmatrix}\begin{pmatrix}d_0\\d_1\\d_2\\d_3\end{pmatrix}.$$

According to Example 3.2 in Chapter 3,

$$\begin{pmatrix}d_0\\2d_1\\2d_2\\4d_4\end{pmatrix} = \begin{pmatrix}1&0&0&0\\-1&1&0&0\\-1&0&1&0\\1&-1&-1&1\end{pmatrix}\begin{pmatrix}\varepsilon_0\\ \varepsilon_0\varepsilon_1\\ \varepsilon_0\varepsilon_2\\ \varepsilon_0\varepsilon_1\varepsilon_2\end{pmatrix}.$$

In particular, $d_3 = d_0$ if $\varepsilon_1 = \varepsilon_2 = -1$ and 0 otherwise. If one has the following bifurcation problem

$$(\lambda^2 x, \lambda y, \lambda z),$$

then the set (d_0, d_1, d_2, d_3) goes from $(1, -1, -1, 1)$ for $\lambda < 0$, to $(1, 0, 0, 0)$ for $\lambda > 0$, hence one has a global bifurcation in E^{H_1} and E^{H_2}. Note, however, that the branches may coincide and be in E^Γ. For instance,

$$(\lambda^2 x + y^2 + z^2, \lambda y, \lambda z) = 0$$

has its solutions, either $x = y = z = 0$, the trivial solution, or $\lambda = 0$, $y = z = 0$, in E^Γ. Note also that if one breaks the symmetry, then one may have no bifurcation. For instance $(\lambda^2 x + y^2 + z^2 + \varepsilon(r)x, \lambda y + \varepsilon(r)z, \lambda z - \varepsilon(r)y)$, where $\varepsilon(r) = \varepsilon(x^2+y^2+z^2)$, with $\varepsilon > 0$, has the only solution $x = y = z = 0$.

For maps without parameters between spaces which satisfy hypothesis (H), one has the following result.

Theorem 2.1. *Assume U and W satisfy* (H), *in particular if $U = W$, and let $f(x)$ be an equivariant map from $\bar{\Omega} \subset U$ into W, which is non-zero on $\partial\Omega$. Then*

$$\deg_\Gamma(f(x); \Omega) = \sum d_j [F_j]_\Gamma,$$

where H_j is such that Γ/H_j is finite and, with the usual order, one has

$$\begin{pmatrix} \deg(f^\Gamma; \Omega^\Gamma) \\ \vdots \\ \deg(f^{H_i}; \Omega^{H_i}) \\ \vdots \\ \deg(f^{T^n}; \Omega^{T^n}) \end{pmatrix} = \begin{pmatrix} 1 & & 0 \\ \vdots & & \vdots \\ \beta_{i1} & |\Gamma/H_j| & 0 \\ \vdots & \vdots & \vdots \\ \beta_{s1} & \beta_{sj}|\Gamma/H_j| & |\Gamma/T^n| \end{pmatrix} \begin{pmatrix} d_0 \\ \vdots \\ d_j \\ \vdots \\ d_s \end{pmatrix}$$

as in Theorem 3.4 *and Corollary* 3.1 *of Chapter* 3. *In particular, if $U = W$ then $\beta_{ij} = 1$ if and only if $H_i < H_j$.*

Proof. This follows from Corollary 3.1 of Chapter 3 and the fact that $B_0 = I \times B_R$ with $\deg(2t + \varphi(x) - 1, \tilde{f}(x); I \times B_R) = \deg(f(x); \Omega)$ in this case. □

The above relations imply that the information obtained from the Γ-degree is, in this case, equivalent to the one obtained from the set of all the ordinary degrees on Ω^H, for isotropy subgroups H, with Γ/H finite. The value of the Γ-degree is to prove the above equivalence (in particular that one may forget H's with $\dim \Gamma/H > 0$) and that if Ω is a ball, then the Hopf property implies that two Γ-maps are Γ-homotopic if and only if they have the same set of d_H's. This fact, used in (3) of Proposition 2.2 cannot be proved directly from the equality of the ordinary degrees.

However, *the full strength of the Γ-degree is clearer in case of parametrized problems.* Let $f(\lambda, X) : \bar{\Omega} \to W$ be a Γ-equivariant map, where Ω is an open bounded invariant subset of $V = \mathbb{R}^k \times U$, where U and W satisfy hypothesis (H). Assume that $f^{-1}(0) = (\lambda_0, \Gamma X_0)$ with $\Gamma_{X_0} \equiv H$ such that $\dim \Gamma/H = k$. Then, f has a well-defined Γ-degree with respect to Ω or to any small invariant neighborhood of $f^{-1}(0)$. Furthermore, X_0 has coordinates $z_1^0, \ldots, z_k^0$ which are non-zero and with $H_0 \equiv H_1 \cap \cdots \cap H_k$ such that $\dim \Gamma/H_0 = k$. From Lemma 2.4 in Chapter 1, one may use the action of Γ in order to assume that z_j^0 are real and positive.

Theorem 2.2. *Let* $(\lambda_0, \Gamma X_0)$ *be an isolated orbit with isotropy* H *such that* $\dim \Gamma/H = k$ *and a zero of* $f(\lambda, X) : \mathbb{R}^k \times U \to W$, *where* U *and* W *satisfy hypothesis* (H). *Then*

$$i_\Gamma(f; (\lambda_0, X_0)) = \sum_{\underline{H} \leq K \leq H} d_K[\tilde{F}_K]_\Gamma,$$

where $\underline{H}$ *is the torus part of* H. *Furthermore, for any* K *in the above sum, define* i_K *as the Poincaré index at* (λ_0, X_0) *of* f^K *restricted to* $V^K \cap \{z_j \in \mathbb{R}^+, j = 1, \dots, k\}$. *Then*

$$i_K = \sum_{K \leq L \leq H} \beta_{KL} d_L |H/L|,$$

where, if $(x_1^{l_1}, \dots, x_s^{l_s})$ *is the complementing map from* $(V^L)^\perp \cap V^K$ *into* $(W^L)^\perp \cap W^K$, *then* $\beta_{KL} = \prod l_j$. *In particular,* $\beta_{KK} = 1$ *and* $\beta_{KL} = 1$ *if* $U = W$.

Proof. Choose the tubular neighborhood Ω of the orbit so small that if X_0 has a coordinate $x_j^0 \neq 0$, then x_j is non-zero in Ω. Thus, if

$$\deg_\Gamma(f; \Omega) = [2t + 2\varphi(\lambda, X) - 1, \tilde{f}(\lambda, X)]_\Gamma = [F]_\Gamma,$$

one may construct $\varphi(\lambda, X)$ such that it has value 1 whenever one of the coordinates x_j is 0. This implies that $F|_{V^K} \neq 0$ for any K which is not a subgroup of H (and not only of H_0 as in Theorem 3.4 of Chapter 3). The argument of this last result implies that $d_K = 0$ for such a K and that one gets, in $i_\Gamma(f; (\lambda_0, X_0))$, contributions only from those isotropy subgroups between $\underline{H}$ and H.

Furthermore, from Theorem 3.4 in Chapter 3, one has

$$\deg(F^K; B_k^K) = \sum_{K \leq L \leq H} \beta_{KL} d_L |H_0/L|,$$

where $B_k = \{(t, \lambda, X)$ in $I \times B_R$, with z_j in $\mathbb{R}^+$ for $j = 1, \dots, k\}$. Then, from the product theorem for the ordinary degree, one has

$$\deg(F^K; B_k^K) = \deg(f^K; \Omega_k^K),$$

where $\Omega_k = \Omega \cap B_k$. Now, $|H_0/L| = |H_0/H||H/L|$ and, due to the H_0-action on B_k, as in Theorem 1.2 of Chapter 3, one has that $f^{-1}(0) \cap B_k$ has $|H_0/H|$ points, each with the same index i_K on $V^K \cap B_k$. Hence, one may divide the above equality by $|H_0/H|$ and obtain the result. □

Assume now that f is C^1 in a neighborhood of $(\lambda_0, \Gamma X_0)$. Then according to Properties 3.3 and 3.4 of Chapter 1, it follows that for any $K < H$

$$Df(\lambda_0, X_0)^K = \begin{pmatrix} Df(\lambda_0, X_0)^H & 0 \\ 0 & Df^\perp(\lambda_0, X_0)^K \end{pmatrix}$$

which is H-equivariant. Suppose also that 0 is a regular value of f on Ω, that is $Df(\lambda_0, X_0)$ has maximal rank. Since U and W have the same dimension, from Hypothesis (H), this implies that $Df(\lambda_0, X_0)$ is onto and has a k-dimensional kernel.

Lemma 2.1. *If* 0 *is a regular value of the equivariant map* f *and* $f(\lambda_0, X_0) = 0$, *where* X_0 *has isotropy* H *with* $\dim \Gamma/H = k$, *then* U *and* W *are equivalent* H-*representations,* $\ker Df(\lambda_0, X_0)$ *is* k-*dimensional and is generated by* k *vectors among* $A_1X_0, \ldots, A_nX_0$, *with* $A_jX = \partial(\gamma X)/\partial\varphi_j|_{\gamma=\mathrm{Id}}$, *the infinitesimal generators of the action of* T^n, *the torus part of* Γ. *Furthermore,* $Df(\lambda_0, X_0)|_{B_k}$ *is invertible, where* B_k *is the global Poincaré section, and the Poincaré index* i_K *of* f *at* (λ_0, X_0) *on* $V^K \cap B_k$ *is*

$$i_K = \text{Sign} \det Df(\lambda_0, X_0)^K|_{B_k} = i_H \,\text{Sign} \det Df^{\perp}(\lambda_0, X_0)^K.$$

Proof. By differentiating the relation $f(\lambda_0, \gamma X_0) = 0$ with respect to φ_j, one has that $Df(\lambda_0, X_0)A_jX_0 = 0$, as in Lemma 7.2 of Chapter 1. Furthermore, from Lemma 7.1 in that chapter, one has exactly k among the A_jX_0 which are linearly independent. Since A_jX_0 generate the tangent space to the orbit at (λ_0, X_0) and that one has assumed $z_1^0, \ldots, z_k^0$ to be real and positive, one has that B_k is orthogonal to that tangent space and corresponds to the usual Poincaré section of the orbit at (λ_0, X_0). Hence, $Df(\lambda_0, X_0)$, when restricted to B_k, is invertible and the formula for i_K follows. Finally, since $Df(\lambda_0, X_0)|_{B_k}$ is H-equivariant and invertible, one has that $V \cap B_k$ and W are equivalent H-representations and, since $z_1, \ldots, z_k$ are fixed by H, one gets that U and $V \cap B_k$ are equivalent H-representations. □

Assume then that $U = W$. This implies that β_{KL}, in Theorem 2.2, is always 1.

Theorem 2.3. *Let* $V = \mathbb{R}^k \times W$ *and* 0 *be a regular value of* f *on* Ω *with an isolated orbit* $(\lambda_0, \Gamma X_0)$ *with isotropy* H *such that* $\dim \Gamma/H = k$. *Let* i_K *be the Poincaré indices given in Lemma* 2.1. *Then, the* Γ-*index of the orbit is given by* $(d_H, d_{K_1}, \ldots)$ *such that* $d_H = i_H$, $d_K = (i_K - i_H)/2$, *if* $H/K \cong \mathbb{Z}_2$, d_K *is completely determined by the above integers if* $H/K \cong \mathbb{Z}_2 \times \cdots \times \mathbb{Z}_2$ *with more than one factor, and* $d_K = 0$ *otherwise.*

Proof. The result follows directly from Theorem 2.2, Lemma 2.1 and Proposition 3.1 in Chapter 3, since this last result is purely number theoretical and is based on the fact that $i_K = \pm 1$. □

Another way to prove it, is to see that, on $\Omega \cap B_k$, $f(\lambda, X)$ is H-deformable to

$$(f_\lambda(\lambda_0, X_0)(\lambda - \lambda_0) + f^H_{X^H}(\lambda_0, X_0)(X^H - X_0), f^{\perp}_{X^{\perp}}(\lambda_0, X_0)X^{\perp})$$

and one may compute the H-degree of the linearization $Df(X_0)|_{B_k}$:

$$i_H(Df(\lambda_0, X_0)|_{B_k}; 0) = \sum_{\underline{H} \le K \le H} d'_K [F'_K]_H,$$

where F'_K are the generators of $\Pi^H_{S^{\tilde{W}}}(S^{\bar{W}})$, with $\tilde{W} = V^{\underline{H}} \cap B_k$ and $\bar{W} = W^{\underline{H}}$.

From Proposition 2.1, one obtains that the d'_K's are given by the formulae of the theorem, since $f^{\perp}_{X^{\perp}}$ decomposes into a block diagonal matrix, according to the H-irreducible representations, where each block is a real matrix if H acts as $\mathbb{Z}_2$ and a complex matrix, if H acts as $\mathbb{Z}_m$, $m \geq 3$ (on $\tilde{W}$, H acts as a finite group). By deforming the complex matrices to the identity, one has to consider only K's with $H/K \cong \mathbb{Z}_2 \times \cdots \times \mathbb{Z}_2$, with

$$Df(\lambda_0, X_0)^K|_{B_k} = \operatorname{diag}(A^H, A_1, \ldots, A_s),$$

with $i_H = \operatorname{Sign}\det A^H$, A_j is the matrix $Df^{\perp}|V^{K_j}$, with $H/K_j \cong \mathbb{Z}_2$ and $i_{K_j} = i_H \operatorname{Sign}\det A_j$. Hence $i_K = i_H \prod_{j=1}^{s}(i_{K_j}/i_H)$ and Corollary 3.1 in Chapter 3 gives

$$\sum d'_L |H/L| = i_K - i_H - \sum (i_{K_j} - i_H),$$

where the sum on the left is over all L's with $K \leq L$ and $|H/L| > 2$. This gives, when varying K over all non-maximal isotropy subgroups, i.e., with $|H/K| > 2$, a lower triangular invertible matrix. Since the right hand side is completely determined by $i_H, i_{K_1}, \ldots, i_{K_s}$, over all maximal isotropy subgroup K_j, one obtains the relations of the theorem.

Now, we have seen in Remark 3.2 of Chapter 3, that if

$$\deg_\Gamma(f;\Omega) = \sum_{\underline{H} \leq K \leq H_0} d_K [\tilde{F}_K]_\Gamma,$$

then

$$\deg_{H_0}(f|_{B_k}; \Omega \cap B_k) = \sum_{\underline{H} \leq K \leq H_0} d_K [\tilde{F}|_{B_k}]_{H_0}.$$

But, on one hand we know that $d_K = 0$ if K is not a subgroup of H and, on the other hand, from Corollary 7.1 in Chapter 3, one has

$$\deg_H(f|_{B_k}; \Omega \cap B_k) = \sum |H_0/H| d_K [F'_K]_H,$$

since for the reduction from H_0 to H (with $k = 0$) one has the factor $|H_0/K|/|H/K| = |H_0/H|$. But, $f|_{B_k}$ has $|H_0/H|$ zeros in $\Omega \cap B_k$, all with the same H-index: in fact, if γ in H_0/H sends X_0 to γX_0, one has, from Property 3.3 in Chapter 1,

$$Df(\lambda_0, \gamma X_0) = \gamma Df(\lambda_0, X_0)\gamma^T$$

and, since these two matrices are conjugate, one has the same set of indices i_H, i_K, and, from the previous argument, the same set of d'_K. Thus,

$$\deg_H(f|_{B_k}; \Omega \cap B_k) = |H_0/H| i_H(Df(\lambda_0, X_0)|_{B_k}; 0),$$

proving that $d'_K = d_K$.

Example 2.2. If $\Gamma = S^1$ and $k = 1$, then any strict subgroup H of Γ is of the form $\mathbb{Z}_m$ and H/K cannot be a product. Thus, if $H = \Gamma$, one has $d_\Gamma = i_\Gamma$ and $d_K = 0$ for all K's, while, if $H = \mathbb{Z}_m$, then $d_H = i_H$, $d_K = 0$ unless m is even and $K = \mathbb{Z}_{m/2}$ with $d_K = (i_K - i_H)/2$.

As an abstract application of the preceding theorem, *assume that* $f(\mu, \lambda, X)$ *is a family, parametrized by* μ, *of* Γ*-equivariant functions from* $\mathbb{R}^k \times W$ *into* W, *with* 0 *as a regular value for* $\mu \neq \mu_0$. Assume there is a known curve of zeros of $f(\mu, \lambda, X)$, $\lambda_0(\mu)$, $X_0(\mu)$ with common isotropy H, with $\dim \Gamma/H = k$. Then $i_H(\mu)$ and $i_K(\mu)$ are well defined for $\mu \neq \mu_0$ and $K < H$.

Corollary 2.1. (a) *If* $i_H(\mu)$ *changes sign at* μ_0, *then one has a global bifurcation at* $(\mu_0, \lambda_0(\mu_0), X_0(\mu_0))$ *in* V^H.

(b) *If* $i_H(\mu)$ *remains constant but* $i_K(\mu)$ *changes sign at* μ_0 *for some* K *with* $H/K \cong \mathbb{Z}_2$, *then there is global bifurcation in* V^K, *i.e., with a period doubling. Topologically all bifurcations are in maximal isotropy subgroups, i.e., with* $H/K \cong Z_2$.

Proof. This is clear from our previous results on bifurcation. The last sentence means that if i_H and i_K's, for all K's with $H/K \cong \mathbb{Z}_2$, do not change, then there will be no other changes for smaller isotropy subgroups. *This does not hold for non-abelian actions.* Note that the isotropy of the bifurcating solution is at least H is case (a) and at least K in the second case and one may construct examples where, in case (b), this isotropy is H. □

Example 2.3. Assume that S^1 acts on $\mathbb{C}^2$ as $(e^{i\varphi}z_1, e^{2i\varphi}z_2)$ and consider the equivariant map, for μ in a neighborhood of $1/2$:

$$f(\mu, \lambda, z_1, z_2) = \big((1-\mu)z_1 - \mu z_2 \bar{z}_1, (1 - |z_2|^2 + i\lambda)z_2\big).$$

If $z_2 \neq 0$, a zero of f implies $\lambda = 0$, $|z_2| = 1$ and $(1-\mu)|z_1| = \mu|z_1|$, i.e., $\mu = 1/2$ or $z_1 = 0$. Hence, for $\mu \neq 1/2$, the isotropy of the orbit $(0, |z_2| = 1)$ is $H \cong \mathbb{Z}_2$. The only other isotropy subgroup is $K = \{e\}$. One has $i_H = \deg((1 - z_2 + i\lambda); |\lambda| < 1, 1-\varepsilon < z_2 < 1+\varepsilon) = 1$, while i_K, which is constant for $\mu \neq 1/2$, changes from $+1$ for $\mu = 0$ to -1 for $\mu = 1$. Hence, any equivariant perturbation of this map will have a bifurcation on this μ-interval.

Note that the linearization of f at $(\mu, 0, 0, 1)$ is $((1-\mu)z_1 - \mu\bar{z}_1, i\lambda - z_2 - \bar{z}_2)$ which is $\mathbb{Z}_2$-equivariant (changing z_1 into $-z_1$ but keeping z_2 fixed). One has that

$$D^H f = \begin{pmatrix} 0 & -2 & 0 \\ 1 & 0 & 0 \end{pmatrix}, \quad D^K f^\perp = \begin{pmatrix} 1-2\mu & 0 \\ 0 & 1 \end{pmatrix}.$$

It is important to recall that, as usual, index computations are mostly useful in getting degrees of complicated maps after performing deformations. It is clear that if one has a map where one may compute directly the Poincaré index then one could object to the construction of the equivariant degree. In order to convince the reader to the contrary, let us present two simple examples.

Example 2.4. Let S^1 act on $\mathbb{R}^2 \times \mathbb{C}^2$ as $(x, y, e^{i\varphi}z_1, e^{i\varphi}z_2)$ and consider the map

$$(|z_1|^2 + |z_2|^2 - 2, (x + iy)(|z_1|^2 - 1)^2 z_1, (x + iy)z_2)$$

with zeros on the 3-dimensional sphere and $x = y = 0$. Hence, one may not compute directly any Poincaré index. However, the deformation $(1 - \tau)(|z_1|^2 - 1) + i\tau$ is admissible, as well as the rotation

$$\begin{pmatrix} (1-\tau)(x+iy) & -\tau \\ \tau(x+iy)^2 & (1-\tau)(x+iy) \end{pmatrix} \begin{pmatrix} z_1 \\ z_2 \end{pmatrix},$$

i.e., after another rotation, and simple deformations, one may compute the S^1-degree of

$$(|z_1|^2 - 1, (x + iy)^2 z_1, z_2).$$

For this last map, one may compute the Poincaré index, which is 2. Thus,

$$\deg_{S^1}(f; B) = 2[F_e]_{S^1},$$

where B is a big ball and $F_e = (|z_1|^2 - 1, (x + iy)z_1)$ is the generator of Π^{S^1} in this case. Accordingly, any S^1-perturbation of the map will have a zero in B. However, the non-equivariant perturbation

$$(|z_1|^2 + |z_2|^2 - 2, (x + iy)(|z_1|^2 - 1)^2 z_1 + \tau\bar{z}_2, (x + iy)z_2 - \tau\bar{z}_1)$$

has no zeros for $\tau \neq 0$ (write the last two equations as a linear system in z_1 and $\bar{z}_2$ by conjugating the last equation).

Note that the first map is S^1-deformable to

$$(1 - x^2 - y^2, (x + iy)(|z_1|^2 - 1)^2 z_1, (x + iy)z_2).$$

In order to compute the S^1-degree of this map, one may either deform $|z_1|^2 - 1$ to 1 as above, or use the deformation $|z_1|^2 - \tau$. In both cases, the S^1-index of the orbit $(|z_1| = 1, z_2 = 0)$ is 0.

Example 2.5. Consider the pair of averaged Van der Pol's equations, that is for integro-differential equations. Look for 2π-periodic solutions to

$$\begin{aligned} x'' - x'\Big(1 - \frac{1}{2\pi}\int_0^{2\pi}(x^2 + y^2)\,dt\Big) + (1 + \nu)x &= f(x, y) \\ y'' - y'\Big(1 - \frac{1}{2\pi}\int_0^{2\pi}(x^2 + y^2)\,dt\Big) + (1 + \nu)y &= g(x, y). \end{aligned}$$

If $x(t)=\sum x_n e^{int}$ and $y(t)=\sum y_n e^{int}$, one has, for $n\geq 0$ and denoting by ρ^2 the integral term,

$$(-n^2-in(1-\rho^2)+(1+\nu))x_n=0.$$

Thus, if ν is close to 0, the only non-trivial solutions will be $x_n=y_n=0$ for $n\neq 1$, $\nu=0$, $|x_1|^2+|y_1|^2=1$, corresponding to $x(t)=\alpha\cos(t+\varphi)$, $y(t)=\beta\cos(t+\psi)$, with $\alpha^2+\beta^2=2$.

In order to compute the S^1-degree of the non-trivial solution, one takes $[-1/2,1/2]\times B$, where B is a big ball containing these solutions. For $n\neq 1$, one may deform the coefficients to 1 and the S^1-degree is that of

$$((\nu-i(1-\rho^2))x_1,(\nu-i(1-\rho^2))y_1),$$

or, after a rotation as in the preceding example, the S^1-index of $(\nu-i(1-|x_1|^2))^2x_1$ near $|x_1|=1$, which is 2. Hence, any small *autonomous* perturbation of the system will have solutions near $\nu=0$, $|x_1|^2+|y_1|^2=1$.

On the other hand consider the $\mathbb{Z}_2$-perturbation

$$\begin{aligned}f(x,y)\ &+\ \tau(3\cos 2t\ y+\sin 2t\ y')=0\\ g(x,y)\ &-\ \tau(3\cos 2t\ x+\sin 2t\ x')=0.\end{aligned}$$

On 2π-periodic functions, the system is only $\mathbb{Z}_2$-equivariant and is equivalent, on Fourier coefficients, to the system

$$\begin{aligned}\big(-n^2-in(1-\rho^2)+1+\nu\big)x_n\ &+\ \frac{\tau}{2}\big((n+1)y_{n-2}-(n-1)y_{n+2}\big)=0\\ \big(-n^2-in(1-\rho^2)+1+\nu\big)y_n\ &-\ \frac{\tau}{2}\big((n+1)x_{n-2}-(n-1)x_{n+2}\big)=0.\end{aligned}$$

For $n=1$, one has

$$\begin{aligned}(\nu-i(1-\rho^2))x_1+\tau\bar{y}_1\ &=0\\ (\nu-i(1-\rho^2))y_1-\tau\bar{x}_1\ &=0,\end{aligned}$$

whose only solution, for $\tau\neq 0$, is $x_1=y_1=0$ (conjugate the second equation and treat the system as a linear system in x_1 and $\bar{y}_1$). Then, the remaining equations form a closed system with, for ν close to 0 and τ small, dominant diagonal terms, hence with a unique solution $x_n=y_n=0$. Thus, for τ small and non-zero, the only solution is $x=y=0$.

These last two examples are illustrations of the restriction map from

$$\Pi^{S^1}_{S^{\mathbb{R}\times W}}(S^W)\cong\mathbb{Z}_2\times\mathbb{Z}\times\cdots$$

to

$$\Pi^{\mathbb{Z}_2}_{S^{\mathbb{R}\times W}}(S^W)\cong\mathbb{Z}_2\times\mathbb{Z}_2,$$

where, according to Proposition 7.3 in Chapter 3, one has

$$d'_{\Gamma_0} = \sum d_{2k} \mod 2, \quad d'_{\{e\}} = \sum d_{2k+1} \mod 2.$$

Here $d_j = 0$ for $j \neq 1$ and $d_1 = 2$. We leave to the reader the task of building other examples.

Let us continue to study the generic case of 0 as regular value of $f(\lambda, X)$ and relate the Γ-index to the "Floquet multipliers" for a "hyperbolic orbit". We shall take the following setting: $V = \mathbb{R}^k \times W$, $f(\lambda, X) = X - F(\lambda, X)$, from V into W, is C^1 and $F(\lambda, X)$ is a compact map with $f(\lambda_0, X_0) = 0$ for X_0 with isotropy H such that $\dim \Gamma/H = k$. As before, we choose an orientation of W in such a way that the first variables $z_1, \ldots, z_k$ have an isotropy subgroup H_0, with $\dim \Gamma/H_0 = k$ and, on the orbit, z_j^0 is real and positive.

Definition 2.2. Let $K < H$. Then (λ_0, X_0) is said to be *K-hyperbolic* if and only if

(a) $\dim \ker(I - F_X(\lambda_0, X_0))^K = k$

(b) $F_\lambda(\lambda_0, X_0) : \mathbb{R}^k \to W$ is one-to-one

(c) Range $F_\lambda(\lambda_0, X_0) \cap \text{Range}(I - F_X(\lambda_0, X_0))^K = \{0\}$.

Similarly, (λ_0, X_0) is said to be *K-simply hyperbolic* if (λ_0, X_0) is K-hyperbolic and the algebraic multiplicity of 0 as eigenvalue of $(I - F_X(\lambda_0, X_0))^K$ is k.

Note that, since X_0 is in V^H, it follows that $f(\lambda, X_0)$ is in W^H, and thus, $F_\lambda(\lambda_0, X_0)\mu$ belongs to W^H. Similarly, since $\Gamma X_0 \subset V^H$, $A_j X_0$ belongs to V^H. Furthermore, since $F_X^K(\lambda_0, X_0)$ has the diagonal structure

$$\begin{pmatrix} F_X^H(\lambda_0, X_0) & 0 \\ 0 & F_X^{\perp K}(\lambda_0, X_0) \end{pmatrix}$$

one obtains the following result.

Proposition 2.3. *(λ_0, X_0) is K-hyperbolic if and only if (λ_0, X_0) is H-hyperbolic and $I - F_X^{\perp K}$ is invertible.*

Note that the above notions depend only on the orbit and not on the representative X_0. This follows easily from the relations

$$\begin{aligned} I - F_X(\lambda_0, \gamma X_0) &= \gamma(I - F_X(\lambda_0, X_0))\gamma^{-1} \\ F_\lambda(\lambda_0, \gamma X_0) &= \gamma F_\lambda(\lambda_0, X_0). \end{aligned}$$

Example 2.6 (Autonomous differential equations). Consider the problem of finding 2π-periodic solutions to

$$\frac{dX}{dt} - g(X, \nu) = 0, \quad X \text{ in } \mathbb{R}^N,$$

for instance with $g(X, \nu) = g(X)/\nu$ coming from the system $\frac{dX}{d\tau} = g(X)$. In order to set the problem as above, let $W = H^1(S^1)$ and consider the operator

$$\tilde{K} : L^2(S^1) \to H^1(S^1)$$

defined on the Fourier series $X(t) = \langle X\rangle + \sum X_n e^{int}$ as

$$\tilde{K}(\langle X\rangle + \sum_{n\neq 0} X_n e^{int}) = \langle X\rangle + \sum_{n\neq 0} X_n/(in)e^{int}.$$

Then, $\tilde{K}X' = (\tilde{K}X)' = X - \langle X\rangle$ and the above equation is equivalent to

$$X - \langle X\rangle - \tilde{K}g(X, \nu) = 0,$$

where $\tilde{K}g$ is a compact map on $H^1(S^1)$.

If $X_0(t)$ is a solution, with minimal period $2\pi/p$, for some ν_0, and $g(X, \nu)$ in C^1 in a neighborhood of $(X_0(t), \nu_0)$, then one gets the linearization

$$X - F_X(X_0, \nu_0)X - F_\nu(X_0, \nu_0)\mu = X - \langle X\rangle - \tilde{K}g_X X - \tilde{K}g_\nu\mu = \tilde{K}(X' - g_X X - g_\nu\mu).$$

Here H is the space of $(2\pi/p)$-periodic functions, or else those Fourier series with n a multiple of p. Now, X_0' is solution of

$$X' - g_X(X_0, \nu_0)X = 0,$$

hence, if $X_0(t)$ is non-constant, the first condition of K-hyperbolicity means that X_0' is the only $(2\pi/p')$-periodic solution of the last equation, for p' dividing p, while the second condition means that $g_\nu(X_0, \nu_0)$ is non-zero. For the special case of $g(X)/\nu$, then $g_\nu = -g(X)/\nu_0^2 = -X_0'/\nu_0$ and condition (b) is met if X_0 is non-constant. The third condition is equivalent to say that the equation

$$X' - g_X X = g_\nu\mu$$

has no $\frac{2\pi}{p'}$-periodic solution for $\mu \neq 0$. In the case of $g(X)/\nu$, taking $\mu = -\nu_0$, then $X' - g_X X = X_0'$ cannot have solution. But this means that $\ker\left(\frac{d}{dt} - g_X\right)^2$ is generated by X_0', that is 0 is a simple eigenvalue of the operator $\frac{d}{dt} - g_X$.

Proposition 2.4. *Let $(\nu_0, X_0(t))$ be a $(2\pi/p)$-periodic solution of $\nu X' - g(X) = 0$. Then, if W^K is the subspace of $H^1(S^1)$ consisting of $(2\pi/p')$-periodic functions, with p' dividing p, (ν_0, X_0) is K-hyperbolic if and only if 0 is a simple eigenvalue of the operator $\nu_0\frac{d}{dt} - g_X(X_0)$ in W^K, that is 1 is a simple Floquet multiplier of $\Phi(2\pi/p')$, where $\Phi(t)$ is the fundamental matrix of the linear system.*

Proof. There remains only to see the equivalence with Floquet theory. This is done in Appendix B. □

Since $\Phi(2\pi) = \Phi(2\pi/p)^p$ and X_0' is always an eigenvector, then (ν_0, X_0) is $\{e\}$-hyperbolic if and only if 1 is a simple Floquet multiplier of the first return map $\Phi(2\pi/p)$ and this matrix has no other eigenvalues which are p-th roots of unity. This is the usual definition of hyperbolicity.

In the general case, $\{K\}$-hyperbolicity means that X_0' is the only $(2\pi/p')$-periodic solution of $X' - g_X X = 0$, that $g_\nu(X_0, \nu_0) \neq 0$ and $\int_0^{2\pi} g_\nu \cdot Z(t)\,dt \neq 0$, where $Z' = -g_X^T Z$ is the solution of the adjoint problem (see Appendix B).

Returning to the abstract setting of $f(\lambda, X)$, with (λ_0, X_0) an H-hyperbolic orbit, note that Range $F_\lambda(\lambda_0, X_0)$ has the right dimension to complement Range$(I - F_X(\lambda_0, X_0)^H)$ in W^H. In order to compute the Γ-index of the orbit, we shall introduce an auxiliary operator. Recall that $z_1, \dots, z_k$ are the first variables in W with isotropy $H_0 > H$ and $\dim \Gamma/H_0 = k$.

Definition 2.3. Let the compact linear H-equivariant operator $\mathcal{K}$, from V into itself, be defined by

$$\mathcal{K}(\mu, Y) = (\mu_1 - \operatorname{Im} z_1, \dots, \mu_k - \operatorname{Im} z_k, F_\lambda(\lambda_0, X_0)\mu + F_X(\lambda_0, X_0)Y).$$

In particular, $\mathcal{K}$ maps V^K into V^K for any $K < H$.

Proposition 2.5. *(λ_0, X_0) is K-hyperbolic if and only if $(I - \mathcal{K})^K$ is invertible, for $K < H$.*

Proof. If $(I - \mathcal{K}^K)(\mu, Y) = 0$, then $\operatorname{Im} z_j = 0$ and $(I - F_X)Y = F_\lambda \mu$. Thus, if (λ_0, X_0) is K-hyperbolic, one needs $\mu = 0$ and Y belongs to $\ker(I - F_X)^H$, that is $Y = \sum_{l=1}^k \alpha_l A_l X_0$. Considering the first k coordinates, one has

$$\operatorname{Im} z_j = \Big(\sum_{l=1}^{k} \alpha_l n_l^j\Big) z_j^0,$$

since z_j^0 has been taken real and positive. But, from Lemma 2.4 of Chapter 1, the matrix $(n_l^j)_{lj}$ is invertible. Then, since $\operatorname{Im} z_j = 0$, one has $\alpha_l = 0$ and $I - \mathcal{K}^K$ is one-to-one. Since $\mathcal{K}$ is compact, one has that $I - \mathcal{K}^K$ is invertible.

Conversely, if $\dim\ker(I - F_X^K) > k$, let Y_0 be in this kernel and linearly independent from $A_j X_0$. Let B be the invertible matrix given by $B_{jl} = n_l^j$ and, if $y_1, \dots, y_k$ are the first k variables of Y_0, define $\alpha_1, \dots, \alpha_k$ through the relations $(B\alpha)_j = \operatorname{Im} y_j / z_j^0$. Replacing then Y_0 by

$$Y_0 - \sum \alpha_l A_l X_0,$$

one may assume that $\operatorname{Im} y_j = 0$. Thus, $(0, Y_0)$ is in $\ker(I - \mathcal{K}^K)$, which is not possible, unless $Y_0 = 0$, if $I - \mathcal{K}^K$ is invertible. Similarly, if $F_\lambda(\lambda_0, X_0)\mu = 0$, then $(\mu, 0)$ is in $\ker(I - \mathcal{K}^H)$. Hence, if $I - \mathcal{K}^H$ is invertible, one has $\mu = 0$ and condition (b) is verified. Finally, if $F_\lambda \mu = (I - F_X^K)Y$, then

$$\left(\mu, Y - \sum \gamma_l A_l X_0\right)$$

is in $\ker(I - \mathcal{K}^K)$ if γ_l are defined by $(B\gamma)_j = \operatorname{Im} y_j / z_j^0$. But then, if $I - \mathcal{K}^K$ is invertible, one has $\mu = 0$, $Y = \sum \gamma_l A_l X_0$ and condition (c) is met. □

Thus, if (λ_0, X_0) is K-hyperbolic, then it is an isolated zero of the H-map

$$(I - \mathcal{F})(\lambda, X) \equiv (\operatorname{Im} z_1, \dots, \operatorname{Im} z_k, X - F(\lambda, X))$$

from V^K into itself, since its linearization is $I - \mathcal{K}^K$.

Recall, from Theorem 2.2, that i_K is the index, at (λ_0, X_0), of the map $X - F(\lambda, X)$ when restricted to $V^K \cap \{\operatorname{Im} z_j = 0, \operatorname{Re} z_j > 0, j = 1, \dots, k\}$.

For the same reason, the Leray–Schauder index of $(I - \mathcal{F})^K$ is also defined at (λ_0, X_0), and clearly both indices are related.

Lemma 2.2. $i_K = (-1)^{k(3k+1)/2} \operatorname{Index}((I - \mathcal{F})^K; (\lambda_0, x_0))$.

Proof. The natural orientation of $\mathbb{R}^k \times W^K$ is given by

$$(\lambda_1, \dots, \lambda_k, \operatorname{Re} z_1, \operatorname{Im} z_1, \dots, \operatorname{Re} z_k, \operatorname{Im} z_k, \dots).$$

Via a series of permutations, this identity map is homotopic to

$$((-1)^{k+1} \operatorname{Im} z_1, (-1)^{k+2} \operatorname{Im} z_2, \dots, (-1)^{2k} \operatorname{Im} z_k, \lambda_1, \dots, \lambda_k, \operatorname{Re} z_1, \dots, \operatorname{Re} z_k, \dots).$$

Hence, from the product theorem, $\operatorname{Index}(I - \mathcal{F}^K) = (-1)^{\sum_{k+1}^{2k} j} i_K$, giving the result. □

Now, if (λ_0, X_0) is K-hyperbolic, one may approximate $(I - \mathcal{F})^K$ by its linearization at (λ_0, X_0), i.e., by $(I - \mathcal{K}^K)(\mu, Y)$ and compute the index of this linear H-equivariant map at $(0, 0)$. Here $\mu = \lambda - \lambda_0$ and $Y = X - X_0$, since $\operatorname{Im} z_j^0 = 0$. From the fact that F_X^H is a compact linear operator, one has the decomposition

$$W^H = \ker(I - F_X^H)^\alpha \oplus \operatorname{Range}(I - F_X^H)^\alpha,$$

where the first term is the *generalized eigenspace*, whose dimension is the *algebraic multiplicity* m of 1 as eigenvalue of F_X^H and α is the *ascent*. Then, $I - F_X^H$ leaves each subspace invariant and one may write

$$Y^H = u \oplus v$$

with u in the generalized eigenspace. Furthermore, since $A_1 X_0, \dots, A_k X_0$ generate $\ker(I - F_X^H)$, one may choose a basis for the generalized eigenspace in such away that $I - F_X^H$ is in Jordan form on it. Thus, there are exactly k Jordan blocks, of size $m_1, \dots, m_k$ such that $\sum m_l = m$ and $\max m_l = \alpha$. On the l-th block, corresponding to $A_l X_0$, one has

$$(I - F_X^H)u_l = J_l u_l,$$

where

$$J_l = \begin{pmatrix} 0 & 1 & & \\ & \ddots & \ddots & \\ & & \ddots & 1 \\ & & & 0 \end{pmatrix}$$

is an $m_l \times m_l$ matrix and u_l is the projection of u on the block (recall that, on the generalized kernel, the matrix $I - F_X^H$ is nihilpotent). Thus, on this basis, $A_l X_0$ has coordinates $(1, 0, \dots, 0)$ on the j-th block and 0 on the others. Let $(x_{l1}, \dots, x_{lm_l})$ be the coordinates of u_l, then

$$u = \sum x_{l1} A_l X_0 + w,$$

where w corresponds to the other variables. Let $F_\lambda \mu$ be written as

$$F_\lambda \mu = (F_1 \mu, \dots, F_l \mu, \tilde{F} \mu),$$

where $F_l \mu$ is the projection on the l-th block with components

$$(F_{l1}\mu, \dots, F_{lm_l}\mu)$$

and $\tilde{F}\mu$ is the projection on Range$(I - F_X^H)^\alpha$. Then, $(I - \mathcal{K}^H)$ has the following form

$$\begin{aligned}(I - \mathcal{K}^H)(Y, \mu) = \Big(\Big\{\Big(\sum_{l=1}^{k} n_l^j x_{l1}\Big) z_j^0 + \operatorname{Im}(w_j + v_j)\Big\}_{j=1,\dots,k}, \\ \{J_l u_l - F_l \mu\}_{l=1,\dots,k}, (I - F_X^H)v - \tilde{F}\mu\Big),\end{aligned}$$

where $\sum_{l=1}^{k} n_l^j x_{l1}$ will be written, as before, $(Bx)_j$ and the components of $J_l u_l - F_l \mu$ are $(x_{l2} - F_{l1}\mu, \dots, x_{ln_l} - F_{lm_{l-1}}\mu, F_{lm_l}\mu)$.

In particular, if Y_l has coordinates $(0, \dots, 0, 1)$ on the l-th block and 0 on the others, i.e., if Y_l generates, for $l = 1, \dots, k$, the kernel of the adjoint matrix, one has

$$F_{lm_l}\mu = \Big(\sum \frac{\partial F}{\partial \mu_j}\mu_j, Y_l\Big).$$

Let Λ be the $k \times k$ matrix with l-th row given by $\big(\frac{\partial F}{\partial \mu_j}, Y_l\big)$.

Lemma 2.3. *The matrix Λ is invertible.*

Proof. Assume this is not true and that some μ belongs to $\ker \Lambda$. Then taking $v = (I - F_X^H)^{-1}\tilde{F}\mu$, and $x_{li+1} = F_{li}\mu$, for $i = 1, \dots, m_l - 1$, i.e., for the coordinates of w, and $(x_{11}, \dots, x_{k1})$ solving the system

$$Bx = -\begin{pmatrix} \mathrm{Im}(w_1 + v_1)/z_1^0 \\ \vdots \\ \mathrm{Im}(w_k + v_k)/z_k^0 \end{pmatrix},$$

one obtains an element of $\ker(I - \mathcal{K})$, which is impossible, unless $\mu = 0$. □

Theorem 2.4. *Let* (λ_0, X_0) *be* K*-hyperbolic and* Λ *be the above matrix. Then:*

(a) $i_H = (-1)^{k(k+1)/2}(-1)^{n_H}$ Sign det Λ Sign det B, *where* n_H *is the number of eigenvalues of* F_X^H, *counted with algebraic multiplicity, which are larger than or equal to* 1.

(b) $i_K = (-1)^{n'_K} i_H$, *where* n'_K *in the number of eigenvalues of* $F_X^{\perp K}$, *counted with algebraic multiplicity, which are larger than* 1.

Proof. In order to compute i_H or the index of $I - \mathcal{K}^H$ at 0, one may deform linearly to 0 the terms $\tilde{F}\mu$ and F_{lj}, for $j = 1, \dots, m_l - 1$ and $l = 1, \dots, k$, i.e., those concerning w and v. Then, one may also deform $\mathrm{Im}(w_j + v_j)$ to 0 and, later, z_j^0 to 1. Using the two compositions $\mu \to \Lambda\mu$ and $x \to Bx$, and since the permutation $(\mu_j, x_j) \to (x_j, -\mu_j)$ has index 1, one is left with the map

$$\begin{aligned}&(x_{11}, x_{12}, \dots, x_{1m_1}, x_{21}, \dots, x_{k1}, \dots, v)\\ &\qquad\to (x_{12}, x_{13}, \dots, x_{1m_1}, x_{11}, x_{22}, x_{2m_2}, x_{22}, \dots, (I - F_X^H)v).\end{aligned}$$

Via permutations, the x-part of this map contributes $(-1)^{m-k}$ to the index. Hence,

$$\mathrm{Index}(I - \mathcal{K}^H) = (-1)^{m-k}\,\mathrm{Sign}\det\Lambda\,\mathrm{Sign}\det B\,\mathrm{Index}\,((I - F_X^H)v).$$

One may decompose $\mathrm{Range}(I - F_X^H)^\alpha$ into

$$\bigoplus \ker(I - \lambda_j F_X^H)^{\alpha_j} \oplus \tilde{W},$$

where λ_j are the characteristic values (i.e., inverses of eigenvalues) of F_X^H between 0 y 1 with algebraic multiplicity m_j and ascent α_j. In fact, the generalized eigenspaces are disjoint, since if $(I - \lambda_1 F_X^H)^{\alpha_1}x = 0 = (I - \lambda_2 F_X^H)^{\alpha_2}x$, then if $y = (I - \lambda_1 F_X^H)^{\alpha_1 - 1}x$, one has $y = \lambda_1 F_X^H y$ and $(I - \lambda_2 F_X^H)^{\alpha_2} y = 0 = (1 - \lambda_2/\lambda_1)^{\alpha_2} y$. Thus, $y = 0$ and if $z = (I - \lambda_1 F_X^H)^{\alpha_1 - 2}$ one may proceed to prove $x = 0$. Furthermore, since $(I - \lambda_j F_X^H)$ commutes with $(I - F_X^H)^\alpha$, one has the above decomposition with a finite number of subspaces due to the fact that the compact operator F_X^H has only a finite number of eigenvalues larger than 1.

The operator $I - F_X^H$ preserves each of these subspaces. One may choose bases on $\ker(I - \lambda_j F_X^H)^{\alpha_j}$ so that the nihilpotent matrix $I - \lambda_j F_X^H$ is in Jordan blocks, i.e., of the form J as above. Hence, on such a block, $I - F_X^H$ will have the form $(1 - \lambda_j^{-1})I + \lambda_j^{-1} J$, which is deformable to $-I$. Hence, each generalized kernel will contribute $(-1)^{m_j}$ to the index. On the other hand, on $\tilde{W}$, the operator $I - F_X^H$ is deformable to I. Thus,

$$\mathrm{Index}((I - F_X^H)v) = (-1)^{\sum m_j}.$$

Using Lemma 2.2, one obtains

$$i_H = (-1)^{k(3k+1)/2}(-1)^{m-k}(-1)^{\sum m_j} \operatorname{Sign} \det \Lambda \operatorname{Sign} \det B.$$

Since $n_H = m + \sum m_j$ and $k(3k+1)/2 + k = 3k(k+1)/2$ has the parity of $k(k+1)/2$, one obtains the first part of the theorem.

For (b), it is enough to recall the block diagonal structure of $I - F_X^K$. Thus,

$$\mathrm{Index}(I - \mathcal{K}^K) = \mathrm{Index}(I - \mathcal{K}^H)\, \mathrm{Index}(I - F_X^{\perp K}),$$

where $I - F_X^{\perp K}$ is invertible in $W^K \cap (W^H)^\perp$. Decomposing this last space in $\bigoplus \ker(I - \lambda_j F_X^{\perp K})^{m_j} \oplus \tilde{W}$ as before, one obtains the contribution $(-1)^{n'_K}$ to the index. □

Remark 2.2. Note that if one has a set of equivalent irreducible H-representations, where H acts as S^1 or as $\mathbb{Z}_m$, $m \geq 3$, then, since $F_X^{\perp K}$ preserves these representations, the map $I - F_X^{\perp K}$ can be seen on them as a real operator of the form

$$(A + iB)(X + iY) = \begin{pmatrix} A & -B \\ B & A \end{pmatrix} \begin{pmatrix} X \\ Y \end{pmatrix}.$$

If

$$P = \begin{pmatrix} I & I \\ -iI & iI \end{pmatrix} \quad \text{and} \quad P^{-1} = \frac{1}{2i} \begin{pmatrix} iI & -I \\ iI & I \end{pmatrix},$$

one has

$$\begin{pmatrix} A & -B \\ B & A \end{pmatrix} = P \begin{pmatrix} A + iB & 0 \\ 0 & A - iB \end{pmatrix} P^{-1},$$

and it follows that

$$\det \begin{pmatrix} A - \lambda I & -B \\ B & A - \lambda I \end{pmatrix} = |\det(A - \lambda I + iB)|^2 > 0.$$

Hence, the algebraic multiplicity of any real eigenvalue is even. Similarly, if $\begin{pmatrix} X \\ Y \end{pmatrix}$ is an eigenvector with real eigenvalue, then $\begin{pmatrix} Y \\ -X \end{pmatrix}$ is also an eigenvector and the geometric multiplicity is even. Thus, in the computation of i_K one has to take into account only the representations in $(V^H)^\perp$ where H acts as $\mathbb{Z}_2$, since, on the others, n'_K will conserve its parity. This gives another proof, for the case of hyperbolicity, of this part of Theorem 2.3.

Corollary 2.2. *If (λ_0, X_0) is $\{e\}$-hyperbolic with isotropy H, then the Γ-index is given by $(d_H, d_K, \dots)$, for $K < H$, where*

$$d_H = i_H = (-1)^{k(k+1)/2}(-1)^{n_H} \operatorname{Sign}\det \Lambda \operatorname{Sign}\det B,$$

where n_H is the number of generalized eigenvalues of F_X^H which are larger than or equal to 1, and B and Λ are defined above. If $K/H \cong \mathbb{Z}_2$, then

$$d_K = (i_K - i_H)/2, \quad \text{with } i_K = (-1)^{n'_K} i_H,$$

where n'_K is the number of generalized eigenvalues of $F_X^{\perp K}$ which are larger than 1. The integer d_K is completely determined by the above integers if K/H is a product of $\mathbb{Z}_2$'s and $d_K = 0$ otherwise.

Proof. This is just a rephrasing of Theorems 2.3 and 2.4. □

Example 2.6 (continued). Let us return to the system

$$\nu X' - g(X) = 0, \quad X \text{ in } \mathbb{R}^N,$$

with a hyperbolic solution (ν_0, X_0), i.e., if $A(t) = g_X(X_0(t))/\nu_0$ and $\Phi(t)$ is the fundamental matrix of the linearization

$$LX = X' - A(t)X$$

then 1 is a simple eigenvalue of $\Phi(2\pi) = \Phi(2\pi/p)^p$, with X'_0 as only solution of $LX = 0$, where $2\pi/p$ is the least period of $X_0(t)$.

Now, the operator $I - F_X$ of Theorem 2.4 has the form $\tilde{K}(X' - A(t)X)$ and its characteristic values, i.e., such that $\ker(I - \lambda F_X)$ is non trivial, correspond to non-trivial solutions of $X' - \lambda A(t)X = 0$. However, since $\tilde{K}$ and $A(t)$ do not commute unless A is constant, the generalized kernels are difficult to relate. Hence, we shall use another way in order to compute i_H and i_K, a way which is related to standard Floquet multipliers.

Proposition 2.6. *If 1 is a simple eigenvalue of $\Phi(2\pi)$, let σ_+ be the number of real eigenvalues, counted with algebraic multiplicity, of $\Phi(2\pi/p)$ which are larger than 1 and let σ_- be the number of real eigenvalues of $\Phi(2\pi/p)$ which are less than -1, then, on $W^K = \{X(t),$ in $H^1(S^1)$, which are $2\pi/p'$-periodic, p' dividing $p\}$,*

$$\begin{aligned} i_K &= -(-1)^{\sigma_+} && \text{if } p/p' \text{ is odd} \\ i_K &= -(-1)^{\sigma_- + \sigma_+} && \text{if } p/p' \text{is even.} \end{aligned}$$

2 In particular, the S^1-index has at most two non-zero components: $d_H = i_H$ and, if p is even, $d_K = (i_K - i_H)/2$, for $|H/K| = 2$ or $p' = p/2$, corresponding to period doubling.

Proof. As seen in Lemma 2.2, $i_K = \text{Index}(I - \mathcal{K}^K)$, where

$$(I - \mathcal{K})(\mu, X) = (\operatorname{Im} z_1, \tilde{K}(X' - AX + X_0'\mu/\nu_0)),$$

with z_1 the component of X on a mode m (a multiple of p) for which X_0 has its corresponding z_1^0 real and positive. Now, from Proposition 2.5, one has that $I - \mathcal{K}$ is invertible on V^K, hence this will be also the case for $I - \mathcal{K} + \lambda\tilde{K}$ for small λ and this small compact perturbation does not alter the index.

Now, since the matrix $\Phi(2\pi)$ has N eigenvalues, the number $e^{2\pi\lambda}$ will not be one of them for small, *strictly positive* λ. Thus, from Appendix B, the Fredholm operator, from $H^1(S^1)$ into $L^2(S^1)$,

$$L_\lambda X = X' - AX + \lambda X$$

will be invertible. In particular, the solution of the equation $L_\lambda X = -\tau\mu X_0'/\nu_0$ is $X = -\tau\mu X_0'/\lambda\nu_0$, with corresponding $\operatorname{Im} z_1 = -\tau\mu m z_1^0/\lambda\nu_0$. Hence, one may deform linearly $(I - \mathcal{K} + \lambda\tilde{K})(\mu, X)$ to $(-\mu, \tilde{K}(X' - AX + \lambda X))$ and

$$i_K = -\text{Index}\left(\tilde{K}(X' - AX + \lambda X)|_{W^K}\right).$$

Increasing λ, one will get a possible change of index at a point λ_0 such that $\tilde{K}L_\lambda$ is not invertible in W^K, i.e., if $e^{2\pi\lambda_0/p'}$ is a Floquet multiplier of $\Phi(2\pi/p')$, p' dividing p.

Although $\tilde{K}L_\lambda$ and L_λ have the same kernel, their generalized kernels do not coincide in general. However, as seen in Remark 2.1, one may detect the change of index by looking at the bifurcation equation $B(\lambda)$, for both operators. For

$$\tilde{K}L_{\lambda_0} + \lambda\tilde{K} = \tilde{K}L_{\lambda_0+\lambda} = \tilde{K}\left(\frac{d}{dt} - A + (\lambda_0 + \lambda)I\right),$$

defined on $H^1(S^1)$ or on W^K, one has

$$\tilde{B}(\lambda) = \lambda(I - \tilde{Q})\tilde{K}(I + \lambda\tilde{R}\tilde{Q}\tilde{K})^{-1}P,$$

where P is a projection from $H^1(S^1)$ onto $\ker L_{\lambda_0}$, the operator $\tilde{Q}$ is a projection from W^K onto Range $\tilde{K}L_{\lambda_0}$ and $\tilde{R}$ is the pseudo-inverse of $\tilde{K}L_{\lambda_0}$ defined by $\tilde{R}\tilde{K}L_{\lambda_0}(I - P) = I - P$ and $\tilde{K}L_{\lambda_0}\tilde{R}\tilde{Q} = \tilde{Q}$. On the other hand, for the Fredholm operator from $H^1(S^1)$ into $L^2(S^1)$, or from W^K into $L^2(S^1)^K$, defined by $\frac{d}{dt} - A + (\lambda_0 + \lambda)I = L_{\lambda_0+\lambda} = L_{\lambda_0} + \lambda I$, one has

$$B(\lambda) = \lambda(I - Q)(I + \lambda RQ)^{-1}P,$$

where Q projects L^2 onto Range L_{λ_0} and $L_{\lambda_0}RQ = Q$, $RL_{\lambda_0}(I-P) = I-P$. Note that the identity in $I + \lambda RQ$ is that of $H^1(S^1)$.

Now, for a given $\tilde{Q}$ one may choose $Q = \left(\frac{d}{dt} + P_0\right)\tilde{Q}\tilde{K}$, where P_0 is the projection on the constants, hence $\tilde{K}\left(\frac{d}{dt} + P_0\right)\tilde{K} = I_{H^1}, \left(\frac{d}{dt} + P_0\right)\tilde{K} = I_{L^2}$. One may take in this case $R = \tilde{R}\tilde{K}$. Conversely, for a given Q, one may take $\tilde{Q} = \tilde{K}Q\left(\frac{d}{dt} + P_0\right)$ and $\tilde{R} = R\left(\frac{d}{dt} + P_0\right)$: in fact, it is easy to see that Q, as defined above, maps L^2 into Range L_{λ_0} and that $Q^2 = Q$. Similarly, $\tilde{Q}$ maps H^1 into Range $\tilde{K}L_{\lambda_0}$ and $\tilde{Q}^2 = \tilde{Q}$. Furthermore, it is immediate to check that R and $\tilde{R}$ have the right properties. Moreover, $I_{H^1} + \lambda RQ = I_{H^1} + \lambda\tilde{R}\tilde{Q}\tilde{K}$ and $I_{L^2} - Q = \left(\frac{d}{dt} + P_0\right)(I_{H^1} - \tilde{Q})\tilde{K}$. Thus,

$$B(\lambda) = \left(\frac{d}{dt} + P_0\right)\tilde{B}(\lambda).$$

Hence, there is a change in the sign of the determinant of $\tilde{B}(\lambda)$ if and only if there is a change of sign in the determinant of $B(\lambda)$. The later will be the case if and only if λ_0 is an eigenvalue of $\frac{d}{dt} - A$ of odd algebraic multiplicity (see Remark 2.1). That is, at each Floquet multiplier of $\Phi(2\pi/p')$ one has a change of the index of $(I - F_X + \lambda\tilde{K})^K$ equal to $(-1)^n$, where n is the algebraic multiplicity of the multiplier itself.

Now consider, for $\lambda > 0$, the deformation

$$X - \tau\langle X\rangle + \tau(\lambda\tilde{K}X - \tilde{K}AX), \quad \tau \in [0, 1].$$

Applying $\frac{d}{dt} + P_0$ to this deformation, one obtains the equation

$$X' + (1-\tau)\langle X\rangle + \tau(\lambda X - AX).$$

Multiplying this equation by X^T and taking $\lambda > N^{1/2}\|A\|$, where $\|A\| = \max|A_{ij}(t)|$, one gets, after integrating on $[0, 2\pi]$, that

$$(1-\tau)\langle X\rangle^2 + \tau(\lambda\|X\|^2 - (X, AX)) \geq (1-\tau)\langle X\rangle^2 + \tau(\lambda - N^{1/2}\|A\|)\|X\|^2.$$

Hence, one has a valid deformation for λ sufficiently large. Clearly, for $\tau = 0$, the index is 1 and the index of $(I - F_X)^K$ is $(-1)^{\sum m_j}$, where m_j are the algebraic multiplicities of the eigenvalues of $\Phi(2\pi/p')$ which are real and larger than 1.

Finally, since $\Phi(2\pi/p') = \Phi(2\pi/p)^{p/p'}$, the spectrum of $\Phi(2\pi/p')$ is made of the (p/p')-powers of the eigenvalues of $\Phi(2\pi/p)$. But, if μ is an eigenvalue of $\Phi(2\pi/p)$ then $\bar{\mu}$ will also be an eigenvalue. Hence, non real eigenvalues come in pairs with the same algebraic multiplicity and will not contribute to the index. For real negative μ, one will have $\mu^{p/p'} > 0$ only if p/p' is even.

Then, Theorem 2.2 gives the final part of the proof. □

Remark 2.3 (S^1-degree and Fuller degree). One may define the S^1-degree for the equation $\nu X' - g(X)$ in the form $X - \langle X\rangle - \tilde{K}g(X)/\nu$, in $H^1(S^1)$, by making the following hypothesis:

Assume that there is an open bounded subset $\tilde{\Omega}$ of $\mathbb{R} \times \mathbb{R}^N$ such that the differential equation has no 2π-periodic solution $(\nu, X(t))$ which touches $\partial\tilde{\Omega}$ for some t. This assumption ensures that if $(\nu, \tilde{X})$ belongs to $\tilde{\Omega}$ for some $\tilde{X}$ on a periodic solution, then the whole orbit stays in $\tilde{\Omega}$. Moreover, if $\tilde{X}$ is a stationary solution, that is $g(\tilde{X}) = 0$, then $(\nu, \tilde{X})$ is also a solution for all ν. Thus, since $\tilde{\Omega}$ is bounded, the set $\tilde{\Omega}$ cannot contain stationary solutions. Furthermore, for the integral equation, one needs that $\nu \geq \delta > 0$ on $\tilde{\Omega}$.

Now, if (ν, X) is in $\tilde{\Omega}$ and belongs to a periodic orbit, then $g(X)/\nu$ is bounded in $\tilde{\Omega}$ and X' will be bounded in L^2 and $\|X\|_1 < R$ for some constant R. Let

$$\Omega = \left\{(\nu, X) \text{ in } \mathbb{R} \times H^1(S^1) : \|X\|_1 \leq R, (\nu, X(t)) \in \tilde{\Omega}\right\}.$$

Since any function in $H^1(S^1)$ is continuous, we have that if (ν, X) is close to (ν_0, X_0) in H^1 then $X(t)$ will be close to $X_0(t)$ for all t. Thus, the set Ω is open. Clearly, Ω is invariant under the S^1-action and any periodic solution $(\nu, X(t))$ in $\tilde{\Omega}$ will give exactly one solution (ν, X) in Ω of the integral equation and conversely. In particular, $X - F(\nu, X) \neq 0$ on $\partial\Omega$ and its S^1-degree is well defined.

Since $g(X) \neq 0$ in $\tilde{\Omega}$, the invariant part of the S^1-degree is 0 and

$$\deg_{S^1}(X - F(\nu, X); \Omega) = \sum d_H[F_H]_{S^1},$$

where d_H is in $\mathbb{Z}$ and H runs over all the Fourier modes m, with $|H| = m > 0$. (Since F is compact all but a finite number of the d_H's are 0).

In this case, Fuller has defined a rational number which turns out to be

$$\sum d_H/|H|.$$

For a hyperbolic orbit, of least period $2\pi/p$, this number is

$$-(-1)^{\sigma_+}/p, \text{ if } p \text{ is odd},$$
$$-(-1)^{\sigma_+}/p - (((-1)^{\sigma_+ + \sigma_-} - (-1)^{\sigma_+})/2)/(p/2) = -(-1)^{\sigma_+ + \sigma_-}/p, \text{ if } p \text{ is even}.$$

Thus, in both cases, the Fuller index is $I_{\{e\}}/p$.

Example 2.7 (Differential equations with fixed period). Consider now the autonomous differential equation

$$\frac{dX}{dt} - g(X, \nu) = 0, \quad X \in \mathbb{R}^N,$$

where ν is not necessarily the frequency. Assume $(\nu_0, X_0(t))$ is a $2\pi/p$-periodic solution which is K-hyperbolic, for K corresponding to $(2\pi/p')$-periodic functions, with p' dividing p. This means that multiples of X_0' are the only non-trivial solutions of the linearized equation

$$\frac{dX}{dt} - A(t)X = 0,$$

where $A(t) = g_X(X_0(t), \nu_0)$. Furthermore, if $Z(t)$ generates the kernel of the adjoint equation

$$\frac{dZ}{dt} + A^T Z = 0,$$

with fundamental matrix $\Phi^{-1T}(t)$, then $g_\nu(X_0, \nu_0)$ is not L^2-orthogonal to $Z(t)$.

Proposition 2.7. *Let k be the algebraic multiplicity of 1 as eigenvalue of $\Phi(2\pi/p')$ and let e_{k-1} be a vector in $\mathbb{R}^N$ such that $(I - \Phi(2\pi/p'))^{k-1}e_{k-1} = X_0'(0) = e_0$. Then, if n_K is the sum of the algebraic multiplicities of real eigenvalues of $\Phi(2\pi/p')$ which are larger than or equal to 1, one has*

$$i_K = -\operatorname{Sign}(e_{k-1} \cdot Z(0)) \operatorname{Sign}\left(\int_0^{2\pi} g_\nu \cdot Z(t)\, dt\right)(-1)^{n_K}.$$

Proof. The argument is parallel to the one used in Proposition 2.6:

$$i_K = \operatorname{Index}(I - \mathcal{K}^K),$$

where

$$(I - \mathcal{K})(\mu, X) = (\operatorname{Im} z_1, \tilde{K}(X' - AX - g_\nu \mu)).$$

Now, one may replace g_ν by any $Z_1(t)$, in W^K, which has $(Z_1, Z)_{L^2}$ of the same sign as $(g_\nu, Z)_{L^2}$: in fact, the whole segment $\tau Z_1 + (1-\tau) g_\nu$ is not in Range $\left(\frac{d}{dt} - A\right)$. Hence, under the deformation, if one has a zero, one needs $\mu = 0$ and X is a multiple of X_0', which has $z_1 = imz_1^0$, where m is the mode of z_1. Hence, the only zero is for $X = 0$.

Now, let e_j be orthogonal to e_0 and such that $(I - \Phi(2\pi/p'))e_j = e_{j-1}$ for $j = 1, \ldots, k-1$, i.e., the generators of $\ker(I - \Phi(2\pi/p'))^K$. Thus, $e_j = (I - \Phi(2\pi/p'))^{k-1-j}e_{k-1}$, with $e_j \cdot Z(0) = 0$ for $j = 0, \ldots, k-2$, while $e_{k-1} \cdot Z(0) \neq 0$ (if not the algebraic multiplicity would be more than k, recalling that $Z(0)$ generates $\ker(I - \Phi^{-1T}(2\pi/p'))$, see Appendix B).

If η is the product of the two signs in the proposition, define

$$Z_1(t) = \eta \Phi(t) \left(\sum_0^{k-1} \left(\frac{tp'}{2\pi}\right)^{k-1-j} e_j \right).$$

Since $\Phi(2\pi/p') \sum_0^{k-1} e_j = \Phi(2\pi/p') \sum_0^{k-1} (I - \Phi(2\pi/p'))^{k-1-j} e_{k-1} = (I - (I - \Phi(2\pi/p'))^k) e_{k-1} = e_{k-1}$, one has $Z_1(2\pi/p') = Z_1(0)$ and $Z_1(t)$ belongs to W^K. (The above sum is of the form $(I - B)\sum_0^{k-1} B^j = I - B^k$ as a geometric sum, with $B = I - \Phi(2\pi/p')$.

Furthermore, since $Z(t) = \Phi^{-1T} Z(0)$ and $e_j \cdot Z(0) = 0$ for $j \leq k-2$, one has $(Z_1, Z)_{L^2} = \eta(e_{k-1} \cdot Z_0)$ with the right sign.

Now, the $\frac{2\pi}{p'}$-periodic solution of $\left(\frac{d}{dt} - (A - \lambda)\right)X = Z_1$ is then

$$X_\lambda(t) = \eta e^{-\lambda t} \Phi(t) \left(C + \int_0^t e^{\lambda s} \sum_0^{k-1} \left(\frac{sp'}{2\pi}\right)^{k-1-j} e_j ds \right),$$

where C is chosen such that $X_\lambda\left(\frac{2\pi}{p'}\right) = X_\lambda(0)$, i.e.,

$$C = e^{-2\pi\lambda/p'}\Phi(2\pi/p')(I - e^{-2\pi\lambda/p'}\Phi(2\pi/p'))^{-1}\int_0^{2\pi/p'} e^{\lambda s}\sum\left(\frac{sp'}{2\pi}\right)^{k-1-j} e_j\,ds,$$

which is possible if λ is small.

Next, we claim that, if z_1 is the component of X_λ on the m-th mode, then $\operatorname{Im} z_1$ has the sign of $\eta(-1)^{k-1}$ for λ small and positive.

In fact, if μ denotes $e^{-2\pi\lambda/p'}$, Φ denotes $\Phi(2\pi/p')$ and $B = \frac{-\mu}{1-\mu}(I - \Phi)$, then we have $I - B = (I - \mu\Phi)/(1 - \mu)$, and from $(I - B)\sum_0^j B^{j-l} = I - B^{j+1}$ one obtains that $(I - B)\sum_0^j \left(\frac{-\mu}{1-\mu}\right)^{j-l} e_l = (I - B)\sum_0^j B^{j-l}(I - \Phi)^{k-1-j}e_{k-1} = (I - B^{j+1})(I - \Phi)^{k-1-j}e_{k-1} = e_j$, since $B^k e_{k-1} = 0$ and $e_j = (I - \Phi)^{k-1-j}e_{k-1}$. Thus,

$$(I - \mu\Phi)^{-1}e_j = (1 - \mu)^{-1}\sum_0^j \left(\frac{-\mu}{1 - \mu}\right)^{j-l} e_l$$

and $\lim_{\mu\to 1}(1 - \mu)^k(I - \mu\Phi)^{-1}e_j = 0$ unless $j = k - 1$, where it is $(-1)^{k-1}e_0$, since j goes from 0 to $k - 1$, hence, $k - j - 1 + l > 0$, except for $j = k - 1$ and $l = 0$. Thus,

$$\lim_{\lambda\to 0^+}(1 - e^{-2\pi\lambda/p'})^k X_\lambda(t) = \eta(-1)^{k-1}(2\pi/p')X_0'(t),$$

since $(1 - \mu)^k C$ tends to $\Phi(2\pi/p')e_0 = (2\pi/p')e_0$ and $X_0'(t) = \Phi(t)e_0$.

But the component z_1 for X_0' is imz_1^0. This proves the claim.

The next step is to make the deformation

$$(\tau \operatorname{Im} z_1 + (1 - \tau)(-1)^{k-1}\eta\mu, \tilde{K}(X' - AX + \lambda X - \tau\mu Z_1)),$$

which is valid since a zero, in W^k, in the second term gives $X = \tau\mu X_\lambda$, with a corresponding z_1 of the sign of $(-1)^{k-1}\eta\mu$. One obtains, for $\tau = 0$, a product and

$$i_K = (-1)^{k-1}\eta \operatorname{Index}\left(\tilde{K}(X' - AX + \lambda X)|_{W^K}\right),$$

for λ small and positive. But, in Proposition 2.6, we have proved that this last index is $(-1)^{n_K - k}$, proving the proposition.

Note that, if $g(X, \nu) = g(X)/\nu$, then $g_\nu = -X_0'/\nu_0$ and $k = 1$. Then, since $X_0' = \Phi(t)e_0$ and $Z(t) = \Phi^{-1T}(t)Z(0)$, the scalar product in η is $-2\pi/\nu_0(e_0 - Z(0))$ and $\eta = -1$. Note also that k is also the algebraic multiplicity of 1 as eigenvalue of $\Phi(2\pi/p)$ since $\Phi(2\pi/p') = \Phi(2\pi/p)^{p/p'}$ and $\Phi(2\pi/p)$ has no eigenvalues, except 1, which are p/p'-roots of unity.

Example 2.8 (Differential equations with first integrals). This case can be translated into an instance of the last proposition. In fact, assume that the equation

$$X' = g(X),\ X \textit{ in } \mathbb{R}^N,\ \textit{has a first integral } V(X).$$

This means that $V(X(t))$ remains constant on solutions of the equation, or equivalently that

$$\nabla V(X) \cdot X' = \nabla V(X) \cdot g(X) = 0.$$

Consider the problem of finding 2π-periodic solutions to the equation

$$X' = g(X) + \nu \nabla V(X) = g(X, \nu).$$

If $X_0(t)$ is such a solution, then $X_0' \cdot \nabla V(X_0) = \nu \|\nabla V(X_0)\|^2 = \frac{d}{dt} V(X_0(t))$.

Integrating over a period, one has $\nu \|\nabla V(X_0(t))\| \equiv 0$, thus $\nu = 0$ if, on the orbit $\nabla V(X_0) \not\equiv 0$, or $\nabla V(X_0) \equiv 0$ on the orbit and, in both cases, $X_0(t)$ is a 2π-periodic solution of the original problem.

Let then denote by $A(t)$ the matrix $Dg(X_0(t))$ and let $\Phi(t)$ be the fundamental matrix for the variational equation $X' - A(t)X$. Then, if $X(t)$ is solution to the initial value problem

$$\begin{aligned} X' &= g(X) = g(X_0(t)) + A(t)(X(t) - X_0(t)) + R(X - X_0) \\ X(0) &= X_0(0) + W, \end{aligned}$$

then,

$$X(t) = X_0(t) + \Phi(t)W + \Phi(t) \int_0^t \Phi^{-1}(s) R(X(s) - X_0(s))\, ds.$$

Hence, linearizing the identity $V(X(t)) = V(X(0))$, one obtains

$$\nabla V(X_0(t)) \cdot \Phi(t) W = \nabla V(X_0(0)) \cdot W, \text{ for all } W \text{ in } \mathbb{R}^N.$$

Thus, if $\nabla V(X_0(0)) = 0$, one has that $\nabla V(X_0(t))$ is orthogonal to all $\Phi(t)W$ and, since $\Phi(t)$ is invertible, the only possibility is that $\nabla V(X_0(t)) = 0$ on the orbit of X_0, that is, if $\nabla V(X_0(t))$ is non-zero at some time t, it will remain so for all t's.

In general, for a 2π-periodic orbit, one has $X_0(2\pi) = X_0(0)$ and $\nabla V(X_0(0))$ is orthogonal to Range$(I - \Phi(2\pi))$. In other words, $\nabla V(X_0(0))$ belongs to $\ker(I - \Phi(2\pi)^T)$ and generates it if it is non-zero and if $\ker(I - \Phi(2\pi))$ is generated only by $X_0'(0)$. Furthermore, in this case the algebraic multiplicity has to be more than one: in fact, since $X_0'(0)$ is orthogonal to $\nabla V(X_0(0))$, then $X_0'(0)$ belongs to Range$(I - \Phi(2\pi))$. Hence, there is another vector in $\ker(I - \Phi(2\pi))^2$ besides $X_0'(0)$.

Then, if $Z(t) = \Phi^{-1T}(t) \nabla V(X_0(0))$, is the 2π-periodic solution of

$$Z' + A(t)^T Z = 0,$$

and, since $g_\nu(X_0(t), \nu) = \nabla V(X_0(t))$, one has

$$\begin{aligned} \int_0^{2\pi} g_\nu \cdot Z(t) dt &= \int_0^{2\pi} \nabla V(X_0(t)) \cdot \Phi(t) \Phi^{-1}(t) \Phi^{-1T}(t) \nabla V(X_0(0))\, dt \\ &= \int_0^{2\pi} \nabla V(X_0(0)) \cdot \Phi^{-1}(t) \Phi^{-1T}(t) \nabla V(X_0(0))\, dt \\ &= \int_0^{2\pi} \|Z(t)\|^2\, dt, \end{aligned}$$

where one has used the relation $\nabla V(X_0(t)) \cdot \Phi(t)W = \nabla V(X_0(0)) \cdot W$.

Since $\Phi^{-1}\Phi^{-1T}$ is a positive definite matrix, the integrand is positive and the only condition for hyperbolicity in this case is that $\dim\ker(I - \Phi(2\pi)) = 1$, or else, whenever X is a 2π-periodic solution of $X' - A(t)X = 0$, then X is a multiple of X_0'.

Proposition 2.8. *Let $X_0(t)$ be a non-constant $(2\pi/p)$-periodic solution of $X' = g(X)$ such that $\nabla V(X_0(0)) \neq 0$ and X_0' is the generator of $\ker(I - \Phi(2\pi))$. Let k, e_{k-1} and n_K be as in Proposition 2.7, then*

$$i_K = -\operatorname{Sign}(e_{k-1} \cdot \nabla V(X_0(0)))(-1)^{n_K}.$$

In particular, the S^1-index of $X' - g(X) - \nu\nabla V(X)$ at $(0, X_0(t))$ has at most two non-zero components $d_H = i_H = \eta(-1)^{\sigma_+}$, where $\eta = -(-1)^k \operatorname{Sign}(e_{k-1} \cdot \nabla V(X_0(0)))$ and σ_+ is the number of real Floquet multipliers of $\Phi(2\pi/p)$ counted with algebraic multiplicity, which are larger than 1, while $d_K = (i_K - i_H)/2$, for $|H/K| = 2$ (hence for p even) and $i_K = \eta(-1)^{\sigma_+ + \sigma_-}$, where σ_- is the number of real Floquet multipliers of $\Phi(2\pi/p)$, which are less than -1.

Proof. It is enough to apply Propositions 2.6 and 2.7. □

Remark 2.4. (a) Given an open bounded set $\tilde{\Omega}_1$ in $\mathbb{R}^N$ such that no periodic solution of $X' = g(X)$ touches $\partial\tilde{\Omega}_1$ and such that $\nabla V(X) \neq 0$ on 2π-periodic solution in $\tilde{\Omega}_1$ (including stationary ones), one may take $\tilde{\Omega} = \{(\nu, X) : |\nu| < \varepsilon, X \in \tilde{\Omega}_1\}$ and Ω as in Remark 2.3. Thus, $\deg_{S^1}(\tilde{K}(X' - g(X) - \nu\nabla V(X)); \Omega)$ is well defined.

(b) For the Hamiltonian system $X' - J\nabla H(X)$, we have that $H(X)$ is a first integral. The augmented system looks like

$$X' - (J - \nu I)\nabla H(X),$$

which may also be written as, on solutions of the equations

$$JX' + \nabla H(X) + \nu X' = 0.$$

But, this is exactly what is obtained when one studies $JX' + \nabla H(X)$ as an orthogonal map. This approach has the advantage of considering also stationary solutions. This will be done in the next section.

(c) One may have several first integrals to the systems $X' - g(X)$ and one could look at the augmented system

$$X' - g(X) + \sum \lambda_j \nabla V_j(X) = 0.$$

Taking the scalar product of this equation with $\sum \lambda_j \nabla V_j(X)$ and integrating on $[0, 2\pi]$ for a 2π-periodic solution of the augmented system one obtains

$$\|\sum \lambda_j \nabla V_j(X)\|_{L^2} = 0.$$

Thus, if (λ, X) is a 2π-periodic solution of the augmented system, then $X(t)$ is a 2π-periodic solution of $X' = g(X)$ and $\sum \lambda_j \nabla V_j(X) = 0$. Note that, as before, if this relation holds at some t_0 it will hold for all t's. In order to have a well-defined equivariant degree one needs to conclude that all λ_j's are 0, i.e., that the vector fields $\nabla V_j(X)$ are linearly independent on the orbit. Since $\nabla V_j(X(0))$ are in $\ker(I - \Phi(2\pi)^T)$, this implies a high dimensional kernel for $I - \Phi(2\pi)$. If S^1 is the only group acting, then the computation of the S^1-degree of the augmented system may be quite involved. On the other hand, one has a nonlinear equivalent of orthogonal maps. In fact, if $g(X) = J\nabla H(X)$, where H is Γ_0-invariant and $A_j X$ are the infinitesimal generators of the symplectic action of Γ_0, then for the equation

$$JX' + \nabla H + \sum \lambda_j A_j X,$$

one may define $V_j(X) = \frac{1}{2}(A_j JX, X)$. Since $A_j J$ is self-adjoint, one has that $\nabla V_j(X) = JA_j X$: see Proposition 9.1 in Chapter 1. This particular case will be studied in the next section.

Example 2.9 (Time dependent equations). Consider the problem of finding 2π-periodic solutions to the problem

$$\frac{dX}{dt} = f(X, t),$$

where $f(X, t)$ is $2\pi/p_0$-periodic in t. Then, as seen in § 9 of Chapter 1, one has a natural $\mathbb{Z}_{p_0}$ action on $C^1_{2\pi}(S^1)$. If $X_0(t)$ is a $2\pi/p$-periodic solution of the equation, with p dividing p_0, then the linearization of the equation at X_0 will be

$$\frac{dX}{dt} - A(t)X,$$

where $A(t) = Df(X_0(t), t)$ is $2\pi/p$-periodic.

Proposition 2.9. *If $W^K = \{X(t)$ in $H^1(S^1)$ which are $2\pi/p'$-periodic, where p' divides $p\}$, then $X_0(t)$ is K-hyperbolic if and only if $\frac{dX}{dt} - A(t)X = 0$ has no $2\pi/p'$-periodic solutions. If σ_+ and σ_- are the number of real eigenvalues, counted with multiplicity, of $\Phi(2\pi/p)$ which are larger than 1, respectively less than -1, then*

$$i_K = \begin{cases} (-1)^{\sigma_+} & \text{if } p/p' \text{ is odd} \\ (-1)^{\sigma_+ + \sigma_-} & \text{if } p/p' \text{ is even.} \end{cases}$$

In this case the $\mathbb{Z}_{p_0}$-index of $X_0(t)$ is $d_H = i_H$ and, if p is even, $d_K = (i_K - i_H)/2$ for $|H/K| = 2$.

Proof. In this case, applying Theorem 2.4, the number k is 0 and the argument follows the proof of Proposition 2.6. Recall that

$$\Pi^{\mathbb{Z}_{p_0}}_{S^W}(S^W) = \mathbb{Z} \times \mathbb{Z} \times \cdots$$

with one $\mathbb{Z}$ for each divisor of p_0. □

Clearly, this example corresponds to the classical situation where one can use Poincaré sections. The purpose of including it here is to contrast it with the situation of the following examples.

Example 2.10 (Symmetry breaking for differential equations). Assume that the autonomous equation $\frac{dX}{dt} = g(X, \nu)$ has an $\{e\}$-hyperbolic solution $(\nu_0, X_0(t))$ of least period $2\pi/q$. Consider the problem of finding 2π-periodic solutions to the equation

$$\frac{dX}{dt} = g(X, \nu) + \tau h(t, X, \nu),$$

for small τ and where h is $2\pi/p_0$-periodic in t. Hence, the S^1-symmetry is broken to a $\mathbb{Z}_{p_0}$-symmetry, for $\tau \neq 0$. This is an *entrainment* or *phase locking* problem and solutions of the perturbed problem are called *p_0-subharmonics*.

Let p be the largest common divisor of q and p_0, then the isotropy subgroup of X_0 is $\mathbb{Z}_p$, with $W^{\mathbb{Z}_p}$ corresponding to $2\pi/p$-periodic functions.

Proposition 2.10. *If (ν_0, X_0) is a $2\pi/q$-periodic solution of the autonomous equation $X' = g(X, \nu)$ such that X_0' generates the kernel of $LX = X' - Dg(X_0, \nu_0)X$ and such that the non-homogeneous equation $LX = g_\nu(X_0, \nu_0)$ has no solution, then the equation $X' = g(X, \nu) + \tau h(t, X, \nu)$, where h is $2\pi/p_0$-periodic in t, has a global continuum of $(2\pi/p)$-periodic solutions (ν, X) going through (ν_0, X_0) and parametrized by τ, where p is the largest common divisor of q and p_0, provided, in the case $p_0 = p$ and q/p_0 even, one has that the sum of the algebraic multiplicities of real eigenvalues of $\Phi(2\pi/q)$ which are less than -1 is even.*

Proof. Recall that $\Pi^{\mathbb{Z}_{p_0}}_{S^{\mathbb{R}\times W}}(S^W) = \bigoplus \Pi(H)$, where $\Pi(H) \cong \mathbb{Z}_2 \times \Gamma/H$ for each isotropy subgroup of $\mathbb{Z}_{p_0}$ (see Theorem 5.5 in Chapter 3), hence if $H \cong \mathbb{Z}_{p_0/p'}$, where p' divides p_0, then $\Gamma/H \cong \mathbb{Z}_{p'}$ and $\Pi(H) \cong \mathbb{Z}_2 \times \mathbb{Z}_{p'}$, if p' is even and $\mathbb{Z}_{2p'}$, if p' is odd. We have proved, in Proposition 7.3 of Chapter 3, that if

$$[F]_{S^1} = \sum d_m [\eta_m]_{S^1},$$

then

$$P_*[F] = \sum d'_{p_0/p'} [\eta'_{p_0/p'}]_{\mathbb{Z}_{p_0}},$$

where

$$d'_{p_0/p'} = \sum n_j \Big(\sum_{k \geq 0} d_{m_j p_0/p' + kp_0} \Big),$$

with $|n_j|$ odd, $n_j m_j \equiv 1$, modulo p', and $1 \leq m_j < p'$ with m_j and p' relatively prime (one may take $|n_j| < p'$).

The number $d'_{p_0/p'}$ is in $\mathbb{Z}_{p'}$ if p' is even and in $\mathbb{Z}_{2p'}$ if p' is odd. For $p' = 1$, corresponding to $H = \mathbb{Z}_{p_0}$, one has $d'_{p_0} = \sum_{k\geq 0} d_{kp_0}$ in $\mathbb{Z}_2$.

Here, $d_m = 0$ except for $m = q$ and $m = q/2$, if q is even, where one has

$$\begin{aligned} d_q &= \eta(-1)^{\sigma_+} \\ d_{q/2} &= -\eta(-1)^{\sigma_+}(1-(-1)^{\sigma_-})/2, \text{ if } q \text{ is even,} \end{aligned}$$

where $\eta = -\operatorname{Sign}(e_{k-1}\cdot Z(0)) \operatorname{Sign} \int_0^{2n} g_\nu \cdot Z(t)dt$ and σ_+ is the sum of the algebraic multiplicities of real eigenvalues of $\Phi(2\pi/q)$ which are larger than or equal to 1, while σ_- is the corresponding sum for real eigenvalues of $\Phi(2\pi/q)$ which are less than -1: see Proposition 2.7.

Then, if $p_0 = p_1 p$ and $q = q_1 p$ with p_1 and q_1 relatively prime, one has $m_j p_0/p' = q$ if and only if $m_j p_1 = q_1 p'$, that is $p' = kp_1$ and $m_j = kq_1$, but, since p' and m_j are also relatively prime, one has $p' = p_1$ and $m_j = q_1$ with $p_0/p' = p$. In this case $n_j = n$ is such that $nq_1 = 1 + kp_1$, with $|n| < p_1$. On the other hand, if q is even, one has $m_j p_0/p' = q/2$, if and only if $2m_j p_1 = q_1 p'$, that is $p' = kp_1$ and $2m_j = kq_1$ which implies $k = 1$ or 2.

If $k = 1$, then $p' = p_1, 2m_j = q_1$ and $p_0/p' = p$ and $n_j = n'$ is such that $n'q_1/2 = 1 + k'p_1$. While, if $k = 2$, then $p' = 2p_1, m_j = q_1$ which must hold since p' is even and $p_0/p' = p/2$, and $n'q_1 = 1 + 2k'p_1$ (in this case, since q is even but q_1 is odd, one has that p is even).

Thus, if q is odd, one has $d'_p = nd_q$ with $nq_1 \equiv 1, [p_1]$. While, if q is even, then either p_0 is odd, which implies that the only possibility for $d_{q/2}$ is $k = 1$ and $d'_p = nd_q + n'd_{q/2}$, or p_0 is even. If p_0 is even and q_1 is odd, then $k = 2$ and $d'_p = nd_q, d'_{p/2} = n'd_{q/2}$, while if q_1 is even, then $k = 1$ and $d'_p = nd_q + n'd_{q/2}$.

Since $nq_1 = 1 + kp'$ and $n'm_j = 1 + k'p'$ the numbers n and n' are non-zero modulo p', hence the only case where the new invariants may be 0 is when $k = 1$ and $d'_p = 0$, modulo $2p_1$. This is possible only if σ_- is odd and $n' - n = 2\tilde{k}p_1$. But then, multiplying by q_1, one would have

$$(2\tilde{k}q_1 - 2k' + k)p_1 = 1.$$

Hence, $p_1 = 1$, in which case $n = n' = 1, k = q_1 - 1, k' = q_1/2 - 1$ and $\tilde{k} = 0$ and q_1 is even.

The global continuum going through (ν_0, X_0) is then given by Theorem 5.1 of Chapter 2, by taking any bounded S^1-invariant set Ω in $\mathbb{R} \times W$ which intersects the slice $\tau = 0$ in a neighborhood of X_0, where X_0 is the only solution. Since $d'_p \neq 0$ one has that the solutions are in $W^{\mathbb{Z}_p}$, i.e., they are $2\pi/p$-periodic. □

Remark 2.5. It would be quite interesting to construct an autonomous system with an isolated π-periodic solution, with σ_- odd, such that with a 2π-periodic perturbation one looses the solution at some value of the parameter. Since $\det \Phi(t) =$

$\exp(\int_0^t \text{Trace}\, A(s)ds)$ is positive, σ_- odd implies that there is at least 3 eigenvalues of $\Phi(\pi)$: 1 and two negative, one less than -1 and one between -1 and 0, hence the system must be at least three-dimensional. In this case one would have $q = 2$ and $p_0 = 1$.

On Fourier series one may take the system, from $\mathbb{R} \times \mathbb{C}^2$ into $\mathbb{C}^2$,

$$f(\nu, z_2, z_1) = ((\nu + i(|z_2| - 1))z_2, z_2\bar{z}_1)$$

where S^1 acts on z_1 as $e^{i\varphi}$ and on z_2 as $e^{2i\varphi}$. If $z_2 \neq 0$, the only solution is $\nu = 0$, $z_1 = 0$, $|z_2| = 1$. If $H \cong \mathbb{Z}_2$ is the isotropy subgroup of z_2, the index of $f(\nu, z_2, z_1)^H$ at $(0, 1, 0)$ is 1 while the index of $f(\nu, z_2, z_1)$, for z_2 real and positive at $(0, 1, 0)$ is -1. Hence, from Theorem 2.1, one has $d_2 = 1$ and $d_1 = -1$, the situation of the last proposition, where Ω_0 is a small neighborhood of the orbit.

If one adds the parameter τ one may look at the open bounded set

$$\Omega = \{(\tau, \nu, z_2, z_1), |\nu| < 1, |z_1| < 1, |\tau| < 2, 1/4 - \tau^2 < |z_2| < 4 - \tau^2\}$$

which, in the (τ, z_2) space and $\tau \geq 0$ is the region between two paraboloids, the first of vertex $(\tau = 1/2, z_2 = 0)$ and basis $(\tau = 0, |z_2| = 1/4)$ and the second of vertex $(\tau = 2, z_2 = 0)$ and basis $(\tau = 0, |z_2| = 4)$. Now, if $\lambda_\tau = \nu + i(|z_2| - 1 + \tau^2)$, the non-equivariant map

$$(\lambda_\tau z_2 + \tau z_1, z_2\bar{z}_1 - \tau\bar{\lambda}_\tau)$$

which, after conjugation of the second component, can be written as

$$\begin{pmatrix} z_2 & \tau \\ -\tau & \bar{z}_2 \end{pmatrix} \begin{pmatrix} \lambda_\tau \\ z_1 \end{pmatrix}$$

has the only solution, for $\tau \neq 0$, the point $\nu = 0$, $z_1 = 0$, $|z_2| = 1 - \tau^2$, which disappears at $\tau = 1$, without touching the boundary of Ω.

Example 2.11 (Twisted orbits). Consider the problem of finding 2π-periodic solutions to the problem

$$\frac{dX}{dt} = g(X, \nu), \quad X \text{ in } \mathbb{R}^N,$$

where ν could be the frequency and g is equivariant with respect to the abelian group Γ_0. The preceding examples were particular cases with Γ_0 trivial. If $(\nu_0, X_0(t))$ is a solution, we have seen, in § 9 of Chapter 1, that $X_0(t)$ may be a time-stationary solution, or a rotating wave or a truly time periodic solution. The first two cases correspond to a Hopf bifurcation and will be studied in the last section of this chapter or, if one fixes ν, to invariants which involve only the stationary part, i.e., for the equation $g(X, \nu) = 0$. Thus, let us assume that $X_0(t)$ is a truly periodic solution, that is $X_0(t)$ is non-constant and $(2\pi/p)$-periodic, with isotropy $H = \mathbb{Z}_p \times H_0$, where $\dim \Gamma_0/H_0 = 0$. (The case of more parameters and higher dimensional orbit will be studied, for orthogonal maps, in the next section). Assume that $X_0(t)$ is $\{e\}$-hyperbolic, i.e., $X_0'(t)$ generates the kernel of the linearization $X' - A(t)X$, with $A(t) = Dg(X_0(t), \nu_0)$ and the equation $X' - A(t)X = g_\nu(X_0(t), \nu_0)$ has no 2π-periodic solution. Then, according to Theorem 2.3, the only relevant isotropy subgroups of $\Gamma = S^1 \times \Gamma_0$ are H and K's, with $K < H$ and $H/K \cong \mathbb{Z}_2$. Since $\dim \Gamma/H = 1$, the torus part of H (and K) is the torus part of H_0 and that of Γ_0: see Lemma 9.2 in Chapter 1. If T^k is this torus part, all the relevant information will be given by orbits which lie in $V_0 \equiv (\mathbb{R}^N)^{T^k}$ for all time. We have seen, in Lemma 9.4 of Chapter 1, that V^H is the space of all $2\pi/p$-periodic functions with $X(t)$ in $V_0^{H_0}$ for all t and $X(t) = \gamma_0 X(t + 2\pi/q)$, where $\gamma_0^{q_0}$ is in H_0 and $q = pq_0$. The element γ_0 of Γ_0 and the integer q are determined by $X_0(t)$.

Furthermore, for each K_j, with $H/K_j \cong \mathbb{Z}_2$, one has a subgroup K_{0j} of H_0 such that $H_0/K_{0j} \cong \mathbb{Z}_2$ or $H_0 = K_0$, with $V_j = V_0^{K_{0j}} = V_j^+ \oplus V_j^-$ where $\gamma_0^{q_0}$ acts as $\pm\,\mathrm{Id}$ on $V_j^\pm$. We have seen, in Lemma 9.4 and Remark 9.4 of Chapter 1, that $V_0^{H_0} = V_j^+$ if and only if $V_j^- \neq \{0\}$ and that the elements of V^{K_j} are those 2π-periodic functions $X(t)$, with $X(t)$ in V_j for all t, and $X(t) = \gamma_0^2 X(t + 4\pi/q)$.

Now, the matrix is H-equivariant and since $g(X, \nu_0)$ is Γ_0-equivariant and $X_0(t)$ is in $V_0^{H_0}$ for all t, one has, for any δ in Γ_0,

$$\delta Dg(X_0, \nu_0) = Dg(\delta X_0, \nu_0)\delta.$$

In particular, for δ in H_0, the matrix $A(t)$ is H_0-equivariant for each t. Thus, on V_j one has

$$A(t) = \begin{pmatrix} A_0(t) & 0 \\ 0 & A_j(t) \end{pmatrix},$$

where A_0 corresponds to $V_0^{H_0}$ and $A_j(t)$ to V_j^- or to the complement of $V_0^{H_0}$ in V_j^+ (if $K_{0j} = H_0$, the matrix A_j is not present). If $\Phi(t)$ is the fundamental matrix for the problem $X' - A(t)X$, one has $\Phi(t) = \mathrm{diag}(\Phi_0(t), \ldots, \Phi_j(t), \ldots)$ on the decomposition of $\mathbb{R}^N$ into irreducible representations of H_0.

Lemma 2.4. *One has the following relations*

$$\gamma_0 A(t + 2\pi/q) = A(t)\gamma_0, \quad \gamma_0 \Phi(t + 2\pi/q) = \Phi(t)\gamma_0\Phi(2\pi/q).$$

In particular, for any integer s,

$$\Phi(2\pi s/q) = \gamma_0^{-s}(\gamma_0\Phi(2\pi/q))^s.$$

Proof. For any γ in Γ one has $\gamma Dg(X, \nu_0) = Dg(\gamma X, \nu_0)\gamma$. In particular, for $\gamma_0 X_0(t + 2\pi/q) = X_0(t)$, one obtains $\gamma_0 A(t + 2\pi/q) = A(t)\gamma_0$. Then, $\Phi'(t + 2\pi/q) = A(t + 2\pi/q)\Phi(t + 2\pi/q) = \gamma_0^{-1}A(t)\gamma_0\Phi(t + 2\pi/q)$, that is $\gamma_0\Phi(t + 2\pi/q)$ is also a fundamental matrix and as such, one has

$$\gamma_0\Phi(t + 2\pi/q) = \Phi(t)C, \text{ with } C = \gamma_0\Phi(2\pi/q).$$

Then, $\gamma_0^s\Phi(2\pi s/q) = \gamma_0^{s-1}(2\pi(s-1)/q)\gamma_0\Phi(2\pi/q) = (\gamma_0\Phi(2\pi/q))^s$ for $s > 0$ and $\gamma_0 = \Phi(-2\pi/q)\gamma_0\Phi(2\pi/q)$ gives the result for $s < 0$. □

Proposition 2.11. *Let* $(\nu_0, X_0(t))$ *be a hyperbolic solution of* $X' = g(X, \nu)$ *and define* $\eta = \text{Sign}(e_{k-1} \cdot Z(0)) \, \text{Sign}(g_\nu(X_0, \nu_0), Z(t))_{L^2}$, *as in Proposition* 2.7, *where k is the algebraic multiplicity of* 1 *as eigenvalue of* $\Phi_0(2\pi/p)$. *If* $g(X, \nu) = g(X)/\nu$ *then* $\eta = -1$.

Let $\sigma_j^\pm$ *be the number of real eigenvalues, counted with algebraic multiplicity, of* $\gamma_0\Phi_j(2\pi/q)$ *which are larger than* 1, *for* σ_j^+, *or less than* -1, *for* σ_j^-, *where* $j = 0$ *for H and* $j \geq 1$ *for each* K_j *with* $H/K_j \cong \mathbb{Z}_2$. *Then*

$$i_H = (-1)^{k-1}\eta(-1)^{\sigma_0^+}$$

$$i_H i_{K_j} = \begin{cases} (-1)^{\sigma_j^+} & \text{if } q \text{ is odd and} V_j^- = \{0\} \\ (-1)^{\sigma_j^-} & \text{if } q \text{ is odd and} V_j^- \neq \{0\} \\ (-1)^{\sigma_0^-} & \text{if } q \text{ is even, } p \text{ is odd and } V_j^- \neq \{0\} \\ (-1)^{\sigma_0^- + \sigma_j^+ + \sigma_j^-} & \text{if } q \text{ is even, } p \text{ is odd and } V_j^- = \{0\} \text{ or } p \text{ is even.} \end{cases}$$

Proof. Recall first that X_0' is the only generator of $\ker(X' - A(t)X)$ and that $g_\nu(X_0, \nu_0)$ does not belong to the range of this operator. In particular, k depends only on $\Phi(2\pi/p)$ and it is the algebraic multiplicity of $X' - A(t)X$ on $H^1(S^1)$. Furthermore, since the operator $X' - A(t)X + \lambda X$ and $\tilde{K}$ are H-equivariant, the arguments of Propositions 2.6 and 2.7 remain valid, that is the index i_K is given by

$$i_K = (-1)^{k-1}\eta(-1)^{\sigma_K},$$

where σ_K is the number of real eigenvalues $\lambda > 0$, including algebraic multiplicity, of $X' - A(t)X + \lambda X = 0$ in V^K. But, from Appendix B, X satisfies this equation in $H^1(S^1)$ if and only if $X(t) = e^{-\lambda t}\Phi(t)W$, with W in $\ker(\Phi(2\pi) - e^{2\pi\lambda}I)$. Thus, $X(t)$ will be in V^H if and only if $X(t)$ lies in $V_0^{H_0}$ and $\gamma_0 X(t + 2\pi/q) = X(t)$, while $X(t)$ will be in V^{K_j} if and only if $X(t)$ lies in $V_0^{K_j} = V_j$ and $\gamma_0^2 X(t + 4\pi/q) = X(t)$. Thus, for V^H, one needs

$$e^{-\lambda 2\pi/q}\Phi(t)\gamma_0\Phi(2\pi/q)W = \Phi(t)W,$$

and, since $\Phi(t)$ is invertible and $\Phi(t)|_{V_0^{H_0}} = \Phi_0(t)$, .

$$\gamma_0\Phi_0(2\pi/q)W = e^{2\pi\lambda/q}W.$$

Conversely, if W satisfies this last relation then

$$\gamma_0^{q_0}\Phi_0(2\pi/p)W = (\gamma_0\Phi_0(2\pi/q))^{q_0}W = e^{2\pi\lambda/p}W$$

and, since $\gamma_0^{q_0} = \text{Id}$ on $V_0^{H_0}$, one obtains $\Phi(2\pi)W = e^{2\pi\lambda}W$.

For the generalized kernel, one has that, if $\left(\frac{d}{dt} - A + \lambda I\right)^\alpha X = 0$, then $X(t) = e^{-\lambda t}\Phi(t)\sum_0^{\alpha-1} W_l t^l/l!$, with W_l in $\ker(\Phi(2\pi) - e^{2\pi\lambda}I)^{\alpha-l}$ uniquely determined by W_0. The relation $\gamma_0 X(t + 2\pi/q) = X(t)$ leads to

$$B^{-1}\Big(\sum_{l=0}^{\alpha-1} W_l(t + 2\pi/q)^l/l!\Big) = \sum_{l=0}^{\alpha-1} W_l t^l,$$

where $B^{-1} = e^{-2\pi\lambda/q}\gamma_0\Phi(2\pi/q)$. This polynomial equality is satisfied if and only if all k-derivatives at $t = 0$ are equal, that is

$$B^{-1}\Big(\sum_{l=k}^{\alpha-1} W_l(2\pi/q)^{l-k}/(l-k)!\Big) = W_k.$$

But these are the relations given in Appendix B, with 2π replaced by $2\pi/q$. Thus, $(B - I)^k W_{\alpha-k} = 0$, i.e., W_l is in $\ker(\gamma_0\Phi(2\pi/q) - e^{2\pi\lambda/q}I)^{\alpha-l}$ and is completely determined by W_0. The converse is clear, hence the algebraic multiplicity of $\frac{d}{dt} - A + \lambda I$ on V^H is that of $e^{2\pi\lambda/q}$ as eigenvalue of $\gamma_0\Phi(2\pi/q)$. In particular, $\sigma_H = \sigma_0^+$.

For K_j, the relation $\gamma_0^2 X(t + 4\pi/q) = X(t)$ leads to

$$\gamma_0^2\Phi(4\pi/q)W = (\gamma_0\Phi(2\pi/q))^2 W = e^{4\pi\lambda/q}W,$$

for any $X(t) = e^{-\lambda t}\Phi(t)W$, with W in $\ker(\Phi(2\pi) - e^{2\pi\lambda}I)$.

Conversely, if $\gamma_0\Phi(2\pi/q)W = \varepsilon e^{2\pi\lambda/q}W$, with $\varepsilon = \pm 1$, then $\gamma_0^q\Phi(2\pi)W = \varepsilon^q e^{2\pi\lambda}W$, for W in V_j. Writing $W = (W_0, W_j)$ on $V_0^{H_0} \oplus (V_0^{H_0})^\perp \cap V_j$ and $\Phi(2\pi) = \operatorname{diag}(\Phi_0(2\pi), \Phi_j(2\pi))$, one has

$$\Phi_0(2\pi)W_0 = \varepsilon^q e^{2\pi\lambda}W_0 \quad \text{and} \quad (\gamma_0^{q_0})^p\Phi_j(2\pi)W_j = \varepsilon^q e^{2\pi\lambda}W_j,$$

where $\gamma_0^{q_0} = \mathrm{Id}$ if $V_j^- = \{0\}$ and $\gamma_0^{q_0} = -\mathrm{Id}$ if $V_j^- \neq \{0\}$.

Hence, for Φ_0, one has $\varepsilon = \pm 1$ if q is even, while only $\varepsilon = 1$ is possible if q is odd. For Φ_j and $V_j^- = \{0\}$, then $\gamma_0^{q_0} = \mathrm{Id}$ and ε is as above, while if $V_j^- \neq \{0\}$ then $\gamma_0^{q_0} = -\mathrm{Id}$. In this case, if p is even, then $\varepsilon = \pm 1$ (q is also even). If p is odd and q even, then $W_j = 0$ (in fact, we have seen in Lemma 2.4 of Chapter 1 that, in this case, $X(t)$ is in $V_j^+ = V_0^{H_0}$). Finally, if q is odd, then $\varepsilon = -1$.

The argument for the generalized kernel is then as before, with $(\gamma_0\Phi(2\pi/q))^2 - e^{4\pi\lambda/q}I$, with no further restriction. This proves the proposition. Note that one may have $H_0 = K_0$: in this case $H/K \cong \mathbb{Z}_2$ only if q is even (if q is odd then $X(t)$ is in V^H) and the contribution to i_K is $(-1)^{\sigma_0^-}$. □

Remark 2.6. On may also look at 2π-periodic solutions, in V_j for all t, of the equation

$$Y' - A(t)Y + \lambda Y = 0, \quad \lambda > 0,$$

for functions which satisfy the relations

$$Y(t+2\pi/p) = aY(t), \quad \gamma_0 Y(t+2\pi/q) = bY(t),$$

where $a = \pm 1, b = \pm 1$.

In fact, $b = 1$ corresponds to $y = X_+$, in Lemma 9.5 of Chapter 1, with components in V_j^+ if $a = 1$ or in V_j^- if $a = -1$. While, if $b = -1$ then $Y(t) = X_-(t)$ with components in V_j^+, if $a = 1$ and q_0 even or if $a = -1$ and q_0 odd, or in V_j^-, if $a = -1$ and q_0 even or $a = 1$ and q_0 odd.

Now, the requirement that $Y(t)$ is 2π-periodic implies that $a^p = 1$ and $\gamma_0^{q_0} Y(t + 2\pi/p) = \gamma_0^{q_0} aY(t) = b^{q_0} Y(t)$, that is $a\gamma_0^{q_0} = b^{q_0} I$, where $\gamma_0^{q_0} = I$ on V_j^+ and $\gamma^{q_0} = -I$ on V_j^- and $V_j^- \neq \{0\}$ if and only if $V_j^+ = V_0^{H_0}$.

Thus, by writing $Y(t) = e^{-\lambda t}\Phi(t)W$, with $W = (W_0, W_j)$, W_0 in $V_0^{H_0}$ and W_j in $(V_0^{H_0})^\perp \cap V_j$ and $\Phi(t) = \operatorname{diag}(\Phi_0, \Phi_j)$, one has the following spectral problem: find $\lambda > 0$, $W \neq 0$, such that

$$\begin{aligned} (\gamma_0\Phi_0)^{q_0} W_0 &= a e^{2\pi\lambda/p} W_0 &\quad \text{and} \quad \gamma_0\Phi_0 W_0 &= b e^{2\pi\lambda/q} W_0, \\ (\gamma_0\Phi_0)^{q_0} W_j &= a e^{2\pi\lambda/p} \gamma_0^{q_0} W_j &\quad \text{and} \quad \gamma_0\Phi_0 W_j &= b e^{2\pi\lambda/q} W_j, \end{aligned}$$

with the restrictions $a^p = 1$, $(a - b^{q_0})W_0 = 0$, $(a - b^{q_0})W_j = 0$ if $V_j^- = \{0\}$ or $(a + b^{q_0})W_j = 0$ if $V_j^- \neq \{0\}$.

Hence, one has the following cases:

1. $K = H$, i.e., $a = b = 1$, $W_j = 0$, with a contribution of σ_0^+.

2. $H/K \cong \mathbb{Z}_2$ and p odd, then $a = 1$. If q_0 is odd, one has a contribution from W_0 (only if $b = 1$) of σ_0^+ and from W_j of σ_j^+ (if $V_j^- = \{0\}$ and $b = 1$) or of σ_j^- (if $V_j^- \neq \{0\}$ and $b = -1$). While, if q_0 is even, one has a contribution from W_0 of $\sigma_0^+ + \sigma_0^-$ (for $b = \pm 1$), and from W_j of $\sigma_j^+ + \sigma_j^-$ only if $V_j^- = \{0\}$.

3. $H/K \cong \mathbb{Z}_2$ and p even. If q_0 is odd, one has a contribution from W_0 only if a and b have the same sign. Thus, if $V_j^- = \{0\}$ one has $\sigma_0^+ + \sigma_j^+$ (for $a = b = 1$) and $\sigma_0^- + \sigma_j^-$ (for $a = b = -1$) for a total of $\sigma_0^+ + \sigma_0^- + \sigma_j^+ + \sigma_j^-$. If $V_j^- \neq \{0\}$, then a and b have the same sign for W_0 and opposite signs for W_j, giving $\sigma_0^+ + \sigma_0^-$ for W_0 and $\sigma_j^+ + \sigma_j^-$ for W_j with the same total as above. While, if q_0 is even, then if $V_j^- = \{0\}$ one needs $a = 1, b = \pm 1$ and a total contribution of $\sigma_0^+ + \sigma_0^- + \sigma_j^+ + \sigma_j^-$. If $V_j^- \neq \{0\}$ then, for W_0 one has $a = 1$ and a contribution of $\sigma_0^+ + \sigma_0^-$, and for W_j one has $a = -1$ and a contribution of $\sigma_j^+ + \sigma_j^-$. Thus, for p even one gets $\sigma_0^+ + \sigma_0^- + \sigma_j^+ + \sigma_j^-$, confirming Proposition 2.11.

Remark 2.7. Since $\gamma_0^{2q_0} = \text{Id}$ on V_j, one has an action of the cyclic group generated by γ_0, which couples two variables x_l and y_l, giving them a complex structure and an action of the form $e^{2\pi i a_l/2q_0}(x_l + iy_l)$, where a_l is even on V_j^+ and odd on V_j^-. Now, it is important to note that $A(t)$ does not preserve this complex structure, unless it is γ_0-equivariant. Thus, since $A(t)$ is a real matrix, the action of γ_0 on the couple (x_l, y_l) has to be represented by the rotation $R_{a_l/2q_0}$. Then, since $A(t)$ is $2\pi/p$-periodic, one has the Fourier series expansion for $A(t)$:

$$A(t) = \sum A_m e^{imt},$$

where m is a multiple of p. The relation $\gamma_0 A(t + 2\pi/q)\gamma_0^{-1} = A(t)$ leads to $\gamma_0 A_m e^{2\pi i m/q}\gamma_0^{-1} = A_m$ and, if A_m is decomposed in 2×2 matrices A_m^{kl} corresponding to the couples (x_l, y_l) and on the coordinates (x_k, y_k), one obtains

$$A_m^{kl} = e^{2\pi i m/q} R_{a_k/2q_0} A_m^{kl} R_{-a_l/2q_0}.$$

If, for some (k, l) and fixed m, one has that $\det A_m^{kl}$ is non-zero, then one has that $2m/q$ is an integer. If this integer is even, then A_m is γ_0-equivariant and, as such, has a block diagonal structure, in particular $A_m^{kl} = 0$ unless $a_k = a_l$. If this integer is odd, then A_m is γ_0^2-equivariant and $A_m^{kl} = 0$ unless $a_k = a_l$ or $a_l + q_0/2$: in fact, in this case one needs q_0 even, since $\gamma_0^{q_0} = \pm\,\text{Id}$ and $A_m = -\gamma_0 A_m \gamma_0^{-1}$.

This is the situation if A_m^{kl} has a complex structure, i.e., of the form $(a+ib)(x_l+iy_l)$, with $A_m^{kl} = \begin{pmatrix} a & -b \\ b & a \end{pmatrix}$.

However, consider $\gamma_0 = \begin{pmatrix} \cos 2\pi/3 & -\sin 2\pi/3 \\ \sin 2\pi/3 & \cos 2\pi/3 \end{pmatrix} = R_{2\pi/3}$ and

$$A(t) = \begin{pmatrix} \cos 2t & -\sin 2t \\ -\sin 2t & -\cos 2t \end{pmatrix} = \frac{1}{2}\begin{pmatrix} 1 & i \\ i & -1 \end{pmatrix} e^{2it} + \frac{1}{2}\begin{pmatrix} 1 & -i \\ -i & -1 \end{pmatrix} e^{-2it}.$$

Here, $p = 2$, $q_0 = 3$ and $q = 6$, with $2m/q = \pm 2/3$, which is not an integer.
It is easy to see that $\gamma_0 A(t + 2\pi/3) = A(t + \pi/3) = A(t)\gamma_0$ and $\det A_{\pm 2} = 0$.

4.3 Γ-Index for an orthogonal map

Orthogonal maps give a very rich structure for their orthogonal Γ-degrees, since, if $\Gamma = T^n \times \mathbb{Z}_{m_1} \times \cdots \times \mathbb{Z}_{m_s}$ acts on the finite dimensional space V, with infinitesimal generators $A_1 x, \ldots A_n x$ for the action of T^n, then according to Theorem 6.1 in Chapter 3, one has $\Pi^{\Gamma}_{\perp S^V}(S^V) \cong \mathbb{Z} \times \cdots \times \mathbb{Z}$, with one $\mathbb{Z}$ for each isotropy subgroup of Γ. Hence, if Γx_0 is an isolated zero-orbit of an orthogonal Γ-map $f(x)$, i.e., such that $f(x) \cdot A_j x = 0$ for $j = 1, \ldots, n$ and $f(\gamma x) = \gamma f(x)$ for all γ in Γ, then one should expect an index with many components. In this section we shall compute the orthogonal Γ-index at Γx_0, relating it to the spectral properties of its linearization. Furthermore, we shall apply these computations to the case of differential equations, in particular to Hamiltonian systems and to examples of spring-pendulum mechanical systems.

Let us *assume that* Γx_0 *is an isolated k-dimensional orbit*, with $f(x_0) = 0$ *and* H *the isotropy subgroup of* x_0, that is $\dim \Gamma/H = k$. Then, there are complex coordinates $z_1, \ldots, z_k$ with isotropy $H_0 > H$ and $|H_0/H| < \infty$ and z_j real and positive for x_0. One may choose an invariant neighborhood of Γx_0 such that $z_j \neq 0$ in it, that is $F(t, x) = (2t + 2\varphi(x) - 1, \tilde{f}(x))$ will be non-zero on the set given by $z_j = 0$ for each $j = 1, \ldots, k$. We shall assume that $A_1 x, \ldots, A_k x$ are the linearly independent vectors if x has its coordinates $z_1, \ldots, z_k$ non-zero. Then, according to Proposition 6.1 of Chapter 3, one has

$$[F]_\perp - \sum_{H_j \leq H_0} d_j [F_j]_\perp,$$

where, for $H_j > \underline{H}$ the torus part of H_0 and $B_k^j = B^{H_j} \cap \{z_1, \ldots, z_k \in \mathbb{R}^+\}$, the d_j's are given by

$$\deg\left(\left(F + \sum_1^k \lambda_l A_l x\right)^{H_i}; B_k^i\right) = \sum_{H_i \leq H_j \leq H_0} d_j |H_0/H_j|.$$

Choose the tubular neighborhood Ω of the orbit so small that if x_0 has a coordinate $x_j^0 \neq 0$, then x_j is non-zero in Ω and construct $\varphi(x)$ with value 1 whenever one of these coordinates x_j is 0. Thus, $F|_{V^K} \neq 0$ for any K which is not a subgroup of H. From Theorem 6.1 (2) of Chapter 3, this implies that the corresponding d_K is 0. Hence,

$$[F]_\perp = \sum_{H_j \leq H} d_j [F_j]_\perp.$$

Furthermore, if $\Omega_k^i = \Omega \cap B_k^i$, one has, for $H_j > \underline{H}$, that

$$\begin{aligned}\deg\left(\left(f + \sum_1^k \lambda_l A_l x\right)^{H_i}; \mathbb{R}^k \times \Omega_k^i\right) &= \sum_{H_i < H_j < H} d_j |H_0/H_j| \\ &= |H_0/H| \operatorname{Index}\left(\left(f + \sum_1^k \lambda_l A_l x\right)^{H_i}; (0, x_0)\right),\end{aligned}$$

since the orbit Γx_0 intersects Ω_k in $|H_0/H|$ points, all with the same index. Thus, the argument is, up to here, parallel to Theorem 2.2, except that one may have isotropy subgroups H_j with $\dim H/H_j > 0$.

If f is C^1 at x_0, let D denote $Df(x_0)$. Then, we have seen in Lemma 7.2 of Chapter 1, that D is H-equivariant (and as such it has a block diagonal structure on fixed-point subspaces of subgroups of H and that, for $K < \underline{H}$, then $D =$

$\text{diag}(D^H, D_{\perp \underline{H}}, D'_{\perp K})$, where $D'_{\perp K}$ is complex self-adjoint). Furthermore, $A_j x_0$ are in $\ker D$ and orthogonal to Range D. In particular, if $\dim \ker D = k$, then for $K < H$, the matrix $D_{\perp K}$ is invertible and the algebraic multiplicity of 0 as eigenvalue of 0 is k.

We shall use this information in the following result

Theorem 3.1. *Let Γx_0 be an isolated orbit of dimension k and isotropy H. Assume that* $\dim \ker Df(x_0) = k$. *Then, the orthogonal index is well defined and is equal to the product $i_\perp(f^{\underline{H}}(x_{\underline{H}}); x_0) i_\perp(Df_\perp(x_0)\bar{X}; 0)$, where $\underline{H}$ is the torus part of H and $Df_\perp(x_0)\bar{X}$ is the linearization on $(V^{\underline{H}})^\perp$, which is complex self-adjoint and H-orthogonal. One has*

$$i_\perp(f^{\underline{H}}) = d_H[F_H]_\perp + \sum_{H/H_i \cong \mathbb{Z}_2} d_{H_i}[F_{H_i}]_\perp + \sum_{H/\tilde{H}_i \cong \mathbb{Z}_2 \times \cdots \times \mathbb{Z}_2} d_{\tilde{H}_i}[F_{\tilde{H}_i}]_\perp,$$

with $d_H = \eta(-1)^{n_H}$, where n_H is the number of negative eigenvalues of $Df^H(x_0)$, and $\eta = (-1)^{k(k+1)/2}$ Sign det B, *where $B_{ij} = n_i^j$ is given by the i-th-coordinate of $A_j x_0$, for $i, j \le k$. The integer $d_{H_i} = d_H((-1)^{n_{H_i}} - 1)/2$, where n_{H_i} is the number of negative eigenvalues of $Df_\perp^{H_i}(x_0)$ and $d_{\tilde{H}_i}$ is completely determined by d_H and d_{H_j}. Furthermore,*

$$i_\perp(Df_\perp(x_0)\bar{X}) = [F_\Gamma]_\perp + \sum \eta_i n_i(K_i)[F_{K_i}]_\perp + \sum_{s=2}^{n-k} \Pi n_j(K_j)[F_{\cap K_j}]_\perp,$$

where K_i are the irreducible representations of H in $(V^{\underline{H}})^\perp$, i.e., $H/K_i \cong S^1$ and $Df_\perp(x_0)$, which is block-diagonal on these representations, has a complex Morse number $n(K_i)$. In the second sum one has the product $\eta_j n(K_{i_1}) \dots n(K_{i_s})$ with $\dim H/K_{i_1} \cap \cdots \cap K_{i_s} = s$ *and* $\eta_j = (-1)^{s(s-1)/2}$ Sign det B_j, *where B_j corresponds to the action of Γ on s variables, defining the generator F_{K_j}. For the first sum $\eta_i =$* Sign B_i. *Finally, $[F_{H_i}]_\perp [F_{K_j}]_\perp = [F_{H_i \cap K_j}]_\perp$. If one takes normalized generators $F^*_{H_i}$, $F^*_{K_j}$ then $\eta = (-1)^{k(k+1)/2}$ and $\eta_j = (-1)^{s(s-1)/2}$.*

Since the proof is involved, we shall break it up in several remarks and lemmas together with some illustrative examples.

Remark 3.1. (a) The generators F_H, F_{H_i}, F_{K_i} are those of Theorem 6.1 in Chapter 3 such that if $K < H$ and $\dim \Gamma/K = s \ge k$, with $A_1 x, \dots, A_s x$ linearly independent in V^K, then $F_K(x) + \sum_1^s \lambda_j A_j x$ has index 1 in the fundamental cell and degree, with respect to $\mathbb{R}^s \times B_s$, equal to $|K_0/K|$, where $K_0 \le H_0$ corresponds to the isotropy of s variables such that $\dim \Gamma/K_0 = s$. Thus, if $s = k$, one has $K_0 = H_0$.

The generator F_K has an orientation factor ε designed to compensate the sign of the determinant of the matrix given by $A_1, \dots, A_s$ on those s coordinates. If one uses the normalized generators F^*_K, of Proposition 6.2 in Chapter 3, then the terms Sign det B_j are not present.

(b) In the product, $f^{\underline{H}}(x_{\underline{H}})$ is Γ-orthogonal but $Df_\perp(x_0)\bar{X}$ is H-orthogonal only. The term $i_\perp(Df_\perp(x_0)\bar{X})$ has to be interpreted as the sum in the formula. We shall prove that $f(x)$ is Γ-orthogonally deformable to $(f^{\underline{H}}(x_{\underline{H}}), \Lambda\bar{X})$ where Λ has the same Morse numbers as $Df_\perp(x_0)$. Also, strictly speaking, we should have written $\sum \eta'_j n'_j [F'_{K_j}]_\perp$ for each irreducible representation of Γ on $(V^{\underline{H}})^\perp$, where $\eta'_j = \text{Sign } B_j$, where B_j is the non-zero coefficient of the action of T^n on a variable in that representation and F'_{K_j} is the associated generator, with Λ having n'_j terms with -1 on variables with the same action of Γ. Now, when one takes the product with $[F_{H_i}]_\perp$, one obtains, for the same $H_i \cap K_j = H_i \cap K$, subgroup of H, the sum $\sum \eta'_j n'_j [F_{H_i\cap K_j}]_\perp = \eta_{H_i\cap K}\big(\sum n'_j\big)[F_{H_i\cap K}]_\perp$, with $\sum n'_j = n(K)$ and $\eta_{H_i\cap K} = 1$ if one has chosen normalized generators. In this context it is important to recall that one has to assimilate complex conjugate representations of H, since, as seen in Remark 5.3 of Chapter 1, they are the same real representations. Thus, if H acts as $e^{ln\varphi}$ on $X = (x_1, \dots, x_s)$ and as $e^{-in\varphi}$ on $Y = (y_1, \dots, y_l)$, coordinates of $\bar{X}$, it may happen that $Df_\perp(x_0)$ has a block-diagonal form on (X, Y), with complex Morse numbers n_X and n_Y. Then the n_K of the theorem is $n_X + n_Y$. Note also that if the complex self-adjoint matrix $\mathcal{A} = A + iB$ has an eigenvector $X + iY$, with real eigenvalue λ, then the real matrix $\begin{pmatrix} A & -B \\ B & A \end{pmatrix}$ has eigenvectors (X, Y) and $(-Y, X)$, i.e., the real Morse number is twice the complex Morse number (see Lemma 7.2 of Chapter 1).

(c) The numbers $d_{\tilde{H}_i}$ are given in terms of d_H and d_{H_j}, as in the proof of Theorem 2.3. If H acts on $V^K \cap (V^H)^\perp$ as $\mathbb{Z}_m$, $m \geq 3$, then the algebraic multiplicity of any real eigenvalue is even: See Remark 2.2.

Lemma 3.1. *$i_\perp(f^{\underline{H}}(x_{\underline{H}}))$ is given by the formula of Theorem* 3.1.

Proof. As seen above one has

$$[F^{\underline{H}}]_\perp = \sum_{\underline{H}\leq H_j\leq H} d_j[F_j]_\perp,$$

where

$$i_{H_i} = \text{Index}((f + \sum_1^k \lambda_l A_l x)^{Hi}; (0, x_0)) = \sum_{H_i\leq H_j\leq H} d_j|H/H_j|.$$

Now, $(0, x_0)$ is $\underline{H}$-hyperbolic for the map $I - F \equiv f(x) + \sum_1^k \lambda_l A_l x$, since its linearization with respect to x at $(0, x_0)$ is $Df(x_0)$ with a k-dimensional kernel, while its linearization with respect to λ is the matrix A with columns given by $A_l x_0$ which are linearly independent and orthogonal to Range $Df(x_0)$: see Definition 2.2. Hence, from Theorem 2.4, one has

$$i_H = \eta(-1)^{n'_H}, \quad i_K = i_H(-1)^{n_K},$$

for $\underline{H} \leq K \leq H$ and n'_H is the number of eigenvalues of $F_x^H(0, x_0) = I - Df^H(x_0)$ which are larger than or equal to 1, that is the number of non-positive eigenvalues of $Df^H(x_0)$, while n_K corresponds to eigenvalues of $F_x^{\perp K} = I - Df_\perp^K(x_0)$, and $\eta = \pm 1$ is an orientation factor which depends on the matrix A. According to Theorem 2.4, one has

$$\eta = (-1)^{k(k+1)/2} \text{ Sign det } \Lambda \text{ Sign det } B,$$

where Λ is given in Lemma 2.3, as the projection of F_λ on $\ker(I - F_x)^T$. Here, since the algebraic multiplicity of 0 as eigenvalue of D is k, one has $\ker(I - F_x)^T = \ker D$, generated by $A_1 x_0, \ldots, A_k x_0$. Furthermore, $F_{\lambda_j} = -A_j x_0$, thus, $\Lambda = -I$. Then, Sign det $\Lambda = (-1)^k$ and, since $n'_H = n_H + k$, one has

$$i_H = (-1)^{k(k+1)/2} \text{ Sign det } B(-1)^{n_H}.$$

Thus, Corollary 2.2 gives the result. □

Before computing d_K for K with $\dim H/K > 0$, let us look at some examples.

Example 3.1. Let $\mathbb{Z}_2$ act on y as antipodal map and S^1 act on z as $e^{i\varphi}$. Then, the map $f(y, z) = (-y, (|z|^2 - 1)z)$ is Γ-orthogonal with respect to $\Gamma = \mathbb{Z}_2 \times S^1$. One has the isolated zero-orbit, $y = 0$, $|z| = 1$ with $H = \mathbb{Z}_2$ and $K = \{e\}$. Furthermore,

$$Df(y = 0, z = 1) = \begin{pmatrix} -1 & 0 & 0 \\ 0 & 2 & 0 \\ 0 & 0 & 0 \end{pmatrix}$$

with $\begin{pmatrix} 2 & 0 \\ 0 & 0 \end{pmatrix}$ corresponding to $Df^H(0, 1)$. Hence, $n_H = 0$, $k = 1$ and $B = 1$. Thus, $i_H = -1$ and $i_K = 1$, which coincides with the index of the map $(\lambda, x) \to (|x|^2 - 1)x + i\lambda x$. Note that $f = \nabla\Phi$, with $\Phi(y, z) = -y^2/2 + (|z|^2/2 - 1)|z|^2/2$.

Example 3.2. Let $\Gamma = S^1$ act on (z_1, z_2) by $(e^{i\varphi} z_1, e^{2i\varphi} z_2)$ and let

$$f(z) = f_0(z) - \lambda(z) Az,$$

with $f_0(z) = (z_2 \bar{z}_1, (|z_2|^2 - 1)z_2)$, $Az = (iz_1, 2iz_2)$ and $\lambda(z) = f_0(z) \cdot Az/|Az|^2$. Recall, from Definition 7.1 of Chapter 1, that the scalar product is the real scalar product, i.e., if $f = a + ib$ and $g = c + id$, then $f \cdot g = ac + bd = \mathrm{Re}(f\bar{g})$. Hence, $\lambda(z)$ is real and $\lambda(z) = i(\bar{z}_2 z_1^2 - z_2 \bar{z}_1^2)/(|z_1|^2 + 4|z_2|^2)$.

Clearly $f(z)$ is Γ-orthogonal and the zeros of $f(z)$ are $(z_1, 0)$ and $(0, |z_2| = 1)$. This second set is an isolated orbit for which $Df^H(0, 1) = \begin{pmatrix} 2 & 0 \\ 0 & 0 \end{pmatrix}$. It is then easy to compute the index of $f(z) + \lambda Az$ at $\lambda = 0$, $z_1 = 0$, $z_2 = 1$, for z_2 in $\mathbb{R}^+$, by deforming $\lambda(z)$ to 0, getting $i_H = -1$ and $i_K = 1$ with $H = \mathbb{Z}_2$ and $K = \{e\}$. Here, $k = 1$, $B = 2$, $n_H = 0$ and $n_K = 1$ since $Df_\perp(0, 1) = \begin{pmatrix} 1 & 0 \\ 0 & -1 \end{pmatrix}$ corresponding to $\bar{z}_1$.

Example 3.3. If f is in normal form, then $f^{\perp}(x_H, x_\perp) = x_\perp$ for $|x_\perp| < \varepsilon$, then $Df^{\perp}(x_0, 0) = \mathrm{Id}$ and $i_K = i_H$ for any $K < H$. In this case $d_H = i_H$ and $d_K = 0$ for $\underline{H} \leq K < H$. By choosing Ω contained in the set where $|x_\perp| < \varepsilon$, one has that $f^{\perp}$ acts as a suspension and $\mathrm{Index}_\perp(f; x_0) = d_H[F_H]_\perp$.

We may now go on to the next step of the proof of Theorem 3.1.

Lemma 3.2. *In Ω, the map $f(x)$ is Γ-orthogonally deformable to $(f^{\underline{H}}(x_{\underline{H}}), \Lambda\bar{X})$, where Λ is the diagonalization of $Df_\perp(x_0)|_{(V^{\underline{H}})^\perp}$, hence with the same Morse number.*

Proof. Recall that one may reparametrize the torus T^n in Γ in such a way that $\underline{H}$ corresponds to $\psi_1 = \cdots = \psi_k \equiv 0, [2\pi]$, and that $\psi_{k+1}, \ldots, \psi_n$ act trivially on $V^{\underline{H}}$: see Lemma 2.4 and Remark 2.1 in Chapter 1. Take then the orthonormal $\tilde{A}_j(x)$ constructed in Theorem 7.1 of Chapter 1, starting the orthogonalization process from $j = n$, i.e., in reverse order. Then, for $j > k$, one has $A_j x_{\underline{H}} = 0$ and $\tilde{A}_j(x)$ is in $(V^{\underline{H}})^\perp$ and orthogonal to $D_{x_\perp} f_\perp(x_H)x_\perp$ since this matrix is H-orthogonal: the infinitesimal generators for the action of H are the derivatives with respect to $\psi_{k+1}, \ldots, \psi_n$. For $j = 1, \ldots, k$ define

$$\lambda_j(x) = D_{x_\perp} f_\perp(x_H)x_\perp \cdot \tilde{A}_j(x),$$

and define, as in Theorem 7.1 of Chapter 1, the Γ-equivariant map

$$\tilde{f}_\perp(x) = D_{x_\perp} f_\perp(x_H)x_\perp - \sum_1^k \lambda_j(x)\tilde{A}_j(x).$$

The Γ-equivariance follows from Property 3.3 of Chapter 1 and the fact that x_H is left as a variable. Furthermore, $\tilde{f}_\perp(x)$ is Γ-orthogonal. From Lemma 7.5 of Chapter 1, one has, for $j \leq k$, that $\tilde{A}_j(x) = \tilde{A}_j(x_H) + O(x_\perp)$ and, since $\tilde{A}_j(x_H)$ is in V^H, while the matrix $D_{x_\perp} f_\perp(x_H)$ maps into $(V^H)^\perp$, one has that $\lambda_j(x) = 0(|x_\perp|^2)$.

Consider then the Γ-orthogonal homotopy

$$(f^H(x_H, \tau x_\perp), \tau f_\perp(x_H, \tau x_\perp)) + (1 - \tau^2)\tilde{f}_\perp(x),$$

on the tubular neighborhood of the orbit Γx_0, which may be taken of the form $\Omega = \{(x_H, x_\perp) : \mathrm{dist}(x_H, \Gamma x_0) < \eta, |x_\perp| < \varepsilon\}$. Since the homotopy reduces, for $x_\perp = 0$, to $(f^H(x_H, 0), 0)$ which is non-zero on the boundary of Ω (since Γx_0 is isolated), one may choose ε so small that the second component is non-zero for $|x_\perp| = \varepsilon$: in fact, by linearizing $\tau f_\perp(x_H, \tau x_\perp)$, one has the approximation $D_{x_\perp} f_\perp(x_H)x_\perp + \tau^2 o(x_\perp) + (1 - \tau^2)0(|x_\perp|^3)$ and since $D_{x_\perp} f_\perp(\Gamma x_0)$ is invertible, one may choose η so small that $D_{x_\perp} f_\perp(x_H)$ is invertible in Ω.

Now, $D_{x_\perp} f_\perp(x_H)$ has the form $\mathrm{diag}(B(x_H), \bar{B}(x_H))$, where $\bar{B}$ is complex self-adjoint and has a block diagonal structure on the equivalent irreducible representations of H. On each block, $\bar{B}(x_H)$ is similar to a diagonal real matrix $\Lambda(x_H)$ with a well-defined Morse index n_K which is constant in Ω^H, since $\bar{B}(x_H)$ is invertible there. Furthermore, if v is an eigenvector of $\bar{B}(x_H)$ then, from Property 3.3 of Chapter 1,

$$\bar{B}(\gamma x_H)\gamma = \gamma B(x_H)$$

and γv is an eigenvector of $\bar{B}(\gamma x_H)$ with the same eigenvalue. Hence, if $\bar{B}(x_H) = U(x_H)\ \Lambda(x_H)U^*(x_H)$, with U unitary, then $U(\gamma x_H) = \gamma U(x_H)\gamma^*$, $\Lambda(\gamma x_H) = \gamma\Lambda(x_H)\gamma^* = \Lambda(x_H)$ will diagonalize $\bar{B}(\gamma x_H)$ since Λ and γ are diagonal, hence commute. Note that $U(x_H)$ is continuous in x_H if the eigenvalues of $\bar{B}(x_0)$ are simple. In general, for x_H close to x_0 and in $\mathcal{C}_H$, the fundamental cell for H, define $\tilde{U}(x_H) = U(x_0)$ and $\tilde{U}(\gamma x_H) = \gamma U(x_0)\gamma^*$ (γx_H is not in $\mathcal{C}_H$ by construction). Let $\tilde{\Lambda}(x_H) = \tilde{U}^*(x_H)\bar{B}(x_H)\tilde{U}(x_H)$, and $\tilde{\Lambda}(\gamma x_H) = \gamma\tilde{\Lambda}(x_H)\gamma^*$. Then, $\tilde{\Lambda}(\gamma x_H)$ is close to $\Lambda(x_0)$, for x_H close to x_0, but not necessarily diagonal. Now, the space of unitary complex matrices is path-connected, hence one may choose a path $U_\tau(x_0)$ from $U(x_0)$ to I, hence a path from $\tilde{U}(\gamma x_H)$ to I and from $\bar{B}(x_H)$ to $\tilde{\Lambda}(x_H)$, which is linearly deformable to $\Lambda(x_0) \equiv \Lambda$. By modifying $\lambda_j(x)$ along the deformations, one obtains an equivariant Γ-orthogonal homotopy to

$$(f^H(x_H), B(x_H)X, \Lambda\bar{X}) - \sum_1^k \tilde{\lambda}_j(x)\tilde{A}_j(x),$$

where $x_\perp$ is written as $X + \bar{X}$. Now, since Λ is real and diagonal, it is orthogonal to A_jx for all j and to the corresponding components of $\tilde{A}_j(x)$, hence $\tilde{\lambda}_j(x) = B(x_H)X \cdot \tilde{A}_j(x)$. Since $\Lambda\bar{X} \cdot \tilde{A}_j(x) = 0$, one may deform $\bar{X}$ to 0 in $\lambda_j(x)\tilde{A}_j(x)$ and still get a Γ-orthogonal homotopy. Hence, one has arrived at the map

$$(f^H(x_H), B(x_H)X, \Lambda\bar{X}) - \sum_1^k \tilde{\lambda}_j(x_{\underline{H}})\tilde{A}_j(x_{\underline{H}}).$$

Since $\tilde{A}_j(x_{\underline{H}})$ are in $V^{\underline{H}}$, by letting $\bar{X} = 0$, one has a Γ-orthogonal homotopy of the last map restricted to $V^{\underline{H}}$ to $f^{\underline{H}}(x_{\underline{H}})$, that is a Γ-orthogonal homotopy of $f(x)$ to $(f^{\underline{H}}(x_{\underline{H}}), \Lambda\bar{X})$. Note that, if one had linearized f at $x_{\underline{H}}$, instead of x_H, then the matrix $Df_{\bar{X}}(x_{\underline{H}})$ would be $\underline{H}$-equivariant and would give larger blocks, however the final result would be the same. □

Since $(f^{\underline{H}}(x_{\underline{H}}), \Lambda\bar{X})$ is a product, Theorem 3.1 will be essentially proved once the orthogonal index of $\Lambda\bar{X}$ at 0 is computed.

Lemma 3.3. *The orthogonal index of $\Lambda\bar{X}$ at 0 is given by*

$$i_\perp(\Lambda\bar{X}) = [F_\Gamma]_\perp + \sum \eta_j n_j [F_{K_j}]_\perp + \sum_{s>1} \eta_j \Big(\prod_{i=1}^{s} n_{j_i}\Big)[F_{K_{j_1}\cap\cdots\cap K_{j_s}}]_\perp,$$

where the first sum is over all K_j isotropy subgroups of the coordinates of $\bar{X}$, n_j is the number of these coordinates for which Λ is negative and the second sum is over those $K_j = K_{j_1} \cap \cdots \cap K_{j_s}$, intersection of s of the previous isotropy subgroups for which dim $\Gamma/K_j = s$. *The orientation factor η_j is $(-1)^{s(s-1)/2}$* Sign det B_j, *where B_j corresponds to the invertible matrix of the action of Γ on these s coordinates.*

Proof. It is clear that Λ may be deformed to blocks of the form $(-I, I)$, where one deforms linearly each eigenvalue to -1 or 1 according to its sign. The I-part acts as a suspension and does not affect the degree while, on a ball of the form $\{z_1, \dots, z_s : |z_j| < 2\}$, one may change any $-z_j$ to $(1-|z_j|^2)z_j$ and one gets the sum of the degrees on sets of the form $\{z_1, \dots, z_l : |z_j| < \frac{1}{2}; z_{l+1}, \dots, : \frac{1}{2} < |z_{l+j}| < 2\}$. For $|z_j| < 1/2$, one may deform back to z_j and obtain a suspension. Hence, one is reduced to compute the orthogonal degree on sets of the form $\tilde{\Omega} \equiv \{z_1, \dots, z_l : 1/2 < |z_j| < 2\}$ of the map $(\dots, (1-|z_j|^2)z_j, \dots)$. Let H_j be the isotropy subgroup of z_j (by construction $\Gamma/H_j \cong S^1$) and let $K = \bigcap_1^l H_j$ with $\dim \Gamma/K = s$ and let K_0 be the intersection of s of the H_j such that $\dim \Gamma/K_0 = s$, say the first s variables. Then, from Proposition 6.1 in Chapter 3, the orthogonal degree with respect to $\tilde{\Omega}$ is given by

$$[F]_\perp - \sum_{K<K_j<K_0} d_j [F_j]_\perp,$$

where, since $\underline{K} = K$, the d_j are given by the relations

$$\deg\Big([((1-|z_1|^2)z_1, \dots, (1-|z_l|^2)z_l) + \sum_1^s \lambda_k A_k z]^{K_i}; \tilde{\Omega}_s^{K_i}\Big) = \sum_{K_i<K_j<K_0} d_j |K_0/K_j|,$$

as a map defined on $\{\lambda_1, \dots, \lambda_s, z_1 > 0, \dots, z_s > 0, z \in \tilde{\Omega}_s^{K_i}\}$ and where $A_1 z, \dots, A_s z$ are the s linearly independent vector fields. Since on $\tilde{\Omega}$ all z_j are non-zero, it follows that the degree on the left is 0, except for $K_i = K$.

For K, since $A_1 z, \dots, A_s z$, restricted to V^{K_0}, are linearly independent, one may deform $A_k z_j$ to 0, for $j > s$, if of course $s < l$. In this case one may add $i\tau z_l$ to $(1-|z_l|^2)z_l$, giving a non-zero map in $\tilde{\Omega}$. Hence, if $s < l$, all d_j are 0.

The only case left is when $s = l$, $K = K_0$, where one has to compare the indices of the following two maps

$$F_{K_0}(\lambda, z) = \Big(2t + 1 - 2\prod |z_i|, \{\varepsilon_j(|z_j|^2 - 1)z_j\}_{j=1}^s\Big) + \sum_1^s \lambda_k A_k z$$

$$F(\lambda, z) = \big(2t + 2\varphi(z) - 1, \{(1-|z_j|^2)z_j\}_{j=1}^s\big) + \sum_1^s \lambda_k A_k z,$$

where $\varepsilon_1 = \dots = \varepsilon_{s-1} = 1$ and ε_s is chosen in such a way that the index of $F_{K_0}|_{B_s}$ at $\lambda_j = 0$, $z_j = 1$ and $t = 1/2$ is 1: see Theorem 6.1 of Chapter 3. On the other hand, $\varphi(z)$ is 1 if one of the z_j has norm less that $1/4$ and is 0 if all z_j have norm larger than $1/2$.

For z_j all real and positive, one may easily deform the first map to

$$\Big(2t - 1, \varepsilon_j(z_j - 1) + i\sum \lambda_k n_k^j\Big)$$

and the second map is deformed to

$$\left(2t-1, 1-z_j+i\sum \lambda_k n_k^j\right).$$

The degree of the first map is $\varepsilon_s(-1)^{s(s+1)/2}$ Sign det $B_s = 1$ (determining ε_s), while the degree of the second map is $(-1)^s(-1)^{s(s+1)/2}$ Sign det $B_s = (-1)^s\varepsilon_s$ (here B_s is the $s \times s$ matrix given by n_k^j of the action of Γ on $z_1, \dots, z_s$). Thus,

$$d_{K_0} = (-1)^s\varepsilon_s \text{ and } [F]_\perp = d_{K_0}[F_{K_0}]_\perp.$$

When $s = 0$, the only contribution is from the set where all z's are small, giving the generator $F_\Gamma = (2t-1, \bar{X})$. While, for $s = 1$, i.e., with only one z_j, we may collect all sets giving the same F_{K_j}, i.e., with the same action of Γ, giving a total contribution of $-n_j\varepsilon_j$, where $\varepsilon_j = -\operatorname{Sign} n_1^j$ (if $A_1z_j \neq 0$) and n_j is the number of sets, i.e., of coordinates with the same K_j and where Λ is -1. In terms of the normalized generators of Proposition 6.2 of Chapter 3, one has $[F]_\perp = d_{K_0}$ Sign det $B_s[F^*_{K_0}]_\perp$, i.e.,

$$[F]_\perp = (-1)^{s(s-1)/2}[F^*_{K_0}]_\perp.$$

In particular, for coordinates with the same action of Γ, one obtains the sum of the contributions. Furthermore, as seen in Proposition 6.3 of Chapter 3, on conjugate representations one has the same normalized generator. For $s > 1$, one has to collect all sets with exactly s coordinates ($s = l$) and the same isotropy $K_j = K_{j_1} \cap \dots \cap K_{j_s}$ with $\dim \Gamma/K_j = s$. Since $\dim \Gamma/K_{j_i} = 1$ one cannot have two coordinates with the same isotropy and, if K_j is as above, one will have $n_{j_1} \dots n_{j_s}$ sets with the same contribution. Note that here $s \leq n$. □

Proof of Theorem 3.1. It is enough to use the formula for the product given in Proposition 7.7 of Chapter 3. Here, $\tilde{H}_1^0 = H_0$ and $\tilde{H}_2^0 = H_2 = K_j$ or $K_{j_1} \cap \dots \cap K_{j_s}$. Since $A_jx = 0$ for $j > k$ and x in $V^{\underline{H}}$, the condition $\dim(\Gamma/(H_i \cap K_j)) = k + s$ implies that one has to take into account only those K_j's for which there are s among $A_{j+1}x, \dots, A_nx$ which are linearly independent on the s coordinates of $(V^{\underline{H}})^\perp$. By construction of $V^{\underline{H}}$ this is clearly true for $s = 1$ and any coordinate of $(V^{\underline{H}})^\perp$. Furthermore, if $x_1, \dots, x_k$ are the coordinates of V^{H_0}, defining the matrix B, and if $x'_1, \dots, x'_s$ are the coordinates of $(V^{\underline{H}})^\perp$ defining B_j, with the vectors $A_{j+1}x, \dots, A_{j+s}x$ (for instance), then on the union of those $k + s$ coordinates one has the matrix $\begin{pmatrix} B & 0 \\ C & B_j \end{pmatrix}$. Since B and B_j may be deformed to $\operatorname{diag}(\varepsilon, I)$ and $\operatorname{diag}(I_j, I)$, the above matrix has a determinant with sign equal to $\varepsilon\,\varepsilon_j$.

Now, $|H_0/H_i| = \prod k_l$ in the decomposition of Γ/H_i on coordinates of V^{H_i}. Then, for $|H_0 \cap K_j/H_i \cap K_j|$, one has the same product, by ordering the coordinates of $V^{H_i \cap K_j}$ by taking first $(x_1, \dots, x_k)$, then the rest of V^{H_i}, then $(x'_1, \dots, x'_s)$ and finally any other coordinate. One will get the decomposition

$$\Gamma/H_i \cap K_j = (\Gamma/H_0)(H_0/H_i)(H_i/H_i \cap K_j).$$

Since $A_{j+1}x, \dots, A_{j+s}x$ are linearly independent on $(x'_1, \dots, x'_s)$, one has that the k_l's for these s variables are ∞ and $k_l = 1$ for any other coordinate in the tail. Thus,

$$[F_{H_i}]_\perp [F_{K_j}]_\perp = [F_{H_i \cap K_j}]_\perp .$$

Furthermore, from the above argument for $\varepsilon\varepsilon_j$, this equality is also true for normalized generators. Note that one may have several K_j's giving the same $H_i \cap K_j$ and their contributions have to be summed according to Propositions 6.2 and 6.3 of Chapter 3, with a direct sum if one takes normalized generators. In particular, since $H_i < H$, the K_j's coming from equivalent H-representations in $(V^{\underline{H}})^\perp$ give the same $H_i \cap K_j = H_i \cap K$, with $\sum n_j = n(K)$, the Morse number of $Df_\perp(x_0)$ on $(V^{\underline{H}})^\perp \cap V^K$. For $K_j = K_{j_1} \cap \dots \cap K_{j_s}$, the count of the sets of Lemma 3.3 gives $\prod_{i=1}^s n(K_{j_i})$ sets. □

Remark 3.2. Another way to prove Theorem 3.1 is the following: since one has $A_l x_{\underline{H}} = 0$ for $l > k$, consider the map

$$\tilde{f}(\lambda_1, \dots, \lambda_k, x) = f(x) + \sum_1^k \lambda_l A_l x_{\underline{H}},$$

on $\mathbb{R}^k \times \Omega_k$, i.e., with $x_1, \dots, x_k$ real and positive. It is clear that $\tilde{f}$ is H-equivariant (in fact it is H_0-equivariant) and, since the projection of $\tilde{f}$ on $(V^{\underline{H}})^\perp$ is that of f with $\varphi_1 = \dots = \varphi_k = 0$ for H, the map $\tilde{f}$ is in fact H-orthogonal. As in Lemma 3.2, one may prove that $\tilde{f}$ is H-orthogonally homotopic to

$$\Big(f^{\underline{H}}(x_{\underline{H}}) + \sum_1^k \lambda_l A_l x_{\underline{H}},\ B_1(x_H)X_1, \dots, B_m(x_H)X_m\Big),$$

where B_j are the blocks of $Df_\perp$ on $(V^{\underline{H}})^\perp$. In fact, in this case one does not need to orthogonalize the linearization. As before, each block may be deformed to $\operatorname{diag}(-I_j, I)$, where I_j has the dimension of the Morse number of $B_j(x_0)$. One may replace $-I_jX_j$ by terms of the form $(1 - |z_j|^2)z_j$, as in Lemma 3.3, and compute the sum of the H-orthogonal degrees on the different sets, arriving at

$$\text{Index}_{\perp H}(\tilde{f}; (0,0)) = \sum d'_j [F'_j]_{\perp H},$$

where the d'_j are given by the same formula as the d_j's: in fact, for $\tilde{f}^{\underline{H}}$, the formula of Proposition 6.1 of Chapter 3 gives this result. However, the problem is to show that the homotopy of H-orthogonal maps lifts to a homotopy of Γ-orthogonal maps and that $F'_j - F_j + \sum_1^k \lambda_l A_l x_{\underline{H}}$: see Remark 3.2 of Chapter 3 and the proof of Theorem 3.4 in that Chapter. This operation may be done but it is delicate: it requires to prove a different version of Lemma 1.1 in Chapter 3 and to examine carefully the generators. We leave this task to the reader.

Let us consider a particular case of Theorem 3.1, that of a Γ-orthogonal linear map, which is useful for bifurcation.

Corollary 3.1. *Assume that B is an invertible Γ-orthogonal matrix. Hence, $B = \operatorname{diag}(B^\Gamma, B_i^{\mathbb{R}}, B_l, B_j)$, where $B_i^{\mathbb{R}}$ stands for B restricted on a set of real coordinates, each with the same isotropy H_i with $|\Gamma/H_i| = 2$, while B_l stands for B on coordinates with action of Γ as $\mathbb{Z}_m$, $m \geq 3$, and $B_j = B_j^*$ stands for B on complex coordinates with the same isotropy K_j with $\dim \Gamma/K_j = 1$. Let $\sigma_0 = \operatorname{Sign}\det B^\Gamma$, $\sigma_i = \operatorname{Sign}\det B_i^{\mathbb{R}}$ and n_j be the complex Morse number of B_j. Then*

$$i_\perp(Bx) = (-1)^{\sigma_0}\Big([F_\Gamma^*]_\perp + \textstyle\sum_{\Gamma/H_i\cong\mathbb{Z}_2} \frac{((-1)^{\sigma_i}-1)}{2}[F_{H_i}^*]_\perp + \cdots\Big)$$
$$\times\Big([F_\Gamma^*]_\perp + \textstyle\sum_{\Gamma/K_j\cong S^1} n_j[F_{k_j}^*]_\perp + \cdots\Big),$$

where the unspecified terms are completely determined by σ_0, σ_i, n_j and are given in Theorem 3.1.

Proof. Here, $k = 0$, $s = 1$ and since we have chosen the normalized generators, there are no more signs to take into account. □

Example 3.4 (Bifurcation). Assume that one has a family $f(\lambda, x)$ of Γ-orthogonal C^1 maps, with $f(\lambda, 0) = 0$, $\lambda \in \mathbb{R}$, $x \in V$. As seen in Lemma 7.2 of Chapter 1, if one writes

$$f(\lambda, x) = B(\lambda)x + R(\lambda, x),$$

where $B(\lambda) = Df(\lambda, 0)$, then $B(\lambda)x$ and $R(\lambda, x)$ are Γ-orthogonal and $B(\lambda)$ has the structure given in Corollary 3.1.

Assume $B(\lambda)$ is invertible for $\lambda \neq 0$ in a neighborhood of 0, then, as seen in Theorem 5.2 of Chapter 2, $\deg_\perp(|x| - \varepsilon, f(\lambda, x); B_{2\rho} \times B_{2\varepsilon})$ is well defined, where $B_{2\rho} = \{\lambda : |\lambda| < 2\rho\}$ and $B_{2\varepsilon} = \{x : |x| < 2\varepsilon\}$. Furthermore, one may deform linearly R to 0 (this is an orthogonal deformation). Then

$$\deg_\perp((|x| - \varepsilon, B(\lambda)x); B_{2\rho} \times B_{2\varepsilon}) = \deg_\perp((\rho^2 - |\lambda|^2, B(\lambda)x); B_{2\rho} \times B_{2\varepsilon})$$

will give the standard results on local and global bifurcation.

For the case of one parameter, the above degree is

$$\operatorname{Index}_\perp(B(-\rho)x) - \operatorname{Index}_\perp(B(\rho)x),$$

see Corollary 5.1 in Chapter 2. Hence, one has to compare the orthogonal indices at 0 of $B(\pm\rho)x$ given in Corollary 3.1.

Proposition 3.1. *Let $f(\lambda, x)$ be C^1-orthogonal, with $f(\lambda, 0) = 0$ and $Df(\lambda, 0)$ invertible for $\lambda \neq 0$ small. Let*

$$\sigma_0(\lambda) = \operatorname{Sign}\det Df^\Gamma(\lambda, 0), \quad \sigma_i(\lambda) = \operatorname{Sign}\det Df_\perp^{H_i}(\lambda, 0),$$

for H_i such that $\Gamma/H_i \cong \mathbb{Z}_2$, and $n_j(\lambda)$ be the complex Morse number of $Df_{\perp}^{K_j}(\lambda, 0)$ for K_j isotropy of a coordinate such that $\Gamma/K_j \cong S^1$. Then one has global bifurcation, i.e., there is a continuum of non-trivial solutions emanating from $(0, 0)$ which is either unbounded or returns to $(\lambda, 0)$ with $\lambda \neq 0$:

- *in V^Γ, if $\sigma_0(\lambda)$ changes sign, or*
- *in V^{H_i} if $\sigma_i(\lambda)$ changes sign, or*
- *in V^{K_j}, if $n_j(\lambda)$ changes.*

Furthermore, if the continuum is bounded and the bifurcation points on it are isolated, then the sum of the jumps of the orthogonal indices is 0. Finally, if $\sigma_0(\lambda)$, $\sigma_i(\lambda)$ and $n_j(\lambda)$ don't change, then there is an orthogonal nonlinearity $\tilde{R}(\lambda, x)$ such that $Df(\lambda, 0)x + \tilde{R}(\lambda, x)$ is zero only at $x = 0$.

Proof. The first part is a direct consequence of Theorem 5.2 in Chapter 2. For the last part, the construction of $\tilde{R}$, we refer to [I.V. 3, Theorem 5.2]. □

Remark 3.3. (a) The case of more parameters is treated in [I] and [IV3]. In that case the real and complex Bott periodicity theorems play a major role.

(b) If $f(\lambda, x) = \nabla\varphi(\lambda, x)$, then the change in the Morse number is sufficient to guarantee local bifurcation, even in the non-equivariant case. However one does not get a continuum. See the bibliographical remarks at the end of this chapter.

(c) If $B_{\perp}^{K_j}(\lambda) = \lambda B$, then n_j changes provided B has a non-zero signature, for example if $B = I$.

(d) For the correct application of Proposition 3.1 it is important to assimilate complex conjugate representations (they are the same as real representations) as the following example shows: Let S^1 act on $\mathbb{C}^2$ as $(e^{i\varphi}z_1, e^{-i\varphi}z_2)$. Consider the orthogonal Γ-map

$$f(\lambda, z) = (\lambda z_1 + t\bar{z}_2, -\lambda z_2 + t\bar{z}_1),$$

with $t = |z_1|^2 + |z_2|^2$. Since $Az = i(z_1, -z_2)$, one has $f(\lambda, z) \cdot Az = \operatorname{Re}(f, \bar{A}z) = 0$. Taking the conjugate of the second component, one has that $f(\lambda, z) = 0$ if and only if $\begin{pmatrix} \lambda & t \\ t & -\lambda \end{pmatrix} \begin{pmatrix} z_1 \\ \bar{z}_2 \end{pmatrix} = 0$, i.e., the map has no zeros except $z_1 = z_2 = 0$, that is there is no bifurcation. When λ goes through 0, the Morse number for z_1 goes from 1 to 0 and that for z_2 goes from 0 to 1, but their sum remains constant.

Example 3.5 (Periodic solutions of Hamiltonian systems). As an illustration of the preceding results, we shall consider the problem of finding 2π-periodic solutions of Hamiltonian systems of first and second order.

For first order systems, one looks at

$$f(X) = JX' + \nabla H(X) = 0,$$

where $X = (Y, Z)$ is in $\mathbb{R}^{2N}$, J is the standard symplectic matrix and H is C^2. Note that by rescaling time, there is no loss of generality when one looks for 2π-periodic solutions instead of a fixed period T.

Assume that the abelian group Γ_0 acts symplectically on $\mathbb{R}^{2N}$, i.e., it commutes with J or, if $X = (Y, Z)$, with Y and Z in $\mathbb{R}^N$, then the action on Y and Z is the same. Assume also that H is Γ_0-invariant and autonomous. Then, if $\Gamma = S^1 \times \Gamma_0$ acts on spaces of 2π-periodic functions with values in $\mathbb{R}^{2N}$ and S^1 acts by time translation, we have seen, in Proposition 9.1 of Chapter 1, that $f(X)$ is Γ-orthogonal with respect to the $L^2(S^1)$ scalar product. Here the infinitesimal generators for Γ will be $AX \equiv X'$ for the action of S^1 and $A_j X$, $j = 1, \ldots, n = \text{Rank}\,\Gamma_0$.

For the second order Hamiltonian equation

$$E(X) \equiv -X'' + \nabla V(X) = 0,$$

for X in $\mathbb{R}^N$ and a C^2 potential V which is Γ_0-invariant, one has that $E(X)$ is Γ-orthogonal with respect to the $L^2(S^1)$ scalar product (Proposition 9.2 in Chapter 1), with infinitesimal generators $AX \equiv X'$, $A_j X$, $j = 1, \ldots, n$.

In order to apply the orthogonal degree, we shall assume that there is an open bounded subset $\tilde{\Omega}$ of $\mathbb{R}^{2N}$ (or $\mathbb{R}^N$) invariant under Γ_0 such that any 2π-periodic solution cannot touch $\partial\tilde{\Omega}$ at any time. Then, as in Remark 2.3, one defines

$$\Omega = \{X \in W : \|X\|_W < R, X(t) \in \tilde{\Omega}\},$$

where $W = H^1(S^1)$ in the first case and $H^2(S^1)$ for the second order system, and R is chosen so large that any periodic solution in $\tilde{\Omega}$ has $\|X\|_W < R/2$, since ∇H and ∇V are bounded on $\tilde{\Omega}$ and $X(t)$ is continuous (or C^1) in W. However, the orthogonal degree has been defined here only for finite dimensional spaces and the extension to infinite dimensional spaces requires either modifying the equations and/or working with intermediate spaces like $H^{1/2}(S^1)$, and many technical difficulties: for instance the operator AX is a Fredholm operator only from H^1 into L^2. Furthermore, the compactness assumptions which we have used in order to define the Γ-degree in infinite dimensions imply that almost all the components of the degree should be 0: this is a result of the suspension isomorphism. However, if one takes $\nabla H(X) = BX$ for a constant matrix B, then the complex Morse index of $inJ + B$ is N for large n. Then, one could look at differences of degrees or differences with respect to a fixed reference map like JX'. However, it is simpler to use the fact that one has a large ball in W and, as in Remark 9.2 of Chapter 1, decompose W, writing any X as $X_1 \oplus X_2$, where $X_1 = PX$ corresponds to modes n, with $|n| \le N_1$ and X_2 to the others. Then, solving for X_2 as a function of X_1, by the global implicit theorem, one is lead to the study of the reduced equation

$$JX_1' + P\nabla H(X_1 + X_2(X_1)) = 0,$$

which is Γ-orthogonal and inherits the gradient structure, for X_1 in the finite dimensional space PW. One may then study $\deg_\perp(Pf(X_1 + X_2(X_1); P\Omega)$. Of course the price one has to pay is that it will be necessary to see how the spectrum of the linearization depends on N_1.

Assume then that ΓX_0 is an isolated solution of the Hamiltonian system $f(X)$, i.e., an orbit of dimension k with $\ker Df(X_0)$ of dimension k and generated by $X_0'(t)$, $A_j X_0(t)$ with exactly k of them linearly independent. Let H be the isotropy of X_0. One will be able to apply Theorem 3.1 provided one identifies $\underline{H}$ and one computes n_H, n_{H_i} and $n(K_i)$ for the reduced equation and for all N_1's large enough.

Remark 3.4. The hyperbolic condition on $Df(X_0)$ prevents it to commute with J, unless $k = 0$. In fact, if this would be the case, then if V belongs to $\ker Df(X_0)$ so does JV which has thus to be a real combination of X_0' and $A_j X_0$.

But, from Proposition 9.1 of Chapter 1, JX_0' is Γ-orthogonal with respect to the L^2-scalar product, i.e., it is orthogonal to X_0' and $A_j X_0$. This is possible only if X_0 is constant. In that case, the relation $JA_1X_0 = \sum_{j=1}^k \lambda_j A_j X_0$, with $X_0 = (Y_0, Z_0)$ leads, for a pair of coordinates (y_i, z_i), with the same action of Γ_0, to

$$-n_1^i z_i = \Big(\sum_l^k \lambda_j n_j^i\Big) y_i$$

$$n_1^i y_i = \Big(\sum_1^k \lambda_j n_j^i\Big) z_i .$$

Thus, either $y_i = z_i = 0$ or n_1^i and $\sum_l^k \lambda_j n_j^i = 0$. Then, if $k > 0$, X_0 is non zero and one arrives at a contradiction. Note that one may have pieces of $Df(X_0)$ which commute with J.

Now, we have seen in §9 of Chapter 1 that X_0 may be of three different types: a time stationary solution, a rotating wave or a truly time periodic solution.

(a) Stationary solution. If X_0 is time stationary, then $H = S^1 \times H_0$, with $H_0 < \Gamma_0$ such that $\dim \Gamma_0/H_0 = k$ and $\underline{H} = S^1 \times T^{n-k}$. Thus, $V^{\underline{H}}$ is contained in $\mathbb{R}^{2N}$, the space of constant functions, $B \equiv Df(X_0)$ has the form $\operatorname{diag}(B^H, B_m^{\mathbb{R}}, B_l^{\mathbb{C}}, B_s^{\mathbb{C}})$, where, on each B_m the group H acts as $\mathbb{Z}_2$, on the complex B_l as $\mathbb{Z}_p$, $p \geq 3$, and on the complex B_s as S^1. Since $B = D^2H(X_0)$, each of these matrices is self-adjoint. Furthermore, B_s is complex self-adjoint and H-orthogonal. Note that since J commutes with Γ_0, J has also a diagonal structure $\operatorname{diag}(J_H, J_m, J_l, J_s)$, each piece coupling a pair of coordinates. The hyperbolicity condition means that $\ker B^H$ has dimension k, that B_m, B_l and B_s are all invertible and that, for $n \neq 0$, $in\, J + B$ is invertible. One has the following result.

Proposition 3.2. *For a stationary hyperbolic orbit, the orthogonal index is given by*

(a) $d_H = \eta(-1)^{n_H}$, *with* n_H *the Morse number of* B^H *and* $\eta = (-1)^{k(k+1)/2}$;

(b) $d_{H_j} = d_H((-1)^{n_j} - 1)/2$, *with* $(-1)^{n_j} = \operatorname{Sign}\det B_j^{\mathbb{R}}$;

(c) *the Morse index of* $inJ + \tilde{B}$ *where* $\tilde{B}$ *is any of the matrices* B^H, $B_m^{\mathbb{R}}$, $B_l^{\mathbb{C}}$ *or* $B_s^{\mathbb{C}}$, *for the mode* $n > 0$ *and the decomposition of* $\mathbb{C}^{2N}$ *(induced by that of* $\mathbb{R}^{2N}$*) into irreducible representations of* H*;*

(d) *the Morse index of* $B_s^{\mathbb{C}}$*.*

Proof. As seen above, the index we have to compute is that of the projection on the modes n with $|n| \leq N_1$ for N_1 large enough, that is

$$(inJ + B)X_n + g_n(\underline{X}_1 + \underline{X}_2(\underline{X}_1)) = 0,$$

where $\underline{X}_1$ corresponds to these modes. As in Theorem 3.1, one has to look at the index of $B^{\underline{H}}X_0 + g_0^{\underline{H}}(X_{\underline{H}})$ and that of $(inJ + B)X_n$ for $n \neq 0$. Now, if $X = X_{\underline{H}}$ in $V^{\underline{H}}$, since $in\,J + B$ are all invertible for $n \neq 0$, one has $\underline{X}_2(X_{\underline{H}}) = 0$. This gives (a) and (b). The rest of the proof comes from Corollary 3.1. □

Remark 3.5. (a) Since $B = D^2H(X_0) = B^T$ is a real self-adjoint matrix and $in\,J+B$ is complex self-adjoint, this last matrix has real spectrum. Furthermore, if X is an eigenvector in $\mathbb{C}^{2N}$ of $inJ+\tilde{B}$, with eigenvalue λ, then $\bar{X}$ is an eigenvector of $-inJ+\tilde{B}$ with the same eigenvalue λ. Hence, the Morse index of $in\,J + \tilde{B}$ is equal to the Morse index of $-inJ + \tilde{B}$. Since one has to assimilate conjugate representations, the Morse index of Corollary 3.1, is, for $n \neq 0$, twice the Morse index of $inJ + \tilde{B}$. For $B_s^{\mathbb{C}}$, which has a complex structure and is complex self-adjoint, its complex Morse index is half its real Morse index, as seen in Remark 3.1. (b).

(b) If one has a family of Hamiltonians $f(\lambda, X)$, with $f(\lambda, X_0) = 0$ for some stationary X_0, which is hyperbolic for λ_1 and λ_2, then, if any of the above numbers change, one has a global bifurcation in the interval from λ_1 to λ_2. This bifurcation will take place in V^K, where $K < H$ is any of the isotropy subgroups for which d_K has changed and V^K is characterized in Lemmas 9.4–9.6 in Chapter 1. In particular, if there is no bifurcation in $V^{\underline{H}}$, then one has a bifurcation from a k-torus ΓX_0 to a $(k + 1)$-torus, either stationary if the Morse index of $B_s^{\mathbb{C}}$ has changed, or, if there is no bifurcation of stationary solutions, to a time-periodic solution, i.e., a pulsating k-torus.

(c) If J commutes with $B_j^{\mathbb{R}}$, then n_j is even and $d_{H_j} = 0$. More generally, if J commutes with $\tilde{B}$, then one may decompose the space into two-dimensional subspaces, invariant under J, $\langle X_k, JX_k\rangle$ corresponding to the eigenvalue λ_k of $\tilde{B}$ with two eigenvectors, orthogonal between them. The eigenvalues of $in\,J + \tilde{B}$ on this subspace are $\lambda_k \pm n$ and the Morse number of $in\,J + \tilde{B}$ is $(a(n) + a(-n))/2$, where $a(n)$ is the number of eigenvalues of $\tilde{B}$ which are less that n: recall that, since $B = D^2H(X_0) = B^T$, the spectrum of B is real and that $a(n)$ is even. In particular, if n is very large (larger that $\|B\|$), then $a(n) = 2N$ and $a(-n) = 0$.

(d) For the system $-X'' + \nabla V(X) = 0$, with $B = D^2V(X_0)$, then the Morse index of $n^2I + \tilde{B}$ is $a(-n^2)$, the number of eigenvalues of $\tilde{B}$ which are less that $-n^2$. Part

(a) of these remarks apply here. Note that, for the system $X' = -Y$, $Y' = \nabla V(X)$, then J commutes with $D^2(V(X) + \|Y\|^2/2)$ only if $B = I$.

(b) Reduction to the stationary case. Assume that X_0 is a rotating wave, i.e., that X_0' is a linear combination of the $A_j X_0$'s. Then, we have seen in Case 9.2 of Chapter 1 that there is a moving coordinates change of variables $A(t)$ such that $Y(t) = A(t)X(t)$ satisfies the equivalent systems

$$\begin{aligned} JY' &- JA'(0)Y + \nabla H(Y) = 0, \\ -Y'' &- A'(0)^2 Y + 2A'(0)Y' + \nabla V(Y) = 0, \end{aligned}$$

which are Γ-orthogonal and $Y_0(t) = A(t)X_0(t)$ is constant in time, i.e., one has frozen the rotating wave. See Remark 9.3 of Chapter 1 for the form of $A(t)$. Then, if $B = D^2H(Y_0)$, respectively $D^2V(Y_0)$, with $\nabla H(Y_0) = JA'(0)Y_0$, respectively $\nabla V(Y_0) = A'(0)^2 Y_0$, one has to look at the Morse numbers of

$$inJ - JA'(0) + \tilde{B},\ n^2 I + 2inA'(0) - A'(0)^2 + \tilde{B},$$

respectively, where $\tilde{B}$ is one of the pieces of B due to the action of Γ_0.

(c) Truly periodic solutions. Assume that X_0 is a $(2\pi/p)$-periodic hyperbolic solution of any of the above Hamiltonian systems, with isotropy H and $\dim \Gamma/H = k$, such that X_0', $A_1X_0, \dots, A_{k-1}X_0$ are linearly independent. One has that $H = \mathbb{Z}_p \times H_0$, the torus part of H is $\underline{H} = \underline{H}_0$, $V^{\underline{H}} = \{X(t) \in V_0 = (\mathbb{R}^N)^{\underline{H}_0}\}$. Furthermore, according to Lemmas 9.4–9.6 of Chapter 1, one has

$$V^H = \{X(t) \in V_0^{H_0},\ (2\pi/p)\text{-periodic},\ X(t) = \gamma_0 X(t + 2\pi/q))\},$$

where $\gamma_0^{q_0}$ is in H_0 and $q = pq_0$. See Example 2.11.

Also, for each K_j, with $H/K_j \cong \mathbb{Z}_2$, one has a subgroup K_{0j} of H_0 such that $H_0/K_{0j} \cong \mathbb{Z}_2$ or $H_0 = K_{0j}$ with $V_j = V_0^{K_{0j}} = V_j^+ \oplus V_j^-$ where $\gamma_0^{q_0}$ acts as $\pm$ Id on $V_j^\pm$ and $V_0^{H_0} = V_j^+$ if and only if $V_j^- \neq \{0\}$. Then,

$$V^{K_j} = \{X(t) \in V_j, 2\pi\text{-periodic},\ X(t) = \gamma_0^2 X(t + 4\pi/q)\}.$$

Finally, for each set of equivalent irreducible representations $\tilde{V}_l$ of H_0 in $V_0^\perp$, with complex coordinates $X^0, \dots, X^r$ and action of γ_0 on X^j as $e^{2\pi i\alpha_j}$, then for each $n_0 = 0, \dots, q-1$, there is a different set of equivalent irreducible representations of H, with isotropy K_{n_0}, in $(V^{\underline{H}})^\perp$, and

$$V^{K_{n_0}} = \{X(t) = (X^0(t), \dots, X^r(t)),\ R_{-2\pi(n_0/q+\alpha_0)}\gamma_0 X^j(t + 2\pi/q) = X^j(t)\},$$

when R_φ is a rotation of angle φ of the coordinates of X^j. A complete description of $V^{K_{n_0}}$ is given in Lemma 9.6 of Chapter 1.

Let $B(t) = D^2H(X_0(t))$ which is symmetric, $(2\pi/p)$-periodic and H_0-equivariant for each t. Furthermore, as in Lemma 2.4, one has

$$B(t) = \operatorname{diag}(B_0(t),\ B_j(t),\ \tilde{B}_l(t)),$$

where B_0 corresponds to $V_0^{H_0}$, B_j to V_j^- or $(V_j^+)\cap(V_0^{H_0})^\perp$ and $\tilde{B}_l$ to $\tilde{V}_l$. Furthermore,

$$\gamma_0 B(t + 2\pi/q) = B(t)\gamma_0.$$

Now, recall that $LX = JX' + B(t)X$ is a bounded Fredholm operator of index 0, from $H^1(S^1)$ into $L^2(S^1)$ and self-adjoint on $L^2(S^1)$, with kernel generated by $\{X_0', A_1X_0, \dots, A_{k-1}X_0\}$. (For $-X'' + B(t)X$ the domain is $H^2(S^1)$).

Hence, one has the decompositions

$$\begin{aligned} H^1(S^1) &= \ker L \oplus \text{Range } L \cap H^1 \\ L^2(S^1) &= \ker L \oplus \text{Range } L, \end{aligned}$$

where this last decomposition is L^2-orthogonal, and one has a bounded pseudo-inverse K from Range L onto Range $L \cap H^1$.

Furthermore, the reduction to finite dimensions, on V_{N_1} generated by all modes n with $|n| \le N_1$, is done by using the implicit function theorem on the higher modes to solve the equation

$$J\tilde{X}'_{N_1} + (I - P_{N_1})\nabla H(X_{N_1} + \tilde{X}_{N_1}) = 0,$$

for $\tilde{X}_{N_1}$ in $V_{N_1}^\perp$ and reduce to

$$JX'_{N_1} + P_{N_1}\nabla H(X_{N_1} + \tilde{X}_{N_1}(X_{N_1})) = 0,$$

which is a finite dimensional Γ-orthogonal map. It is not difficult to see that the linearization of this last equation is of the form

$$L_{N_1}X_{N_1} = JX'_{N_1} + P_{N_1}B(t)(X_{N_1} + \tilde{X}_{N_1}) = 0,$$

where $\tilde{X}_{N_1}$ is the unique solution, for N_1 large enough, in $V_{N_1}^\perp$ of the equation

$$\tilde{L}_{N_1}\tilde{X}_{N_1} = J\tilde{X}'_{N_1} + (I - P_{N_1})B(t)(X_{N_1} + \tilde{X}_{N_1}) = 0.$$

Thus, $\|\tilde{X}_{N_1}\|_1 \le C\|X_{N_1}\|_0$ and $\|\tilde{X}_{N_1}\|_0 \le \|\tilde{X}_{N_1}\|_1/N_1$, where C depends only on $\sup|B(t)|$, defining a continuous operator $\tilde{X}_{N_1}(X_{N_1})$ into $H^1(S^1)$.

Lemma 3.4. *If N_1 is large enough, the operator L_{N_1} is self-adjoint, with* ker $L_{N_1} =$ P_{N_1}(ker L) *of dimension k and* Range L_{N_1} = Range $L \cap V_{N_1}$. *The pseudo-inverse K_{N_1} of L_{N_1} is $P_{N_1}KP_{N_1}$, with $\|K_{N_1}\| \le \|K\|$.*

Proof. One may use the gradient structure of the linearization of the reduction, or see directly that

$$\begin{aligned}(L_{N_1}X_{N_1}, Z_{N_1})_{L^2} - (X_{N_1}, L_{N_1}Z_{N_1})_{L^2} &= (B\tilde{X}_{N_1}, Z_{N_1}) - (X_{N_1}, B\tilde{Z}_{N_1}) \\ &= (X, BZ_{N_1}) - (X_{N_1}, BZ),\end{aligned}$$

using the symmetry of B. But, since $J\tilde{X}'_{N_1} = -(I - P_{N_1})BX$, then

$$(J\tilde{X}'_{N_1}, \tilde{Z}_{N_1}) = -(BX, \tilde{Z}_{N_1}) = (\tilde{X}_{N_1}, J\tilde{Z}'_{N_1}) = -(BZ, \tilde{X}_{N_1}).$$

Then, the above difference is $(Z, BX) - (X, BZ) = 0$. Since

$$LX = L_{N_1}X_{N_1} \oplus \tilde{L}_{N_1}\tilde{X}_{N_1}, \quad X = X_{N_1} \oplus \tilde{X}_{N_1},$$

where $\tilde{L}_{N_1}\tilde{X}_{N_1} = 0$ in the definition of L_{N_1}, then clearly $\ker L_{N_1} = P_{N_1}(\ker L)$ with $\dim\ker L_{N_1} \leq k$, equality coming from linear independence and the definition of $\tilde{L}_{N_1}$. Furthermore, if $L_{N_1}X_{N_1} = Z_{N_1}$ then $L(X_{N_1} + \tilde{X}_{N_1}) = Z_{N_1}$ and $\text{Range}\, L_{N_1} = \text{Range}\, L \cap V_{N_1}$. From $LKZ = Z$, for $Z = Z_{N_1}$ in V_{N_1}, one has that $K_{N_1} = P_{N_1}KP_{N_1}$ and, as operator from L^2 into H^1, one has $\|K_{N_1}\| \leq \|K\|$ and $L_{N_1}K_{N_1} = \text{Id}_{V_{N_1}}$. □

Note that, if P is the projection onto $\ker L$ and $I - P$ that on $\text{Range}\, L$, one has that $P_{N_1}P$ will map onto $\ker L_{N_1}$.

Now, since the inclusion of H^1 in L^2 is compact, the operator $L - \lambda I$ is also a Fredholm operator of index 0, from H^1 into L^2, self-adjoint in L^2 and K, as an operator from L^2 into L^2, is compact. Hence, the spectrum of L, $\sigma(L)$, is discrete.

Recall that JX' is strongly indefinite, i.e., its spectrum goes to $\pm\infty$, while $-X''$ is an elliptic operator which is non-negative.

Lemma 3.5. *Let $\mathcal{K}$ be a compact interval in $\mathbb{R}$. Then:*

(a) *If $\sigma(L) \cap \mathcal{K} = \phi$, then $\sigma(L_{N_1}) \cap \mathcal{K} = \phi$ for N_1 large enough.*

(b) *If $\sigma(L) \cap \mathcal{K} = \lambda_0$ with $\dim\ker(L - \lambda_0 I) = d \leq 2N$, then for N_1 large enough $\sigma(L_{N_1}) \cap \mathcal{K}$ has d eigenvalues (counted with multiplicity).*

Proof. Assume λ is not in $\sigma(L)$, let K_λ be the inverse of $L - \lambda I$, then $\|(L - \lambda I)X\|_0 \geq \|K_\lambda\|^{-1}\|X\|_1 \geq \|K_\lambda\|^{-1}\|X\|_0$. If $\tilde{X}_{N_1}$ is defined by $\tilde{L}_{N_1}\tilde{X}_{N_1} = 0$, then one has

$$(L - \lambda I)(X_{N_1} + \tilde{X}_{N_1}) = (L_{N_1} - \lambda I)X_{N_1} - \lambda\tilde{X}_{N_1}.$$

Thus, one gets the estimate

$$\|(L_{N_1} - \lambda I)X_{N_1}\|_0 \geq \|K_\lambda\|^{-1}\|X_{N_1}\|_0 - |\lambda|\|\tilde{X}_{N_1}\|_0 > (\|K_\lambda\|^{-1} - C|\lambda|/N_1)\|X_{N_1}\|_0.$$

Hence, for N_1 large enough, λ is not in $\sigma(L_{N_1})$. For the compact set $\mathcal{K}$, it is easy to see that one will get an upper bound for $\|K_\lambda\|$ on $\mathcal{K}$ and $L_{N_1} - \lambda I$ will be invertible, for all λ in $\mathcal{K}$, provided N_1 is large enough.

For (b), write

$$(L_{N_1} - \lambda I)X_{N_1} = (L - \lambda_0 I)(X_{N_1} + \tilde{X}_{N_1}) + (\lambda_0 - \lambda)X_{N_1} + \lambda_0 \tilde{X}_{N_1},$$

and treat this problem as a (linear) bifurcation problem by projecting on $\ker(L - \lambda_0 I)$, via P_0, and on Range$(L - \lambda_0 I)$, via $I - P_0$:

$$\begin{aligned}(L_{N_1} - \lambda I)X_{N_1} = (L - \lambda_0 I)[(I - P_0)(X_{N_1} + \tilde{X}_{N_1} \\ + K_{\lambda_0}(I - P_0)((\lambda_0 - \lambda)X_{N_1} + \lambda_0 \tilde{X}_{N_1})] \\ \oplus (\lambda_0 - \lambda)P_0 X_{N_1} + \lambda_0 P_0 \tilde{X}_{N_1}.\end{aligned}$$

Then, $(L_{N_1} - \lambda I)X_{N_1} = 0$ gives two equations. The first one is uniquely solvable for $(I - P_0)X_{N_1}$ in terms of $P_0 X_{N_1}$, as a linear operator:

$$(I - P_0)X_{N_1} = -(I + (\lambda_0 - \lambda)K_{\lambda_0})^{-1}(I + \lambda_0 K_{\lambda_0})(I - P_0)\tilde{X}_{N_1},$$

for λ such that $|\lambda - \lambda_0| \leq \|K_{\lambda_0}\|^{-1}/2$ and N_1 large enough. In fact,

$$\|(I - P_0)X_{N_1}\|_0 \leq \tilde{C}\|\tilde{X}_{N_1}\|_0 \leq \tilde{C}(\|P_0 X_{N_1}\|_0^2 + \|(I - P_0)X_{N_1}\|_0^2)^{1/2}/N_1,$$

that is, $\|(I - P_0)X_{N_1}\|_0 \leq \tilde{C}\|P_0 X_{N_1}\|_0/N_1$, for this range of λ's. The second term is of the form

$$((\lambda_0 - \lambda)I + C(\lambda))P_0 X_{N_1} = 0,$$

where $C(\lambda)$ is a $d \times d$ matrix, analytic in λ and symmetric (as it is easy to see) with $\|C(\lambda)\| \leq C/N_1$: since L_{N_1} is symmetric, its spectrum is real. Then, for $N_1 \geq 2C\|K_{\lambda_0}\|$, the spectrum of $C(\lambda)$ is completely contained in the interval $|\mu| \leq \|K_{\lambda_0}\|^{-1}/2$ and gives d curves parametrized by λ (this is due to the fact that $C(\lambda)$ is symmetric) and the line $\lambda - \lambda_0 = \mu$ intersects these curves in d points. From (a), in $\mathcal{K}\backslash\{\lambda : |\lambda - \lambda_0| \leq \|K_{\lambda_0}\|^{-1}/2\}$, one has no eigenvalues of L_{N_1}, for N_1 large enough. (Note that, if $LX = \lambda X$, then $X(t)$ is a 2π-periodic solution of the differential equation, hence $d \leq 2N$). □

The above information is enough to prove the following

Proposition 3.3. *If $X_0(t)$ is a hyperbolic $(2\pi/p)$-truly periodic solution of the system $-X'' + \nabla V(X) = 0$, then, for N_1 large enough, the truncated system $-X''_{N_1} + P_{N_1}\nabla V(X_{N_1} + \tilde{X}_{N_1}) = 0$ has an orthogonal index given by the following:*

1. *$d_H = \eta(-1)^{n_H}$, where n_H is the Morse number of $-X'' + B_0(t)X$, with $B_0 = D^2V(X_0)$ restricted on $V_0^{H_0}$ and X in V^H, i.e., $X(t)$ is in $V_0^{H_0}$, $(2\pi/p)$-periodic and $X(t) = \gamma_0 X(t + 2\pi/q)$, with $\gamma_0^{q_0}$ in H_0 and $q = pq_0$. Here $\eta = (-1)^{k(k+1)/2}$.*

2. *$d_{K_j} = d_H((-1)^{n_{K_j}} - 1)/2$, where n_{K_j} is the Morse number of $-X'' + B_j(t)X$, with $B_j = D^2V(X_0)$ restricted to $V_j \cap (V_0^{H_0})^\perp$ and X in V^{K_j}, i.e., $X(t)$ is in this subspace and $X(t) = \gamma_0^2 X(t + 4\pi/q)$.*

3. $d_{K_{n_0}}$ *is the complex Morse index of* $-X'' + \tilde{B}_l(t)X$, *for each* $n_0 = 0, \ldots, q-1$ *and* $X(t)$ *in* $V^{K_{n_0}}$, *i.e.,* $X(t)$ *is in* $\tilde{V}_l$ *and* $R_{-2\pi(n_0/q+\alpha_0)}\gamma_0 X^j(t+2\pi/q) = X^j(t)$ *for* $j = 1, \ldots, r$ *the complex coordinates of* $X(t)$ *in* $\tilde{V}_l$, *and* $\tilde{B}_l(t)$ *is* $D^2V(X_0)$ *restricted to* $\tilde{V}_l$.

Proof. Note first that the projection on the modes is compatible with the decomposition of $B(t)$. Hence, the orthogonal index is given by the different Morse numbers of $L_{N_1}X = -X''_{N_1} + P_{N_1}\tilde{B}(X_{N_1} + \tilde{X}_{N_1})$. Since $LX = -X'' + \tilde{B}(t)X$ is an elliptic operator, its spectrum is bounded from below, because

$$(LX - \lambda X, X)_{L^2} \geq \|X'\|^2_{L^2} - (M+\lambda)\|X\|^2_{L^2} \geq 0 \quad \text{for } \lambda \leq -M = -\|\tilde{B}\|_{C^0}.$$

Then, the finite number of strictly negative eigenvalues of L will give the same number of negative eigenvalues of L_{N_1}, for N_1 large enough, due to Lemma 3.5: this gives the result for n_{K_j} and $d_{K_{n_0}}$ since L is invertible on V^{K_j} and $V^{K_{n_0}}$. For V^H, Lemma 3.4 implies that $\ker L_{N_1}$ is k-dimensional, i.e., none of the zero eigenvalues of L escapes from the origin when dealing with the approximation L_{N_1}. This finishes the argument. □

For the system $JX' + \nabla H(X) = 0$, the situation is slightly different and requires the following preliminary result

Lemma 3.6. *For the system* $LX = JX' + \tilde{B}(t)X$, *the Morse numbers* $n(L_{N_1})$ *of* L_{N_1} *and* $X(t)$ *in* $\tilde{V} = V_0^{H_0}$ *or* $V_j \cap (V_0^{H_0})^{\perp}$ *or* $\tilde{V}_l$ *are such that, for* N_2, *the next integer after* N_1 *where one has new modes, one has* $n(L_{N_2}) = n(L_{N_1}) + \dim \tilde{V}$, *where* $\tilde{V}$ *has even dimension.*

Proof. Note first that in V^H all functions are $2\pi/p$-periodic, hence $N_2 = N_1 + p$, while in V^{K_j} and $V^{K_{n_0}}$ one has $N_2 = N_1 + q$ (see Lemmas 9.5 and 9.6 in Chapter 1). Hence, if $X_{N_2} = X_{N_1} \oplus Y_{N_1}$, one has that $Y_{N_1} = (X_m, X_{-m} = \bar{X}_m)$ for one mode m (a multiple of p for V^H), with $X_m = (\tilde{X}, \tilde{Y})$ in $\mathbb{C}^{2r}$, where $2r = \dim \tilde{V}$: because of J, the space $\tilde{V}$ is even dimensional. Then,

$$\begin{aligned} L_{N_2}X_{N_2} = {} & L_{N_1}X_{N_1} + P_{N_1}B(Y_{N_1} + \tilde{X}_{N_2} - \tilde{X}_{N_1}) \\ & \oplus JY'_{N_1} + (P_{N_2} - P_{N_1})B(X_{N_1} + Y_{N_1} + \tilde{X}_{N_2}). \end{aligned}$$

But, since $\tilde{X}_{N_1} = \tilde{X}_{N_2} \oplus \tilde{Y}_{N_1}$, with $J\tilde{Y}'_{N_1} + (P_{N_2} - P_{N_1})B(X_{N_1} + \tilde{X}_{N_1}) = 0$, one has

$$\begin{aligned} L_{N_2}X_{N_2} = {} & L_{N_1}X_{N_1} + P_{N_1}B(Y_{N_1} - \tilde{Y}_{N_1}) \\ & \oplus J(Y'_{N_1} - \tilde{Y}'_{N_1}) + (P_{N_2} - P_{N_1})B(Y_{N_1} - \tilde{Y}_{N_1}). \end{aligned}$$

Now, since $L_{N_2}X_{N_2}$ and $L_{N_1}X_{N_1} \oplus JY'_{N_1}$ are self-adjoint, this is also the case for the linear deformation $L^{\tau}_{N_2}$, where B is replaced by τB in the above expression.

Then, for $m > \|B\|$, if one has $L^{\tau}_{N_2} X_{N_2} = 0$ then $Y_{N_1} - \tilde{Y}_{N_1} = 0$ and $L_{N_1} X_{N_1} = 0$, i.e., $L_{N_2} X_{N_2} = 0$. Thus, the kernel of L_{N_2} is preserved and the other eigenvalues of L_{N_2} do not cross 0. Similarly, the deformation $J(Y'_{N_1} - \tau \tilde{Y}'_{N_1})$ will not introduce a new eigenvector with eigenvalue 0, hence the Morse indices are related by

$$n(L_{N_2}) = n(L_{N_1}) + n(JY'_{N_1}).$$

Now, if $JY'_{N_1} = \lambda Y_{N_1}$, with $Y_{N_1} = \cos mt X + \sin mt Y$, where X and Y are in $\mathbb{R}^{2r}$, then $mJY = \lambda X$, $-mJX = \lambda Y$ and, multiplying by J, $m^2 Y = -\lambda m JX = \lambda^2 Y$, that is $\lambda = \pm m$, each eigenvalue with a $(2r)$-dimensional eigenspace. □

Note that the matrices L_{N_1} and $P_{N_1} L P_{N_1}$ are such that

$$\|L_{N_1} X_{N_1} - P_{N_1} L X_{N_1}\|_0 = \|P_{N_1} B \tilde{X}_{N_1}\|_0 \leq C\|X_{N_1}\|_0 / N_1,$$

hence, for N_1 large enough,, the two matrices have their spectra close, but the 0 eigenvalue of L_{N_1} may split into k eigenvalues for $P_{N_1} L P_{N_1}$.

Proposition 3.4. *If $X_0(t)$ is a hyperbolic $(2\pi/p)$-truly periodic solution of the system $JX' + \nabla H(X) = 0$, then, for N_1 large enough, $P_{N_1} X_0$ gives an orthogonal index for $JX'_{N_1} + P_{N_1} \nabla H(X_{N_1} + \tilde{X}_{N_1})$ equal to*

1. *$d_H = \eta(-1)^{n_H}$, where n_H is the Morse number of L_{N_1} restricted to V^H, with $\eta = (-1)^{k(k+1)/2}$,*
2. *$d_{K_j} = d_H((-1)^{n_{K_j}} - 1)/2$, where n_{K_j} is the Morse number of L_{N_1} restricted to $V^{K_j} \cap (V^H)^{\perp}$,*
3. *$d_{K_{n_0}}$ the Morse number of L_{N_1} restricted to $V^{K_{n_0}}$,*

where V^H, V^{K_j} and $V^{K_{n_0}}$ are defined in Proposition 3.3. *For N_1 large enough, the numbers n_H and n_{K_j} have a constant parity while $d_{K_{n_0}}$ increases by the even number* dim $\tilde{V}_l$, *when N_1 is replaced by $N_1 + q$.*

Proof. It is enough to note that dim $\tilde{V} = 2r$ and apply Corollary 3.1. □

Remark 3.6. If $J\tilde{B} = \tilde{B}J$ for some block $\tilde{B}$ in B, let $\Phi(t)$ be the fundamental matrix for $X' = J\tilde{B}X$, with $\Phi(0) = I$. Since $\Phi' = J\tilde{B}\Phi = \tilde{B}J\Phi$, then $J\Phi$ and ΦJ are also fundamental matrices and, being equal for $t = 0$, one has that J and Φ commute. Now, $(\Phi^T J \Phi)' = 0$, hence $\Phi^T J \Phi = J$ and, since J and Φ commute, one has that Φ is an orthogonal matrix and hence with spectrum on the unit disk. If $JX' + \tilde{B}X = \lambda X$, then $X(t) = e^{-\lambda J t}\Phi(t)X(0)$ and $X(2\pi) = X(0)$ if and only if $X(0)$ is in $\ker(I - e^{-2\pi\lambda J}\Phi(2\pi))$. Furthermore, $e^{\lambda J t}$ preserves the generalized eigenspaces of $\Phi(t)$. Then, if $\Phi(2\pi)W = \mu W$, one has $(I - e^{-2\pi\lambda J}\Phi(2\pi))W = 0$ if and only if $e^{2\pi\lambda J} W = \mu W = (\cos 2\pi\lambda I + \sin 2\pi\lambda J)W$, that is $\mu = e^{\pm 2\pi i \lambda}$.

Note also that if $JX' + \tilde{B}X = \lambda X$, then $Y(t) = e^{-Jt}X(t)$ satisfies $JY' + \tilde{B}Y = (\lambda + 1)Y$ and is 2π-periodic if $X(t)$ is 2π-periodic. Similarly, if $X(t)$ belongs to V^H, V^{K_j} or $V^{K_{n0}}$ then $Y(t) = e^{-qJt}X(t)$ belongs to the same subspace: this is due to the facts that $V_0^{H_0}$, V_j and $\tilde{V}_l$ are invariant under J, that e^{-qJt} is $(2\pi/q)$-periodic and that Γ_0 commutes with J (see Lemmas 9.4–9.6 in Chapter 1). Thus, if one knows the eigenvalues of $JX' + \tilde{B}X = \tilde{L}X$ which are in $[-q, 0]$, then the whole spectrum of $\tilde{L}$ will be given by translations of multiples of q. Note also that, if X is an eigenvector of $\tilde{L}$, then JX is also an eigenvector with the same eigenvalue. Hence, all eigenvalues of $\tilde{L}$ have even multiplicity (see also Remark 3.4). In this case, one has that $\tilde{X}_{N_1}(JX_{N_1}) = J\tilde{X}_{N_1}(X_{N_1})$, by uniqueness of the solution to $\tilde{L}_{N_1}\tilde{X}_{N_1} = 0$, and L_{N_1} commutes with J. Hence, the corresponding Morse number is even and, if $\tilde{B}$ is based on V^{K_j}, one has $d_{K_j} = 0$.

Note finally that one may relate the spectra of $\Phi(2\pi/q)$ to that of $\Phi(2\pi)$, as in Lemma 2.4 and Proposition 2.11.

However, if J and $\tilde{B}$ don't commute then the spectra of $\tilde{L}$ and of Φ need not be related.

Example 3.6 (Spring-pendulum systems). We shall give now an illustration of how the equivariant degree for orthogonal maps, in particular for Hamiltonian systems, may be used to show bifurcation from an S^1-orbit to a T^2-orbit in two spring-pendulum apparatus.

The first system consists of a spring, moving vertically only, with a rigid pendulum suspended at the end, free to move in any direction. If one pulls the pendulum downwards slightly, one obtains a stable harmonic oscillation. For a stronger pull, this oscillation looses its stability and one has an oscillation in a plane. For a still stronger pull, one gets an oscillation of the pendulum with a triangular pattern in space. Stronger pulls seem to lead to more complicated patterns.

The second apparatus is a pendulum with an elastic shaft. The same succession of patterns is observed and follow the behavior predicted by the study we shall present.

For the first system, the spring has length l_0 at rest and a constant k. It is suspended at the origin, with a mass M at the end, i.e., at the point $(0, 0, l)$, orienting the z-axis downwards. From this mass, one attaches a rigid pendulum, of length r_0, with a mass m at its end, of coordinates (x, y, z). The kinetic and potential energies are

$$\begin{aligned} T &= \frac{1}{2}M\dot{l}^2 + \frac{1}{2}m(\dot{x}^2 + \dot{y}^2 + \dot{z}^2) \\ K &= \frac{1}{2}k(l - l_0)^2 - Mgl - mgz \end{aligned}$$

with the relation $r_0^2 = x^2 + y^2 + (z - l)^2$.

Instead of using a Lagrange multiplier for this holonomic relation, we shall write

$$l = z - r = z - (r_0^2 - x^2 - y^2)^{1/2},$$

assuming thus that $0 \leq l \leq z$, i.e., that the pendulum does not reach the horizontal position. The Euler equations for $x(t)$, $y(t)$, $z(t)$, denoted $x_j(t)$, $j = 1, 2, 3$, are

$$\frac{d}{dt}\left(\frac{\partial}{\partial \dot{x}_j}(T - K)\right) = \frac{\partial}{\partial x_j}(T - K)$$

and give the following system of equations

$$\mathcal{M}\begin{pmatrix} \ddot{x} \\ \ddot{y} \\ \ddot{z} \end{pmatrix} + \tilde{C}\begin{pmatrix} x/r \\ y/r \\ 1 \end{pmatrix} + \begin{pmatrix} 0 \\ 0 \\ -mg \end{pmatrix} = 0,$$

where $\tilde{C} = C + k(z - r - l_0) - Mg$, with

$$C = M\left(\frac{\dot{x}^2 + \dot{y}^2}{r} + \frac{(x\dot{x} + y\dot{y})^2}{r^3}\right)$$

$$\mathcal{M} = \begin{pmatrix} m + M\frac{x^2}{r^2} & M\frac{xy}{r^2} & M\frac{x}{r} \\ M\frac{xy}{r^2} & m + M\frac{y^2}{r^2} & M\frac{y}{r} \\ M\frac{x}{r} & M\frac{y}{r} & m + M \end{pmatrix}.$$

If one defines

$$Z = z - r_0 - l_0 - (m + M)g/k,$$

then $\tilde{C}$ is transformed into $C + B + A$, where

$$A = mg, \quad B = k(Z + r_0 - r).$$

Since we are looking at periodic solutions of the system, of unknown frequency ν, we shall scale the time by $\tau = \nu t$, and get the system

$$f(\nu, x, y, z) \cong \nu^2\mathcal{M}\begin{pmatrix} x'' \\ y'' \\ Z'' \end{pmatrix} + A\begin{pmatrix} x/r \\ y/r \\ 0 \end{pmatrix} + (B + \nu^2 C)\begin{pmatrix} x/r \\ y/r \\ 1 \end{pmatrix} = 0,$$

where, in C, we have changed the derivative with respect to t by the derivative with respect to τ.

Note that $\det \mathcal{M} = m^2(m + M + M(x^2 + y^2)/r^2)$, that is, $\mathcal{M}$ is an invertible, symmetric, positive definite matrix. Furthermore,

$$\nu^{-2}\mathcal{M}^{-1}f = X'' + \tilde{A}\begin{pmatrix} x/r \\ y/r \\ -M(x^2 + y^2)/(m + M)r^2 \end{pmatrix} + (\tilde{B} + \nu^2 C_1)\begin{pmatrix} x/r \\ y/r \\ 1 \end{pmatrix},$$

where $X^T = (x, y, Z)$, $\tilde{A} = g(m + M)/(m + M + M(x^2 + y^2)/r^2)$, $\tilde{B} + \nu^2 C_1 = (B + \nu^2 C)/(m + M + M(x^2 + y^2)/r^2)$.

Notice that, for $x = y = 0$, the system reduces to $\nu^2(m+M)Z''+kZ = 0$, with 2π-periodic solutions of the form $a\cos(n\tau+\varphi)$, with $n = \nu_0/\nu$ and $\nu_0 = (k/(m+M))^{1/2}$ is the natural frequency of the spring when it oscillates vertically.

The mapping $f(\nu, X)$ is continuous from $C^2_{2\pi}$ into $C^0_{2\pi}$ and satisfies the properties of the following lemma.

Lemma 3.7. (a) *The mapping $f(\nu, X)$ is $S^1 \times S^1$-orthogonal, that is*

1. $f(\nu, T_{\varphi,\psi}X) = T_{\varphi,\psi}f(\nu, X)$

2. $(f(\nu, X), X')_{L^2} = 0, \quad (f(\nu, X), AX)_{L^2} = 0,$

where $T_{\varphi,\psi}X(\tau) = R_\psi(X(\tau+\varphi))$ and R_ψ is a rotation of angle ψ and axis the Z-axis, hence with infinitesimal generator $AX = (-y, x, 0)$.

(b) *$f(\nu, X)$ is also reversible, in the sense that $f(\nu, \tilde{R}_\varepsilon X) = \tilde{R}_\varepsilon f(\nu, X)$, where*

$$\tilde{R}_\varepsilon(x(\tau), y(\tau), Z(\tau)) = (x(-\tau), \varepsilon y(-\tau), Z(-\tau)), \varepsilon = \pm 1.$$

Proof. The equivariance with respect to the time shift, $X(\tau+\varphi)$, follows from the fact that the system is autonomous. Furthermore, since in C one has terms of the form $X\cdot X'$ and $X'\cdot X'$, it is clear that C is invariant under R_ψ and the equivariance of $f(\nu, X)$ with respect to R_ψ reduces to that of $\mathcal{M}X''$. Since $\mathcal{M} = mI + MD(X)$ and it is easy to check directly that $D(R_\psi X) = R_\psi D(X) R_\psi^{-1}$, one has $D(R_\psi X)R_\psi X = R_\psi D(X)X$, i.e., that $f(\nu, R_\psi X) = R_\psi f(\nu, X)$.

For the orthogonality one has, by direct calculation, that $f(\nu, X)\cdot X' = \frac{d}{d\tau}(K+T)$ which integrates to 0 on periodic functions, i.e., one has conservation of energy on solutions.

On the other hand, $f(\nu, X)\cdot AX = \nu^2 m(xy'' - x''y)$, which integrates to 0 on periodic functions.

Finally, the reversibility is easily checked. □

Remark 3.7. (a) One has $(xx' + yy')/r = (r_0 - r)'$ and $(r_0 - r)'' = C/M + (xx'' + yy'')/r$, hence one may rewrite the systems in the form

$$m\nu^2U'' + (mg + kV + M\nu^2V'')U/r = 0$$
$$m\nu^2Z'' + kV + M\nu^2V'' = 0,$$

where $U = x + iy$ and $V = Z + r_0 - r$. In this form the equivariance with respect to $R_\psi U = e^{i\psi}U$ is clear, as well as that with respect to conjugation, equal to $\tilde{R}_{-1}\tilde{R}_1$. Furthermore, one derives easily the conservation of the energy. One may also use the second equation in order to write the first one as

$$\nu^2U'' + (g - \nu^2Z'')U/r = 0,$$

but with the loss of the orthogonality.

(b) It is easy to check that $\tilde{R}_\varepsilon T_{\varphi,\psi} = T_{-\varphi,\varepsilon\psi}\tilde{R}_\varepsilon$, i.e., these actions don't commute. As a matter of fact, we should have stated that $f(\nu, X)$ is $O(2) \times O(2)$-orthogonal, where the first component $O(2)$ acts as T_φ and $\tilde{R}_1$, while the second component acts as T_ψ and the conjugation $\tilde{R}_{-1}\tilde{R}_1$. We have chosen, since the theory developed in this book is for abelian actions, to put together the non-abelian part in $\tilde{R}_\varepsilon$ and we shall use this information to characterize further the bifurcated solutions. Note here that if $X(\tau)$ is a solution, then $\tilde{R}_{-1}\tilde{R}_1 X$ and $R_\pi X$ are also solutions.

The second system consists of an elastic spring, of length at rest r_0 and suspended at the origin (the z-axis is again oriented downwards), with a mass m at its end. One has

$$\begin{aligned} T &= \frac{1}{2}m(\dot{x}^2 + \dot{y}^2 + \dot{z}^2) \\ K &= \frac{1}{2}k(r - r_0)^2 - mgz, \end{aligned}$$

where $r = (x^2 + y^2 + z^2)^{1/2}$. The Euler equations are

$$m\ddot{X} + k(r - r_0)X/r - mg(0, 0, 1)^T = 0.$$

Let $Z = z - (r_0 + mg/k)$ and $\tau = \nu t$, then one gets the system

$$g(\nu, X) = m\nu^2 X'' + kX + k\frac{r_0}{r}\begin{pmatrix} -x \\ -y \\ (x^2 + y^2)/(Z + r + r_0 + mg/k) \end{pmatrix} = 0,$$

where the last term in the third component comes from $k(r_0/r)(r - z)$ and $r - z = (r^2 - z^2)/(r + z)$, with $r = (x^2 + y^2 + (Z + r_0 + mg/k)^2)^{1/2}$. Thus, the problem is to find 2π-periodic solutions to $g(\nu, X) = 0$. It is easy to see that $x = 0$, $y = 0$, $Z = a\cos(n\tau + \varphi)$, with $n = \nu_0/\nu$ and $\nu_0 = (k/m)^{1/2}$, the natural frequency of the spring when it oscillates vertically, is a solution.

Note that $g(\nu, X)$ is a continuous map from $C^2_{2\pi}$ into $C^0_{2\pi}$, provided $r \neq 0$, i.e., if the spring does not collapse.

Lemma 3.8. *The mapping $g(\nu, X)$ is $S^1 \times S^1$-orthogonal with respect to the action defined in Lemma* 3.7 *and reversible as well.*

Proof. The equivariance with respect to the time shift and the rotation around the z-axis are easy to prove. The orthogonality to X' follows from the conservation of energy and that to AX is immediate. The reversibility follows as in the previous system. □

As noted above, $X_0(\tau) = (0, 0, a\cos(n\tau + \varphi))$ is a solution, with $\nu = \nu_0/n$, of $f(\nu, X) = 0$. Let us linearize $f(\nu, X)$ around this solution, with $\nu = \nu_0/n + \mu$ and

$Z = Z_0 + z$ and obtain

$$\begin{aligned}
&(\nu_0/n)^2 m x'' + m(g + (k/(m+M))Z_0)(x/r_0)\\
&(\nu_0/n)^2 m y'' + m(g + (k/(m+M))Z_0)(y/r_0)\\
&(\nu_0/n)^2 (m+M)z'' + kz - 2k(n/\nu_0)Z_0
\end{aligned}$$

where we have used the fact that $(\nu_0/n)^2(m+M)Z_0'' + kZ_0 = 0$. Equivalently, with $Z_0 = a\cos(n\tau+\varphi)$ and $\nu_0 = (k/(m+M))^{1/2}$, one has

$$L_n(\mu, X) = X'' + n^2 \begin{pmatrix} (\alpha + 2\beta\cos(n\tau+\varphi))x \\ (\alpha + 2\beta\cos(n\tau+\varphi))y \\ z + \mu\gamma\cos(n\tau+\varphi) \end{pmatrix}$$

where $\alpha = g/(\nu_0^2 r_0) = g(m+M)/(kr_0)$, $\beta = a/(2r_0)$, $\gamma = -2an/\nu_0$.

That is, the two first equations are Mathieu's equations and the third is a resonant harmonic oscillation. The amplitude a of Z_0 plays the role of an extra parameter.

As before we shall fix the phase φ of the one-dimensional orbit Z_0, at 0. Notice the complex structure induced by the action of R_ψ.

Lemma 3.9. *One has* $\dim\ker L_n = 2, 4$ *or* 6, *with eigenvectors* $\mu = 0$, $z = \cos n\tau$ *or* $\sin n\tau$, *x and y are Mathieu functions corresponding to analytic curves $\alpha_{k/n}(\beta)$, $\tilde{\alpha}_{k/n}(\beta)$ passing through the point ($\alpha = (k/n)^2$, $\beta = 0$). Solutions on $\alpha_{k/n}(\beta)$ are even in τ and those on $\tilde{\alpha}_{k/n}(\beta)$ are odd. Furthermore, these curves are symmetric with respect to the α-axis, except $\alpha_{(2k+1)/2}(-\beta) = \tilde{\alpha}_{(2k+1)/2}(\beta)$. Also, $\alpha_{k/n}(\beta) = \tilde{\alpha}_{k/n}(\beta)$ if $k/n \neq k_1/2$, where k_1 is an integer, while $\alpha_{k_1/2}(\beta)$ and $\tilde{\alpha}_{k_1/2}(\beta)$ intersect only at $\beta = 0$. Also, $\alpha_{k/n}(\beta)$ tends to $-\infty$ when $|\beta|$ goes to ∞ and $\alpha_0(\beta) = \alpha_0(-\beta) < 0$, while $\tilde{\alpha}_0(\beta)$ does not exist. Moreover, $\alpha_{k/n}(\beta)$ foliate the region between the curves bifurcating from two consecutive half-integers, i.e., those curves do not intersect and are dense. In this region any solution of $L_2X = 0$ (not necessarily periodic) is bounded, while in the complementing region (the Arnold's tongues), the solutions are unbounded, as well as the other solution on the transition curves $\alpha_{k_1/2}(\beta)$ and $\tilde{\alpha}_{k_1/2}(\beta)$.*

If $x_n(\tau)$ is a 2π-periodic solution for $\alpha_{k/n}(\beta)$ and $k/n = k_1/n_1$, with k_1 and n_1 relatively prime, then $x_n(\tau) = x_{n_1}(n\tau/n_1)$, in particular $x_n(\tau)$ is $(2\pi n_1/n)$-periodic. The solutions $x_n(\tau)$ on $\alpha_{k/n}(\beta)$ have $2k$ simple internal zeros in $(0, 2\pi)$ and on $\tilde{\alpha}_{k/n}(\beta)$ the solution $\tilde{x}_n(\tau)$ has $2k-1$ internal zeros.

Proof. On the space of 2π-periodic functions one needs that the last equation has bounded solutions, hence it cannot be resonant, then $\mu = 0$ and $z(\tau)$ is a combination of $\cos n\tau$ and $\sin n\tau$.

For the Mathieu equation, if $x(\tau)$ is a 2π-periodic solution, then so is $x(-\tau)$ and $x(\tau) \pm x(-\tau)$. Hence one may assume that $x(\tau)$ has a definite parity. Furthermore, from the uniqueness of the initial value problem, for a given (α, β), one has at most one even and one odd solution. Similarly, $x(\tau + \pi/n)$ is a solution for $(\alpha, -\beta)$ and

$x(\tau+2\pi/n)$ is a solution for (α, β). In particular, $x(\tau+\pi/n)\pm x(-\tau+\pi/n)$ are two linearly independent solutions for $(\alpha, -\beta)$ since of different parity, unless one of them is 0. If one has started with $x(\tau)$ even, if $x(\tau+\pi/n) = x(-\tau+\pi/n) = x(\tau-\pi/n)$, then $x(\tau)$ is $(2\pi/n)$-periodic, while, if $x(\tau+\pi/n) = -x(-\tau+\pi/n)$, then $x(\tau)$ is $(2\pi/n)$-antiperiodic and $(4\pi/n)$-periodic (this implies that n is even). On the other hand, if $x(\tau)$ is odd then the situation is reversed. Thus, if $x(\tau)$ is not $(4\pi/n)$-periodic, one has two periodic solutions for $(\alpha, -\beta)$ and also for (α, β). Note that if $x(\tau)$ is $(4\pi/n)$-periodic, then $y(\tau) = x(2\tau/n)$ is a 2π-periodic solution of $L_2 y(\tau) = 0$, the classical Mathieu equation, i.e., with $n = 2$.

Assume now that, for (α_0, β_0), one has a solution $x_0(\tau)$ of a definite parity. Consider then the self-adjoint operator L_n in the spaces of periodic functions with that parity, say from H^1 into L^2, with a one-dimensional kernel generated by x_0. Take $x = ax_0 + x_1$, with x_1 being L^2-orthogonal to x_0. Then, the classical Ljapunov–Schmidt reduction implies that $L_n x = 0$ is equivalent to a unique analytic solution $x_1(a, \alpha, \beta) = ax_1(1, \alpha, \beta)$, with $x_1(1, \alpha_0, \beta_0) = 0$, and a solution to the bifurcation equation, coming from the L^2-projection on x_0 (see Section 9 of Chapter 1)

$$\alpha - \alpha_0 + 2(\beta - \beta_0) \int_0^{2\pi} (x_0^2 + x_0 x_1) \cos n\tau \, d\tau = 0,$$

after normalizing x_0 to have norm 1 in L^2. The implicit function theorem implies that this equation has a unique analytic solution $\alpha(\beta)$, with

$$\alpha'(\beta_0) = -2 \int_0^{2\pi} x_0^2 \cos n\tau \, d\tau = \int_0^{2\pi} (\alpha_0 x_0^2 - x_0'^2/n^2)/\beta_0 \, d\tau.$$

Since for $\beta = 0$, one has $\alpha = (k/n)^2$ one obtains the curves $\alpha_{k/n}(\beta)$ and $\tilde{\alpha}_{k/n}(\beta)$ characterized by the parity of the solutions and defined for all β's. These curves are monotonous for $\alpha_0 < 0$. Furthermore, the number of zeros on each curve is conserved (by uniqueness of the initial value problem, the zeros are simple, since the L^2-norm of x_0 is 1) and are those of $\cos k\tau$ for $\alpha_{k/n}(\beta)$ or of $\sin k\tau$ for $\tilde{\alpha}_{k/n}(\tau)$. It is then easy to see that $\alpha'_{k/n}(0) = 0$, except if $k/n = 1/2$, with $\alpha'_{1/2}(0) = -1 = -\tilde{\alpha}'_{1/2}(0)$.

Furthermore, if $k/n = k_1/n_1$, then if $x_{n_1}(\tau)$ is a 2π-periodic solution of $L_{n_1} x = 0$, then $x(\tau) = x_{n_1}(n\tau/n_1)$ is a $(2\pi n_1/n)$-periodic solution of $L_n x = 0$, with the parity of x_{n_1}, that is $x(\tau)$ belongs to the unique curve which goes through $(k/n)^2$. This implies that these curves are correctly labelled by k/n and that, conversely, the solutions on $\alpha_{k/n}(\beta)$ are $(2\pi n_1/n)$-periodic. In particular, if $n_1 \geq 3$, one has that $x(\tau)$ is not $(4\pi/n)$-periodic (if it were $y(\tau) = x(2\tau/n)$ would belong to the curve for L_2 going through k_1/n_1, for $\beta = 0$, and hence $n_1 = 1$ or 2) and, as seen above, $\alpha_{k/n}(\beta) = \tilde{\alpha}_{k/n}(\beta)$, curves which are symmetric with respect to the α-axis.

For $n = 2$, if $k/n = k_1$, i.e., $n_1 = 1$, then $x(\tau)$ is π-periodic and $x(\tau + \pi/2)$, solution for $(\alpha, -\beta)$, has the parity of $x(\tau)$, that is the curves are symmetric with respect to the α-axis. On the other hand, if k is odd, then we have seen that $x(\tau+\pi)$ is a solution for (α, β) of the same parity as $x(\tau)$, hence from the uniqueness, $x(\tau+\pi) =$

$ax(\tau)$. Evaluation at $\tau = 0$ and $\tau = \pi$, leads to $a^2 = 1$. Furthermore, it is clear that the sign of a is invariant on the curves $\alpha_{(2k+1)/2}(\beta)$ and $\tilde{\alpha}_{(2k+1)/2}(\beta)$ and it is -1 at $\beta = 0$. Thus, $x(\tau + \pi) = -x(\tau)$ and $x(-\tau + \pi/2) = -x(-\tau - \pi/2)$ has the opposite parity of $x(\tau)$, that is $\alpha_{(2k+1)/2}(-\beta) = \tilde{\alpha}_{(2k+1)/2}(\beta)$. For $\alpha = \beta = 0$, only the constant solutions (hence even) exist, thus $\tilde{\alpha}_0(\beta)$ does not appear.

Now, if $n = 2$, two curves may intersect at $(\alpha_0, \beta_0 \neq 0)$ only if one has $\alpha_0 = \alpha_{k_1/2}(\beta_0) = \tilde{\alpha}_{k_1/2}(\beta_0)$: in fact, within the same parity, the implicit function theorem prevents intersections and if $\alpha_{k_1/2}(\beta_0) = \tilde{\alpha}_{k_1'/2}(\beta_0)$ for $k_1' \neq k_1$, with solutions x_0 and $\tilde{x}_0$, then, from the separation of the zeros (i.e., between two consecutive zeros of x_0 one has exactly one zero of $\tilde{x}_0$: if not, if $\tilde{x}_0$ is not 0 on this interval, then $(x_0/\tilde{x}_0)' = W(x_0, \tilde{x}_0)/\tilde{x}_0^2$, with a wronskian $|W| \equiv 1$, must have a zero) and from the conservation of zeros along the curves, one gets a contradiction. Thus, if k_1 is even (hence x_0 and $\tilde{x}_0$ are π-periodic), one has

$$x_0(\tau) = A_0/2 + \sum_{m\geq 1} A_{2m} \cos 2m\tau$$

$$\tilde{x}_0(\tau) = \sum_{m\geq 1} B_{2m} \sin 2m\tau,$$

with the recurrence relations

$$\begin{aligned}(\alpha/2)A_0 &+ \beta A_2 = 0\\ (\alpha - m^2)A_{2m} &+ \beta(A_{2m-2} + A_{2m+2}) = 0, \quad m \geq 1,\\ (\alpha - m^2)B_{2m} &+ \beta(B_{2m-2} + B_{2m+2}) = 0, \quad m \geq 1,\ B_0 = 0.\end{aligned}$$

Now, for $m > m_0$ large enough, one may solve this system in terms of βA_{2m_0}, in particular $A_{2m_0+2} = \beta a(\alpha, \beta)A_{2m_0}$ and $B_{2m_0+2} = \beta a(\alpha, \beta)B_{2m_0}$, reducing to two tri-diagonal systems: one for $X = (A_0, \dots, A_{2m_0})$ of the form

$$\mathcal{A}X = \begin{pmatrix} \alpha/2 & \beta \\ \beta & \mathcal{B} \end{pmatrix} X = 0,$$

and one for $Y = (B_1, \dots, B_{2m})$ of the form

$$\mathcal{B}Y = 0,$$

where the last line of $\mathcal{B}Y$ is $\beta B_{2m_0-2} + (\alpha - m_0^2 + \beta^2 a(\alpha, \beta))B_{2m_0}$.

If the two systems have non-trivial solutions, then $\det \mathcal{A} = \det \mathcal{B} = 0$. But $\det \mathcal{A} = (\alpha/2) \det \mathcal{B} - \beta^2 \det \mathcal{B}_{1,1}$, where $\mathcal{B}_{1,1}$ is $\mathcal{B}$ with its first line and first column deleted. Thus, $\det \mathcal{B}_{1,1} = 0$ and, since $\det \mathcal{B} = (\alpha - 1) \det \mathcal{B}_{1,1} - \beta^2 \det \mathcal{B}_{2,2}$, with deleting from $\mathcal{B}_{1,1}$ the first line and first column to get $\mathcal{B}_{2,2}$, one obtains $\det \mathcal{B}_{2,2} = 0$. Continuing this process, one arrives at $\alpha - m_0^2 + \beta^2 a(\alpha, \beta) = 0$, a contradiction if m_0 is large enough (it is easy to prove that $a(\alpha, \beta)$ is a decreasing function of m_0 and tends to 0 when m_0 goes to ∞).

If k_1 is odd (hence x_0 and $\tilde{x}_0$ are π-antiperiodic), one has

$$\begin{aligned} x_0(\tau) &= \sum_{m\geq 0} A_{2m+1}\cos(2m+1)\tau \\ \tilde{x}_0(\tau) &= \sum_{m\geq 0} B_{2m+1}\sin(2m+1)\tau, \end{aligned}$$

with the relations

$$\begin{aligned} (4\alpha - 1 + 4\beta)A_1 + 4\beta A_3 &= 0 \\ (4\alpha - 1 + 4\beta)B_1 + 4\beta B_3 &= 0 \\ (4\alpha - (2m+1)^2)C_{2m+1} + 4\beta(C_{2m-1} + C_{2m+3}) &= 0, \quad m \geq 1 \text{ and } C_k = A_k \text{ or } B_k. \end{aligned}$$

With the same arguments, one obtains the tridiagonal matrices

$$\begin{aligned} \tilde{\mathcal{A}}X &= \begin{pmatrix} 4\alpha - 1 + 4\beta & 4\beta \\ 4\beta & \mathcal{C} \end{pmatrix} X = 0 \\ \tilde{\mathcal{B}}Y &= \begin{pmatrix} 4\alpha - 1 - 4\beta & 4\beta \\ 4\beta & \mathcal{C} \end{pmatrix} Y = 0, \end{aligned}$$

where $\mathcal{C}$ is common. Then $\det\tilde{\mathcal{A}} = (4\alpha - 1 + 4\beta)\det\mathcal{C} - (4\beta)^2\det\mathcal{C}_{1,1}$ and $\det\tilde{\mathcal{B}} = (4\alpha - 1 - 4\beta)\det\mathcal{C} - (4\beta)^2\det\mathcal{C}_{1,1}$, where both determinants have to be 0. This implies, if $\beta \neq 0$, that $\det\mathcal{C} = \det\mathcal{C}_{1,1} = 0$. One arrives then at the same contradiction.

If, for n and n' greater than 2, one has $\alpha_{k/n}(\beta) = \alpha_{k'/n'}(\beta)$, with k and n relatively prime (respectively k' and n'), then for N the least common multiple of n and n' and $x(\tau)$ solution on $\alpha_{k/n}(\beta)$, then $y(\tau) = x(N\tau/n)$ will be a solution on $\alpha_{k_1/N}(\beta)$ for $k/n = k_1/N$ with $2k_1$ or $2k_1 - 1$ zeros. Hence, one would have four periodic solutions for $L_N y = 0$, two even and two odd, with $2k_1$ and $2k_1' = 2k'N/n'$ zeros, hence independent, which is clearly a contradiction for a second order equation.

With respect to the boundedness of solutions of $L_2 x = 0$, convert the equation into a first order system $X' = A(\tau)X$, with Trace $A = 0$ and $A(\tau + \pi) = A(\tau)$. Then, from Remark B.1, one has that the fundamental matrix satisfies $\Phi(\tau + m\pi) = \Phi(\tau)\Phi(\pi)^m$. Instead of using the full strength of Floquet theory, it is easy to see that $\Phi(\pi)$ has two eigenvalues with product equal to 1. If they are complex conjugate, then $\Phi(\pi)^m$ is bounded as well as $\Phi(\tau)$, for τ in $[0, 2\pi]$. Hence, in this case any solution is bounded. While, if the eigenvalues are real, then $\Phi(\pi)^m$ is similar to $\begin{pmatrix} \lambda^m & 0 \\ 0 & \lambda^{-m} \end{pmatrix}$, giving unbounded solutions when τ goes to $\pm\infty$. Finally, if $\lambda = 1$, since there is only one periodic solution on $\alpha_{k/2}(\beta)$, for $\beta \neq 0$, $\Phi(\pi)^m$ is similar to $\begin{pmatrix} 1 & m \\ 0 & 1 \end{pmatrix}$, giving an unbounded second solution.

Notice that, if $x(\tau)$ belongs to $\alpha_{k/n}(\beta)$, then $y(\tau) = x(2\tau/n)$ is a $n\pi$-periodic solution of $L_2 y(\tau) = 0$. Thus, for $n \geq 3$, $\Phi(\pi)^n = I$ and the eigenvalues of

$\Phi(\pi)$ are $\exp(\pm 2k' i\pi/n)$. But Trace $\Phi(\pi) = 2\cos 2k'\pi/n$ is analytic on $\alpha_{k/n}(\beta)$, hence constant and equal to its value at $\alpha_{k/n}(0) = (k/n)^2$, that is for $\Phi(\tau) = \begin{pmatrix} \cos 2k\tau/n & k^{-1}\sin 2k\tau/n \\ -k\sin 2k\tau/n & \cos 2k\tau/n \end{pmatrix}$, or else $k' \equiv k, mod\, n$. The rational k/n is called the *rotation number*. Now, Trace $\Phi(\pi)$ is an analytic function of α and β and, on each stable region and for fixed β_0, covers the interval from -2 (for $k/n = k_1/2$ with k_1 odd) to 2 (for $k/n = k_1/2$ with k_1 even).

Furthermore, since the curves $\alpha_{k/n}(\beta)$ do not intersect, the points $\alpha_{k/n}(\beta_0)$, where Trace $\Phi(\pi) = 2\cos 2k\pi/n$, are ordered in the same way as $\alpha_{k/n}(0)$. Therefore, Trace $\Phi(\pi)$ cannot be locally constant and is strictly monotone as a function of α. This implies that, arbitrarily close to (α, β), there are points of the form $(\alpha_{k/n}(\beta), \beta)$, giving the foliation.

The last point is the asymptotic behavior, when β goes to ∞: let $x(\tau)$ belong to $\alpha_{k/2}(\beta)$, hence an even solution of

$$x'' + 4(\alpha + 2\beta - 4\beta\sin^2\tau)x = 0,$$

with $2k$ zeros in $[-\pi, \pi]$. Multiplying the equation by $x(\tau)$, integrating over $[-\pi, \pi]$ and using the periodicity of $x(\tau)$, one needs that $\alpha + 2\beta \geq 0$. Let $y(\tau) = x(\tau/(2\beta^{1/4}))$, then $y(\tau)$ is a $(4\pi\beta^{1/4})$-periodic, even solution of

$$y'' + (A - (2\beta^{1/4}\sin\tau/(2\beta^{1/4}))^2)y = 0,$$

with $A = (\alpha + 2\beta)/\beta^{1/2}$. Since $4\beta^{1/2}\sin^2(\tau/(2\beta^{1/4})) \leq \tau^2$, the solution $y(\tau)$ will be compared to solutions of the equation

$$z'' + (A - \tau^2)z = 0.$$

Let $H(\tau) = z(\tau)e^{\tau^2/2}$, then $H(\tau)$ satisfies Hermite's equation

$$H'' - 2\tau H' + (A - 1)H = 0.$$

A series solution $H(\tau) = \sum h_n\tau^n$ yields the recurrence relation

$$(n + 2)(n + 1)h_{n+2} = (2n + 1 - A)h_n.$$

In particular, if $A = 2N + 1$, one solution will be a polynomial $H_N(\tau)$ of degree N, with the parity of N and normalized so that the leading coefficient is τ^N. Now, it is easy to verify that $\tau H_N - H'_N/2$ satisfies the equation for H_{N+1} and, having a leading coefficient equal to τ^{N+1}, one has that

$$H_{N+1} = \tau H_N - H'_N/2.$$

This implies that between two consecutive zeros of H_N (hence with H'_N of different signs) one has a zero of H_{N+1}. Furthermore, if τ_N is the largest zero of H_N (with $H'_N(\tau_N) > 0$, since the leading term of H_N is τ^N), then $H_{N+1}(\tau_N) < 0$ and one has a zero of H_{N+1} to the right of τ_N. Due to the parity of H_{N+1}, this last zero generates its symmetric, for $\tau < 0$. An easy induction argument implies that $H_N(\tau)$ has exactly N zeros. Now, $z_N(\tau) = e^{-\tau^2/2}H_N(\tau)$ is bounded and as such has all its N zeros confined to the interval $|\tau| \leq (2M+1)^{1/2}$: in fact, if z_N has its last zero, τ_N, outside this interval, then $z_N(\tau) > 0$, $z_N''(\tau) > 0$ for $\tau > \tau_N$ and, since $z_N'(\tau_N) > 0$, one has that $z_N(\tau)$ is increasing and convex, contradicting the boundedness.

We shall then use the following *comparison principle*:

If $y(\tau)$ and $z(\tau)$ are solutions to the equations

$$y'' + p(\tau)y = 0 \quad \textit{and} \quad z'' + q(\tau)z = 0,$$

with $p(\tau) \geq q(\tau)$, then between two zero of z there is at least one zero of y.

(If $y(\tau)$ is not 0, say positive, between τ_1 ant τ_2, two consecutive zeros of z with, say, $z(\tau) > 0$ and $z'(\tau_1) > 0$, $z'(\tau_2) < 0$ (if not change z to $-z$), then $\int_{\tau_1}^{\tau_2}(z''y - zy'')\,d\tau = \int_{\tau_1}^{\tau_2}(p-q)yz\,d\tau = z'y - zy'|_{\tau_1}^{\tau_2}$ leads to a contradiction, since the integral is non-negative while the last term is strictly positive).

Hence, if $A \geq 2N+1$, then $y(\tau)$ has at least $N-1$ zeros for $|\tau| \leq (2N+1)^{1/2}$, since $z_N(\tau)$ has N zeros in this interval. But $y(\tau)$ has $2k$ zeros for $|\tau| \leq 2\pi\beta^{1/4}$. Taking $N = 2k+2$, one would arrive, if $A \geq 4k+5$, at a count of at least $2k+1$ zeros for $y(\tau)$ for $|\tau| \leq (4k+5)^{1/2}$, which is not possible if $2\pi\beta^{1/4} > (4k+5)^{1/2}$. Then, if $\beta > (4k+5)^2/(2\pi)^4$ one has that $0 \leq A = (\alpha + 2\beta)\beta^{-1/2} \leq 4k+5$. This implies that $\alpha_{k/2}(\beta)$ cannot be bounded from below.

In fact, one may prove, by looking more closely at the distribution of zeros of $y(\tau)$, that

$$\alpha = -2\beta + (4k+1)\beta^{1/2}/2 + 0(1). \qquad \square$$

Remark 3.8. One may prove that, for $k/n \neq 1/2$, one has

$$\alpha''_{k/n}(0) = \tilde{\alpha}''_{k/n}(0) = \frac{4}{4(k/n)^2 - 1}.$$

In fact, since x_0 is analytic in β, one has that

$$\alpha''(\beta) = -4\int_0^{2\pi} x_0 x_{0\beta} \cos n\tau \, d\tau$$

and, by differentiating the Mathieu equation, that at $\beta = 0$, one has that $x_{0\beta}$ is a solution (of the same parity of x_0) of

$$y'' + k^2 y = -2n^2 \cos n\tau x_0,$$

where one has used that $\alpha(0) = (k/n)^2$ and $\alpha'(0) = 0$. Then, for $x_0 = \cos k\tau/\sqrt{\pi}$, one has that $x_{0\beta}$ is orthogonal to x_0 and

$$x_{0\beta} = \frac{n}{\sqrt{\pi}}\left(\frac{\cos(n+k)\tau}{n+2k} + \frac{\cos(n-k)\tau}{n-2k}\right)$$

and the result follows.

For the second system, one has that $X_0(\tau) = (0, 0, a\cos(n\tau + \varphi))$ is a solution, with $\nu = \nu_0/n$, of $g(\nu, X) = 0$, where $\nu_0 = (k/m)^{1/2}$. Let us linearize $g(\nu, X)$ around that solution, with $\nu = \nu_0/n + \mu$ and $Z = Z_n + z$ and obtain $M_n X = 0$, or else

$$x'' + n^2\left(\frac{\alpha + \beta\cos(n\tau + \varphi)}{1 + \beta\cos(n\tau + \varphi)}\right)x = 0$$
$$y'' + n^2\left(\frac{\alpha + \beta\cos(n\tau + \varphi)}{1 + \beta\cos(n\tau + \varphi)}\right)y = 0$$
$$z'' + n^2 z = 2n^3\nu_0^{-1}\cos(n\tau + \varphi)\mu,$$

where $\alpha = mg/(mg + kr_0)$ and $\beta = ak/(mg + kr_0)$.

In this linearization, we have taken $r = Z_n + r_0 + mg/k$, i.e., that $|\beta| \le 1$. Note that $|\beta| = 1$ corresponds to $a = r_0 + mg/k$, i.e., to $Z_n = 0$ for $n\tau + \varphi = \pi$, that is to a spring totally collapsed.

Note also that $0 < \alpha < 1$. Thus, we shall work in the rectangle $0 \le \alpha \le 1$, $|\beta| \le 1$. The first two equations are singular Hill's equations, while the third will have a non-resonant solution only for $\mu = 0$ and $z = \varepsilon\cos(n\tau + \psi)$. Note that, as before, one may fix the phase φ of the one-dimensional orbit Z_0, at 0.

Lemma 3.10. *One has* dim ker $M_n = 2, 4$ *or* 6, *with eigenvectors* $\mu = 0$, $z = \cos n\tau$ *or* $\sin n\tau$, *x and y corresponding to analytic curves* $\alpha_{k/n}(\beta)$, $\tilde{\alpha}_{k/n}(\beta)$ *passing through* $\alpha_{k/n}(0) = \tilde{\alpha}_{k/n}(0) = (k/n)^2$. *Solutions on* $\alpha_{k/n}(\beta)$ *are even in* τ *and those on* $\tilde{\alpha}_{k/n}(\beta)$ *are odd. Furthermore, these curves are symmetric with respect to the* α*-axis and equal except for* $n = 2k$, *where* $\alpha_{1/2}(-\beta) = \tilde{\alpha}_{1/2}(\beta)$. *The region, for constant* β, *between these two curves, which intersect only at* $\beta = 0$, $\alpha = 1/4$, *is a region of instability, while the regions, for constant* α, *between* $\alpha_{1/2}(\beta)$ *and* $\tilde{\alpha}_{1/2}(\beta)$ *is foliated by the curves* $\alpha_{k/n}(\beta)$. *Also,* $\alpha_0(\beta) \equiv 0$, *with unique solution* $1 + \beta\cos n\tau$ *(while* $\tilde{\alpha}_0(\beta)$ *does not exist) and* $\alpha_1(\beta) = \tilde{\alpha}_1(\beta) = 1$ *with solution* $\cos n\tau$ *and* $\sin n\tau$. *The solutions on* $\alpha_{1/2}(\beta)$ *and* $\tilde{\alpha}_{1/2}(\beta)$ *are* $(2\pi/n)$*-antiperiodic and, if* $k/n = k_1/n_1$, *with* k_1 *and* n_1 *relatively prime, then* $x_n(\tau) = x_{n_1}(n\tau/n_1)$ *and has period* $2\pi n_1/n$. *On* $\alpha_{k/n}(\beta)$, *the solution* $x_n(\tau)$ *has* $2k$ *internal zeros and on* $\tilde{\alpha}_{k/n}(\beta)$ *the solution* $\tilde{x}_n(\tau)$ *has* $2k - 1$ *internal zeros.*

Finally, $\alpha_{k/n}(\beta)$ *goes to* 0 *if* $|\beta|$ *goes to* 1, *if* $k/n < 1/2$ *and to* 1, *if* $k/n > 1/2$. *In fact,* $\alpha_{1/2}(\beta)$ *is monotone, decreasing from* $(0, 1)$, *where it is vertical, to* $(1, -1)$ *where it is horizontal.*

Proof. Fortunately, most of the arguments in the proof of Lemma 3.9 did not depend on the special form of $\alpha + 2\beta\cos n\tau$ but only on the fact that this function is even. Thus, one has to concentrate on the complement of these arguments. The first one is the Ljapunov–Schmidt reduction to the bifurcation equation

$$h(\alpha, \beta) = \int_0^{2\pi}\frac{(\alpha - \alpha_0 + (\beta - \beta_0 + \alpha\beta_0 - \alpha_0\beta)\cos n\tau)(x_0^2 + x_0 x_1)}{(1 + \beta\cos n\tau)(1 + \beta_0\cos n\tau)}\,d\tau = 0,$$

where x_0 and x_1 have the same meaning as before, that is $x = ax_0 + x_1$, with $x_1(a, \alpha, \beta) = ax_1(1, \alpha, \beta)$ and $x_1(1, \alpha_0, \beta_0) = 0$: $h(\alpha, \beta)$ is just the projection on x_0 of $x'' + P(\alpha, \beta)x$. Then,

$$h_\alpha(\alpha_0, \beta_0) = \int_0^{2\pi} x_0^2(1 + \beta_0 \cos n\tau)^{-1}\, d\tau > 0$$

and

$$h_\beta(\alpha_0, \beta_0) = (1 - \alpha_0) \int_0^{2\pi} x_0^2 \cos n\tau (1 + \beta_0 \cos n\tau)^{-2}\, d\tau.$$

Thus, one has an analytic curve $\alpha(\beta)$, for $|\beta| < 1$, which must cross the α-axis at some $\alpha(0) = (k/n)^2$, for some $0 \leq k \leq n$.

In particular, for $n = 1$, there are only two curves, $\alpha_0(\beta) = 0$ with only even solutions (bifurcating from $\beta = 0$ with the constant solution) and $\alpha_1(\beta) = 1$ with solutions of both parities and given in the lemma. The other solution, for $\alpha_0(\beta) = 0$, obtained by reduction of order, is

$$v(\tau) = (1 + \beta \cos n\tau) \int_0^\tau (1 + \beta \cos ns)^{-2}\, ds,$$

which is neither periodic nor bounded.

For $n = 2$ and $\beta = 0$ with $\alpha_{1/2}(0) = \tilde{\alpha}_{1/2}(0) = 1/4$, the solutions are $\cos \tau$ and $\sin \tau$ and one has $\alpha'_{1/2}(0) = -3/8$ and $\tilde{\alpha}'_{1/2}(0) = 3/8$. Furthermore, solutions on $\alpha_{1/2}(\beta)$ are even and π-antiperiodic and those on $\tilde{\alpha}_{1/2}(\beta)$ are odd and π-antiperiodic. Hence, if these curves intersect at some (α, β) one has

$$x(\tau) = \sum_0^\infty A_{2n+1} \cos(2n + 1)\tau$$

$$\tilde{x}(\tau) = \sum_0^\infty B_{2n+1} \sin(2n + 1)\tau$$

with the recurrence relations

$$(m^2 - 4\alpha)C_m + (\beta/2)((m - 2)^2 - 4)C_{m-2} + ((m + 2)^2 - 4)C_{m+2}) = 0,$$

where $m = 2n + 1 \geq 0$, C_m stands for A_{2n+1} or B_{2n+1}, with $A_1 = A_{-1}$ and $B_{-1} = -B_1$. If one defines $D_m = (m^2 - 4)C_m$, with $\sum D_m^2 < \infty$, since the solution is in H^2, one has to solve the system

$$\begin{aligned} ((4\alpha - 1)/3 + \varepsilon\beta/2)D_1 + (\beta/2)D_3 &= 0 \\ (\mathcal{A} + 4(1 - \alpha)\mathcal{M})X + (\beta/2)D_1 &= 0, \end{aligned}$$

where $\varepsilon = 1$ for $x(\tau)$ and $\varepsilon = -1$ for $\tilde{x}(\tau)$, $X = (D_3, D_5, \dots)$ is in L^2, the operator $\mathcal{A}$ is symmetric and tridiagonal, with 1 on the diagonal and $\beta/2$ on the two off-diagonals, and $\mathcal{M}$ is a diagonal operator with terms $1/(m^2 - 4)$, $m = 3, 5, \dots$ Now,

$$(\mathcal{A}X, X) = \|X\|^2 + \beta \sum D_i D_{i+1} \geq |\beta| D_3^2/2 + (1 - |\beta|)\|X\|^2,$$

then, for $|\beta| < 1$ and $\alpha \leq 1$, the operator $\mathcal{B} = \mathcal{A} + 4(1-\alpha)\mathcal{M}$ is invertible with $\|\mathcal{B}^{-1}\| \leq 1/(1-|\beta|)$. In particular, $D_3 = -(\beta/2)a(\alpha,\beta)D_1$, where $a(\alpha,\beta) = (\mathcal{B}^{-1}(1,0,\dots),(1,0,\dots)) > 0$, since $\mathcal{B}$ and $\mathcal{B}^{-1}$ are positive operators. Hence, the problem is reduced to

$$f(\alpha,\beta) = (4\alpha-1)/3 + \varepsilon\beta/2 - (\beta^2/4)a(\alpha,\beta) = 0.$$

This implies that the two transition curves meet only at $\beta = 0$ and $\alpha = 1/4$.

Note that $a(1,\beta) = \lim_{\beta\to\infty} \det(\mathcal{A}_{p-1})/\det(\mathcal{A}_p)$, where $\mathcal{A}_p$ is the truncation of $\mathcal{A}$ to p modes. As $\det\mathcal{A}_p = \det\mathcal{A}_{p-1} - (\beta^2/4)\det\mathcal{A}_{p-2}$ for $p \geq 3$, with $\det\mathcal{A}_1 = 1$, $\det\mathcal{A}_2 = 1-\beta^2/4$, one may use the generating function $g(z) = \sum_1^\infty \det\mathcal{A}_p z^p$ to get

$$g(z) = -1 + ((\beta^2/4)z^2 - z + 1)^{-1} = -1 + 4/\beta^2(z_1 - z_2)\sum_0^\infty (z_2^{-p-1} - z_1^{-p-1})z^p,$$

where

$$z_{1,2} = (2/\beta^2)(1 \pm (1-\beta^2)^{1/2}),\ z_1 z_2 = 4/\beta^2 \text{ and } z_1 - z_2 = (1-\beta^2)^{1/2}4/\beta^2.$$

Thus, $\det\mathcal{A}_p = [(1+(1-\beta^2)^{1/2})^{p+1} - (1-(1-\beta^2)^{1/2})^{p+1}]/2^{p+1}(1-\beta^2)^{1/2}$.

From this relation it follows that $a(1,\beta) = 2/(1+(1-\beta^2)^{1/2})$.

Now

$$f_\alpha = 4/3 - (\beta^2/4)a_\alpha,$$

with $\alpha_\alpha = 4(\mathcal{B}^{-1}\mathcal{M}\mathcal{B}^{-1}(1,0,\dots),(1,0,\dots))$ and, using $\mathcal{M} = (\mathcal{B}-\mathcal{A})/4(1-\alpha)$, one has

$$a_\alpha = (1-\alpha)^{-1}(a - (\mathcal{A}\mathcal{B}^{-1}(1,0,\dots),\mathcal{B}^{-1}(1,0,\dots))).$$

Since $(\mathcal{A}\mathcal{B}^{-1}(1,0,\dots),\mathcal{B}^{-1}(1,0,\dots)) \geq |\beta|a^2/2$, one obtains

$$(1-\alpha)f_\alpha \geq 4(1-\alpha)/3 - (\beta^2/4)a + (|\beta|^3/8)a^2.$$

Then, if $f(\alpha,\beta) = 0$, one has $a = 4\beta^{-2}(1+\varepsilon\beta/2 - x)$, with $x = 4(1-\alpha)/3$ and, after a short computation,

$$2|\beta|(1-\alpha)f_\alpha \geq (2+\varepsilon\beta - |\beta| - 2x)^2 + |\beta|(2+\varepsilon\beta-|\beta|).$$

Thus, since $|\beta| \leq 1$, the left-hand side is strictly positive, unless $\varepsilon\beta = -1$ and $x = 0$, i.e., for $\alpha = 1$. We have recovered the fact that the transition curves can be parametrized by β. Note that, since $\mathcal{M}$ is a positive operator, one has $a_\alpha > 0$ and $0 \leq f_\alpha \leq 4/3$.

On the other hand,

$$f_\beta = \varepsilon/2 - \beta a/2 - \beta^2 a_\beta/4,$$

with $a_\beta = -(\mathcal{B}^{-1}\mathcal{A}_\beta\mathcal{B}^{-1}(1,0,\dots),(1,0,\dots))$. Since $\beta\mathcal{A}_\beta = \mathcal{B} - I - 4(1-\alpha)\mathcal{M}$, one obtains

$$\beta f_\beta = \varepsilon\beta/2 - \beta^2 a/4 - (\beta^2/4)(\|\mathcal{B}^{-1}(1,0,\dots,)\|^2 \\ + 4(1-\alpha)(\mathcal{M}\mathcal{B}^{-1}(1,0,\dots),\mathcal{B}^{-1}(1,0,\dots)).$$

In particular, on the transition curves, one gets

$$-\beta f_\beta = (4\alpha-1)/3 + (\beta^2/4)(\|\mathcal{B}^{-1}(1,0)\|^2 + 4(1-\alpha)(\mathcal{M}\mathcal{B}^{-1}(1,0),\mathcal{B}^{-1}(1,0))).$$

Hence, if $\alpha \geq 1/4$, one has $\beta f_\beta < 0$ and in this range of α's one may parametrize the curves by α. This implies that, for $1/4 \leq \alpha \leq 1$, the curve $\alpha_{1/2}(\beta)$ is decreasing, with $\beta < 0$, while $\tilde{\alpha}_{1/2}(\beta)$ is increasing from $(1/4, 0)$ to $(1, 1)$. Note that $\mathcal{B}^{-1}(1,0,\dots) = 2(1,-1,1,-1,\dots)$ for $\alpha = 1$, $\beta = 1$ and $f_\alpha = 4/3 - \sum_1^\infty((2n+1)^2-4)^{-1}$ is positive and finite, that is, the transition curves arrive horizontally at $(1, 1)$.

Now, if one denotes by $\mathcal{C}$ the diagonal operator $I + 4(1-\alpha)\mathcal{M}$ and by X the vector $\mathcal{B}^{-1}(1,0,\dots)$, one has

$$\beta a_\beta = -a + (\mathcal{C}X, X) \\ f_\beta = \varepsilon/2 - \beta a/4 - (\beta/4)(\mathcal{C}X, X),$$

then, using the relation $\beta\mathcal{A}_\beta = \mathcal{B} - \mathcal{C}$, one obtains

$$f_{\beta\beta} = -(1/2)(\mathcal{B}^{-1}\mathcal{C}X, \mathcal{C}X) < 0,$$

since $\mathcal{B}^{-1}$ is a positive operator, hence f_β is a decreasing function. Now, for $\alpha = 0$, $\beta = -1$, it is easy to verify that $X = 2(1/3, 1/5, 1/7, \dots)$, noting that, for $|\beta| \leq 1$, one has $(\mathcal{B}X, X) \geq |\beta|a^2/2$, i.e., $\mathcal{B}$ is one-to-one. Thus, $a(0,-1) = 2/3$ and, for $\varepsilon = -1$, one has $f(0,-1) = 0$, that is the curve $\tilde{\alpha}_{1/2}$ starts at $\beta = -1$, $\alpha = 0$. Furthermore, for that X, one has

$$(\mathcal{C}X, X) = 4\sum_1^\infty (2n+1)^{-2}(1 + 4/((2n+1)^2 - 4)) \\ = 4\sum_1^\infty 1/(2n-1)(2n+3) = 4/3,$$

by using partial fractions. Thus, $f_\beta(0,-1) = 0$ for $\varepsilon = -1$. This implies that $f_\beta < 0$ and that one may parametrize the curves $\tilde{\alpha}_{1/2}$ by α, starting vertically at $(0,-1)$ and arriving horizontally at $(1, 1)$.

The rest of the properties (regions of stability, foliations) follow the same lines of proof as in the preceding case, since they do no depend on the particular form of the equation. □

Remark 3.9. By normalizing x_0 in such a way that $h_\alpha(\alpha_0, \beta_0) = 1$, then $\alpha'(\beta) = -h_\beta$. Hence, for $k/n \neq 1/2$ and $\beta_0 = 0$, one has $x_0(\tau) = \pi^{-1/2} \cos k\tau$ and $\alpha'(0) = 0$, with

$$\alpha''(0) = -2(1 - (k/n)^2) \int_0^{2\pi} x_0 \cos n\tau (x_{0\beta} - x_0 \cos n\tau)\, d\tau,$$

where $x_{0\beta}$ is a solution (by differentiating the Hill equation) of

$$y'' + k^2 y = (k^2 - n^2) x_0 \cos n\tau,$$

thus,

$$x_{0\beta}(\tau) = \frac{n^2 - k^2}{2n\sqrt{\pi}} \left(\frac{\cos(n+k)\tau}{n+2k} + \frac{\cos(n-k)\tau}{n-2k} \right).$$

From this it is easy to prove that

$$\alpha''_{k/n}(0) = 3(k/n)^2(1 - (k/n)^2)(4(k/n)^2 - 1)^{-1}.$$

Remark 3.10. For a given n, one may compute numerically the curves $\alpha_{k/n}(\beta)$ by combining a path following method with a numerical integration of the solution: in fact, if (α_0, β_0) is on this curve, one may take a point at a certain distance on the tangent and, on an orthogonal line, test for periodicity by looking at the Poincaré return map of the solution for (α, β): these are obtained by Runge–Kutta of high order. For the transition curves ($n = 1$ or 2), one may also use the fact that the solutions in the Arnold tongues are unstable. It is interesting to see the foliation phenomenon and that the curves corresponding to high rotation numbers are easier to follow than the transition curves. For Mathieu's equation, the regions of stability decay very fast as $|\beta|$ increases.

Let us turn now to the non-linear systems and the application of the orthogonal degree. Since the arguments are similar for both systems, we shall treat them simultaneously. Thus, fix n and assume that the vertical line, corresponding to a fixed α, crosses the line $\alpha_{k/n}(\beta)$ at (α_0, β_0). Since the points of tangency are finite, we may assume that the crossing is transversal. On that line the nonlinear systems have the solution $\nu = \nu_0/n$, $x = y = 0$, $Z_n = a\cos(n\tau + \varphi)$, where a is proportional to β, that is a family, parametrized by β, of one-dimensional orbits. If a_0 corresponds to β_0, let Ω be the following tubular neighborhood of $(\nu_0/n, 0, 0, a_0\cos(n\tau + \varphi))$:

$$\Omega = \{(\nu, x, y, a\cos(n\tau + \varphi) + \tilde{Z}) : \tilde{Z} \text{ is } L^2\text{-orthogonal to} \cos n\tau \text{ and } \sin n\tau,$$
$$|\nu - \nu_0/n| < 2\varepsilon,\ \|x\|^2 + \|y\|^2 < 4\varepsilon^2,\ |a - a_0| < 2\rho,\ \|\tilde{Z}\| < 2\varepsilon\},$$

where $\|x\|$ is the H^2-norm of $x(\tau)$. Consider, from $\mathbb{R} \times H^2(S^1)$ into $\mathbb{R} \times L^2(S^1)$, the following pair

$$f_\varepsilon(\nu, X) = (d^2(\nu, X) - \varepsilon^2, -f(\nu, X)),$$

(or $-g(\nu, X)$ for the second system), where $d^2(\nu, X) = (\nu - \nu_0/n)^2 + \|x\|^2 + \|y\|^2 + \|\tilde{Z}\|^2$ is the distance to the plane $\nu = \nu_0/n$, $x = y = 0$, $Z = a\cos(n\tau + \varphi)$, which will be called the trivial solution.

Choose ε so small that the only ν in Ω, of the form ν_0/m, is for $m = n$. In particular, any zero in Ω of $f_\varepsilon(\nu, X)$ must have, from the form of the equations, $\|x\|^2 + \|y\|^2 > 0$: in fact, if $x = y = 0$, then $f(\nu, X)$ reduces to $\nu^2 Z'' + \nu_0^2 Z = 0$. Furthermore, for $|a - a_0| = 2\rho$ small enough, the (x, y) part of the linearization is invertible, since one is off the curve $\alpha_{k/n}(\beta)$, and the only solution, for ε small enough, will be on the plane, i.e., with $d = 0$. Thus, $f_\varepsilon(\nu, X)$ is non-zero on $\partial\Omega$ and its orthogonal degree is well defined, more precisely, the orthogonal degree of the projection on N_1 modes, after solving for the other modes in Ω, as in Example 3.5; we leave these details to the reader.

Choosing ρ and ν appropriately, one may assume that, whenever $f(\nu, X) = 0$ in Ω and $|a - a_0| > \rho/2$, then $x = y = 0$ and $d(\nu, X) = 0$. Then, one may perform the orthogonal deformation $(\lambda(d^2 - \varepsilon^2) + (1 - \lambda)(\rho^2 - (a - a_0)^2), -f(\nu, X))$ on $\partial\Omega$. Then,

$$\deg_\perp(f_\varepsilon; \Omega) = i_-(f_\varepsilon) + i_+(f_\varepsilon),$$

where $i_\pm(f_\varepsilon)$ is the orthogonal index of $(\rho^2 - (a - a_0)^2, -f)$ at $\nu = \nu_0/n$, $x = y = 0$, $a = a_0 \pm \rho$, $Z_n = a\cos(n\tau + \varphi)$, with isotropy $H = \mathbb{Z}_n \times S^1$. For $\varphi = 0$, the linearization of the pair, at $Z_n = (a_0 \pm \rho)\cos n\tau$, will be

$$Df_\pm(\mu, X) = (\mp 2\rho\varepsilon_1, -\mathcal{M}_0 L_n(\mu, X)),$$

where $\nu = \nu_0/n + \mu$, $Z = Z_n + z$, with $z = \varepsilon_1 \cos n\tau + \varepsilon_2 \sin n\tau + \tilde{Z}$, hence $a^2 = (a_0 \pm \rho + \varepsilon_1)^2 + \varepsilon_2^2$. Here, $\mathcal{M}_0 = (\nu_0/n)^2 \operatorname{diag}(m, m, m + M)$ comes from the matrix $\mathcal{M}$ evaluated at $x = y = 0$. For the second system, one has $-m(\nu_0/n)^2 M_n(\mu, X)$. Hence, the kernel of the linearization is generated by $\mu = 0$, $x = y = 0$, $z = \varepsilon_2 \sin n\tau$, i.e., by X_0'. Thus, both indices may be computed from the results given for the case of truly periodic solutions of 2nd order Hamiltonian systems of Proposition 3.3, modified by the ν-variable.

Since $H = \mathbb{Z}_n \times S^1$, one has $H_0 = S^1$, $V_0^{H_0} = \{(\nu, 0, 0, Z)\}$, the torus part $\underline{H} = \{e\} \times S^1$ with $V^{\underline{H}} = \{(\nu, 0, 0, Z(\tau))\}$ and $V^H = \{(\nu, 0, 0, Z(\tau))$, with $Z(\tau)$ being $(2\pi/n)$-periodic$\}$, i.e., with modes which are multiples of n: here $p = q = n$ and $\gamma_0 = \mathrm{Id}$. Furthermore, if K is such that $H/K \cong \mathbb{Z}_2$ then n is even, $K \cong \mathbb{Z}_{n/2} \times S^1$ and V^K corresponds to $(\nu, 0, 0, Z(\tau))$ where $Z(\tau)$ is $(4\pi/n)$-periodic, i.e., with modes which are multiples of $n/2$. Thus, d_H and d_K are given by the number of negative eigenvalues λ of the system

$$(\mp\varepsilon_1 - \lambda\mu, -z'' - n^2 z + bZ_n\mu - \lambda z),$$

in the spaces of $(2\pi/n)$ and $(4\pi/n)$-periodic functions, where $b = 2n^3/\nu_0$, that is b is positive.

Since, for $\lambda < 0$, the second equation $z'' + (n^2 + \lambda)z = b\mu a \cos n\tau$ is non-resonant, its particular solution has to be $z = \varepsilon_1 \cos n\tau$, with $\varepsilon_1 = b\mu(a_0 \pm \rho)/\lambda = \mp\lambda\mu$. Since $b > 0$, one has a contribution, for $\mu \neq 0$, only at $a_0 - \rho$ with $\lambda = -(b(a_0 - \rho))^{1/2}$. For $\mu = 0$, one has non-trivial solutions only for $n^2 + \lambda = k^2$, hence, in the spaces under consideration, only for $\lambda = (m^2 - 1)n^2$ for the case of $(2\pi/n)$-functions, i.e., with modes mn and, for $\lambda < 0$, only for $m = 0$, while for the case of $(4\pi/n)$-functions, i.e., with modes $mn/2$, for $\lambda = (m^2/4 - 1)n^2$ and, for $\lambda < 0$, only for $m = 0$ and 1. For this last case, with $m = 1$, one has a two-dimensional kernel, while for $m = 0$, the kernel is one-dimensional. Hence, $n_H(a_0 + \rho) = 1, n_H(a_0 - \rho) = 2$, while $n_K(a_0+\rho) = n_K(a_0-\rho) = 2$, since we have to look at modes in $V^K \cap (V^H)^\perp$. Note that, if one restricts the study to even functions $z(\tau)$, then the linearization is invertible at $a_0 \pm \rho$, the Morse number n_H is the same but n_K is lowered by one. Thus, taking into account that $\eta = -1$, since the orbit is one-dimensional, one has

$$d_H(a_0 + \rho) = 1, \quad d_H(a_0 - \rho) = -1,$$

and, for n even, one has $d_K(a_0 \pm \rho) = 0$, according to Proposition 3.3.

Hence, by writing $i_\pm(f_\varepsilon)$ as $i_|^\pm(f^{\underline{H}}) \times i_|^\pm(Df_\perp X)$, as in Theorem 3.1, we have,

$$i_\perp^\pm(f^{\underline{H}}) = \pm[F_H^*]_\perp,$$

where F_H^* is the normalized generator. From Proposition 3.1 in Chapter 3, the other elements in $i_\perp(f^{\underline{H}})$ are 0.

It remains to identify the irreducible representations of H in $(V^{\underline{H}})^\perp$, that is for x and y only, their isotropy K_{n_0}, the operators $\tilde{B}_l$ and their Morse numbers as well as the isotropy subspaces.

Lemma 3.11. *There are n different irreducible representations of H in $(V^{\underline{H}})^\perp$, with $K_{n_0} \equiv \{(l, \psi = -2\pi n_0 l/n), l = 0, \dots, n-1\}$, for $n_0 = 0, \dots, n-1$. The space $V^{K_{n_0}}$ is spanned by functions $x(\tau), y(\tau), z(\tau)$ with the property that*

$$R_{2\pi n_0/n} \begin{pmatrix} x(\tau) \\ y(\tau) \end{pmatrix} = \begin{pmatrix} x(\tau + 2\pi/n) \\ y(\tau + 2\pi/n) \end{pmatrix}$$

and $z(\tau)$ is $2\pi/n$-periodic. More precisely, $x(\tau) = \operatorname{Re} X(\tau)$, $y(\tau) = \operatorname{Im} X(\tau)$, where

$$X(\tau) = \sum_{-\infty}^{\infty} x_m e^{in_0\tau} e^{imn\tau}.$$

Furthermore, if $(x(\tau), y(\tau))$ is in $V^{K_{n_0}}$, then $(x(\tau), -y(\tau))$ is in $V^{K_{n-n_0}}$.

Proof. This is just a straight application of Lemma 9.6 of Chapter 1, after one notices that, since $\gamma_0 = \mathrm{Id}$, then $\alpha_0 = \alpha_j = a_j = 0$ and that H acts on $z(\tau)$ only by the time shift of $2\pi l/n$ for $l = 0, \dots, n-1$, i.e., that $z(\tau)$ is $(2\pi/n)$-periodic. Recall that the action of H on $(x(\tau), y(\tau))$ is by the time shift and a rotation R_ψ. Hence, on the mode m, one has

$$R_\psi \begin{pmatrix} x_m \\ y_m \end{pmatrix} e^{2\pi ilm/n} = P^{-1} \begin{pmatrix} e^{i\psi} & 0 \\ 0 & e^{-i\psi} \end{pmatrix} P \begin{pmatrix} x_m \\ y_m \end{pmatrix} e^{2\pi ilm/n},$$

where

$$2P = \begin{pmatrix} 1 & i \\ 1 & -i \end{pmatrix}.$$

Hence, $\begin{pmatrix} x_m \\ y_m \end{pmatrix}$ will be fixed if either $\psi \equiv -2\pi lm/n$, $[2\pi]$, and $y_m = -ix_m$ or $\psi \equiv 2\pi lm/n$, $[2\pi]$, and $y_m = ix_m$, since $R_\psi \begin{pmatrix} x_m \\ y_m \end{pmatrix} e^{2\pi ilm/n} = \begin{pmatrix} x_m \\ y_m \end{pmatrix}$ leads to

$$\begin{aligned} e^{i(\psi+2\pi lm/n)}(x_m + iy_m) &= x_m + iy_m \\ e^{i(-\psi+2\pi lm/n))}(x_m - iy_m) &= x_m - iy_m. \end{aligned}$$

Thus, for each n_0 fixed, $0 \leq n_0 < n$, one has K_{n_0}, as in the statement of the lemma, and modes $m = n_0 + kn$, with $y_{n_0+kn} = ix_{n_0+kn}$, and also modes $m = -n_0 + kn = -n_0 - \tilde{k}n$, with $y_{-n_0-\tilde{k}n} = ix_{-n_0-\tilde{k}n}$. The first set of modes gives $x_1(\tau)$ and the second, with the condition that $x(\tau) = x_1(\tau) + x_2(\tau)$ must be real, gives $x_2(\tau) = \bar{x}_1(\tau)$, with $y_1(\tau) = -ix_1(\tau)$, $y_2(\tau) = ix_2(\tau)$. Note that, for $n_0 = 0$ or $n/2$, the modes in x_1 and x_2 are the same: for $n_0 = 0$, the functions $x(\tau)$ and $y(\tau)$ are $(2\pi/n)$-periodic, while, for $n_0 = n/2$, they are $(2\pi/n)$-anti-periodic. Note also that, for $m = n_0$, the elements of $V^{K_{n_0}}$ are $(\cos n_0\tau, \sin n_0\tau)$ and $(-\sin n_0\tau, \cos n_0\tau)$. □

In order to compute the Morse numbers at $a_0 \pm \rho$, one has to look at negative eigenvalues of $-L_nX$ (respectively $-M_nX$), when restricted to $V^{K_{n_0}} \cap (V^{\underline{H}})^{\perp}$, i.e., only for the functions $x(\tau)$ and $y(\tau)$: the parts corresponding to the variables ν and $z(\tau)$ are in $V^{\underline{H}}$.

Hence, one has to consider the eigenvalue problem

$$\begin{aligned} x'' + n^2(\alpha + 2\beta \cos n\tau)x &= \lambda x \\ y'' + n^2(\alpha + 2\beta \cos n\tau)y &= \lambda y, \end{aligned}$$

for $\lambda > 0$ and x and y in $V^{K_{n_0}}$, with an analogous linear system in the second case. One could plug in the Fourier series of Lemma 3.11 and arrive at an infinite system of equations, as in Lemmas 3.9 and 3.10. However, it is simpler to see that this Morse number is constant in the regions separated by the curves $\alpha_{k/n}(\beta)$: since the operators are self-adjoint no eigenvalue may change sign without going through 0. In particular, one may compute them for $\beta = 0$. Furthermore, since the linearization is H-equivariant, this argument can be done for each $V^{K_{n_0}}$ separately. Also, since $i_{\perp}^{\pm}(f^{\underline{H}}) = \pm[F_H^*]_{\perp}$ and $\deg_{\perp}(f_\varepsilon; \Omega) = i_-(f_\varepsilon) + i_+(f_\varepsilon)$, the only n_0's which will count in this last degree are those for which there is a change when crossing $\alpha_{k/n}(\beta)$, at (α_0, β_0), when β varies, that is, those for which L_nX is not invertible in $V^{K_{n_0}}$. Hence, one may identify these n_0's by looking, as above, at the kernel of L_n at $\beta = 0$, $\alpha = (k/n)^2$.

Lemma 3.12. *Let $k = k_0 + \tilde{k}n$, with $0 \leq k_0 < n$, then the only n_0's for which there is a change in the Morse number are $n_0 = k_0$ and $n_0 = n - k_0$. If $n_{n_0}(a_0 \pm \rho)$ is this Morse number, one has $n_{k_0}(a_0 \pm \rho) = n_{n-k_0}(a_0 \pm \rho)$. Furthermore, if $\varepsilon_0 =$ Sign $\alpha'_{k/n}(\beta_0)$,*

(a) *if $2k/n$ is not an integer, then the complex Morse numbers are*

$$n_{k_0}(a_0 + \varepsilon_0\rho) = [2k/n]$$
$$n_{k_0}(a_0 - \varepsilon_0\rho) = [2k/n] + 1,$$

where $[2k/n]$ is the integer part of $2k/n$;

(b) *if $2k/n = k_1$, then, on the left transition curve, one has*

$$n_{k_0}(a_0 + \varepsilon_0\rho) = k_1 - 1$$
$$n_{k_0}(a_0 - \varepsilon_0\rho) = k_1,$$

while, on the right transition curve, one has

$$n_{k_0}(a_0 + \varepsilon_0\rho) = k_1$$
$$n_{k_0}(a_0 - \varepsilon_0\rho) = k_1 + 1.$$

Proof. It is enough to look at the spectrum of $x'' + \alpha n^2 x$, for $\beta = 0$, near $\alpha = (k/n)^2$ and in $V^{K_{n_0}}$, that is for x (and y) as in Lemma 3.11. One gets, for $X(\tau) = \sum_{-\infty}^{\infty} x_m e^{in_0\tau} e^{inm\tau}$, and $\lambda \geq 0$,

$$(-(n_0 + mn)^2 + \alpha n^2 - \lambda)x_m = 0.$$

For $\lambda = 0, \alpha = (k/n)^2, k = k_0 + \tilde{k}n$, the only non-zero modes are such that $n_0 = \pm k_0 + (\pm\tilde{k} - m)n$, that is $n_0 = k_0$ and $m = \tilde{k}$, i.e., $n_0 + mn = k$, and $n_0 = n - k_0$ and $m = -\tilde{k} - 1$, i.e., $n_0 + mn = -k$.

For $\lambda > 0$ and $\alpha = (k - \varepsilon)^2/n^2$, one has a contribution of the mode $mn + n_0$, i.e., with $x_m \neq 0$, only if $\lambda = (k_0 + \tilde{k}n - \varepsilon)^2 - (k_0 + mn)^2 = -(mn - \tilde{k}n + \varepsilon)(2k_0 + mn + \tilde{k}n - \varepsilon) > 0$, that is for all integers m between $-\tilde{k} - 2k_0/n + \varepsilon/n$ and $\tilde{k} - \varepsilon/n$. Taking ε small enough, it is easy to see that the number of m's in this interval is $[2k/n]$, if $\varepsilon > 0$, and $2k/n$ not an integer, $[2k/n] + 1$, if $\varepsilon < 0$, and $2k/n$ not an integer, while, if $2k/n = k_1$, one has $k_1 - 1$ such m's, if $\varepsilon > 0$, and $k_1 + 1$, if $\varepsilon < 0$.

If $2k/n$ is not an integer the y component is completely determined by $x_1(\tau)$ which is complex, i.e., the real Morse number is twice the complex Morse number. Furthermore, if $a_0 > 0$, that is, if $\beta_0 > 0$, then the point $(\alpha_0, a_0 - \rho)$ is to the left of the curve $\alpha_{k/n}(\beta)$, if $\varepsilon_0 = -1$, and to the right, if $\varepsilon_0 = 1$, while $(\alpha_0, a_0 + \rho)$ has the inverse collocation. Being to the left means $\varepsilon > 0$ and to the right means $\varepsilon < 0$. On the other hand, if $a_0 < 0$, that is, if $\beta_0 < 0$, then $(\alpha_0, a_0 - \rho)$ is to the right of $\alpha_{k/n}(\beta)$, if $\varepsilon_0 = 1$, and to the left, if $\varepsilon_0 = -1$.

If $2k/n = k_1$, a crossing of the transition curve will increase (or decrease according to the sign of ε_0) the real Morse number by 1, but then the y-component will give a similar contribution, with a total complex Morse number changing by 1. In particular, on the left transition curve, $a_0 - \rho$ will be with $\varepsilon > 0$, if $\varepsilon_0 = -1$ (and the complex Morse number for $a_0 + \rho$ is increased by 1), while $a_0 + \rho$ will be with $\varepsilon > 0$, if

$\varepsilon_0 = 1$, with an increase of 1 in the Morse number for $a_0 - \rho$. On the right transition curve, if $\varepsilon_0 = 1$, then $a_0 - \rho$ corresponds to $\varepsilon < 0$, while, if $\varepsilon_0 = -1$, then $a_0 + \rho$ corresponds to $\varepsilon < 0$. Crossing the transition curve will decrease the Morse number by 1.

Finally, since $V^{K_{n_0}}$ and $V^{K_{n-n_0}}$ are isomorphic the Morse numbers are equal. □

Since Γ/H has dimension one, one has (for the relevant isotropy subgroups) that, according to Theorem 3.1,

$$i_\perp(\Lambda \bar{X}) = [F_\Gamma]_\perp + n_{k_0}[F^*_{k_0}]_\perp + n_{n-k_0}[F^*_{n-k_0}]_\perp,$$

since Γ is two-dimensional, hence, in the formula of Theorem 3.1, one has only $s = 1$. Note that, if $2k/n = k_1$, then there is only one $F^*_{k_0}$.

Theorem 3.2. *The orthogonal degree for the spring-pendulum system is*

$$\deg_\perp(f_\varepsilon; \Omega) = -\text{ Sign } \alpha'_{k/n}(\beta_0)([F^*_{K_{n_0}}]_\perp + [F^*_{K_{n-n_0}}]_\perp),$$

where $k \equiv n_0, [n]$, and only one generator if $2k/n = k_1$, with $n_0 = 0$ or $n/2$. From $(\beta_0, \alpha_{k/n}(\beta_0))$ there is a global bifurcation, in $V^{K_{n_0}}$ and $V^{K_{n-n_0}}$, of a branch of non-trivial solutions which is either unbounded in $(\nu, x, y, \tilde{Z})$ or returns to another intersection of the line $\alpha = \alpha_{k/n}(\beta_0)$ with the curve $\alpha_{k/n}(\beta)$ with an opposite sign of $\alpha'_{k/n}$.

Solutions (x, y) on the branch are not identically zero (except at $\nu = \nu_0/n, \beta = \beta_0$, $\alpha = \alpha_{k/n}(\beta_0)$), have $2k$ zeros in $[0, 2\pi)$ and satisfy the symmetry of Lemma 3.11, i.e., for $V^{K_{n_0}}$

$$R_{2\pi n_0/n}\begin{pmatrix} x(\tau) \\ y(\tau) \end{pmatrix} = \begin{pmatrix} x(\tau + 2\pi/n) \\ y(\tau + 2\pi/n) \end{pmatrix},$$

$z(\tau)$ is $(2\pi/n)$-periodic and $(x(\tau)), -y(\tau))$ is in $V^{K_{n-n_0}}$.

Proof. Using the product formula of Theorem 3.1, one has

$$i_\pm(f_\varepsilon) = \pm([F^*_H]_\perp + n_{k_0}(a_0 \pm \rho)([F^*_{K_{n_0}}]_\perp + [F^*_{K_{n-n_0}}]_\perp)).$$

Since $\deg_\perp(f_\varepsilon; \Omega) = i_+(f_\varepsilon) + i_-(f_\varepsilon)$, one gets

$$\deg_\perp(f_\varepsilon; \Omega) = (n_{k_0}(a_0 + \rho) - n_{k_0}(a_0 - \rho))([F^*_{K_{n_0}}]_\perp + [F^*_{K_{n-n_0}}]_\perp).$$

The difference of Morse numbers is -1 if $\varepsilon_0 = 1$ and 1 if $\varepsilon_0 = -1$, and the same argument works for $2k/n = k_1$.

The argument for the global bifurcation comes from Theorem 5.2 in Chapter 2 and Proposition 3.1, in particular, that the bifurcation takes place in V^{Kn_0} and V^{Kn-n_0}. The relation between the two branches is given in Remark 3.7 and the isomorphism $(x, y) \to (x, -y)$ of Lemma 3.11, between the two isotropy subspaces. The nodal properties follow from the fact that the equations for x and y are of the form $x'' + f(\tau)x = 0$, hence, the number of zeros on the branch is conserved. Furthermore, the branch cannot return to a point on a curve $\alpha_{k'/n}(\beta)$, with $k' \neq k$, since, when x goes to 0 and near a bifurcation point, the number of zeros is determined by k: Lemmas 3.9 and 3.10.

Since (x, y) is in $V^{K_{n_0}}$, then, if $x(\tau) \equiv 0$, one has $y(\tau) \equiv 0$, unless $2n_0/n = k_1$. Hence, if (x, y) tends to $(0, 0)$ on the branch, one has to go to a bifurcation point, i.e., with $\nu = \nu_0/\tilde{n}$, $Z = a\cos(\tilde{n}\tau + \varphi)$ and $\alpha = \alpha_{\tilde{k}/\tilde{n}}(\beta)$. In the limit, the elements on that curve need to have $2k$ zeros, thus, $k = \tilde{k}$ and, from the periodicity of $Z(\tau)$, $\tilde{n}$ should be a multiple of n. From the fact that on each curve $\alpha_{\tilde{k}/\tilde{n}}$ there are only two linearizations which are not invertible, corresponding to $V^{K_{\tilde{n}_0}}$ and $V^{K_{\tilde{n}-\tilde{n}_0}}$, the above argument is reversible and $n = \tilde{n}$. Thus, the only $(x, y) = (0, 0)$ on the branch are the bifurcation points from the trivial solutions. Of course, this argument may also be given directly from the fact that the bifurcation is in $V^{K_{n_0}}$. □

Remark 3.11. If one varies α, one obtains "surfaces" bifurcating from the curve $\alpha_{k/n}(\beta)$, following the arguments of Γ-epi maps of [I.M.V.0].

In particular, for any segment, in the (α, β)-plane, which is transversal to $\alpha_{k/n}(\beta)$, either the branch is unbounded over the segment (in (x, y, Z)), or it covers one of the end points. For instance, if one has a-priori bounds, then the branch covers one of the components of the complement of the curve: See [I. p.395].

Remark 3.12. If one wishes to use the reversibility, then one may restrict the study to fixed point subspaces of $\tilde{R}_\varepsilon$ and of $R_\pi \tilde{R}_\varepsilon$, i.e., for $Z(\tau)$ even, $x(\tau)$ and $y(\tau)$ of equal parity ($\varepsilon = 1$) or opposite parity ($\varepsilon = -1$).

(a) Fixing the parity will destroy the equivariance with respect to the time shift. On the other hand, if $x(\tau)$ and $y(\tau)$ have the same parity, that is $\varepsilon = 1$, one will keep the equivariance with respect to R_ψ and the equations are still orthogonal. In this case $\Gamma = S^1$, acting via R_ψ, and the isotropy of $(x = 0, y = 0, Z_n = a\cos n\tau)$ is $H = S^1$, with only one strict subgroup $K = \{e\}$. In order to compute $i_\perp^\pm(f^H)$, one has to count the negative eigenvalues λ of the linearization in $V^H = \{(\nu, 0, 0, Z(\tau))$, with $Z(\tau)$ even and 2π-periodic$\}$,

$$Df_\pm(\mu, z) = (\mp\varepsilon_1 - \lambda\mu, -z'' - n^2 z + bZ_n\mu - \lambda z),$$

where $z(\tau) = \varepsilon_1 \cos n\tau + \tilde{Z}(\tau)$, with $\tilde{Z}(\tau)$ even. As before, for $\mu \neq 0$, one has a contribution only at $a_0 - \rho$. On the other hand, if $\mu = 0$, then $\varepsilon_2 = 0$ and $\tilde{Z}'' + (n^2 + \lambda)\tilde{Z} = 0$ will give 2π-periodic even solutions for $\lambda = k^2 - n^2$ with $0 \leq k < n$ and $\tilde{Z}(\tau) = \cos k\tau$. Thus, $d_H(a + \rho) = (-1)^n$ and $d_H(a - \rho) = (-1)^{n+1}$. As before, one has

$$i_\pm(f_\varepsilon) = \pm(-1)^n [F^*_{S^1}]_\perp \times ([F^*_{S^1}]_\perp + n(a_0 \pm \rho)[F^*_{\{e\}}]_\perp)$$

and, as a consequence,

$$\begin{aligned}\deg_\perp(f_\varepsilon;\Omega) &= (-1)^n(n(a_0+\rho)-n(a_0-\rho))[F^*_{\{e\}}]_\perp\\ &= -(-1)^n \operatorname{Sign}\ \alpha'_{k/n}(\beta_0)[F^*_{\{e\}}]_\perp,\end{aligned}$$

since the difference, in the complex Morse numbers, is – Sign $\alpha'_{k/n}(\beta_0)$, if one crosses the curve $\alpha_{k/n}$ (for $x(\tau)$ and $y(\tau)$ even) or the curve $\tilde{\alpha}_{k/n}$ (for $x(\tau)$ and $y(\tau)$ odd). Solutions on the branch conserve the parity, by construction, and the nodal properties. However, since the periodicity of $Z(\tau)$ on the branch is only 2π, there is no topological argument to prevent the branch coming out of $(\beta_0, \alpha_{k/n}(\beta_0))$ to go to a point $(\beta_1, \alpha_{k/\tilde{n}}(\beta_1) = \alpha_{k/n}(\beta_0))$, for a $\tilde{n}$ different from n. Furthermore, one may have $x(\tau) \equiv 0$ or $y(\tau) \equiv 0$ on the bifurcating branch.

In fact, if one puts $y(\tau) \equiv 0$, then the system reduces to two equations and one may use the standard Leray–Schauder degree theory on spaces of functions $(x(\tau), Z(\tau))$, with are 2π-periodic, with $Z(\tau)$ even and $x(\tau)$ of a given parity, either even or odd. On these spaces, the kernels of the linearization of f_ε are one-dimensional (due to the parity of $Z(\tau)$) on the curve $\alpha_{k/n}(\beta)$, with a change in the index from $a_0 - \rho$ to $\alpha_0 + \rho$. Hence, one has the same bifurcation results, but now of planar solutions. It is likely that these solutions, rotated by R_ψ, generate the solutions obtained by the reversibility argument. However, except for the case where $2k/n$ is an integer, they are different from the ones given in $V^{K_{n0}}$. Hence, one has a double bifurcation from $(\beta_0, \alpha_{k/n}(\beta_0))$, if $2k/n$ is not an integer, of planar and non-planar solutions.

(b) For the case of opposite parity, then the equivariance with respect to R_ψ is also destroyed. However, the subspaces $V^{K_{n0}}$ remain fixed by the action of $\tilde{R}_{-1}$, that is, if

$$X(\tau) = \sum_{-\infty}^{\infty} x_m e^{i(n_0+nm)\tau},$$

with $x(\tau) = \operatorname{Re} X(\tau)$ and $y(\tau) = \operatorname{Im} X(\tau)$, then, if all x_m's are taken real, one has

$$x(\tau) = \sum_{-\infty}^{\infty} x_m \cos(n_0+nm)\tau \quad \text{and} \quad y(\tau) = \sum_{-\infty}^{\infty} x_m \sin(n_0+nm)\tau,$$

and $x(\tau)$ is even, while $y(\tau)$ is odd. On the other hand by taking all x_m's pure imaginary, then $x(\tau)$ will be odd and $y(\tau)$ will be even. Thus, by decomposing functions in $V^{K_{n0}}$ as sums of the form ($x(\tau)$ even, $y(\tau)$ odd) and ($x(\tau)$ odd, $y(\tau)$ even), one may study the equations in the above subspaces. One will have a jump of one eigenvalue when crossing $\alpha_{k/n}(\beta)$ and one may use the Leray–Schauder theory in that space. However, this invariance property of $V^{K_{n0}}$ is not clear a priori, while the existence of $V^{K_{n0}}$, coming from the analysis of twisted orbits is natural.

(c) From the stability in the complement of the Arnold's tongues, it seems likely that the first bifurcation will correspond to a crossing of a transition curve, i.e., with $2k/n$ an integer and a planar solution.

4.4 Γ-Index of a loop of stationary points

In this last section of the book, we shall study the case of an isolated loop of stationary solutions, for problems with one extra parameter, with the main intention of applying the results to different kinds of Hopf bifurcation. More precisely, let $F : \mathbb{R} \times U \to W$ be a Γ-equivariant map such that F has a simple loop P of zeros in $\mathbb{R} \times U^{\Gamma}$, on which F is regular, with the usual compactness if U is infinite-dimensional. Hence DF has a one-dimensional kernel, at each point of P, generated by the tangent vector to P. This situation forces U and W to be equivalent representations (see § 5 in Chapter 1). Then, if Ω is a small invariant neighborhood of P such that F^{Γ} has only P as zeros in Ω^{Γ}, one may define $\deg_{\Gamma}(F; \Omega)$. Furthermore, if X, in $\mathbb{R} \times U$, is written as $X^{\Gamma} \oplus X_{\perp}$ and F as $F^{\Gamma} \oplus F^{\perp}$, the regularity implies that $D_{X_{\perp}} F^{\perp}$ is invertible, hence, as it has been done already several times

$$\deg_{\Gamma}(F; \Omega) = \deg_{\Gamma}((F^{\Gamma}(X^{\Gamma}), D_{X_{\perp}} F^{\perp}(X^{\Gamma})X_{\perp}); \Omega).$$

We have in mind the special case of the Hopf bifurcation, or variations of it, that is $U = \mathbb{R} \times V$, and X is written as $(\mu, \nu, X_0, X_{\perp})$, with X_0 in V^{Γ}, while $F^{\Gamma}(\mu, \nu, X_0) = (\rho^2 - \mu^2 - \nu^2, F_0(\mu, \nu, X_0))$, with $F_0(\mu, \nu, 0) = 0$. Thus, the loop P is the circle $\rho^2 = \mu^2 + \nu^2$, $X_0 = X_{\perp} = 0$.

If we assume that $D_{X_0} F_0(\mu, \nu, 0)$ is invertible on the loop, then one may simplify further the computation of the Γ-degree

$$\deg_{\Gamma}(F; \Omega) = \deg_{\Gamma}((\rho^2 - \mu^2 - \nu^2, D_{X_0} F_0(\mu, \nu, 0)X_0, D_{X_{\perp}} F^{\perp}(\mu, \nu, 0)X_{\perp}); \Omega)$$

and we take Ω to be $\{(\mu, \nu, X_0, X_{\perp}) : \mu^2 + \nu^2 < 4\rho^2, \|X_0\|, \|X_{\perp}\| < 2\varepsilon\}$.

As we have seen in Corollary 5.2 of Chapter 1, one has

$$\deg_{\Gamma}(F; \Omega) = \Sigma_0 J^{\Gamma}(D_{X_0} F_0, D_{X_{\perp}} F^{\perp}),$$

where Σ_0 is the suspension by $2t - 1$ and J^{Γ} is the J^{Γ}-homomorphism, or Whitehead map, from the set of all Γ-homotopic classes from S^1 into $\mathrm{GL}^{\Gamma}(V)$ into the group $\Pi^{\Gamma}_{S^{\mathbb{R}\times V}}(S^V)$:

$$\begin{array}{rcl} [S^1 \to \mathrm{GL}^{\Gamma}(V)]_{\Gamma} & \xrightarrow{J^{\Gamma}} & \Pi^{\Gamma}_{S^{\mathbb{R}\times V}}(S^V) \\ A(\mu, \nu) & \longrightarrow & (\|X\| - \varepsilon, A(\mu, \nu)X), \end{array}$$

recalling that the homotopy $\tau(\rho^2 - \mu^2 - \nu^2) + (1 - \tau)(\|X\| - \varepsilon)$ is valid on $\partial\Omega$, once $A(\mu, \nu) = \operatorname{diag}(D_{X_0} F_0, D_{X_{\perp}} F^{\perp})|_{\mu^2+\nu^2=\rho^2}$ has been extended, to all (μ, ν), by defining $A(\mu\tau, \nu\tau) = \tau A(\mu, \nu)$, for $\tau \geq 0$ and $\mu^2 + \nu^2 = \rho^2$. Note that, from Definition 8.1 of Chapter 1, the group $\Pi^{\Gamma}_{S^{\mathbb{R}\times V}}(S^V)$ requires an extra variable, here given by $t = 2\mu - 1$. Recall also that Σ_0 is an isomorphism provided $\dim V^{\Gamma} \geq 3$, see Corollary 7.1 of Chapter 3.

Example 4.1 (The classical Hopf bifurcation). In order to motivate the study described above, consider the problem of finding 2π-periodic solutions to the autonomous system

$$(\nu_0 + \nu)\frac{dX}{dt} - L(\mu)X - f(X, \mu) = 0, \quad X \text{ in } \mathbb{R}^N,$$

where $f(X, \lambda) = 0(\|X\|^2)$. Thus, $X = 0$ is a solution for all (μ, ν). The problem is equivalent, on Fourier series, to

$$in(\nu_0 + \nu)X_n - L(\mu)X_n - f_n(X, \mu) = 0, \quad n \geq 0,$$

where f_n is S^1-equivariant. Clearly, a necessary condition for the existence of solutions with $X \not\equiv 0$, is that $in(\nu_0 + \nu)I - L(\mu)$ is not invertible. Hence, assume that $L(0)$ has $\pm i\nu_0$ as eigenvalues, but that $in\nu_0 I - L(0)$ is invertible for $n \neq 1$ (*non-resonance condition*), then this will be case for (μ, ν) close to $(0, 0)$ and one may solve, by the implicit function theorem for instance, for X_n in terms of (μ, ν, X_1) and one is reduced to

$$(i(\nu_0 + \nu)I - L(\mu))X_1 - \tilde{f}_1(X_1, \mu, \nu) = 0,$$

where, due to the uniqueness of X_n, one has $\tilde{f}_1(e^{i\varphi}X_1, \mu, \nu) = e^{i\varphi}\tilde{f}_1(X_1, \mu, \nu)$. If, furthermore, $i\nu_0$ is a *simple* eigenvalue of $L(0)$, then the Ljapunov–Schmidt reduction leads to

$$(i\nu - a(\mu))x - \tilde{f}(x, \mu, \nu) = 0,$$

where $a(0) = 0$ and $L(\mu)$ has the eigenvalue $i\nu_0 + a(\mu)$, the variable x is now in $\mathbb{C}$ and $\tilde{f}$ is S^1-equivariant and $0(|x|^2)$. Using this last fact for $x = re^{i\varphi}$, one is finally reduced to

$$r(i\nu - a(\mu) - g(r, \mu, \nu)) = 0,$$

with $g(r, \mu, \nu) = 0(r)$. If $\operatorname{Re} a'(0) \neq 0$ (*non-zero speed crossing*), then one may solve uniquely, again by the implicit function theorem, these two equations for (μ, ν) in terms of r, giving periodic solutions.

In this derivation, the S^1-equivariance was used, at the last step, to reduce the dimension of the domain. Now, if one has resonances or a non-simple eigenvalue, or more symmetries, the argument does not work anymore. But, on the other hand, the Γ-degree (an S^1-degree in the case of classical Hopf bifurcation) can be computed.

Let us return to the general situation described at the beginning of this section, i.e., to $\Sigma_0 J^{\Gamma}(DF)$, where DF is the linearization at $(\mu, \nu, 0)$ of $(F_0, F^{\perp})$, with respect to $(X_0, X_{\perp})$. Using a Ljapunov–Schmidt reduction, we may assume that V is finite dimensional. Since DF is Γ-equivariant, it has a block diagonal structure (Theorem 5.3 of Chapter 1)

$$DF = \operatorname{diag}(D_{X_0}F_0, D_{Y_j}F_j^{\perp}, \ldots, D_{Z_l}F_l^{\perp}, \ldots, D_{Z_k}F_k^{\perp}),$$

where Y_j is made of real coordinates with an action of Γ as $\mathbb{Z}_2$, while Γ acts on the complex coordinate Z_l as $\mathbb{Z}_{m_l}$ and on Z_k as S^1. Each of the pieces of DF depends on (μ, ν) and is invertible on the circle $\mu^2 + \nu^2 = \rho^2$. In particular, the determinant of each piece has a constant sign on the circle, positive for the complex matrices.

Now, we have seen in Theorem 8.3 of Chapter 1, that if

$$A(\mu, \nu) = \mathrm{diag}(A_0, A_j, B_l, C_k),$$

where A_0 corresponds to X_0, A_j to Y_j with $j = 1, \dots, r$ and isotropy H_j with $\Gamma/H_j \cong \mathbb{Z}_2$, and B_l or C_k correspond to Z_l or Z_k with action of Γ as $\mathbb{Z}_{m_l}$ or S^1, is such that A_0 and A_j have positive determinant, then J^Γ is a morphism of groups, i.e.,

$$\Pi_1(\mathrm{GL}_+^\Gamma(V)) \xrightarrow{J^\Gamma} \Pi^\Gamma_{S^{\mathbb{R}\times V}}(S^V)$$

is such that

$$J^\Gamma(AB) = J^\Gamma(A) + J^\Gamma(B),$$

where all pieces of A and B have positive determinants.

In order to compute $\deg_\Gamma(F; \Omega)$, we shall use this property of J^Γ by relating first $J^\Gamma(A)$ to $J^\Gamma(A^*)$, where A^* is obtained from A by changing the sign of one row in case A has a negative determinant, in which case A^* has a positive determinant (of course this will be done for each piece A_0, A_j of A).

Let I_0 be the linear map which changes the first component of X_0 into its opposite and I_j the similar map for Y_j. Since the addition in $\Pi^\Gamma_{S^{\mathbb{R}\times V}}(S^V)$ is defined on t, the maps I_0 and I_j induce two morphisms on this group by

$$\begin{aligned} I_j^*[f(X)]_\Gamma &= [f(I_j X)]_\Gamma \\ I_j'^*[f(X)]_\Gamma &= [I_j f(X)]_\Gamma, \end{aligned}$$

for $j = 0$ (and I_0), and $j = 1, \dots, r$.

Lemma 4.1. *The morphisms I_j^* and $I_j'^*$ have the following properties:*

(a) $I_j^{*2} = I_j'^{*2} = \mathrm{Id}$,

(b) $I_j^* I_k^* = I_k^* I_j^*$, $I_j'^* I_k'^* = I_k'^* I_j'^*$, $I_j^* I_k'^* = I_k'^* I_j^*$,

(c) $I_0^*[f]_\Gamma = -[f]_\Gamma$.

(d) *If* $\dim V^\Gamma \geq 3$, *then* $I_0^*[f]_\Gamma = I_0'^*[f]_\Gamma$.

(e) *If* $\dim\{Y_j\} \geq 3$, *then* $I_j^*[f]_\Gamma = I_j'^*[f]_\Gamma$.

(f) *If* $\dim\{Y_j\} = 1$, *then* $[J^\Gamma A_j(\mu, \nu)]_\Gamma = 0$.

(g) *If* $\dim\{Y_j\} = 2$, *then* $I_j^*[J^\Gamma A_j]_\Gamma = -I_j'^*[J^\Gamma A_j]_\Gamma$ *and* $2I_j^*[J^\Gamma A_j]_\Gamma = 2[J^\Gamma A_j]_\Gamma$.

(h) *If* $\dim\{Y_j\} \geq 3$, *then* $2[J^\Gamma A_j]_\Gamma = 0$, *where, for* $j = 0$, $\{Y_0\}$ *stands for* V^Γ.

Proof. Since $I_j^2 = \mathrm{Id}$, it follows that I_j^* and $I_j'^*$ are involutions. Furthermore, the commutativity is immediate. Notice that, via a rotation, one has that $(2t-1, -x_0)$ is homotopic to $(1-2t, x_0)$. This proves (c).

Now, if one suspends by y_j, with isotropy H_j (taking again y_0 with $H_0 = \Gamma$), one has

$$[I_j f, y_j] = [f, -y_j] = [f(I_j), y_j],$$

by performing the rotations between the components or the variables. Hence,

$$\Sigma_j I_j'^*[f]_\Gamma = \Sigma_j I_j^*[f]_\Gamma.$$

But, from Corollary 7.1 of Chapter 3, Σ_j is an isomorphism if $\dim\{Y_j\} \geq 3$, proving (d) and (e).

Finally, if $\dim\{Y_j\} = 1$, then $A_j(\mu, \nu)$ is a non-zero scalar, deformable to a constant. Hence, $[J^\Gamma A_j]_\Gamma = [|y_j|-1, \pm y_j]_\Gamma = 0$. On the other hand, if $\dim\{Y_j\} \geq 3$, then $2[A_j] = 0$ if $\det A_j > 0$ (see Theorem 8.3 in Chapter 1), or $2[I_j A_j]_\Gamma = 0$ if $\det A_j < 0$. For the case $\dim\{Y_j\} = 2$, let I_j be the matrix $\begin{pmatrix} -1 & 0 \\ 0 & 1 \end{pmatrix}$, then, if $\det A_j > 0$, one has that A_j is homotopic to λ^d, where $\lambda = \mu + i\nu$ and Y_j is written as $y_1 + iy_2$. Then, according to Theorem 5.1 of Chapter 3, one gets

$$[J^\Gamma A_j]_\Gamma = [|Y_j| - 1, A_j Y_j]_\Gamma = d\eta,$$

where η is the Hopf map and a generator of $\Pi(H_j)$.

Recall that, in this case

$$\Pi(H_j) \cong \mathbb{Z} \times \mathbb{Z}_2,$$

generated by $\eta = (|Y_j| - 1, \lambda Y_j)$, which is such that its degree (as a map from $(\mu, \nu, y_1 > 0)$ into $\mathbb{R}^3$) is 1, and by η_1 with $2\eta_1 = 0$. Then,

$$I_j^*[J^\Gamma A_j]_\Gamma = dI_j^*\eta = d[|Y_j| - 1, -\lambda \bar{Y}_j]_\Gamma = d(\eta + d_1\eta_1),$$

since the degree of $(y_j - 1, -\lambda y_1)$, for $y_1 > 0$, is also 1. Since $2\eta_1 = 0$, one has part of the answer, with $I_j^*[J^\Gamma A_j]_\Gamma = d([J^\Gamma A_j]_\Gamma + d_1\eta_1)$.

On the other hand, if $\det A_j < 0$, then $A_j I_j$ has a positive determinant and homotopic to $\lambda^d Y_j$, for some d. Then,

$$I_j^*[J^\Gamma A_j I_j]_\Gamma = dI_j^*\eta = d(\eta + d_1\eta_1) = [J^\Gamma A_j]_\Gamma,$$

from the above result and since $I_j^{*2} = \mathrm{Id}$. One gets $I_j^*[J^\Gamma A_j]_\Gamma = d\eta = [J^\Gamma A_j]_\Gamma - dd_1\eta_1 = d([J^\Gamma A_j]_\Gamma + d_1\eta_1)$, since $2\eta_1 = 0$.

Furthermore, since $I_j \lambda I_j = \bar{\lambda}$, one has, if $\det A_j > 0$ and using the fact that the map $\lambda \to \bar{\lambda}$ is homotopic to I_0^*,

$$[J^\Gamma I_j A_j I_j]_\Gamma = I_j'^* I_j^*[J^\Gamma A_j]_\Gamma = -[J^\Gamma A_j]_\Gamma,$$

hence $I_j'^*[J^\Gamma A_j]_\Gamma = -I_j^*[J^\Gamma A_j]_\Gamma$. If, on the contrary, $\det A_j < 0$, then $A_j I_j$ has positive determinant and one has

$$I_j'^* I_j [J^\Gamma A_j]_\Gamma = I_j'^*[J^\Gamma A_j I_j]_\Gamma = -I_j^*[J^\Gamma A_j I_j]_\Gamma = -[J^\Gamma A_j]_\Gamma,$$

with the same result. □

Remark 4.1. We shall prove below, in Lemma 4.2, that, after one suspension, one has

$$\Sigma_j(I_j^*\eta) = \Sigma_j(\eta - \eta_1) = \Sigma_j(\eta + d_1\eta_1),$$

that is $(1 + d_1)\Sigma_j\eta_1 = 0$, where $\Sigma_j\eta$ and $\Sigma_j\eta_1$ generate $\Pi(H_j) \cong \mathbb{Z}_2 \times \mathbb{Z}_2$ in this dimension. Thus, $1 + d_1$ is even, or else d_1 is odd and may be taken to be -1.

Let us now return to the matrix

$$A(\mu, \nu) = \operatorname{diag}(A_0, A_j, B_l, C_k),$$

where the different pieces have the same meaning as before. Let $\varepsilon_j = \operatorname{Sign}\det A_j$, for $j = 0, 1, \ldots, r$, that is, for the pieces of V^Γ and $V^{H_j} \cap (V^\Gamma)^\perp$, where $\Gamma/H_j \cong \mathbb{Z}_2$. Define $A_j^* = A_j I^{\alpha_j}$, where $\alpha_j = (1 - \varepsilon_j)/2$ and let

$$A^* = \operatorname{diag}(A_0^*, A_j^*, B_l, C_k).$$

Then $A^*(\mu, \nu)$ belongs to $\Pi_1(\mathrm{GL}_+^\Gamma(V))$ and can be written as a product of matrices of the form

$$\operatorname{diag}(A_0^*, I, I, I)\operatorname{diag}(I, A_j^*, I, I)\operatorname{diag}(I, I, B_l, I)\operatorname{diag}(I, I, I, C_k).$$

Thus, from the group morphism property of J^Γ, one has

$$J^\Gamma[A^*] = \Sigma^\Gamma J^\Gamma[A_0^*] + \sum_j \Sigma^\Gamma J^\Gamma[A_j^*] + \sum_l \Sigma^\Gamma J^\Gamma[B_l] + \sum_k \Sigma^\Gamma J^\Gamma[C_k],$$

where Σ^Γ is the suspension by the corresponding identity. Hence, one has

$$J^\Gamma[A] = \Big(\prod_{j=0}^{r} I_j^{*\alpha_j}\Big)[I_0^{*\alpha_0}\Sigma^\Gamma J^\Gamma[A_0] + \sum_j I_j^{*\alpha_j}\Sigma^\Gamma J^\Gamma[A_j] + \sum_l \Sigma^\Gamma J^\Gamma[B_l] + \sum_k \Sigma^\Gamma J^\Gamma[C_k]].$$

It remains to identify the action of I_j^* on each term and to compute $J^\Gamma[A_j]$, $J^\Gamma[B_l]$ and $J^\Gamma[C_k]$ in terms of the generators of $\Pi^\Gamma_{S^{\mathbb{R}\times V}}(V)$, as given in Theorems 5.1 and 5.2 in Chapter 3.

Assume, for simplicity, the following dimension conditions:

$$\text{(H1)}\qquad \begin{aligned} &\dim V^{\Gamma} \geq 3 \\ &\dim V_j \geq 3, \quad j = 1, \dots, r \\ &\dim_{\mathbb{C}} V_l \geq 2, \quad \dim_{\mathbb{C}} V_k \geq 1 \end{aligned}$$

where V_j is spanned by $Y_j = (y_1, \dots, y_n)$, each with isotropy H_j with $\Gamma/H_j \cong \mathbb{Z}_2$, the space V_l is spanned by $Z_l = (z_1, \dots, z_n)$, each with isotropy H_l with $\Gamma/H_l \cong \mathbb{Z}_p$ and action of Γ on z_s as $\exp(2\pi i m_s/p)$, where m_s and p are relatively prime. The coordinates $Z_k = (z_1, \dots, z_n)$, giving V_k, have isotropy H_k with action of Γ as $\exp(2\pi i m_k \varphi)$, including conjugates.

From Theorem 7.1 of Chapter 3, any suspension of $\Pi^{\Gamma}_{S^{\mathbb{R}\times V}}(S^V)$ is one-to-one and any suspension by one of the variables present in V is an isomorphism, in particular Σ_0: recall that $\deg_{\Gamma}(F;\Omega) = \Sigma_0 J^{\Gamma}[DF]_{\Gamma}$ is in $\Pi^{\Gamma}_{S^{\mathbb{R}^2\times V}}(S^{\mathbb{R}\times V})$. Now, according to Theorems 3.2 and 5.5 of Chapter 3, this group is a product of $\Pi(H)$'s, with $\Pi(H) \cong \mathbb{Z}$, if $\dim \Gamma/H = 1$, and $\Pi(H) \cong \mathbb{Z}_2 \times \Gamma/H$, if Γ/H is finite. Here, since DF is diagonal on equivalent irreducible representations of Γ, only those H's corresponding to coordinates in V will be concerned in the first computation of the Γ-degree of $J^{\Gamma}[DF]$.

Let us write $\lambda = \mu + i\nu$ and, in $V_0 = V^{\Gamma}, V_j, V_l$ or V_k, single out a complex coordinate z (made up of two real coordinates in the case of V_0 or V_j) and write X_0 as (z, X_0'), Y_j as (z, Y_j') and Z_l or Z_k as (z, Z_l') or (z, Z_k'). As functions of $(t, \mu, \nu, X_0, Y_j, Z_l, Z_k)$, consider the following generators

$$\begin{aligned} F_0 &= (2t-1, |z|^2-1, \lambda z, X_0', Y_j, Z_l, Z_k) \\ F_j &= (2t-1, |z|^2-1, X_0, \lambda z, Y_j', Z_l, Z_k) \\ F_l &= (2t-1, |z|^2-1, X_0, Y_j, \lambda z, Z_l', Z_k) \\ F_k &= (2t-1, |z|^2-1, X_0, Y_j, Z_l, \lambda z, Z_k'). \end{aligned}$$

The map F_0 is the suspension of the Hopf map and generates $\Pi(\Gamma)$: Lemma 5.1 of Chapter 3. The map F_j has an ordinary degree, for z in $\mathbb{R}^+$, equal to $(-1)^{\varepsilon_j}$, where $\varepsilon_j = \dim V_0 + \sum_{i<j} \dim V_i$, and, as such, can be taken as one of the two generators of $\Pi(H_j)$: see Theorem 5.1 in Chapter 3. The same argument yields an ordinary degree of $(-1)^{\varepsilon_{r+1}}$ for F_l and F_k. Since $\Pi(H_l)$ is $\mathbb{Z}_2 \times \mathbb{Z}_p$, if p is even, or $\mathbb{Z}_{2p}$ if p is odd, from Corollary 5.1 in Chapter 3, one may choose F_l as generator (see Lemma 5.4 in Chapter 3). For F_k one uses Theorem 3.3 in Chapter 3.

Recall that DF has a diagonal structure on equivalent irreducible representations of Γ, in particular on V_l, with $\Gamma/H_l \cong \mathbb{Z}_p$, one has $DF|_{V_l} = \operatorname{diag}(A_1, \dots, A_n)$, where the action of Γ on Z_{ls} is as $\exp(2\pi i m_s/p)$ and A_s corresponds to all coordinates with the same m_s.

Theorem 4.1. *Assume hypothesis* (H1) *holds and that* DF *is invertible on the loop* $|\lambda| = \rho$, *then, for* ε *small enough, one has*

$$\deg_\Gamma((\|X\|^2 - \varepsilon^2, F(\mu, \nu, X_0, Y_j, Z_l, Z_k); \Omega)$$
$$= \Big(\prod_{j=0}^{r} I_j^{*\alpha_j}\Big)(d_0[F_0]_\Gamma + \sum_j d_j[F_j]_\Gamma + \sum_l \Big(\sum_s n_s d_s)[F_l]_\Gamma + \sum_k d_k[F_k]_\Gamma\Big),$$

where

$$DF = \operatorname{diag}(D_{X_0}F^\Gamma, D_{Y_j}F_j^\perp, D_{Z_{ls}}F_{ls}^\perp, D_{Z_k}F_k^\perp)$$

and

$$\alpha_j = (1 - \operatorname{Sign}\det D_{Yj}F_j^\perp)/2$$

on the loop, for $j = 0, 1, \dots, r$. *If* η *is the generator of* $\Pi_1(\mathrm{GL}_+(V_j))$, *i.e.,* $\eta = \operatorname{diag}(\lambda, I)$, *for* $j = 0, 1, \dots, r$, *then*

(a) $d_0\eta$ *is the class of* $(D_{X_0}F^\Gamma)I_o^{\alpha_o}$ *in* $\Pi_1(\mathrm{GL}_+(V^\Gamma)) \cong \mathbb{Z}_2$,

(b) $d_j\eta$ *is the class of* $(D_{Y_j}F_j^\perp)I_j^{\alpha_j}$ *in* $\Pi_1(\mathrm{GL}_+(V_j)) \cong \mathbb{Z}_2$,

(c) *For* Z_{ls}, *with action as* $\exp(2\pi i m_s/p)$, *with* m_s *and* p *relatively prime, the number* $|n_s|$ *is an odd integer such that* $n_s m_s \equiv 1$, *mod* p, *and* d_s *is the winding number of* $\det(D_{Z_{ls}}F_{ls}^\perp)$, *as a mapping from the loop onto* $\mathbb{C}\backslash\{0\}$,

(d) *Finally,* d_k *is the winding number of* $\det(D_{Z_k}F_k^\perp)$, *where* Γ *acts as* $\exp(\pm i m_k\varphi)$.

Proof. The only point to check is the computation of $J^\Gamma[D_{Z_{ls}}F_{ls}^\perp]_\Gamma = d_s[F_{ls}]_\Gamma$, where the generator F_{ls} is built on the model of F_l but with action of Γ on z given by $\exp(2\pi i m_s/p)$. Thus, one has to relate $[F_{ls}]_\Gamma$ to $[F_l]_\Gamma$, where the action on z for F_l is given by a fixed m_s, for instance 1. This computation was done in Proposition 5.1 of Chapter 3:

$$[F_{ls}]_\Gamma = n_s[F_l]_\Gamma,$$

where $|n_s|$ is odd and $n_s m_s \equiv 1$, modulo p. For H_k, use Theorem 8.3 of Chapter 1. □

Remark 4.2. The reader should notice that there is a slight inconsistency in our statement of Theorem 4.1: whereas we have assimilated complex conjugate representations in V_k, with action as $\exp(\pm i m_k\varphi)$, we did not do so in V_l, where $\exp(2\pi i m_s/p)$ and $\exp(2\pi i(p - m_s)/p)$give the same equivalent real representations (see Remark 5.3 of Chapter 1). Furthermore, in general $DF|_{V_l}$ is not necessarily block diagonal on conjugate representations. However, in the cases of our applications, to symmetry breaking of differential equations, on one side one may eliminate negative modes (i.e., $m_k < 0$) and, on the other side, one has this block diagonal structure. Note that $-n_s(p - m_s) \equiv 1$, modulo p, if $n_s m_s \equiv 1$ (see the examples of Proposition 7.3 in Chapter 3). On the other hand, if F_k is built on z with action $\exp(i m_k\varphi)$ and

F'_k on a z with a conjugate action, we have seen, in Remark 3.1 of Chapter 3, that $[F'_k]_\Gamma = -[F_k]_\Gamma$. Hence, if $DF|_{V_k}$ is made of two blocks, one should have, in Theorem 4.1, the contribution $(d_k - d'_k)[F_k]_\Gamma$, where d_k is the winding number for the modes with action $\exp(im_k\varphi)$ and d'_k the winding number for the conjugates. If the two blocks are of the form $\pm im_k\nu I - L(\mu)$, as in the Hopf bifurcation, with $L(\mu)$ real, then the blocks are conjugates one of the other and $d'_k = -d_k$, giving $2d_k[F_k]_\Gamma$: then, there is no loss by considering only positive modes.

It remains to study the effect of the isomorphisms I_j^* on each of the generators.

Lemma 4.2. *Let F_u, $u = 0, j, k, l$ denote any of the above generators, then*

(a) $I_0^*[F_u]_\Gamma = -[F_u]_\Gamma$.

(b) $I_j^*[F_u]_\Gamma = [F_u]_\Gamma - [F_{uj}]_\Gamma$, *where F_{uj}, with z in V_u and y_j in V_j, is the map*

$$(2t - 1, |y_j| \cdot |z| - 1, X_0, Y_i, (y_j^2 - 1)y_j, \lambda z, \dots).$$

(c) *If $j_1 \neq j_2$, then $I_{j_2}^*[F_{uj_1}]_\Gamma = [F_{uj_1}]_\Gamma - [F_{uj_1j_2}]_\Gamma$, where the map $F_{uj_1j_2} = (2t - 1, |y_{j_1}| \cdot |y_{j_2}| \cdot |z| - 1, X_0, Y_i, (y_{j_1}^2 - 1)y_{j_1}, (y_{j_2}^2 - 1)y_{j_2}, \lambda z, \dots)$, while $I_j^*[F_{uj}]_\Gamma = -[F_{uj}]_\Gamma$.*

Proof. Write F_u as $(2t - 1, |z|^2 - 1, X_0, y_j, \lambda z, \dots)$, on the ball

$$B = \{0 \le t \le 1, |\lambda| \le 2, |y_j| \le 2, |z| \le 2, \|X_0\|, \dots \|Z_k\| \le 2\},$$

then $[F_u]_\Gamma = \deg_\Gamma(F_u; B)$, by using the fact that the suspension Σ_0 in the definition of the Γ-degree is an isomorphism. Then, the deformation $y_j(1 - \tau + \tau(y_j^2 - 1))$ is valid on ∂B. But then,

$$\deg_\Gamma(F_u; B) = \deg_\Gamma(F_u; B \cap \{|y_j| < 1/2\}) + \deg_\Gamma(F_u; B \cap \{|y_j| > 1/2\}).$$

For the first degree, one may deform y_j^2 to 0 and obtain $I_j^*[F_u]_\Gamma$.

For the second degree, one may use, on the set $\{|y_j| > 1/2\}$, the deformation $(1 + (1 - \tau)|z|)(|z|(1 - \tau + \tau|y_j|) - 1)$, since, there, a zero of $y_j(y_j^2 - 1)$ implies $|y_j| = 1$. For $\tau = 1$, one gets F_{uj}. Since (a) was already proved in Lemma 4.1, one obtains (b).

By using $I_j^{*2} = \text{Id}$, it is easy to see that $I_j^*[F_{uj}]_\Gamma = -[F_{uj}]_\Gamma$. Furthermore, by repeating the above argument, one has $[F_{uj_1}]_\Gamma = I_{j_2}^*[F_{uj_1}]_\Gamma + [F_{uj_1j_2}]_\Gamma$, as stated in (c). Further applications of I_j^* are built on the same scheme. □

Finally, one may identify F_{uj} with some of the remaining generators of $\Pi^\Gamma_{S^{\mathbb{R}^2 \times V}}(S^{\mathbb{R} \times V})$.

Lemma 4.3. *Let $H_u = \Gamma_z$ and $H_j = \Gamma_{y_j}$, with $\Gamma/H_j \cong \mathbb{Z}_2$ and $\Gamma/H_u = \{e\}$, $\mathbb{Z}_p$ or S^1.*

(a) *If $H_j \leq H_u$, then $[F_{uj}]_\Gamma$ is the second generator of $\Pi(H_j) \cong \mathbb{Z}_2 \times \mathbb{Z}_2$.*

(b) *If $H_u < H_j$, then $\Gamma/H_u \cong \mathbb{Z}_{2k}$ and*

$$[F_{uj}]_\Gamma = 2[F_u]_\Gamma + d[\tilde{F}_u]_\Gamma,$$

where $d = 1$ if k is odd and $\tilde{F}_u$ generates the second part of $\Pi(H_u) \cong \mathbb{Z}_{2k} \times \mathbb{Z}_2$, with $I_j^*[\tilde{F}_u]_\Gamma = [\tilde{F}_u]_\Gamma$.

(c) *If H_u is not a subgroup of H_j and $\Gamma/H_u \cong \mathbb{Z}_p$, then $[F_{uj}]_\Gamma$ is a generator of $\Pi(H_u \cap H_j) \cong \mathbb{Z}_p \times \mathbb{Z}_2 \times \mathbb{Z}_2$, with $p([F_{uj}]_\Gamma + [\tilde{F}_{uj}]_\Gamma) = 0$, with $2[\tilde{F}_{uj}]_\Gamma = 0$. The third generator is $[F_{ju}]_\Gamma$. If, furthermore, H_u is not a subgroup of H_{j_1}, then $[F_{ujj_1}]_\Gamma$ is a generator for $\Pi(H_u \cap H_j \cap H_{j_1})$, with $p([F_{ujj_1}]_\Gamma + [\tilde{F}_{ujj_1}]_\Gamma) = 0$, where $2[\tilde{F}_{ujj_1}]_\Gamma = 0$.*

(d) *If $\Gamma/H_u \cong S^1$, then $[F_{uj}]_\Gamma$ generates $\Pi(H_u \cap H_j) \cong \mathbb{Z}$ and $[F_{ujj_1}]_\Gamma$ generates $\Pi(H_u \cap H_j \cap H_{j_1})$.*

The action of I_{j2}^* follows from the above.

Proof. If H_u is not a subgroup of H_j, then there is an h in H_u such that $hy_j = -y_j$, in which case $\Gamma/(H_u \cap H_j) \cong (\Gamma/H_u) \times \mathbb{Z}_2$, since h^2 is in H_u and acts as the identity on y_j. On the other hand, if $H_u < H_j$ and $\Gamma/H_u \cong S^1$, then the action of Γ on z is given by $\exp i(\langle N, \Phi\rangle + 2\pi\langle K, L/M\rangle)$ (see Lemma 1.1 in Chapter 1). Hence, for any L there is a Φ_0 such that the exponential is 1, that is (Φ_0, L) is in H_u, since N is not 0. On y_j, the action of Γ is given by $\exp(\pi i\langle K_j, L\rangle)$. Then, if $H_u < H_j$, this last expression should be 1 for any L, which is impossible, since $\Gamma/H_j \cong \mathbb{Z}_2$. Thus, the only case where H_u is a subgroup of H_j is for $\Gamma/H_u \cong \mathbb{Z}_p$ with a generator γ_0 such that $\gamma_0 z = \exp(2\pi i/p)$ and $\gamma_0 y_j = -y_j$ (if $\gamma_0 y_j = y_j$, then any γ in Γ is of the form $\gamma = \gamma_0^k h$, with h in $H_u < H_j$ and one would have $H_j = \Gamma$). Since γ_0^p is in $H_u < H_j$, this implies that p is even.

Now, if $H_j \cong H_u$, i.e., $H_u = H_j$ or Γ, then the fundamental cell for $\Pi(H_j)$ can be taken as $\{0 < y_j < 2\}$ and F_{uj} is non-zero on its boundary and its class, on this boundary, is the suspension of the Hopf map, hence, from Theorem 5.1 of Chapter 3, F_{uj} is the second generator of $\Pi(H_j)$, proving (a).

Now, if H_u is not a subgroup of H_j and $\Gamma/H_u \cong \mathbb{Z}_p$, then the fundamental cell for $\Pi(H_u \cap H_j)$ is $\{(z, y_j, \dots) : 0 \leq \operatorname{Arg} z < 2\pi/p,\ 0 < y_j < 2\}$, where the order for y_j and z is irrelevant. Hence, from Theorems 5.3 and 5.4 of Chapter 3, one has $\Pi(H_u \cap H_j) = \mathbb{Z}_p \times \mathbb{Z}_2 \times \mathbb{Z}_2$, with generators η_1, η_2 and $\tilde{\eta}$ satisfying the relations $p(\eta_1 + \tilde{\eta}) = 0, 2\eta_2 = 0, 2\tilde{\eta} = 0$, and F_{uj} is given on η_1 and η_2 by $d_1 = \deg(F_{uj}; B \cap \operatorname{Arg} z = 0)/2$ and $d_2 = \deg(F_{uj}; B \cap \{y_j > 0, y_j' = 0\})/p$, where y_j' is one of the twins of y_j. By deforming y_j' to ε, it is clear that $d_2 = 0$, while

$d_1 = (-1)^{\dim V_0+1}$. Then, one may choose F_{uj} as the generator, instead of η_1. The other generators will be, according to Theorem 5.4 of Chapter 3

$$F_{ju} = (2t-1, |z|\cdot|z_l|-1, X_0, \lambda z, (z_l^p-1)z_l, \dots)$$

$$\tilde{F}_{uj} = (2t-1, |z||z_l^p-1|-1, (z^2-1)z, \lambda(z_l^p-1)z_l, \dots),$$

where $z = (y_j + iy_j')$ and z_l belongs to V_l.

Similarly, if H_u does not contain H_j nor H_{j_1}, with H_j different from H_{j_1}, so that $V^{H_j} \cap V^{H_{j_1}} = V^\Gamma$, one has the same situation for $\Pi(H_u \cap H_j \cap H_{j_1})$ and one may take F_{ujj_1} as a generator for this group, with the same relation as above. This proves (c).

On the other hand, if $\Gamma/H_u \cong S^1$, then $\dim \Gamma/(H_u \cap H_j) = 1$ and $\Pi(H_u \cap H_j) \cong \mathbb{Z}$. Then, one may compute the extension degree of F_{uj} on the fundamental cell $\mathcal{C} = \{(z, y_j, \dots) : 0 \le y_j \le 2,\ z \in \mathbb{R}^+\}$, which is $(-1)^{\dim V_0+1}$, as above. Hence, we may choose F_{uj} as a generator of the group. This proves (d).

Finally, if $H_u < H_j$, with $\Gamma/H_u \cong \mathbb{Z}_{2k}$, one may construct a fundamental cell for $\Pi(H)$ in two different ways: the first one, as the set characterized by $\{z : 0 \le \operatorname{Arg} z < \pi/k\}$, with the generators $[F_u]_\Gamma$ and $[\tilde{F}_u]_\Gamma$ and the relations $2k[F_u]_\Gamma = 0$, $2[\tilde{F}_u]_\Gamma = 0$. The second one, with a fundamental cell characterized by $\{(y_j, z) : 0 \le y_j < 2,\ 0 \le \operatorname{Arg} z < 2\pi/k\}$, with the generators

$$\eta_1 = (2t-1, |y|\cdot|z|-1, X_0, \lambda y, (\bar{y}z^k - |y|)z),$$

$$\eta_2 = [F_{uj}]_\Gamma$$

$$\tilde{\eta} = [\tilde{F}_u]_\Gamma = (2t-1, |z^{2k}-i|-1/2, X_0, y_j, \lambda(z^{2k}-i)z, \dots),$$

where $y = y_j + iy_j'$, with the relations

$$2\eta_1 + d_2\eta_2 + \tilde{d}\tilde{\eta} = 0, \quad k(\eta_2 + \tilde{\eta}) = 0, \quad 2\tilde{\eta} = 0,$$

see Theorem 5.2 of Chapter 3: on the set $B \cap \{y_j > 0, y_j' = 0\}$, the map η_1 has a degree equal to $(-1)^{\dim V_0}k$ and, according to Lemma 5.3 in Chapter 3, it may be taken as a generator, since d_2 and $\tilde{d}$ are 0 or 1.

Note that, according strictly to Lemma 5.4 of Chapter 3, $\tilde{\eta}$ should be the map

$$\tilde{\eta} = (2t-1, |y_j||y_jz^k - i| - 1/2, X_0, (y_j^2-1)y_j, \lambda(y_jz^k - i)z),$$

with zeros at $t = 1/2$, $X_0 = 0$, $y_j = \pm 1$, $\lambda = 0$, $|z^k \pm i| = 1/2$. In particular, $\tilde{\eta}$ is non-zero on the boundary of the first fundamental cell, where one has $0 \le \operatorname{Arg} z^k < \pi$ and z^k real on the boundary. Furthermore, in the cell, the only zero is for $y_j = 1$ and $|z^k - i| = 1/2$. From here, it is easy to see that the class of $\tilde{\eta}$, on this fundamental cell, is the Hopf map, that is $\tilde{\eta} = [F_u]_\Gamma$.

Now, on the set $B \cap \{z : \operatorname{Arg} z = 0\}$, one has the following relations between the ordinary degrees:

$$\deg(I_j^*[F_u]) = -\deg([F_u]), \quad \deg([F_{uj}]) = 2\deg([F_u]).$$

Hence, according to Lemma 5.4 in Chapter 3, one has

$$I_j^*[F_u]_\Gamma = -[F_u]_\Gamma + d\tilde{\eta}, \quad [F_{uj}]_\Gamma = 2[F_u]_\Gamma + d_1\tilde{\eta}.$$

Since, $I_j^*[F_u]_\Gamma + [F_{uj}]_\Gamma = [F_u]_\Gamma$, from Lemma 4.2, one has $d_1 + d = 0$, that is, in $\mathbb{Z}_2$, $d_1 = d$. Furthermore, in the map η_1, one may perform the equivariant rotation

$$(((1-\tau)\lambda - \tau(\bar{y}z^k - |y|))y, (\tau\lambda + (1-\tau)(\bar{y}z^k - |y|))z).$$

For $\tau = 1$, the term $|y|y - |y|^2 z^k$ is deformed linearly to $|z|y - z^k$ (on a zero of the map, one has $|y| \cdot |z| = 1$). Then, $|y| \cdot |z| - 1$ is deformed linearly to $|z|^k - 1$ and, next, to $|z|^2 - 1$. Finally, $|z|y - z^k$ is deformed (since on a zero one has $|z| = 1$), to $y - z^k$ and then to y. Thus, $\eta_1 = [F_u]_\Gamma$.

From $[F_{uj}]_\Gamma = \eta_2 = 2[F_u]_\Gamma + d\tilde{\eta} = 2\eta_1 + d\tilde{\eta}$, one obtains $d_2 = -1$ and $d = \tilde{d}$. Since $2k[F_u]_\Gamma = 2k\eta_1 = 0$ and $k(\eta_2 + \tilde{\eta}) = 0$, one gets $dk\tilde{\eta} = -k\tilde{\eta}$, that is, if k is odd, one needs $d = 1$. Note that $\tilde{\eta}$ has the class of the Hopf map on the fundamental cell and that $I_j^*\tilde{\eta} = \tilde{\eta}$ (since $2\tilde{\eta} = 0$). As in Propositions 7.2, 7.6 and 7.9 of Chapter 3, we leave out the determination of d, when k is even.

Note that, from Lemma 4.2 and from what we have proved in the present lemma, the effect of subsequent applications of I_{j1}^*, I_{j2}^*, and so on, may be easily derived. Remark that part of this theorem was proved, in Example 7.4 in Chapter 3, by using products. □

Corollary 4.1 (Abstract Hopf bifurcation). *Assume hypothesis* (H1) *and let* $F(\mu, \nu, X)$ *be a* C^1 *map from* $\mathbb{R}^2 \times U$ *into* U *(of the form Identity – Compact, if* U *is infinite dimensional) such that* $F(\mu, \nu, 0) = 0$ *and* $D_X F$ *is invertible for* $0 < \mu^2 + \nu^2 < 4\rho^2$ *and* $X = 0$. *Then, there is a global continuum* $\mathcal{C}$ *of zeros of* F, *with* $X \neq 0$, *bifurcating from* $(0, 0, 0)$, *which is either unbounded or returns to* $(\mu_1, \nu_1, 0) \neq (0, 0, 0)$, *if one of the following numbers is non-zero:*

$d_0 \mod 2$, and $\mathcal{C}$ is in U^Γ;

$d_j \mod 2$, and $\mathcal{C}$ is in U^{H_j};

$\sum n_s d_s \mod p$ if p is even and mod $2p$ if p is odd, and $\mathcal{C}$ is in U^{H_l};

d_k and $\mathcal{C}$ is in U^{H_k}.

If $\mathcal{C}$ *is bounded and* $D_X F$ *is invertible in punctured neighborhoods of the return points* $(\mu_j, \nu_j, 0)$, *then the sum of the* Γ*-degrees in Theorem* 4.1 *is* 0.

If all the numbers are 0, *then there is a* Γ*-map* $\tilde{F}(\mu, \nu, X)$, *with* $D_X\tilde{F}(\mu, \nu, 0) = D_X F(\mu, \nu, 0)$, *for* $\mu^2 + \nu^2 \leq 4\rho^2$, *such that* $\tilde{F}(\mu, \nu, X) = 0$ *only for* $X = 0$.

Proof. It is enough to apply Theorem 5.2 of Chapter 2 and, for the last part, the results of [*I*]. The proof relies only on the fact that for any $\rho > 0$, small enough, there is an $\varepsilon(\rho)$ such that $F(\mu, \nu, X) = 0$, for $\|X\| \leq 2\varepsilon(\rho)$ and $\rho^2 \leq \mu^2 + \nu^2 \leq 4\rho^2$, then $X = 0$. Then, for $\varepsilon \leq \varepsilon(\rho)$, one defines the set

$$\Omega = \{(\mu, \nu, X) : \mu^2 + \nu^2 \leq 4\rho^2, \ \|X\| \leq 2\varepsilon\},$$

and perform the deformation $(\tau(\|X\| - \varepsilon) + (1 - \tau)(\rho^2 - \mu^2 - \nu^2), F(\mu, \nu, X))$ on $\partial\Omega$, followed by the linearization of F, on the loop $\mu^2 + \nu^2 = \rho^2$, to $DF(\mu, \nu, 0)X$. □

Remark 4.3. There are many possible variations on the hypothesis of invertibility of DF. For instance, that DF is invertible only outside a disk of the form $\mu^2 + \nu^2 \leq \rho_0^2$: the above argument goes through and the bifurcation will take place from this disk. Another hypothesis, which we will use in the case of differential equations, is the following:

(H2) *For some* $\varepsilon_0, \rho_0 > 0$, *if* $\rho_0^2 \leq \mu^2 + \nu^2 \leq 4\rho_0^2$, *one has* $D_{X_\perp} F^\perp(\mu, \nu, 0)$ *invertible and, in the same ring,* $F^\Gamma(\mu, \nu, X_0) \neq 0$ *if* $0 < \|X_0\| \leq 2\varepsilon_0$.

In fact, since $F^\perp(\mu, \nu, X_0, X_\perp) = 0(\|X_\perp\|\|X\|)$, due to the equivariance, in the above disk and for $\|X\| \leq 2\varepsilon_0$, a zero of F is only for $X_\perp = 0$ and with $F^\Gamma(\mu, \nu, X_0) = 0$, i.e., with $X_0 = 0$. Thus, the deformation of $(\|X\| - \varepsilon, F)$ to $(\rho_0^2 - \mu^2 - \nu^2, F)$ is possible on $\partial\Omega$. Then, it is straightforward to deform to $(\rho_0^2 - \mu^2 - \nu^2, F^\Gamma(\mu, \nu, X_0), D_{X_\perp} F^\perp(\mu, \nu, 0)X_\perp)$, since the invertibility of $D_{X_\perp} F^\perp$ at $(\mu, \nu, 0)$ implies its invertibility at (μ, ν, X_0), for X_0 small. Note that, for $\mu^2 + \nu^2 = \rho_0^2$, the index of $F^\Gamma(\mu, \nu, X_0)$ at 0, is well defined and independent of (μ, ν) on this circle, since one may move along the circle, with a constant index. Note also that, if $F^\Gamma(\mu, \nu, X_0)$ is zero only at $X_0 = 0$, for $\mu^2 + \nu^2 \leq 4\rho_0^2$, then one may deform F^Γ, via $F^\Gamma(\tau\mu, \tau\nu, X_0)$, to $F^\Gamma(0, 0, X_0)$, obtaining a product map. In general, one has the following result:

Corollary 4.2. *Assume* (H1) *and* (H2) *hold. Then,* $\deg_\Gamma((\|X\|^2 - \varepsilon_0^2, F); \Omega)$ *is given as in Theorem* 4.1, *with* $d_0 = \deg((\|X_0\|^2 - \varepsilon_0^2, F^\Gamma(\mu, \nu, X_0)); \Omega^\Gamma)$, *in* $\mathbb{Z}_2$, *and* $I_0^{*\alpha_0}$ *replaced by* Index$(F^\Gamma(\mu_0, \nu_0, X_0); 0)$, *for any* (μ_0, ν_0) *on the loop* $\mu^2 + \nu^2 = \rho_0^2$. *In particular, if* $F^\Gamma(\mu, \nu, X_0)$ *is zero only at* $X_0 = 0$, *for* $\mu^2 + \nu^2 \leq 4\rho_0^2$, *then* $d_0 = 0$, *there is no bifurcation of stationary solutions, and, if* Index$(F^\Gamma(0, 0, X_0); 0)d_u$ *is non-zero, then one has a global bifurcation of non-stationary solutions in* V^u.

Proof. It is clear that one may perform the above deformation for $\Pi I_j^{\alpha_j} F$. Thus, one may assume that each piece of $D_{X_\perp} F^\perp$ has a positive determinant on the loop $|\lambda| = \rho$. Furthermore, on that loop, the piece $I_j^{\alpha_j} A_j$ is homotopic to diag$(\lambda^{d_j}, \mathrm{Id})$. If $G(\tau, \lambda)$

is the homotopy of the family of matrices, for $|\lambda| = \rho$, then $|\lambda| G(\tau, \lambda\rho/|\lambda|)$ is a valid extension to Ω. Thus, one may assume that $\Pi I_j^{\alpha_j} D_{X_\perp} F^\perp$ has this special form.

Let $\varphi(X_\perp)$ be 1 if $\|X_\perp\| < \varepsilon_0/2$ and be 0 if $\|X_\perp\| > \varepsilon_0$. Then, one may replace $D_{X_\perp} F^\perp$ by $(1-\varphi) D_{X_\perp} F^\perp + \varphi \operatorname{Id}^*$, where Id^*, on a component z_j such that $D_{X_\perp} F^\perp$ is $\lambda^{d_j} z_j$, is of the form $-\rho^{d_j} \eta_j^{d_j}$, with $|\eta_j| = 1$ and the d_j roots of the equation $\eta_j^{d_j} = 1$ are different for all j's. Hence, if Ω_1 is the part of Ω with $\|X_\perp\| < \varepsilon_0/2$ and Ω_2 the part with $\varepsilon_0/2 < \|X_\perp\| < 2\varepsilon_0$, one obtains

$$
\begin{aligned}
&\deg_\Gamma((\rho^2 - |\lambda|^2, F^\Gamma, D_{X_\perp} F^\perp); \Omega) \\
&\quad = \deg_\Gamma((\rho^2 - |\lambda|^2, F^\Gamma, \operatorname{Id}^* X_\perp); \Omega_1) \\
&\qquad + \deg_\Gamma(\rho^2 - |\lambda|^2, F^\Gamma, ((1-\varphi) D_{X_\perp} F^\perp + \varphi \operatorname{Id}^*) X_\perp); \Omega_2).
\end{aligned}
$$

In the first degree, one may deform Id^* to Id and obtain the suspension of d_0. For the second degree, the zeros of the map are such that, for some of the pieces of $D_{X_\perp} F^\perp$, one has $(1-\varphi)\lambda^{d_j} z_j - \varphi \rho^{d_j} \eta_j^{d_j} z_j = 0$, with $z_j \neq 0$. Hence, $\varphi = 1/2$ and $\lambda^{d_j} = \rho^{d_j} \eta_j^{d_j}$, something which happens for different values of λ. Hence, one may divide the loop, and subsequently Ω_2, into smaller pieces, where this last relation occurs for just one value λ_j of λ. On each of these pieces of the loop, one may deform $F^\Gamma(\mu, \nu, X_0)$ to $F^\Gamma(\mu_j, \nu_j, X_0)$, obtaining a product of maps and, from Proposition 7.6 of Chapter 3, a product of degrees. In this case, it is obvious that the product of the generators is just the generator of the product. Furthermore, since the index of $F^\Gamma(\mu_j, \nu_j, X_0)$ is constant on the loop, one may factor it and recompose Ω_2 from its pieces and, in fact, return to Ω, without the dependence on X_0. Note that, since $\dim V^\Gamma \cap V_\perp = 0$, technically the hypothesis (H1) is not satisfied for the Γ-degree for $(\lambda, X_\perp)$. But, one may suspend by $\tilde{X}_0$ of large dimension and use Theorem 7.1 of Chapter 3, where one has that, for strict subgroups of Γ, this suspension is an isomorphism, due to the rest of hypothesis (H1). Note also that one could compute directly the Γ-degree on a small neighborhood of $\lambda_j = \mu_j + i\nu_j$, by taking the section z_j in $\mathbb{R}^+$, giving a contribution of Sign d_j, for each root. Also, if $F^\Gamma(\mu, \nu, X_0) \neq 0$, for X_0 non-zero and $\mu^2 + \rho^2 \leq 4\rho_0^2$, then by deforming to $F^\Gamma(0, 0, X_0)$, one may use directly the product theorem. In this case, $\deg((\rho_0^2 - |\lambda|^2, F^\Gamma(0, 0, X_0)); \Omega^\Gamma) = 0$, since one may deform the first component to $\|X_0\|^2 - \varepsilon_0^2$, obtaining a pair without zeros. □

Example 4.2. As an illustration of the last corollary, consider the group $\mathbb{Z}_2$ acting trivially on z_0 in $\mathbb{C}$ and as $-\operatorname{Id}$ on (z_1, z_2) in $\mathbb{C}^2$. Consider the Γ-map

$$
F = (z_0^2 - \lambda t, \lambda z_1 + z_0(z_2 + \bar{z}_2) - t\bar{z}_1, |\lambda| z_2 - (z_0 z_1 + \bar{z}_0 \bar{z}_1) - t\bar{z}_2),
$$

where $\lambda = \mu + i\nu$ and $t = |z_1|^2 + |z_2|^2$. Then, $D_{X_\perp} F^\perp(\lambda, 0) = \operatorname{diag}(\lambda, |\lambda|)$ and $F^\Gamma = z_0^2$, that is (H2) is satisfied, with $\operatorname{Index}(F^\Gamma; 0) = 2$. Furthermore, $d_1 = [D_{X_\perp} F^\perp] = 1$ (the term $|\lambda|$ is deformable to 1). From Corollary 4.2 and taking into account that $2[F_1] = 0$, one has that $\deg_\Gamma(\rho^2 - |\lambda|^2, F; \Omega) = 0$. In fact, if $F(\lambda, X) = 0$, one may write the last two components, each multiplied by t, in the

form
$$\begin{pmatrix} z_0 & t \\ -t & \bar{z}_0 \end{pmatrix} \begin{pmatrix} z_0 z_1 & + & t\bar{z}_2 \\ z_0 z_2 & - & t\bar{z}_1 \end{pmatrix} = 0.$$
If $X \neq 0$, the matrix is invertible and, conjugating the second component, one may write the vector as
$$\begin{pmatrix} z_0 & t \\ -t & \bar{z}_0 \end{pmatrix} \begin{pmatrix} z_1 \\ \bar{z}_2 \end{pmatrix} = 0.$$
Hence the only zero of F is $(\lambda, 0)$, i.e., with no bifurcation.

Remark 4.4. For a correct application of Corollary 4.1, it is important to note that if $d_u \neq 0$, then there is global bifurcation in V^u. But this does not mean that the isotropy of the solution is exactly H_u. Similarly, one may have d_r and d_s non-zero for two subgroups H_r and H_s. Hence, one will have global bifurcation in V^{H_r} and V^{H_s}, but it may happen that these branches are the same in $V^{H_r} \cap V^{H_s}$, with an isotropy H which contains $H_r \cup H_r$. Even if $d_H = 0$, this is not enough to guarantee that the two solutions are distinct.

In fact, consider the action of S^1 on $\mathbb{C}^4$ via
$$e^{i\varphi}(z_1, z_2, z_3, z_4) = (e^{2i\varphi}z_1, e^{3i\varphi}z_2, e^{6i\varphi}z_3, e^{6i\varphi}z_4),$$
and the map $F(\lambda, X) = (X_0, \lambda z_1, \lambda z_2, \lambda z_3 + z_1^3, \bar{\lambda} z_4 + z_2^3)$.

From the linearization, it is easy to see that $d_{\mathbb{Z}_2} = 1 = d_{\mathbb{Z}_3}$ and $d_{\mathbb{Z}_6} = 0$, corresponding to the linear map $\operatorname{diag}(\lambda, \bar{\lambda})$, with winding number equal to 0. However, the non-trivial solutions are for $\lambda = 0$, $z_1 = z_2 = 0$, i.e., in $V^{\mathbb{Z}_6}$.

Similarly, the map $(|\lambda|^2 x_0 + |z_1|^2 + |z_2|^2, \lambda z_1, \lambda z_2)$, with action of S^1 as $e^{i\varphi}(x_0, z_1, z_2) = (x_0, e^{2i\varphi}z_1, e^{3i\varphi}z_2)$, has $d_\Gamma = 0$, $d_{\mathbb{Z}_2} = d_{\mathbb{Z}_3} = 1$, but the non-trivial solutions are for $\lambda = 0 = z_1 = z_2$.

Clearly, if it is known that there is no bifurcation in any V^H for H containing $H_r \cup H_s$ and the numbers d_{H_r} and d_{H_s} are non-zero, then one will have two distinct branches. This is the case if $DF^H(\lambda, 0)$ is invertible, for all such H's, or in the situation of Corollary 4.2.

Example 4.3 (Hopf bifurcation for autonomous differential equations). Let us return to the autonomous system
$$g(\mu, \nu, X) \equiv (\nu_0 + \nu)\frac{dX}{dt} - L(\mu)X - f(X, \mu) = 0, \quad X \text{ in } \mathbb{R}^N,$$
where $f(X, \mu) = 0(\|X\|^2)$, or equivalently, to the infinite system of Fourier series
$$(in(\nu_0 + \nu)I - L(\mu))X_n - f_n(X, \mu) = 0, \quad n \geq 0,$$
where f_n is S^1-equivariant. Note that the equation for $n = 0$, i.e., for the stationary part, is independent of ν.

Proposition 4.1. *Assume that $L(0)$ has eigenvalues $\pm im_1\nu_0, \ldots, \pm im_s\nu_0$ and that, for μ small but non-zero, $L(\mu)$ has its corresponding eigenvalues off the imaginary axis. Assume also that if $L(\mu)X + f(X, \mu) = 0$, for μ and $\|X\|$ small, then $X = 0$. Then, for any ε and ρ small enough, $\deg_{S^1}((\|X\| - \varepsilon, g(\mu, \nu, X)); \{|\lambda| \le 2\rho, \|X\| \le 2\varepsilon\})$ is well defined and equal to*

$$\sum_{j=1}^{s} d_{m_j}[F_{m_j}]_{S^1},$$

where $d_{m_j} = \mathrm{Index}(L(\mu)X + f(X, \mu); 0)\sigma_{m_j}$ and σ_{m_j} is the net crossing number of eigenvalues of $L(\mu)$ at $im_j\nu_0$, that is the number of eigenvalues which cross the imaginary axis at $im_j\nu_0$, when μ goes through 0, from left to right minus the number of those which cross from right to left.

Proof. The hypothesis insures that $in(\nu+\nu_0)I - L(\mu)$ is invertible for $\lambda = \mu + i\nu$ non zero and small, provided $n > 0$. The second part of the hypothesis implies that one may apply Corollary 4.2. From the point of view of the reduction to finite dimension, any contraction argument will lead to considering the modes $m_1, \ldots, m_s$ and 0. Note that the second part of the hypothesis is met if $L(0)$ is invertible, in which case the index of the statement is just the sign of $\det L(0)$. It remains only to compute the winding number of $\det(im_j(\nu_0 + \nu)I - L(\mu))$.

It is enough to recall that one may identify the eigenvalues $\lambda_1(\mu), \ldots, \lambda_N(\mu)$ of $L(\mu)$ in a continuous way (unless the eigenvalue is simple, or $L(\mu)$ is self-adjoint, the corresponding eigenvector is not continuous, although the projection on the generalized eigenspace is continuous: see for instance [K]). Then, if one writes $\lambda_k(\mu) = \alpha_k(\mu) + i\beta_k(\mu)$, it is clear that in the above determinant one may deform to 1, in $\mathbb{C}$, all the terms corresponding to eigenvalues which do not satisfy $\alpha_k(0) = 0$ and $\beta_k(0) = im_j\nu_0$. Thus, the winding number of the determinant is the degree of $\Pi(-\alpha_k(\mu) + im_j(\nu - \gamma_k(\mu))$, where k runs over all eigenvalues corresponding to the generalized eigenspace $\ker(L(0) - im_j\nu_0 I)^{\alpha_j}$, of dimension d, and $\beta_k(\mu) = im_j(\nu_0 + \gamma_k(\mu))$.

Since $\alpha_k(\mu) \neq 0$ for $\mu \neq 0$, and $\gamma_k(0) = 0$, one may deform, on $\mu^2 + \nu^2 = \rho^2$, the term $\gamma_k(\mu)$ to 0. Furthermore, $\alpha_k(\mu)$ may be deformed to Sign $\alpha_k(\mu)$ and then the corresponding factor is deformed to 1, in $\mathbb{C}$, if $\alpha_k(\mu)$ does not change sign. While, if $\alpha_k(\mu)$ changes sign as $\pm\mu$, then one may deform it to $\pm\mu$. Thus, one has to compute the degree of $(-\mu + im_j\nu)^{n_+}(\mu + im_j\mu)^{n_-}$, where $n_\pm$ are the number of eigenvalues which cross the imaginary axis at $im_j\nu_0$ as $\pm\mu$, when μ goes through 0. Hence the degree is $n_- - n_+$. This gives the result up to an orientation factor $(-1)^{N+1}$, given by the change from $n_- - n_+$ to $n_+ - n_-$ and to the degree of $-\mathrm{Id}$ in $\mathbb{R}^N$. This factor is absorbed by the generator F_{m_j}. The fact that $d_0 = 0$ comes from the special case of hypothesis (H2). □

Remark 4.5. Let us return to the point of the type of the solutions. Consider the three-dimensional system

$$(1+\nu)\begin{pmatrix} x_1 \\ x_2 \\ x_3 \end{pmatrix}' = \begin{pmatrix} \mu & 1 & 0 \\ -1 & \mu & 0 \\ 0 & 0 & \mu^2 \end{pmatrix}\begin{pmatrix} x_1 \\ x_2 \\ x_3 \end{pmatrix} + \begin{pmatrix} P(X,\mu) \\ Q(X,\mu) \\ R(X,\mu) \end{pmatrix},$$

where P, Q, R are C^2 functions with vanishing first derivatives at $X = 0$. Here the action is that of S^1 and only two modes are important: $n = 1$, with a crossing of a simple eigenvalue from left to right, and $n = 0$ with a non-negative eigenvalue touching the origin at $\mu = 0$.

(a) If $R(X, \mu) = a(\mu)x_1^2 + b(\mu)x_2^2 + c(\mu)x_1x_2 + x_3^3 + \text{H.O.T.}$, where H.O.T. means terms of order 3 (different from x_3^3) and higher. Since $L(\mu)$ is singular, one has to look at the stationary solutions in order to verify hypothesis (H2). In this case, one may solve the first two equations, for x_1 and x_2 in terms of x_3, with $x_i = 0(x_3^2)$, $i = 1, 2$. The last equation will be of the form $x_3(\mu^2 + x_3^2 + 0(x_3^3))$, which, for x_3 small, has the only solution $x_3 = 0$. Hence, one obtains a global branch of truly periodic solutions, since the index of $L(\mu)X + f(X, \mu)$ is 1, for μ small and non-zero.

(b) If the third equation is replaced by $(1 + \nu)x_3' = R(X, \mu)$, then, as above, the only stationary solution, for μ small, is $X = 0$ and its index is 1 (from the term x_3^3). Thus, one has the same behavior as in the preceding case.

(c) Take the system, with $P = Q = 0$, $R(X, \mu) = x_1^2 + x_2^2$. Then, any periodic solution gives either $\mu = \nu = 0$, or $x_1 = x_2 = 0$. But $x_3' = x_1^2 + x_2^2 \geq 0$ cannot have a periodic solution, unless $x_1 = x_2 = 0$. Thus, $\mu = 0$, $x_1 = x_2 = 0$, x_3 in $\mathbb{R}$ is a global branch of stationary solutions.

Example 4.4 (Hopf bifurcation for autonomous systems with symmetries). We consider the problem of finding (2π)-periodic solutions to the system

$$g(\mu, \nu, X) = (\nu + \nu_0)\frac{dX}{dt} - L(\mu)X - f(X, \mu, \nu) = 0, \quad X \text{ in } \mathbb{R}^N,$$

for (μ, ν) close to $(0, 0)$ and $f(X, \mu, \nu) = o(\|X\|)$. Here we shall assume that $L(\mu)$ and $g(X, \mu, \nu)$ are Γ_0-equivariant. Then the problem is equivalent to the system

$$(in(\nu + \nu_0)I - L(\mu))X_n - f_n(X, \mu, \nu) = 0, \quad n \geq 0,\ X_n \text{ in } \mathbb{C}^N.$$

We shall assume that $L(\mu)$ has the same spectral behavior as in the preceding example and, for simplicity, that $L(0)$ is invertible. Now, if X_{nj} is the j'th coordinate of X_n, then the action of $\Gamma = S^1 \times \Gamma_0$ on X_{nj} is of the form

$$\exp i(\langle N^j, \Phi\rangle + 2\pi\langle K, L^j/M\rangle + n\varphi),$$

as in Section 1 of Chapter 1, with $n \geq 0$. Hence, X_{nj} and X_{kl} belong to the same representation only if $n = k$, $N^j = N^l$ and $L^j \equiv L^l$, mod M. Since $n \geq 0$, conjugates will enter only for $n = 0$, $N^j = -N^l$, $L^j \equiv -L^l$, i.e., for conjugate representations of Γ_0. Furthermore, if H_{nj} is the isotropy of X_{nj}, then Γ/H_{nj} is finite only if $n = 0$ and $N^j = 0$, in which $d_{0j} = 0$, since $L(0)$ is invertible. Similarly, if $\Gamma/H_{0j} \cong S^1$, one obtains also $d_{0j} = 0$.

Now, since $L(\mu)$ is Γ_0-equivariant, one has

$$L(\mu) = \mathrm{diag}(L_0(\mu), L_k(\mu), \ldots, L_l(\mu)),$$

where Γ_0 acts trivially on L_0, as $-\,\mathrm{Id}$ on L_k and as $\mathbb{Z}_m$ or S^1 on L_l.

One gets the following application of Theorem 4.1:

Proposition 4.2. *Assume $L(0)$ is invertible and has eigenvalues $\pm im_1\nu_0, \ldots, \pm im_s\nu_0$, with $0 < m_1 \leq m_2 \leq \cdots \leq m_s$, and with the corresponding eigenvalues of $L(\mu)$ off the imaginary axis, for μ small and non-zero, and $L(\mu) = \mathrm{diag}(L_0, L_k, \ldots, L_l)$, with L_k of real dimension at least 3 and L_l of complex dimension at least 2, then*

$$\deg_\Gamma((\|X\| - \varepsilon, g); \{|\lambda| < 2\rho\} \times \{\|X\| < 2\varepsilon\}) = \prod I_k^{*\alpha_k}\Big(\sum_{n\geq 1} d_{nj}[F_{nj}]_\Gamma\Big),$$

where d_{nj} is the net crossing number of eigenvalues of $in(\nu + \nu_0)I - L_j(\mu)$ in V^{H_j}, where H_j is the isotropy of the piece corresponding to L_j. The generator F_{nj} is, up to an orientation factor, the suspension of $(1 - |z_{nj}|^2, \lambda z_{nj})$. The terms α_0, α_k are $(1 - \mathrm{Sign}\det L_0)/2$ and $(1 - \mathrm{Sign}\det L_k)/2$ respectively. One has $I_k^*[F_{nj}]_\Gamma = [F_{nj}]_\Gamma - [F_{njk}]_\Gamma$, where the last generator corresponds to the resonance of the stationary part $L_k(\mu)$, with action of Γ_0 as $-\,\mathrm{Id}$, on the n-th mode z_{nj}. For I_0, the action of I_0^* is the antipodal one. If $d_{nj} \neq 0$, one has a global bifurcation in V^{H_j}, with solutions $X(t)$ which satisfy

$$X(t) = \gamma_0 X(t + 2\pi/q), \text{ where } \gamma_0 \text{ is in } \Gamma_0, \gamma_0^{q_0} \text{ is in } H_j \text{ and } nq_0 = q.$$

Proof. This follows from Theorem 4.1, Corollaries 4.1 and 4.2 and Lemma 9.4 in Chapter 1. In order to determine α_k, it is enough to see which subgroups of Γ give $\Gamma/H \cong \mathbb{Z}_2$: this is possible only if $n = 0$, $N^j = 0$ and Γ_0 acts as $-\,\mathrm{Id}$. Note that, due to spectral conditions, there are at most $N/2$ possible d_{nj} which may be non-zero. □

Example 4.5 (Hopf bifurcation for time-dependent differential equations). Consider the problem of Hopf bifurcation for the equation

$$g(\mu, \nu, X) = (\nu + \nu_0)\frac{dX}{dt} - L(\mu)X - f(\mu, \nu, X) - \varepsilon_0 h(X, \mu, \nu, t) = 0, \ X \in \mathbb{R}^N,$$

where $X(t)$ is 2π-periodic, (μ, ν) is close to 0, the autonomous term has $f(\mu, \nu, X) = o(\|X\|)$ and the non-autonomous term has $h(0, \mu, \nu, t) = 0$ and is $(2\pi/p)$-periodic in t. Thus, for $\varepsilon_0 = 0$, one has an S^1-action, while for $\varepsilon_0 \neq 0$, the action is reduced to a $\mathbb{Z}_p$-action.

Proposition 4.3. *Assume $L(0)$ is invertible and has eigenvalues $\pm i m_1 \nu_0, \ldots, \pm i m_s \nu_0$, with $0 < m_1 \le m_2 \cdots \le m_s$, and with the corresponding eigenvalues of $L(\mu)$ off the imaginary axis, for μ small and non-zero. Then, for ε_0 small enough, the $\mathbb{Z}_p$-degree of $(\|X\| - \varepsilon, g)$ with respect to $B_{2\rho} \times B_{2\varepsilon}$ is well defined and equal to*

$$\operatorname{Sign}\det L(0)(d_\Gamma[F_\Gamma]_{\mathbb{Z}_p} + \sum_{p'|p} d_H[F_H]_{Z_p}),$$

where, if d_n denotes the net crossing number of eigenvalues of $in\ (\nu + \nu_0)I - L(\mu)$, one has

$$d_\Gamma \equiv \sum_{k=1}^{\infty} d_{kp} \mod 2,$$

$$d_H \equiv \sum_j n_j \sum_{k=1}^{\infty} d_{m_j p/p' + kp} \mod 2p' \text{ if } p' \text{ is odd and} \mod p' \text{ if } p' \text{ is even.}$$

Here, $\Gamma/H \cong \mathbb{Z}_{p'}$, for any divisor p' of p, the sum is over all m_j's, relatively prime to p', with $1 \le m_j < p'$, and $|n_j|$ is odd such that $n_j m_j \equiv 1$, modulo p'. If d_Γ is odd, one obtains Hopf bifurcation of $(2\pi/p)$-periodic solutions, while if d_H is not congruent to 0, one has Hopf bifurcation of $(2\pi p'/p)$-periodic solutions.

Proof. If $h(X, \mu, \nu, t) = A(t)X + \ldots$ one may choose ε_0 so small that the Fredholm operator $(\nu+\nu_0)d/dt - L(\mu) - \varepsilon_0 A(t)$ is invertible, for $\mu^2+\nu^2 = \rho^2$, from the space of 2π-periodic C^1 functions onto the space of 2π-periodic C^0 functions: this comes from the fact that, for $|\mu| \le \rho$, $L(\mu)$ is invertible and has no pure imaginary eigenvalues, for $\mu \ne 0$, close to a multiple of ν_0, hence the operator $(\nu + \nu_0)d/dt - L(\mu)$ is invertible on the loop. Furthermore, one may $\mathbb{Z}_p$-deform $g(\mu, \nu, X)$, on the loop, to $(\nu + \nu_0)X' - L(\mu)X$, considered, when $\varepsilon_0 \ne 0$, as a $\mathbb{Z}_p$-equivariant linear map.

While, for $\varepsilon_0 = 0$, any non-zero winding number d_n of $in(\nu + \nu_0)I - L(\mu)$ will give rise to a Hopf bifurcation of 2π-periodic solutions (not necessarily least periodic), for $\varepsilon_0 \ne 0$, we have to study the isotropy subgroups H of $\mathbb{Z}_p$ for its action on Fourier series, that is as $\exp(2\pi ink/p)$ on X_n, with $0 \le k < p$. Hence, if $n/p = n'/p'$, with n' and p' relatively prime, the isotropy H of X_n will be $H = \{k = 0, p', 2p', \ldots (p/p' - 1)p'\} \cong \mathbb{Z}_{p/p'}$ and $\Gamma/H \cong \mathbb{Z}_{p'}$.

Now, two representations of $\mathbb{Z}_p$ will be equivalent, on X_n and X_m and as complex representations, if and only if $n \equiv m$, modulo p. Furthermore, in order to apply Theorem 4.1, one needs to identify all modes X_m which have exactly H, as above, as isotropy, i.e., such that the action of Γ on X_m is of the form $\exp(2\pi i m_s k/p')$, for $k = 0, \ldots, p' - 1$ and where m_s and p' are relatively prime, with $m/p = m_s/p'$. Then, $m_s = m_j + ap'$, with $1 \le m_j < p'$, and m_j and p' relatively prime, and $m = m_j p/p' + ap$. If p' is prime, then any integer m_j between 1 and $p' - 1$ is allowed. Clearly, if n_j, with $|n_j|$ odd, is such that $m_j n_j \equiv 1$, modulo p', then $m_s n_j \equiv 1$, modulo p'. Also, if $H = \Gamma$, then $m = kp$, since $p' = 1$ and $m_j = n_j = 1$.

Finally, since Γ acts only on the non-trivial modes, I_k^* is not present, except for I_0^* which contributes Sign det $L(0)$. □

Remark 4.6. Note first that this symmetry breaking argument was given, in an abstract form, in Proposition 7.3 in Chapter 3. From the point of view of Hopf bifurcation, note that a mode m belongs to just one p': in fact, if $m = m_1 p/p_1 + k_1 p = m_2 p/p_2 + k_2 p$, then $m_1 p_2 - m_2 p_1 = k p_1 p_2$, where m_j and p_j are relatively prime. But this implies $p_1 = p_2$. Thus, it is convenient to list the divisors of p in increasing order and begin with the smallest (1 corresponds to d_Γ). Then, for a given integer $j < p'$, either j is relatively prime to p' or the corresponding modes $jp/p' + kp$ have already been assigned to a smaller divisor of p. Note also that, if $m_j n_j \equiv 1$, modulo p', with m_j and p' relatively prime, then it is also true for $m_j' = p' - m_j$ and $n_j' = -n_j$: this natural pairing corresponds to conjugation. Finally, note that if p' is an odd prime (if $p' = 2$, then $m_j = n_j = 1$), then, due to the pairing, one has to consider all integers between 1 and $(p' - 1)/2$, with $n_1 = 1, n_2 = (1 + p')/2$, if this number is odd, or $n_2 = (1 - p')/2$ otherwise, and $n_{(p'-1)/2} = p' - 2$.

Finally, for $p \leq 7$, we refer the reader to the examples after Proposition 7.3 in Chapter 3, where d_Γ and d_H are computed in terms of the d_n's.

Remark 4.7. Recall that, if the bifurcation index is 0, then, given a linear part, there is a non-linear part at the level of Fourier series (not necessarily coming from a differential equation) such that there is no bifurcation. Here, we shall give an example, which is parallel to Example 2.5, showing how one may force a linear system which has a Hopf bifurcation with a linear time-periodic perturbation which destroys the bifurcation.

Take p any integer larger than 1 and consider the following system for 2π-periodic functions:

$$\begin{aligned} x'' &- \mu x' + \nu x + 2\varepsilon((p+1)y\cos pt + y'\sin pt) = 0 \\ y'' &- (p-1)\mu y' + (p-1)^2\nu y - 2\varepsilon(p-1)((2p-1)x\cos pt + x'\sin pt) = 0. \end{aligned}$$

For $\varepsilon = 0$, μ close to 0 and ν close to 1, one has a vertical Hopf bifurcation for $(x, 0)$ with $n = 1$ and for $(0, y)$ with $n = p - 1$. The winding numbers are all 0, except $d_1 = d_{p-1} = 1$.

For $\varepsilon \neq 0$, the system is equivalent to

$$\begin{aligned} (-n^2 &- i\mu n + \nu)x_n + \varepsilon((n+1)y_{n-p} - (n-1)y_{n+p}) = 0 \\ (-n^2 &- i\mu n(p-1) + \nu(p-1)^2)y_n \\ &- \varepsilon(p-1)((n+p-1)x_{n-p} - (n-p+1)x_{n+p}) = 0. \end{aligned}$$

Taking the first equation for $n = 1$ and the second for $n = p - 1$, one obtains the pair

$$((\nu - 1 - i\mu)x_1 + 2\varepsilon\bar{y}_{p-1}, (p-1)^2((\nu - 1 - i\mu)y_{p-1} - 2\varepsilon\bar{x}_1)),$$

with only zeros $x_1 = y_{p-1} = 0$, unless $\nu = 1$, $\mu = 0$ and $\varepsilon = 0$. For $\varepsilon \neq 0$, the remaining equations form a closed system with invertible diagonal, that is, the only solution, for ε small and (μ, ν) close to $(0, 1)$, is $x = y = 0$.

For $p = 1$, one takes out the factors $p - 1$, in the second equation, and one has $d_1 = 2$ but the same result holds.

It would be interesting to have similar simple examples for, say, $p = 3, d_1 = 6, d_j = 0$ for $j > 1$, or $p = 5, d_1 = 3, d_2 = -1$ and $d_j = 0$ otherwise.

Remark 4.8 (Global Hopf bifurcation). In this book we have not stressed the aspects of global bifurcation, since [IMPV] and [I] deal with this problem. However, we should warn the reader about the meaning of unboundedness of global branches, in particular for the equation

$$\nu X' = g(X, \mu).$$

As explained in Remark 2.3 of Chapter 3, this equation has to be transformed into an equation of the form Id-compact, in order to apply any degree theory in infinite-dimensional spaces. The integral equation will then have the term ν^{-1}, that is, when ν goes to 0, the equation becomes unbounded. Hence, a natural parameter for the global bifurcation is the period $T = 2\pi/\nu$.

Furthermore, if $g(X, \mu)$ is C^1 and one has a 2π-periodic solution $X(t)$, then,

$$\nu^2|X'(t) - X'(0)|^2 \leq L^2|X(t) - X(0)|^2,$$

where L is a bound for $Dg(X(s), \mu)$ on the orbit. Since $\int_0^{2\pi} X'(t) \cdot X'(0)\, dt = 0$, from the periodicity, one obtains

$$\nu^2\|X'\|^2 \leq \nu^2(\|X'\|^2 + |X'(0)|^2) \leq 2\pi^2L^2\|X'\|^2.$$

Thus, for a non-stationary solution lying in a bounded region of $\mathbb{R}^N$, one has

$$\nu \leq \pi\sqrt{2}L.$$

Also, if $X(t)$ is close to a stationary solution X_0, the Fourier series

$$in\nu X_n - Dg(X_0)X_n = h_n(X),$$

imply that, if $|\nu| > \|Dg(X_0)\|$, then the only solution is a stationary solution and X_0 cannot be a bifurcation point of truly periodic solutions.

Note that, if there are no stationary solutions in some bounded subset of $\mathbb{R}^N$, then $K \leq |g(X)| \leq M$, and $K \leq |\nu|\|X'\| \leq M$.

In particular, any truly periodic solution, in a bounded subset Ω of H^1, will be, from Sobolev inequality, bounded in $\mathbb{R}^N$ and, from the above, with a bounded frequency.

Thus, the global Hopf theorem should be stated as: either the branch of non-trivial solutions is unbounded in $(\mu, T, \|X\|_1)$, or returns to another trivial solution, where trivial solution means either $X = 0$, if one has complemented with $\|X\|_1 - \varepsilon$, or a stationary solution, if the complementing map is $\|X'\| - \varepsilon$: in this last case, one needs that there is no bifurcation of stationary solutions at $(0, 0)$, for instance, if $L(0)$ is invertible.

Example 4.6 (Hopf bifurcation for autonomous systems with first integrals). Consider the problem of finding 2π-periodic solutions to the problem

$$g(X,\mu) = \frac{dX}{dt} - L(\mu)X - f(X,\mu) = 0, \quad X \text{ in } \mathbb{R}^N, \ f(X,\mu) = o(\|X\|),$$

for which one has a family of first integrals $V(X,\mu)$. Thus, $\nabla V(X,\mu)$ is orthogonal to $g(X,\mu)$ for each fixed μ, that is ∇V is orthogonal, on $\mathbb{R}^N$, to $LX+f$ and $\nabla V(X(t),\mu)$ is L^2-orthogonal to $g(X(t),\mu)$, if $X(t)$ is 2π-periodic. As explained in Example 2.8, this problem is equivalent to finding 2π-periodic solutions to the equation

$$\frac{dX}{dt} - L(\mu)X - f(X,\mu) - \nu\nabla V(X,\mu) = 0,$$

where, if one has a solution with $\nabla V(X,\mu) \neq 0$, then $\nu = 0$.

Assume there is a family of stationary solutions $X(\mu)$ such that $g(X(\mu),\mu) = 0$, $\nabla V(X(\mu),\mu) = 0$. Without loss of generality, we may take $X(\mu) \equiv 0$. Let

$$\nabla V(X,\mu) = H(\mu)X + k(X,\mu), \quad \text{with } k(X,\mu) = o(\|X\|).$$

Lemma 4.4. *One has the following relations:*

$$H(\mu) = H^T(\mu), \quad L(\mu)^T H(\mu) + H(\mu)L(\mu) = 0.$$

Proof. The first relation follows immediately from the fact that $H(\mu)$ is the Hessian of V. For the second relation, from the orthogonality

$$(L(\mu)X + f(X,\mu), \nabla V(X,\mu)) = 0,$$

one obtains, dividing by $\|X\|^2$ and taking limits when X goes to 0:

$$(L(\mu)X, H(\mu)X) = 0,$$

and thus, from $(X, (L^T H + HL)X) = 0$, the symmetric matrix $L^T H + HL$ is 0. □

Assume that $L(0)$ has eigenvalues $\pm im_1, \dots, \pm im_s$, with $0 < m_1 \leq \dots \leq m_s$, counted with multiplicities. Let $\lambda_j(\mu) = \alpha_j(\mu) + i\beta_j(\mu)$ be the eigenvalues of $L(\mu)$, for μ close to 0, such that $\alpha_j(0) = 0$, $\beta_j(0) = \beta_j$. We shall impose the following hypothesis:

(H_j) a) *If* $\lambda_j(\mu) = im_j$, *for* μ *close to* 0, *then* $\mu = 0$.
b) $\ker H(0) \cap \ker(im_j I - L(0)) = \{0\}$ *for* $j = 1, \dots, s$.

Proposition 4.4. *Hypothesis* (H_j) *is equivalent to have* $im_j I - L(\mu) - \nu H(\mu)$ *invertible for* $(\mu,\nu) \neq (0,0)$, *but close to* $(0,0)$.

Proof. If $im_jI - L(\mu) - \nu H(\mu)$ is invertible, take $\nu = 0$, $\mu \neq 0$, then one obtains (a). On the other hand, taking $\mu = 0$, $\nu \neq 0$, one gets (b).

Conversely, consider the complex scalar product

$$((L(\mu)+\nu H(\mu)-\lambda I)X, H(\mu)X) = (H(\mu)L(\mu)X, X)+\nu\|H(\mu)X\|^2-\lambda(H(\mu)X, X),$$

where $\lambda = \alpha + i\beta$. The first term on the right is, due to Lemma 4.4, purely imaginary, while the other two are real, since $H(\mu)$ is real. Hence, if λ is an eigenvalue of $L(\mu) + \nu H(\mu)$, with corresponding eigenvector X, one obtains

$$\nu\|H(\mu)X\|^2 = \alpha(H(\mu)X, X).$$

On the other hand, one has, in general,

$$\begin{aligned}
&\|(L(\mu) + \nu H(\mu) - \lambda I)X\|^2 \\
&\quad = \|(L(\mu) - \lambda I)X\|^2 + \nu^2\|H(\mu)X\|^2 + 2\nu \operatorname{Re}((L(\mu) - \lambda I)X, H(\mu)X) \\
&\quad = \|(L(\mu) - \lambda I)X\|^2 + \nu^2\|H(\mu)X\|^2 - 2\nu\alpha(H(\mu)X, X).
\end{aligned}$$

Thus, if $\lambda = im_j$ and X is an eigenvector, then $\alpha = 0$, $\nu\|H(\mu)X\| = 0$ and X is an eigenvector of $L(\mu) - im_jI$. From (a), this implies $\mu = 0$ and, from (b), one needs $H(0)X \neq 0$, thus, $\nu = 0$. □

We shall need some information on the spectral behavior of $L(\mu)$:

Lemma 4.5. *Assume* (H_j) *holds, then, for small* μ, *one has the following.*

(a) *For any* $k \geq 1$, $\ker H(\mu) \cap \ker(L(\mu) - \lambda_j(\mu)I)^k = \{0\}$.

(b) *If* $\lambda = \lambda_j(\mu)$ *is an eigenvalue of* $L(\mu)$ *so are* $-\lambda$ *and* $\pm\bar{\lambda}$, *with the same algebraic multiplicity.*

(c) *If* $i\beta = \lambda_j(\mu)$ *is a simple eigenvalue of* $L(\mu)$ *with corresponding eigenvector* X, *then* $(H(\mu)X, X) \neq 0$.

(d) *If* $i\beta = \lambda_j(\mu)$ *is an eigenvalue of* $L(\mu)$, *with generalized eigenspace* $\ker(L(\mu) - i\beta I)^k$, *then* $H(\mu)$ *induces a non-degenerate quadratic form on this eigenspace, with a well-defined signature* $\sigma_\beta(\mu)$.

Proof. If (a) is false for $k = 1$, then there are sequences μ_n converging to 0, eigenvalues λ_n converging to im_j, eigenvectors X_n, with norm 1, and a subsequence converging to some X, such that $(L(\mu_n) - \lambda_nI)X_n = 0$, $H(\mu_n)X_n = 0$. Taking limits, one will get a contradiction to (H_j). For $k > 1$, let X be such that $(L(\mu) - \lambda I)^kX = 0$ and set $Y = (L(\mu) - \lambda I)^nX$, where n is the largest integer for which $(L(\mu) - \lambda I)^nX \neq 0$, hence $n < k$. Thus, $(L(\mu) - \lambda I)Y = 0$. If $H(\mu)X = 0$, then $H(\mu)Y = (-1)^n(L(\mu)^T + \lambda I)^nH(\mu)X = 0$, from Lemma 4.4. Hence, from the case $k = 1$, one has $Y = 0$, which results in a contradiction.

For (b) one uses the relation $H(\mu)(L(\mu)-\lambda I)^k = (-1)^k(L(\mu)^T+\lambda I)^k H(\mu)$. Since $L(\mu)$ is real, if λ is an eigenvalue, so is $\bar{\lambda}$, with the same algebraic multiplicity. From the above relation, this is also the case for $-\lambda$ (and $-\bar{\lambda}$) as eigenvalue of $L(\mu)^T$, with eigenvector $H(\mu)X$, non-zero because of *(a)*, and hence for $L(\mu)$. Since $H(\mu)$ is a one-to-one morphism from $\ker(L(\mu)-\lambda I)^k$ into $\ker(L(\mu)^T+\lambda I)^k$, the second space is at least as large as the first.

Decompose orthogonally $\mathbb{R}^N$ as $V(\mu)\oplus V(\mu)^\perp$, where $V(\mu)=\ker H(\mu)$. Since $H(\mu)$ is symmetric, the space $V(\mu)^\perp$ is Range $H(\mu)$. From the relation $L(\mu)^T H(\mu)+H(\mu)L(\mu)=0$, one obtains, on this decomposition,

$$L(\mu)=\begin{pmatrix} A(\mu) & B(\mu) \\ 0 & C(\mu)\end{pmatrix},$$

with $C(\mu)^T H(\mu)+H(\mu)C(\mu)=0$. Since $H(\mu)$ is invertible on $V(\mu)^\perp$, one has $C(\mu)=-H(\mu)^{-1}C(\mu)^T H(\mu)$, which implies that dim $V(\mu)^\perp$ is even. From (H_j), it follows that, if $\lambda=\lambda_j(\mu)$, then $A(\mu)-\lambda I$ is invertible and, from the triangular form of $L(\mu)$, one has

$$(L(\mu)-\lambda I)^k=\begin{pmatrix}(A(\mu)-\lambda I)^k & 0 \\ 0 & I\end{pmatrix}\begin{pmatrix} I & 0 \\ 0 & (C(\mu)-\lambda I)^k\end{pmatrix}\begin{pmatrix} I & (A(\mu)-\lambda I)^{-k}D_k \\ 0 & I\end{pmatrix},$$

that is, $\ker(L(\mu)-\lambda I)^k$ is isomorphic to $\ker(C(\mu)-\lambda I)^k$. Furthermore, the relation with $H(\mu)$ gives

$$H(\mu)(C(\mu)-\lambda I)^k=(-1)^k(C(\mu)^T+\lambda I)^k H(\mu),$$

with $H(\mu)$ invertible on this subspace. Thus, $\ker(C(\mu)^T+\lambda I)^k$ has the same dimension as $\ker(C(\mu)-\lambda I)^k$. This implies that $\ker(L(\mu)-\lambda I)^k$ and $\ker(L(\mu)^T+\lambda I)^k$ have the same dimension and that $H(\mu)$ is an isomorphism between them. The equality of the algebraic multiplicities follows from standard arguments.

For (c), if $i\beta$ is a simple eigenvalue of $L(\mu)$, with eigenvector X, then, from *(a)*, $H(\mu)X\neq 0$ and $H(\mu)X$ generates $\ker(L(\mu)^T+i\beta I)$. If $(H(\mu)X, X)=0$, then X would be orthogonal to $H(\mu)X$, hence X would belong to Range$(L(\mu)-i\beta I)$ and the multiplicity of $i\beta$ would be greater than 1.

For (d), if k is the ascent of $L(\mu)-i\beta I$, then one has the (non necessarily orthogonal) decomposition

$$\mathbb{C}^N=\ker(L(\mu)-i\beta I)^k\oplus \mathrm{Range}(L(\mu)-i\beta I)^k.$$

Let P be the orthogonal projection on $\ker(L(\mu)-i\beta I)^k$. Since $H(\mu)$ is an isomorphism from this last space onto $\ker(L(\mu)^T+i\beta I)^k=(\mathrm{Range}(L(\mu)-i\beta I)^k)^\perp$, then $PH(\mu)P$ generates a symmetric bilinear form on $\ker(L(\mu)-i\beta I)^k$. Furthermore, if $PH(\mu)X=0$, for some X in this space, then $(X, H(\mu)Y)=0$, for any Y in this space and, from the above isomorphism, X is orthogonal to $\ker(L(\mu)^T+i\beta I)^k$, hence X belongs to Range$(L(\mu)-i\beta I)^k$, that is $X=0$: hence the quadratic form is non-degenerate. □

Definition 4.1. (a) The signature of a complex self adjoint matrix A, i.e., the number of positive eigenvalues minus the number of negative eigenvalues (A may be singular) will be denoted by $\sigma(A)$.

(b) We shall denote, for $\mu \neq 0$, by $\sigma_j^{\pm}(\mu)$ the sum of the signatures of H on $\bigcup \ker(L(\mu)-i\beta_j(\mu))^k$, for $\beta_j(\mu) > m_j$ and close to m_j (for $\sigma_j^+(\mu)$) and for $\beta_j(\mu) < m_j$ and close to m_j (for $\sigma_j^-(\mu)$). Let $\sigma_j(\mu) = \sigma_j^+(\mu) + \sigma_j^-(\mu)$.

Note that Hypothesis (H_j) implies that $\sigma_j^{\pm}(\mu)$ are well defined for $\mu \neq 0$. We shall prove below that, in fact, they remain constant provided μ does not change sign and that $\sigma_j(\mu) = \sigma_j(0)$. Recall that Sylvester inertial law says that $\sigma(A)$ is independent of the basis.

In order to compute the Hopf bifurcation indices, we shall need the following perturbation result

Lemma 4.6. (a) *For each fixed μ, one may perturb $L(\mu)$ to $\tilde{L}(\mu)$ and $H(\mu)$ to $\tilde{H}(\mu)$, such that the relationship $L(\mu)^T H(\mu) + H(\mu)L(\mu) = 0$ is preserved during the perturbation and $\tilde{L}(\mu)$ has all its purely imaginary eigenvalues, close to im_j, simple and $\sigma_j(H(\mu)) = \sigma_j(\tilde{H}(\mu))$.*

(b) $\sigma_j(\mu) = \sigma_j(0)$ *and* $\sigma_j^{\pm}(\mu)$ *are constant for* $\mu \neq 0$.

Proof. From our previous considerations, it is enough to look at $C(\mu)$, such that $L(\mu) = \begin{pmatrix} A(\mu) & B(\mu) \\ 0 & C(\mu) \end{pmatrix}$, on $(\ker H(\mu))^{\perp}$. In order to lighten the notation, we shall drop the μ dependence. Let $i\beta$ be an eigenvalue of C, close to some im_j and let k be the least integer such that $V(\mu)^{\perp} \cong \mathbb{C}^s$, with s even, and

$$\mathbb{C}^s = \ker(C - i\beta I)^k \oplus \operatorname{Range}(C - i\beta I)^k.$$

Let P be the orthogonal projection onto $\ker(C - i\beta I)^k$ and let

$$F = i^{k-1} PH(C - i\beta I)^{k-1} P = PH\mathcal{A}^{k-1}P.$$

It is easy to check that $H\mathcal{A} = \mathcal{A}^* H$ and $F^* = \bar{F}^T = F$. Furthermore, if $FX = 0$, then $H(C-i\beta I)^{k-1}PX$ would be orthogonal to $\ker(C-i\beta I)^k$, i.e., in $\operatorname{Range}(C^T+i\beta I)^k$, that is $H(C - i\beta I)^{k-1}PX = (C^T + i\beta I)^k Y$, for some Y. Apply $C^T + i\beta I$ to this equality, use the anticommutativity for H and $(C - i\beta I)$ and the fact that PX is in $\ker(C-\beta I)^k$, to conclude that $(C^T+i\beta I)^{k+1}Y = 0$. But, since $\ker(C^T+i\beta I)^{k+1} = \ker(C^T + i\beta I)^k$, by definition of the ascent, this implies that $(C^T + i\beta I)^k Y = 0$, or else, since H is an isomorphism, that $(C - i\beta I)^{k-1}PX = 0$. Since $\ker(C - i\beta I)^{k-1}$ is strictly contained in $\ker(C - i\beta I)^k$, one concludes that F is not identically 0 and that $\ker F = \ker \mathcal{A}^{k\ \ 1}$.

Thus, there is a non-zero X, in $\ker(C-i\beta I)^k$, and $\lambda_1 \neq 0$, such that $FX_1 = \lambda_1 X_1$. Normalize X_1 in such a way that $(FX_1, X_1) = \eta_1 = \operatorname{Sign} \lambda_1$. Let

$$X_j = i^{j-1}(C - i\beta I)^{j-1}X_1 = \mathcal{A}^{j-1}X_1, \quad j = 1, \ldots, k.$$

Then $(HX_j, X_l) = \eta_1$ if $l + j = k + 1$ and 0 if $l + j > k + 1$. Define

$$\begin{aligned} Y_1 &= X_1 + a_2 X_2 + \cdots + a_k X_k \\ Y_j &= i^{j-1}(C - i\beta I)^{j-1} Y_1 = \mathcal{A}^{j-1} Y_1, \quad \text{for } j = 1, \dots, k, \end{aligned}$$

where $a_2, \dots, a_k$ are obtained by setting $(HY_1, Y_j) = 0$, for $j = 1, \dots, k-1$: since $Y_j = \sum a_l \mathcal{A}^{j-1} X_l = \sum a_l X_{j+l-1}$, one has

$$(HY_1, Y_j) = \sum_{l+m \le k-j+2} a_m a_l (HX_m, X_{j+l-1}) = 2\eta_1 a_{k-j+1} + \cdots,$$

where the suspension dots correspond to indices less that $k - j + 1$. From this triangular form, it is clear that one may find $a_2, \dots, a_k$.

Since $Y_k = X_k \neq 0$ (from $(HX_k, X_1) = \eta_1$), and $Y_j = \mathcal{A} Y_{j-1}$, it is standard to see that $\{Y_1, \dots, Y_k\}$ form a sub-basis of $\ker \mathcal{A}^k$ and that $\mathcal{A}$, on this basis, is in Jordan form, with 0 on the diagonal and 1 on the lower diagonal. Furthermore, $(HY_l, Y_j) = (H\mathcal{A}^{l-1} Y_1, Y_j) = (HY_1, \mathcal{A}^{l-1} Y_j) = (HY_1, Y_{j+l-1}) = 0$ if $j + l \le k$, and, since $Y_m = 0$ for $m > k$, this product is also 0 if $j + l > k + 1$. While, $(HY_1, Y_k) = \sum a_l (HX_l, X_k) = \eta_1$. Thus, on the $\{Y_j\}$ basis, the matrix H is 0 everywhere except on the antidiagonal, $l + j = k + 1$, where it is η_1.

Repeat this process for each eigenvalue of F, then replace F by $PH\mathcal{A}^{k-2}P$, on $\ker F$, and so on. The result is a basis and a change of variables T for $\ker \mathcal{A}^k$, for which $\mathcal{A}$ and H are in the above form. More precisely, if $Y_j = Te_j$, then $J \equiv T^{-1}\mathcal{A}T$ and $Q = T^* HT$. Then, on $\ker \mathcal{A}^k$, one has

$$C = i\beta I - iTJT^{-1}.$$

By repeating this Jordan process for all eigenvalues (not necessarily pure imaginary) of C, one gets

$$C = T(\Lambda - iJ)T^{-1},$$

where J corresponds to the Jordan blocks and Λ is a diagonal matrix composed with the eigenvalues of C. Let λ be such an eigenvalue, of algebraic multiplicity k. Then, if X is in $\ker(C - \lambda I)^k$, one has that $T^{-1}X$ is in $\ker(\Lambda - iJ - \lambda I)^k$, while HX is in $\ker(C^T + \lambda I)^k = (\text{Range}(C + \bar{\lambda} I)^k)^\perp$, hence $T^* HX$ belongs to $(\text{Range}(\Lambda - iJ + \bar{\lambda} I)^k)^\perp = \ker(\Lambda^* + iJ^T + \lambda I)^k$. Thus, Q maps $\ker(\Lambda - iJ - \lambda I)^k$ onto $\ker(\Lambda^* + iJ^T + \lambda I)^k$, i.e., associating the generalized kernels of Λ for λ and $-\bar{\lambda}$ (they are the same if $\lambda = i\beta$). Hence, if Λ, on the direct sum of these kernels , is of the form $\operatorname{diag}(\lambda I, -\bar{\lambda} I)$, then Q is of the form $\begin{pmatrix} 0 & A \\ A^* & 0 \end{pmatrix}$, assuming that λ is not pure imaginary. From here, it is easy to see that $Qi\Lambda = (i\Lambda)^* Q$. On the other hand, since $H(iC) = (iC)^* H$, one has $Q(i\Lambda + J) = ((i\Lambda)^* + J^T)Q$. Thus, $QJ = J^T Q$.

Now, take any real number γ and consider the self-adjoint matrix (on $(\ker H)^\perp$), $\gamma H + iHC = H(\gamma I + iC)$. From Sylvester law, one has

$$\sigma(\gamma H + iHC) = \sigma(Q(\gamma I + i\Lambda + J)).$$

Assume that $\gamma \neq \beta$, for any $i\beta$ eigenvalue of C. Then, $Q(\gamma I + i\Lambda + J)$ is invertible and self-adjoint. Hence, any self-adjoint perturbation, which preserves the invertibility, will also preserve the signature. An admissible perturbation is $Q(-\tau J)$, for any τ. Hence,

$$\begin{aligned}\sigma(\gamma H + iHC) &= \sigma(Q(\gamma I + i\Lambda)) \\ &= \sum_{\beta \in \mathbb{R}} \operatorname{Sign}(\gamma - \beta)\sigma_\beta(Q) + \sum_{i\lambda \notin \mathbb{R}} \sigma_{\lambda,-\bar{\lambda}}(Q(\gamma I + i\Lambda)),\end{aligned}$$

where $\sigma_\beta(Q)$ is the signature of Q on $\ker(i\Lambda + J + \beta I)^k$, i.e., the signature of PHP on $\ker(C - i\beta I)^k$, and $\sigma_{\lambda,-\bar{\lambda}}$ corresponds to the pair of eigenvalues λ and $-\bar{\lambda}$. But, if (X, Y) is an eigenvector of $Q(\gamma I + i\Lambda)$ on this pair of eigenspaces, with real eigenvalue ξ, then

$$\xi X = (\gamma - i\bar{\lambda})AY, \xi Y = (\gamma + i\lambda)A^* X,$$

that is, $\xi^2 X = ((\gamma - \beta)^2 + \alpha^2)AA^* X$, is $\lambda = \alpha + i\beta$, with $\alpha \neq 0$. Since Q is invertible (as H on this space), on has that $A^* X \neq 0$ and $\xi \neq 0$, independently of γ. Thus, this part of the signature is independent of γ.

Take then $\gamma_1 < \gamma_2$, with $i\gamma_j$ not an eigenvalue of C, one obtains

$$\sigma(\gamma_2 H + iHC) - \sigma(\gamma_1 H + iHC) = 2 \sum_{\gamma_1 < \beta < \gamma_2} \sigma_\beta(Q).$$

Let us take $\varepsilon > 0$, so small that $\gamma_j H + iHC + \varepsilon I$ is invertible, hence the signatures are unchanged, and consider the invertible matrices $\gamma_j H + iHL + \varepsilon I$, whose signature is $\sigma(\gamma_j H + iHC) + \dim\ker L$. Then, the above difference is valid for $\gamma_j H + iHL + \varepsilon I$ and for $\gamma_j \tilde{H} + i\tilde{H}\tilde{L} + \varepsilon I$ if $\tilde{H}$ and $\tilde{L}$ are sufficiently close to H and L. In particular, $\sum_{\gamma_1 < \beta < \gamma_2} \sigma_\beta(\mu)$ is locally constant, provided $i\gamma_j$ is not an eigenvalue of $L(\mu)$. Then, choosing $\gamma_1 < m_j < \gamma_2$, and γ_1, γ_2 close to m_j, one gets that $\sigma_j(\mu)$ is constant for small μ and $\sigma_j^\pm(\mu)$ remain constant provided μ keeps the same sign. This proves *(b)*.

Now, recall that Q, on a Jordan block associated to $i\beta$, is $\eta_1 I^*$, where I^* is the anti-diagonal. It is easy to see, by induction, that $\det(Q - \lambda I)$ is $(\lambda^2 - \eta_1^2)^m$, if the dimension of the block is $2m$, or $(\lambda^2 - \eta_1^2)^m(\eta_1 - \lambda)$, if the dimension is $2m + 1$. Hence $\sigma(Q)$ is 0, if the dimension of the block is even, and η_1, if the dimension is odd. This argument implies, for *(a)* to be true, that, for such a block, one will have no imaginary eigenvalue for the perturbed problem, if the dimension is even, and only one, if the dimension is odd. Note that, on such a block, $Q^2 = I$.

Let $K = \operatorname{diag}(1, 2, \dots, m, m+1, m, \dots, 2, 1)$, if the block has dimension $2m+1$, or $K = \operatorname{diag}(1, 2, \dots, m, m, \dots, 2, 1)$, if the dimension is $2m$. On this block, define $\tilde{S} = QK$. Then, $\tilde{S}$ is the anti-diagonal matrix with elements $(1, 2, \dots, 2, 1)$ and $QK = KQ$, that is $\tilde{S}$ is self-adjoint. Define $\tilde{S}$, on the generalized eigenspaces with eigenvalue non pure imaginary, as 0 (also on $\ker H$) and let $S = T^{*-1}\tilde{S}T^{-1}$. Define, for ε small enough,

$$\tilde{H} = H + \varepsilon S,\ \tilde{L} = \begin{pmatrix} A & B \\ 0 & (H+\varepsilon S)^{-1}HC \end{pmatrix}.$$

Then, $\ker \tilde{H} = \ker H$, $\tilde{H}$ is self-adjoint and $\tilde{H}(i\tilde{L}) = (i\tilde{L})^*\tilde{H}$. (Note that we are not claiming that S is real. Since we are studying the winding number of complex determinants, S may be complex self-adjoint).

Then, if λ is an eigenvalue of $\tilde{L}$, close to im_j, one has $\det((H+\varepsilon S)^{-1}HC - \lambda I) = 0$, since $A - im_jI$ is invertible. Thus,

$$\det(C - \lambda I - \lambda\varepsilon H^{-1}S) = \det(\Lambda - iJ - \lambda I - \lambda\varepsilon K) = 0.$$

Thus, either λ is a non pure imaginary eigenvalue of C or $\lambda = i\beta/(1+\varepsilon l)$, for $l = 1, \dots, m$ or $m+1$ according to the parity of the dimension of K. Thus, the Jordan block is split into m two-dimensional blocks and one single eigenvalue if the dimension is odd. By choosing different sets of integers for different blocks, one may assume that the Jordan blocks are at most two dimensional.

On such a block, one may take $\tilde{S} = -\eta_1 I$ and

$$\Lambda - iJ - \lambda I - \lambda\varepsilon Q^{-1}\tilde{S} = \begin{pmatrix} i\beta - \lambda & \lambda\varepsilon \\ -i + \lambda\varepsilon & i\beta - \lambda \end{pmatrix},$$

with two eigenvalues, for $\varepsilon > 0$ small, off the imaginary axis, since β, being close to m_j, is positive. Thus, after the deformation, an even dimensional block will give rise to eigenvalues off the imaginary axis, while an odd dimensional block gives a single pure imaginary eigenvalue, below $i\beta$. Then, the stability analysis of the signature will complete the proof of the lemma. □

In order to complete the set of hypothesis needed for the S^1-index computation, we shall assume one of the following two conditions:

(H$_0$)
a) *$L(\mu)$ is invertible for $\mu \neq 0$, small*
b) $\ker H(0) \cap \ker L(0) = \{0\}$,

(H$_0'$) *There are $\varepsilon_0, \rho_0 > 0$, such that if $L(\mu)X + f(X,\mu) = 0$, for $\|X\| \le 2\varepsilon_0$ and $|\mu| \le 2\rho_0$, then either $X = 0$, or $|\mu| < \rho_0$ and $\nabla V(X,\mu) \neq 0$.*

As in Proposition 4.4., (H$_0$) is equivalent to the invertibility of $L(\mu) + \nu H(\mu)$, for $(\mu, \nu) \neq (0,0)$ and small. Furthermore, its clear that (H$_0$) implies (H$_0'$) which, in turn, implies that if $L(\mu)X + f(X,\mu) + \nu\nabla V(X,\mu) = 0$, then, if $\|X\| \le 2\varepsilon_0$ and $\mu^2 + \nu^2 \le 4\rho_0^2$, either $X = 0$ or $\nu = 0$ and $|\mu| < \rho_0$, that is, hypothesis (H2) of Remark 4.3 is verified. We are then in the position of applying Corollary 4.2, where d_0 will be computed later.

Proposition 4.5. *Assume* (H_0') *and* (H_j) *hold for* $j = 1, \ldots, s$. *Then the* S^1*-degree of* $(\|X\|^2 - \varepsilon_0^2, X' - L(\mu)X - f(X,\mu) - \nabla V(X,\mu))$ *on the set* $\{\|X\| \le 2\varepsilon_0, \mu^2 + \nu^2 \le 4\rho_0^2\}$ *is given by* $d_0[F_{S^1}] + \sum_1^s d_j[F_{m_j}]$, *where*

$$d_j = \mathrm{Index}(L(\rho_0)X + f(X,\rho_0); 0)(\sigma_j^+(-\rho_0) - \sigma_j^+(\rho_0)),$$

with $\sigma_j^{\pm}(\rho)$ *are given in Definition* 4.1.

If (H_0) *holds, then* $\mathrm{Index}(L(\rho_0)X + f(X,\rho_0); 0) = \mathrm{Sign}\det L(\rho_0)$.

Proof. From Theorem 4.1, one has to compute the winding number of $\det(L(\mu) + \nu H(\mu) - im_j I)$, on the circle $\mu^2 + \nu^2 = \rho_0^2$. As in Proposition 4.1, this determinant is $\prod_1^N a_k(\mu,\nu)$, where the eigenvalues $a_k(\mu,\nu)$ are chosen to be continuous and counted according to their multiplicity. Since the winding number of the product is the sum of the winding numbers of the factors, it is enough to look at each of them. If $a_k(0,0) \neq 0$, i.e., it corresponds to an eigenvalue λ of $L(0)$ which is not im_j, then $a(\mu,\nu)$ will remain away from the origin and will not wind around 0: one may then deform it to $a_k(0,0) = \lambda$ and then to 1. On the other hand, if $a_k(0,0) = 0$ and $a_k(\mu,\nu) = \alpha(\mu,\nu) + i\beta(\mu,\nu)$, then it corresponds to an eigenvector X, with $H(\mu)X \neq 0$ and $\nu\|H(\mu)X\|^2 = \alpha(H(\mu)X, X)$: see the proof of Proposition 4.4. Thus, for $\nu \neq 0$, one gets $\alpha(\mu,\nu) \neq 0$ and, since it is continuous, it keeps the same sign for all ν's positive (or negative). If, for $\nu = 0$, one has $\alpha(\pm\rho_0, 0) \neq 0$, then $\alpha_k(\mu,\nu)$ stays on the same half complex plane and its winding number is 0.

Hence, $a_k(\mu,\nu)$ crosses the imaginary axis at most twice, for $\nu = 0$ and $\mu = \pm\rho_0$. Assuming one has performed the perturbation of Lemma 4.6, this implies that $a_k(\pm\rho_0, 0)$ is a simple eigenvalue and, by Lemma 4.5, one has $(H(\pm\rho_0)X, X) \neq 0$, for the corresponding eigenvector. Thus, as ν crosses 0 from negative values to positive values, $\alpha(\mu,\nu)$ will cross 0 in the same direction, if $(HX, X) > 0$, and in the other direction, if $(HX, X) < 0$. Note that, in this case, (HX, X) keeps the same sign on the whole loop, by using the continuity of $X(\mu)$, near $\mu = \pm\rho_0$, which is true since $a_k(\pm\rho_0, 0)$ is simple.

Taking the orientation (μ,ν), the loop described by $a_k(\mu,\nu)$ will give a winding number equal to 0 if $\beta(\pm\rho_0, 0)$ have the same sign, and, otherwise, equal to Sign (HX, X), if $\beta(\rho_0, 0) < 0 < \beta(-\rho_0, 0)$, and to – Sign (HX, X), if $\beta(-\rho_0, 0) < 0 < \beta(\rho_0, 0)$.

For $\mu = \pm\rho_0$ and $\eta = \pm 1$, let $n^{\pm}(\mu,\eta)$ be the number of imaginary eigenvalues $i\lambda$ of $L(\mu)$, close to im_j, which are above im_j, that is $\beta(\mu, 0) > 0$ (for n^+), or below im_j, that is $\beta(\mu, 0) < 0$ (for n^-), and which have the simple eigenvector X with Sign $(HX, X) = \eta$. Hence, $\sigma_j^{\pm}(\mu) = n^{\pm}(\mu, 1) - n^{\pm}(\mu, -1)$. Let $a^{\pm}(\eta)$ be the number of eigenvalues of $L(\mu)$, with Sign $(HX, X) = \eta$, which cross im_j from below to above as μ goes from $-\rho_0$ to ρ_0 (i.e., $\beta(\mu,\nu)$ goes from negative to positive), for $a^+(\eta)$, and in the inverse direction for $a^-(\eta)$. Then, then winding number is $a^-(1) - a^+(1) + a^+(-1) - a^-(-1)$.

Let $b^{\pm}(\eta)$ be the number of eigenvalues, with Sign $(HX, X) = \eta$, which remain above im_j (i.e., with $\beta(\mu,\nu) > 0$), for b^+, or below im_j (i.e., with $\beta(\mu,\nu) < 0$), for b^-. Then, one has the relations

$$\begin{aligned} n^+(-\rho_0, \eta) &= a^-(\eta) + b^+(\eta) \\ n^-(-\rho_0, \eta) &= a^+(\eta) + b^-(\eta) \\ n^+(\rho_0, \eta) &= a^+(\eta) + b^+(\eta) \\ n^-(\rho_0, \eta) &= a^-(\eta) + b^-(\eta). \end{aligned}$$

Thus, $a^+(\eta) - a^-(\eta) = n^+(\rho_0, \eta) - n^+(-\rho_0, \eta) = n^-(-\rho_0, \eta) - n^-(\rho_0, \eta)$. We have proved that the winding number is $\sigma_j^+(-\rho_0) - \sigma_j^+(\rho_0) = \sigma_j^-(\rho_0) - \sigma_j^-(-\rho_0)$. □

For instance, if $L(\mu) = (\mu + \lambda_0)L$, with $\lambda_0 > 0$ and $\pm im_j/\lambda_0$ an eigenvalue of L, then $\sigma_j^+(-\rho_0) = 0, \sigma_j^+(\rho_0) = \sigma_j$, the signature of H for im_j/λ_0.

Remark 4.9. Hypothesis (H_0') implies that $X = 0$ is an isolated zero of $L(\mu)X + f(X, \mu) + \nu\nabla V(X, \mu)$, provided $\mu^2 + \nu^2 = \rho_0^2$, and with a constant index on the loop. At first sight this hypothesis could seem awkward and a more elegant hypothesis could have been to ask that $L(\mu)X + f(X, \mu)$ and $\nabla V(X, \mu)$ have 0 as an isolated zero, for $\mu \neq 0$, for the first equation, and for any small μ, for $\nabla V(X, \mu)$.

This happens, for instance, if $H(0)$ is invertible and $L(\mu)$ is also invertible for $\mu \neq 0$ (a stronger hypothesis than (H_0)). But this new hypothesis implies that either N is even or $\text{Index}(L(\rho_0)X + f(X, \rho_0); 0) = 0$, in which case $d_j = 0$, for all j's. In fact, for ρ_0, the maps $L(\rho_0)X + f(X, \rho_0)$ and $\nabla V(X, \rho_0)$ have a well-defined index at 0 (this is not necessarily true for ∇V in case (H_0') holds). Furthermore, since ∇V is orthogonal to $L(\rho_0)X + f(X, \rho_0)$, the index, at 0, of $\tau(L(\rho_0)X + f(X, \rho_0)) \pm (1-\tau)\nabla V(X, \rho_0)$ is well defined and constant. Then $\text{Index}(L(\rho_0)X + f(X, \rho_0); 0) = \text{Index}(\nabla V; 0) = \text{Index}(-\nabla V; 0) = (-1)^N \text{Index}(\nabla V; 0)$.

In order to compute d_0, assume that (H_0) holds. Thus, according to Theorem 4.1, d_0 is the class of $L(\mu) + \nu H(\mu)$ in $\Pi_1(\text{GL}(\mathbb{R}^N)) \cong \mathbb{Z}_2$ (since the change of orientation I^{α_0} does not affect d_0, we may assume that $L(\mu) + \nu H(\mu)$ has positive determinant on the loop $\mu^2 + \nu^2 = \rho^2$).

Decompose $\mathbb{R}^N$ into $\ker H(0) \oplus \text{Range } H(0)$ and write

$$L(\mu) + \nu H(\mu) = \begin{pmatrix} A + \nu H_1 & B + \nu H_2 \\ D + \nu H_2^T & C + \nu H \end{pmatrix},$$

where $H_1(0) = H_2(0) = D(0) = 0$, and, from ($H_0$), the matrices $A(0)$ and $H(0)$ are invertible. The relation $L^T H + HL = 0$, is then

$$\begin{aligned} &H_1 A + H_2 D \text{ and } H_2^T B + HC \text{ are skew symmetric,} \\ &H_2^T A + HD + B^T H_1 + (H_2 C)^T = 0. \end{aligned}$$

Note that, for $\nu \neq 0$, $L(0) + \nu H(0)$ is deformable to $\text{diag}(A(0), C(0) + \nu H(0))$ and to $\text{diag}(A(0), \nu H(0))$. This implies that dim Range $H(0)$ is even (since the sign of $\det(L(\mu)+\nu H(\mu))$ is positive on the loop) and that $\det A$ and $\det H$ have the same sign, for μ small. Let $\tilde{H} = H - (A^{-1}B)^T H_2$, then, for μ small, $\tilde{H}$ is invertible and the matrix $\mathcal{A} = \operatorname{diag}(A^T, \tilde{H})$ is invertible and deformable to $\mathcal{A}(0)$ and then to I. Consider the matrix

$$\mathcal{A}(L(\mu)+\nu H(\mu)) = \begin{pmatrix} A^T A + \nu A^T H_1 & A^T B + \nu A^T H_2 \\ \tilde{H} D + \nu \tilde{H} H_2^T & \tilde{H} C + \nu \tilde{H} H \end{pmatrix} = \begin{pmatrix} \tilde{A} & \tilde{B} \\ \tilde{D} & \tilde{C} \end{pmatrix}.$$

The class of this matrix in $\Pi_1(\mathrm{GL}(\mathbb{R}^N))$ is the class of $L+\nu H$. Since $\tilde{A}$ is invertible, for (μ,ν) small, multiply $\tilde{B}$ and $\tilde{D}$ by $\cos\tau$ and replace $\tilde{C}$ by $\tilde{C} - \sin^2\tau\, \tilde{D}\tilde{A}^{-1}\tilde{B}$: if (X, Y) gives a zero of the deformation, then $X = -\cos\tau\,\tilde{A}^{-1}\tilde{B}Y$ and $(\tilde{C} - \tilde{D}\tilde{A}^{-1}\tilde{B})Y = 0$, hence $Y = 0$ since this last matrix is invertible on the loop. For $\tau = \pi/2$, on obtains the matrix

$$\operatorname{diag}(\tilde{A}, \tilde{C} - \tilde{D}\tilde{A}^{-1}\tilde{B}) = \operatorname{diag}(\tilde{A}, I)\operatorname{diag}(I, \tilde{C} - \tilde{D}\tilde{A}^{-1}\tilde{B}).$$

The matrix $\tilde{A}$ is always invertible, for μ small, and deformable to I. Hence, the class of $L + \nu H$ is the suspension of the class of $\tilde{C} - \tilde{D}\tilde{A}^{-1}\tilde{B}$.

Lemma 4.7. *Under hypothesis* (H_0), *the matrix* $\tilde{C} - \tilde{D}\tilde{A}^{-1}\tilde{B} = \mathcal{C} + \nu H^2 + \nu O(\mu)$, *where* $\mathcal{C}(\mu)$ *is skew-symmetric and invertible for* $\mu \neq 0$.

Proof. The matrix $\tilde{C} - \tilde{D}\tilde{A}^{-1}\tilde{B}$ is, with $\tilde{H} = H - (A^{-1}B)^T H_2$, equal to

$$\begin{aligned} &\tilde{H}C + \nu\tilde{H}H - (\tilde{H}D + \nu\tilde{H}H_2^T)(A^TA + \nu A^T H_1)^{-1}(A^T B + \nu A^T H_1) \\ &\quad = \tilde{H}(C - DA^{-1}B) + \nu H^2 + \nu 0(\mu). \end{aligned}$$

Using the identity $HDA^{-1} = -(H_2^T + (A^{-1}B)^T A^T H_1 A^{-1} + (H_2C)^T A^{-1})$, one obtains $HC + H_2^T B + (A^{-1}B)^T(A^T H_1 + H_2 D)A^{-1}B + (H_2C)^T A^{-1}B - (A^{-1}B)^T H_2 C + \nu H^2 + O(\mu)$.

It is clear now that the first terms are skew symmetric and, since they are equal to $\tilde{H}(C - DA^{-1}B)$, they give a matrix $\mathcal{C}(\mu)$, which is invertible for $\mu \neq 0$ (from (H_0), $L(\mu)$ is invertible for $\mu \neq 0$). □

Now, recall that any invertible skew-symmetric matrix $\mathcal{C}$ can be put in real Jordan form Λ, via an orthogonal change of basis T, where Λ consists of blocks $\begin{pmatrix} 0 & \beta_j \\ -\beta_j & 0 \end{pmatrix}$, $j = 1, \dots, m$, with $\dim \operatorname{Range} H(0) = 2m$: in fact, $\mathcal{C}^2$ is self-adjoint and negative definite, hence with eigenvalues $-\beta_j^2$ and orthonormal eigenvectors $X_1, \dots, X_{2m}$. If Y_j is defined by $\mathcal{C}X_j = \beta_j Y_j$, then $\mathcal{C}Y_j = -\beta_j X_j$, and Y_j is orthogonal to X_j, with $\mathcal{C}^2 Y_j = -\beta_j^2 Y_j$, that is, $-\beta_j^2$ is a double eigenvalue of $\mathcal{C}^2$. Of course, one may choose all β_j's to be positive, but then $\det T$ (which is ± 1), may be negative. On the other hand, one may insist in $\det T$ being positive, but then one may have to take one of the β_j's to be negative.

Definition 4.2. The matrix $\mathcal{C} = T^T \Lambda T$, with $\det T > 0$, will define a *positive complex structure* if all β_j*'s* are positive. Otherwise, if one β_j has to be negative, then the complex structure will be said to be negative.

Proposition 4.6. *If* (H_0) *holds, then the class of* $L(\mu) + \nu H(\mu)$ *will be non trivial if and only if* $\mathcal{C}(\rho)$ *and* $\mathcal{C}(-\rho)$ *define complex structures of different signs.*

Proof. We have seen that the class of $L(\mu) + \nu H(\mu)$ is the class of $\mathcal{C}(\mu) + \nu H^2 + \nu O(\mu)$. One may deform the last term to 0, since if, for some $X \neq 0$, one has a zero of the deformation, then, because $(\mathcal{C}X, X) = 0$, one obtains $\nu(\|HX\|^2 + \tau(0(\mu)X, X)) = 0$. For $|\mu| \leq \rho$, one has $\|HX\|^2 \geq C\|X\|^2$, hence one gets $\nu = 0$ and $\mathcal{C}(\mu)X = 0$, for $\mu = \pm\rho$ on the loop, something which is impossible. Since H^2 is positive definite, one may use the same sort of deformation to arrive at $\mathcal{C}(\mu) + \nu I$. Furthermore, one may replace $\mathcal{C}(\mu)$ by $\mathcal{C}(\tau, \mu)$ defined as

$$(1-\tau)\mathcal{C}(\mu) + \tau(\mu/\rho)^2(\mathcal{C}(\rho) + \mathcal{C}(-\rho)) + \tau(\mu/\rho)(\mathcal{C}(\rho) - \mathcal{C}(-\rho)).$$

In fact, the above matrix is skew-symmetric, hence, if $\mathcal{C}(\tau, \mu)X + \nu X = 0$, taking the scalar product with X, one has $\nu = 0$ and, for $\mu = \pm\rho$, the condition $\mathcal{C}(\pm\rho)X = 0$, which is not possible, unless $X = 0$. For $\tau = 1$, one may perform a linear deformation to

$$\mathcal{C}(\rho) + \mathcal{C}(-\rho) + (\mu/\rho)(\mathcal{C}(\rho) - \mathcal{C}(-\rho)) + \nu I.$$

Finally, since $\mathcal{C}(\rho) = T^T\Lambda(\rho)T$, with $\det T > 0$, one may deform T to I, keeping the deformed matrix skew - symmetric, hence one may replace $\mathcal{C}(\rho)$ by $\Lambda(\rho)$ and this last matrix by $\mathrm{diag}(J, \dots, \pm J)$, where $J = \begin{pmatrix} 0 & 1 \\ -1 & 0 \end{pmatrix}$, by deforming β_j to 1, and with $\pm J$ according to the positive or negative complex structure for $\mathcal{C}(\rho)$. With a similar argument at $\mu = -\rho$, one obtains

$$\mathrm{diag}(2J + \nu I, \dots, \pm 2(\mu/\rho)J + \nu I),$$

where the last component is not present if the complex structures are the same. Since $2J + \nu I$ is deformable to I and since $\mu J + \nu I$ generates $\Pi_1(\mathrm{GL}(\mathbb{R}^2))$, one obtains the result. □

An interesting particular case is the following

Corollary 4.3. *Assume that* (H_0) *holds, with the condition* $\|L(\mu)\| \geq C|\mu|$, *then* $[L(\mu) + \nu H(\mu)]$ *is non-trivial if and only if* $(\dim\ker L(0))/2$ *is odd.*

Proof. The condition $\|L\| \geq C|\mu|$, which implies the invertibility of L, holds also for $\mathcal{A}L$, that is $\|\mathcal{A}L(X, Y)\|^2 \geq C^2\mu^2(\|X\|^2 + \|Y\|^2)$. In particular, for $X = -A^{-1}BY$, one has $\mathcal{A}L(X, Y) = (0, \mathcal{C}Y)$, that is $\|\mathcal{C}Y\| \geq C|\mu|\|Y\|$. Now, if one writes $\mathcal{C}$, on the orthogonal decomposition given by $\ker \mathcal{C}_0 \oplus \text{Range}\, \mathcal{C}_0$,

$$\mathcal{C} = \mathcal{C}_0 + \mu\mathcal{C}_1(\mu) = \begin{pmatrix} \mathcal{C}_0 + \mu C_1 & \mu C_2 \\ -\mu C_2^T & \mu\tilde{\mathcal{C}} \end{pmatrix},$$

one has that $\mathcal{C}_0 = \mathcal{C}(0)$ and $\mathcal{C}_1(\mu)$ are skew-symmetric, as well as C_1 and $\tilde{\mathcal{C}}$. Repeating the above argument, one has that $\|\mu\tilde{\mathcal{C}}\| \geq C|\mu|$, that is, $\tilde{\mathcal{C}}(0)$ is invertible. Now, as we have done in the last proposition, one may deform $\mathcal{C}$ to $\text{diag}(I, \mu(\tilde{\mathcal{C}} + \mu C_2^T(\mathcal{C}_0 + \mu C_1)^{-1}C_2))$ and then, to $\text{diag}(I, \mu\tilde{\mathcal{C}}(0))$. Deforming $\tilde{\mathcal{C}}(0)$ to its Jordan form, the class of $L + \nu H$ is the suspension of the class of $(\mu J + \nu I, \dots, \pm\mu J + \nu I)$, that is $(m/2)$-times the Hopf map, if m is the dimension of $\tilde{\mathcal{C}}$, i.e., of $\ker \mathcal{C}_0$ (note that, since $\tilde{\mathcal{C}} + \mu C_2^T(\mathcal{C}_0 + \mu C_1)^{-1}C_2$ is invertible (here for all μ's) and skew-symmetric, this dimension is even). Now, since $\mathcal{C}_0 = H(0)C(0)$, one has $\ker \mathcal{C}_0 = \ker C(0) = \ker L(0)$. □

Note that in general, i.e., if only one assumes (H_0), the matrix $\tilde{\mathcal{C}}_1 = \tilde{\mathcal{C}} + \mu C_2^T(\mathcal{C}_0 + \mu C_1)^{-1}C_2$, will be invertible for $\mu \neq 0$, but $\tilde{\mathcal{C}}(0)$ may have a non-trivial kernel. Writing $\tilde{\mathcal{C}}_1 = \tilde{\mathcal{C}}(0) + \mu\mathcal{C}_2(\mu)$, one may repeat the above argument, with a decomposition on $\ker \tilde{\mathcal{C}}(0) \oplus \text{Range}\, \tilde{\mathcal{C}}(0)$. On $\text{Range}\, \tilde{\mathcal{C}}(0)$, one will get a contribution of $(\dim \text{Range}\, \tilde{\mathcal{C}}(0))/2$ and, on $\ker \tilde{\mathcal{C}}(0)$, a skew symmetric matrix $\mu^2\tilde{\mathcal{C}}_2$. On $\text{Range}\, \tilde{\mathcal{C}}_2(0)$, one will get a zero contribution (due to μ^2), while, on $\ker \tilde{\mathcal{C}}_2(0)$, one has to look at the third order terms. Thus, for smooth $L(\mu)$ and $H(\mu)$, one may compute the homotopy class in terms of dimension of subspaces.

Remark 4.10. The reader should notice that, for this Hopf bifurcation, we are asking that $\nabla V(0, \mu) = 0$, while in the case of non-stationary solutions (Example 2.8), the condition was $\nabla V(X) \neq 0$. The reason is the following: from the orthogonality condition $g(X) \cdot \nabla V(X) = 0$, one has, after linearization

$$Dg(X)^T \nabla V(X) + H(X)g(X) = 0,$$

where $H(X)$ is the Hessian of $V(X)$. Thus, if $\nabla V(X) = 0$ and one is in the natural situation of an invertible $H(X)$ (hence the point X is isolated in the set of zeros of ∇V), one needs $g(X) = 0$, i.e., a stationary point. Now, if X belongs to an orbit of solutions of the equation $X' = g(X)$, we have seen, in Example 2.8, that $\nabla V(X(t))$ is either identically zero or never zero. Hence, under the hypothesis of a discrete set of zeros of $\nabla V(X)$, one has that either $X(t)$ is constant, i.e., stationary, or $X(t)$ is truly periodic with $\nabla V \neq 0$. On the other hand, if $g(X) = 0$ and $\nabla V(X) \neq 0$, then $\dim \ker Dg(X) > 0$. For $X = 0$ and $Dg(0, \mu) = L(\mu)$, this is incompatible with (H_0). This explains our hypotheses.

On the other hand, one may still study the bifurcation of periodic solutions, from a set of stationary solutions, when $\nabla V(0, \mu) \neq 0$. As seen above, this implies that $L(\mu)^T \nabla V(0, \mu) = 0$. Assume that $\dim \ker L(0) = 1$.

Let $\mathbb{R}^N = \ker L(0) \oplus \text{Range } L^T(0)$ and write $X = aX_0 \oplus Y$, with $L(0)X_0 = 0$. Then, one may linearize the stationary equation $g(X, \mu) + \nu\nabla V(X, \mu) = F(a, Y, \nu, \mu)$ and obtain

$$D_{(Y,\nu)}F(0, 0, 0, 0)(Z, \eta) = L(0)Z + \eta\nabla V(0, 0).$$

Since this is an isomorphism, the equation $F(a, Y, \nu, \mu) = 0$ has a unique local solution $Y(a, \mu)$, $\nu(a, \mu)$, with $\nu(a, \mu) = 0$, from the orthogonality. Furthermore, $Y(0, \mu) = 0$ and $\ker L(\mu)$ is also one-dimensional, generated by $X_0 + Y_a(0, \mu)$.

Let $X(a, \mu) = aX_0 + Y(a, \mu)$ be this unique local stationary solution of $g(X, \mu) = 0$ and let $A(a, \mu) = D_X g(X(a, \mu), \mu)$. Then, $\ker A(a, \mu)$ is one-dimensional and generated by $X_0 + Y_a(a, \mu)$, while, as seen above, $\nabla V(X(a, \mu), \mu)$ generates $\ker A(a, \mu)^T$.

Proposition 4.7. *Assume* $g(0, \mu) = 0$ *and* $\ker D_X g(0, 0) = \{aX_0\}$. *Then, locally,*

(a) $g(X, \mu) = 0$ *if and only if* $X = X(a, \mu)$.

(b) *If* $A(a, \mu) = D_X g(X(a, \mu), \mu)$ *is such that* $inI - A(a, \mu)$ *is invertible for all* $n > 0$ *and* $(a, \mu) \neq 0$, *then the* S^1*-degree of the pair* $(\|Z\|_1 - \varepsilon, X'(t) - g(X(t), \mu) - \nu\nabla V(X(t), \mu))$, *with respect to* $\Omega = \{(X(t) = X(a, \mu) + Y + Z(t), \mu, \nu) : \|Z\|_1 < 2\varepsilon, |\mu| < 2\rho, |\nu| < 2\varepsilon, |a| < 2\rho, \|Y\| < 2\varepsilon$, *where* Y *is orthogonal to* X_a *and* $Z(t)$ *has only non-zero modes*$\}$ *is well defined. (Here* $\|\cdot\|_1$ *is the* H^1*-norm). This* S^1*-degree is given by*

$$d_0 = 0, \quad d_n = \eta\sigma_n,$$

where σ_n *is the winding number of* $\det(A(a, \mu) - inI)$ *and* η *is the sign of* $\det D_{(Y,\nu)}(g + \nu\nabla V)(0, 0)$, *with* Y *in* $\text{Range } A(0, 0)^T$.

Proof. Part *(a)* has already been proved. For part *(b)*, write $X(t) = \sum X_n e^{int} = X_0 + Z(t)$ and, after linearizing at $X(a, \mu)$, the equation

$$\begin{aligned} X' - A(a, \mu)(X - X(a, \mu)) - \nu\nabla V(X(a, \mu), \mu) \\ - \nu H(X(a, \mu), \mu)(X - X(a, \mu)) + \cdots = 0 \end{aligned}$$

is equivalent to the system

$$\begin{aligned} -A(a, \mu)(X_0 - X(a, \mu)) - \nu\nabla V(X(a, \mu), \mu) \\ + 0(\nu(X_0 - X(a, \mu)) + \|X(t) - X(a, \mu)\|_1^2) = 0 \end{aligned}$$

$$(in\, I - A(a, \mu))X_n + 0(\nu X_n + \|X(t) - X(a, \mu)\|_1^2) = 0.$$

Since $\ker A(a, \mu)$ is generated by $X_a(a, \mu)$, let $X_0 = aX_a \oplus Y$, with Y orthogonal to X_a. Then, $X_0 - X(a, \mu) = Y + 0(a^2)$. As before, $A(a, \mu)Y + \nu\nabla V(X(a, \mu))$ is an isomorphism from $\text{Range } A(a, \mu)^T \times \mathbb{R}$ onto $\text{Range } A(a, \mu) \oplus \ker A(a, \mu)^T$ (since $\nabla V(X(a, \mu))$ generates $\ker A(a, \mu)^T$). Thus, one may solve uniquely for

$(Y(a,\mu,Z), \nu(a,\mu,Z)) = 0(a^2 + \|Z(t)\|_1^2)$. Note that, due to the orthogonality, $\nu(a,\mu,Z) = 0$ for periodic solutions. Furthermore, if in is not an eigenvalue of $A(0,0) = L(0)$, which is true for large n, one may solve these equations in terms of the resonant modes, obtaining an H^1-bound for $X(t)$. In particular, if one has a zero of the differential equation in Ω, then $\|Y\|$ and ν are of the order of ε^2 and, if $a^2 + \mu^2 \geq \rho^2$, one will have $\|Z\|_1 = 0(\varepsilon^2)$, i.e., the S^1-degree of the pair is well defined.

As done many times, one may deform the pair to $(\|Z\|_1 - \varepsilon, -A(a,\mu)Y - \nu\nabla V(a,\mu), \{(in\ I - A(a,\mu))X_n\})$ and one may apply Theorem 4.1. Since $A(a,\mu)Y + \nu\nabla V(a,\mu)$ may be deformed to $A(0,0)Y + \nu\nabla V(0,0)$, which gives an invertible matrix, one gets the result: the orientation factor $(-1)^N$ is, as before, absorbed in the generator. □

Remark 4.11 (Global bifurcation). Corollary 4.1. says that, if d_0 or d_j, in Propositions 4.5 and 4.6, are non-zero, then there is a continuum of solutions $\mathcal{C}$, with $X \not\equiv 0$ on $\mathcal{C}$, which is either unbounded in the space $\{X(t), \mu, \nu\}$ or returns to some point $\{0, \mu_1, \nu_1\}$ with, in case of boundedness and a nice local behavior (i.e., (H_j) and (H_0) hold), a sum of S^1-indices equal to 0. Now, near $\{0,0,0\}$, one has good information on the solution set and on $\mathcal{C}$: for instance, if $L(0)$ is invertible, there is no bifurcation of stationary solutions and, near the bifurcation point, the solutions are truly periodic.

However, if $\mathcal{C}$ contains a point $(X(t),\mu)$, with $X(t) \not\equiv 0$ and $\nabla V(X,\mu) = 0$ (recall that this vector is either identically zero or never zero on solutions of $X' = g(X,\mu)$), then $\mathcal{C}$ will be unbounded in the ν-component. Since ν was introduced in an artificial way, this is not a natural result. In order to avoid this situation, introduce the set

$$\begin{aligned} S = \{&(X(t),\mu), \text{ with } X(t) \text{ a periodic (or stationary)} \\ &\text{solution of } X' = g(X,\mu) \text{ and } \nabla V(X(t),\mu) = 0\}. \end{aligned}$$

The conditions $g(0,\mu) = 0$, $\nabla V(0,\mu) = 0$, imply that $\{(0,\mu)\} \subset S$. Complement the equation $X' - g(X,\mu) - \nu\nabla V(X,\mu)$ with the condition $\text{dist}((X(t),\mu); S) - \varepsilon$, where the distance is in the H^1-norm: S is compact on bounded sets in that norm. Any solution in the complement of S will have $\nu = 0$.

Now, if (H_j) and either (H_0) or (H_0') hold, then, in a neighborhood of $(0,0)$, S is just $(0,\mu)$: in fact, if, for some solution of $X' = g(X,\mu)$, one has $\nabla V(X,\mu) = 0$, then for the Fourier series, one has

$$\begin{aligned} (inI - L(\mu))X_n &- f_n(X,\mu) = 0 \\ H(\mu)X_n &- k_n(X,\mu) = 0. \end{aligned}$$

Hence, for any ν

$$(inI - L(\mu) - \nu H(\mu))X_n - f_n - \nu k_n = 0.$$

Taking $\nu = \rho$, Proposition 4.4. implies that the linear part is invertible, for $n \neq 0$, and the non-linear part is of the order of $\|Z\|_1\|X\|_1$, where $X = X_0 + Z(t)$. Hence, for $\|X\|_1$ small enough, one has $Z(t) = 0$ and $X = X_0$ is a stationary solution of $g(X, \mu) = 0$. The same argument for $n = 0$, in case (H_0) holds, or by hypothesis, if (H_0') holds, implies that $X = 0$, if $|\mu| < \rho$.

Hence, near (0, 0), $\operatorname{dist}((X, \mu); S) = \|X\|_1$ and Propositions 4.5 and 4.6 are valid. Thus, if one of the d_j's is non-zero, the global branch will be unbounded, in X or μ, or will meet a point of S, a stationary point if $S \cap \{\mu = \mu_0\}$ is discrete. Further computations of the S^1-degrees, near S, are given in [IMV2], Remark 6.9.

Note that Remark 4.8 holds also here.

Remark 4.12 (First integrals and symmetries). If $g(X, \mu)$ and $\nabla V(X, \mu)$ are equivariant with respect to a group Γ_0, then one may repeat the considerations of Example 4.3: the linearizations $L(\mu)$ and $H(\mu)$ will have a block diagonal structure and the $S^1 \times \Gamma_0$-degree will be given in terms of the spectral behavior of each of the sub-matrices of $inI - L(\mu) - \nu H(\mu)$, as in Proposition 4.2.

Remark 4.13 (Hamiltonian systems). Consider the system

$$\frac{dX}{dt} = J\nabla V(X, \mu), \quad X \text{ in } \mathbb{R}^{2N}, \ J = \begin{pmatrix} 0 & -I \\ I & 0 \end{pmatrix}.$$

As pointed out in Remark 2.4, we have that $V(X, \mu)$ is a first integral and we may either apply the orthogonal degree, as in Example 3.5, or consider the equations

$$X' - (J - \nu I)\nabla V(X, \mu) = 0,$$

or solutions of the equations

$$JX' + \nabla V(X, \mu) + \nu X' = 0.$$

An important special case is when $V(X, \mu) = V(X)/(\mu_0 + \mu)$, where $\mu_0 + \mu$ stands for the frequency.

Assume that $\nabla V(0, \mu) = 0$. Let $H(\mu)$ be the Hessian of V at $(0, \mu)$. Suppose that $JH(0)$ has eigenvalues $\pm i m_1 \pm \ldots, \pm i m_s$, with $0 < m_1 \leq \cdots \leq m_s$.

Then, hypothesis (H_j) is equivalent to asking that $JH(\mu) - im_j I$ is invertible for $\mu \neq 0$, μ close to 0 (this is always true if $V(X, \mu) = V(X)/(\mu_0 + \mu)$). Hypothesis ($H_0'$) is verified provided $X = 0$ is an isolated zero of $\nabla V(X, \mu)$. For hypothesis (H_0) one needs the invertibility of $H(0)$ (in this case $d_0 = 0$).

Then Proposition 4.5 gives the bifurcation index in terms of $\sigma_j^+(-\rho) - \sigma_j^+(\rho)$, while Proposition 3.2 gives it in terms of $\mathcal{M}_j(-\rho) - \mathcal{M}_j(\rho)$, where $\mathcal{M}_j(\mu)$ is the Morse number of $im_j J + H(\mu)$. The factor $\operatorname{Index}(J\nabla V(X, \rho); 0) = \operatorname{Index}(\nabla V(X, \rho); 0)$ is common to both formulations. *The two formulae are the same*: in fact, one has $(H + im_j J)JH + (JH)^T(H + im_j J) = 0$ and $H + im_j J$ is self-adjoint and invertible, if $\mu \neq 0$. Thus, as in Lemma 4.5, $H + im_j J$ maps $\ker(JH - \lambda I)^k$ into $\ker(HJ - \lambda I)^k$. Furthermore, if $i\beta$ is a simple eigenvalue of JH, with eigenvector

X_β, then $(H + im_j J)X_\beta = (\beta - m_j)\beta^{-1} H X_\beta$. In particular, on $\ker(JH - i\beta I)$, the signature of $H + im_j J$ is

$$\sigma_\beta(H + im_j I) = \operatorname{Sign}(\beta - m_j)\beta^{-1}\sigma_\beta(H).$$

For a couple of eigenvalues, $(\lambda, -\bar{\lambda})$, of JH one has that, as in the proof of Lemma 4.6, the matrix $H + im_j J$ has the form $\begin{pmatrix} 0 & A \\ A^* & 0 \end{pmatrix}$, with a zero signature. Hence,

$$\sigma(H + im_j I) = \sum_{\beta \in \mathbb{R}} \operatorname{Sign}(\beta - m_j)\beta^{-1}\sigma_\beta(H).$$

(Note that, since $H(\mu) + im_j I$ is invertible for $\mu \neq 0$, we don't have to worry about the kernel, as in Lemma 4.6). Thus, after perturbing $JH(\pm\rho)$, so that they have simple purely imaginary eigenvalues and using the fact that (HX, X) has a constant sign on the loop $\mu^2 + \nu^2 = \rho^2$, one obtains

$$\begin{aligned} &\sigma(H(\rho) + im_j I) - \sigma(H(-\rho) + im_j I) \\ &\quad = \sum (\operatorname{Sign}(\beta(\rho) - m_j) - \operatorname{Sign}(\beta(-\rho) - m_j)) \operatorname{Sign}(HX, X), \end{aligned}$$

since only those β's close to m_j are involved. Hence, in terms of $a^{\pm}(\eta)$, defined in the proof of Proposition 4.5, the difference is

$$2(a^+(1) - a^-(1) - a^+(-1) + a^-(-1)) = 2(\sigma_j^+(\rho) - \sigma_j^+(-\rho)).$$

Since $\mathcal{M}(H + im_j I) = N - \sigma(H + im_j I)/2$, one gets that $\mathcal{M}_j(-\rho) - \mathcal{M}_j(\rho) = -(\sigma_j^+(-\rho) - \sigma_j^+(\rho))$, the sign being again an orientation factor.

In case $JH = HJ$, one may see this equality in a more direct way: as in Remark 3.5 (c), decompose $\mathbb{C}^{2N}$ into two-dimensional subspaces, invariant under J, $\langle X_k, JX_k \rangle$, $k = 1, \ldots, N$, corresponding to the eigenvalue λ_k of H. On that subspace, $JH - i\beta I = \begin{pmatrix} -i\beta & -\lambda_k \\ \lambda_k & -i\beta \end{pmatrix}$, with eigenvalues, if $\beta = \pm\lambda_k, 0$ and $-2i\beta$, that is β is a simple eigenvalue, with $(HX, X) = \lambda_k \|X\|^2$. Hence, if $\beta > n$ and $\beta = \lambda_k$, the signature of H is 1, while it is -1 if $\beta = -\lambda_k$. Recalling that each λ_k is a double eigenvalue of H, one has that $2\sigma_j^+(\mu)$ is the number of λ_k's larger than m_j (but close to m_j) minus the number of those less than $-m_j$ (but close to $-m_j$). That is, if $a(n)$ is the number of eigenvalues of H, less than n, as in Remark 3.5 (c), and $[\sigma_j^+]$ denotes the jump $\sigma_j^+(-\rho) - \sigma_j^+(\rho)$, one gets

$$2[\sigma_j^+] = -[a(-m_j) + a(m_j)] = -2[\mathcal{M}].$$

Note that, if $V(X, \mu) = V(X)/(\mu_0 + \mu)$, then $H(\mu) = H/(\mu_0 + \mu)$, one has $\sigma_j^+(\rho) = 0$, $\sigma_j^+(-\rho) = \sigma_j$ the signature of H on $\ker(JH - im_j\mu I)^k$.

Example 4.7 (Hopf bifurcation for equations with delays). As a last illustration of the use of the equivariant degree, let us look at a (slightly) different context from ordinary differential equations. Consider the problem of finding periodic solutions to the system

$$\frac{dX}{d\tau} = g(\lambda, X(\tau - r_1), \dots, X(\tau - r_s)), \quad X \text{ in } \mathbb{R}^N,$$

which, after the scaling $t = \nu\tau$, is equivalent to finding 2π-periodic solutions to

$$\nu X'(t) = g(\lambda, X(t - \nu r_1), \dots, X(t - \nu r_s)).$$

Here r_j may be a fixed delay or may be taken as a parameter or even depend on X. The problem is clearly S^1-equivariant. If there is only one delay in g, then the n-th Fourier coefficient of $g(X(t - \nu r))$ is $e^{-in\nu r} g_n(X(t))$, for a 2π-periodic $X(t)$. In particular, if $g(\lambda, 0, \dots, 0) = 0$ and $A_j(\lambda) = D_{X_j} g(\lambda, 0, \dots, 0)$, the problem is equivalent to the system, for $X(t) = \sum X_n e^{int}$,

$$(in\nu I - \sum_1^s A_j(\lambda) e^{-in\nu r_j}) X_n = f_n(X), \quad n \geq 0.$$

The linear parts are called the *indicial equations* and it is not difficult to devise conditions under which they have isolated singularities and non-zero winding numbers.

For instance, if $N = 1$ and the equation is

$$x'(\tau) = -\lambda x(\tau) - \lambda f(x(\tau - r)), \quad f(0) = 0,$$

with $f'(0) = k > 1$, the indicial equations are

$$(in\nu + \lambda + \lambda k e^{-in\nu r}) x_n.$$

For $\lambda > 0$, the possible bifurcation points are such that $\nu = (\nu_0 + 2m\pi)/(nr)$, where $m \geq 0$ and $\pi/2 < \nu_0 < \pi$ is such that $\cos \nu_0 = -k^{-1}$, and $\lambda = (\nu_0 + 2m\pi) r^{-1} (k^2 - 1)^{-1/2}$. By linearizing the equation $in\nu + \lambda + \lambda k e^{-in\nu r}$ around one of these points, it is immediate to see that the winding number, for the orientation (λ, ν), is -1 if $\lambda > 0$ and 1 if $\lambda < 0$. Thus, one has bifurcation from each of these points (of truly periodic solutions, if $\lambda \neq 0$).

Assume $xf(x) > 0$ for $x \neq 0$. This has several consequences:

(a) the only constant solution is $x = 0$ if $\lambda \neq 0$ or $\lambda = 0$ and any constant.

(b) Any periodic solution for $\lambda \neq 0$ must change sign (if of constant sign then it would be monotone and non-periodic).

(c) No branch of solutions may go to $\lambda = 0$, with $\|x\|_1$ and periods bounded (that is $\nu \geq a > 0$): in fact, if $\|x_n\|_1$ is bounded, then there is a convergent subsequence, in C^0, to a solution, with $\lambda = 0$, i.e., a constant solution. If this constant is non-zero, then nearby periodic solutions can not change sign, while, if the constant is 0, then $(x = 0, \lambda = 0)$ would be a bifurcation point, which would contradict the indicial equations.

Hence, the bifurcating branches must go to ∞ (in $\|x\|_1$, $\lambda > 0$, or periods).

If $\mathcal{C}_{n,m}$ is the branch bifurcating from $\nu = (\nu_0 + 2m\pi)/(nr)$, $\lambda = (\nu_0 + 2m\pi)r^{-1}(k^2 - 1)^{-1/2}$, then, since if $(\nu, \lambda, x(t))$ is solution this is also the case for $(\alpha\nu, \alpha\lambda, x(t))$, one has that, on $\mathcal{C}_{n,m}$, $(\nu, \lambda) = \alpha(\nu_0, \lambda_0)$, with $\alpha = -1+2m\pi\nu_0^{-1}$ and (ν_0, λ_0) on $\mathcal{C}_{n,0}$. Also, if $z(t) = x(nt)$, then $(\nu/n)z'(t) = -\lambda z(t) - \lambda f(x(nt - \nu r))$, if $x(t)$ is solution for (ν, λ). Thus, $z(t)$ is solution for $(\nu/n, \lambda)$. This implies that solutions on $\mathcal{C}_{n,m}$ are those of $\mathcal{C}_{1,m}$, rescaled as above. Thus, *it is enough to study one of these branches, for instance* $\mathcal{C}_{1,0}$.

Now, if $x(t)$ is a solution, for (ν, λ), let $y(t) = x(t - \nu r)$, then

$$\nu y'(t) = -\lambda y(t) - \lambda f(x(t - 2\nu r)).$$

Assume that $\nu r = l\pi$, then, from the 2π-periodicity of $x(t)$, one has the system of ordinary differential equations

$$\nu x' = -\lambda x - \lambda f(y), \quad \nu y' = -\lambda y - \lambda f(x).$$

Suppose that, for some t_0, $(x(t_0), y(t_0)) = (a, b)$ holds with $a \neq b$. Then one has $(x(t_0 - \nu r), y(t_0 - \nu r)) = (b, a)$, that is, if (a, b) is on one side of the diagonal in the (x, y)-plane, then (b, a) is on the other side. But then, the path $(x(t), y(t))$ must cross the diagonal at some point, that is, there is a τ, with $x(\tau) = y(\tau)$. From the uniqueness of the initial value problem for the system of O.D.E.'s, one has $x(t) \equiv y(t)$, which should be a 2π-periodic solution of $\nu x' = -\lambda x - \lambda f(x)$, something impossible in dimension one, unless $x(t)$ is constant and $\lambda = 0$. *Thus, $\nu r = l\pi$ are forbidden frequencies.*

In particular, $\mathcal{C}_{1,1}$, which starts at $\nu = (\nu_0 + 2\pi)/r$, with $\frac{\pi}{2} < \nu_0 < \pi$, must stay in the interval $2\pi < \nu r < 3\pi$. Hence, for $\mathcal{C}_{1,0}$, one has that νr has to be in $(2\pi(1 + 2\pi\nu_0^{-1})^{-1}, 3\pi(1 + 2\pi\nu_0^{-1})^{-1})$. *Thus, the periods are bounded on* $\mathcal{C}_{1,0}$.

If, in addition, $f(x)$ *is bounded by* M, then, since any 2π-periodic solution must have a zero, let $x(\tau) = 0$ and write a solution of the equation as

$$x(t) = -\frac{\lambda}{\nu}\int_\tau^t f(x(s - \nu r))e^{-\lambda(t-s)/\nu}\,ds.$$

Thus, for $t \geq \tau$, one gets $|x(t)| \leq M$ and $\nu|x'(t)| \leq \lambda M$. Thus, on $\mathcal{C}_{1,0}$, one has $\|x'\|_1 \leq K|\lambda|$ and $\mathcal{C}_{1,0}$ *goes to infinity in* λ.

Clearly, these conditions are rather particular, but we hope that the reader will be able to study more general situations.

4.5 Bibliographical remarks

There is an uncountable number of applications of classical degree theory. For problems with symmetries, the current literature is more inclined toward variational methods or to generic situations, as in the books [Fi], [B], [GS]. For the case of equivariant degrees, we refer to [KW] and the articles in the References.

Section I is taken from [IV1], for the general case, and from [IV3] for the orthogonal degree. The basic material of Section 2 comes from [IV2]. However some of the examples are taken from [IMV2], [I], and earlier work. The notion of hyperbolicity, which mimics the one for differential equations, was introduced in [IMV0]. The treatment of autonomous differential equations and their periodic solutions is now a standard application of Fourier series, as well as the period doubling phenomenon. Chow and Mallet-Paret were the first to use the Fuller index in this context. Many papers were published later on generalizing Fuller's ideas.

The examples of differential equations with first integrals are taken from [IMV2]. A treatment with Fuller degree is due to Dancer and Toland. The special spectral behavior of Remark 2.6 is similar to the one introduced by Fiedler.

The bulk of Section 3 comes from [IV3]. However, we invite the reader to compare these results with the ones coming from variational methods (in case of gradients or Hamiltonians) so that he may judge by himself the advantages and shortcomings of this degree for orthogonal maps. For the case of S^1-orthogonal maps, we refer also to the papers by Rybicki.

The spring-pendulum was published in part in [I2]. Here, we have given complete proofs of the local behavior of the singular Hill's equation. We refer to the references of this last paper for other special treatments of these systems.

The last section, essentially on Hopf bifurcation, is taken from [I0], [I], [IMV2] and [IV2]. There is a vast literature on the classical Hopf bifurcation. Among the first papers using topological tools, one has to mention [AY], [I0] and [CM-P]. The examples of classical Hopf bifurcation are taken from [IMV0], while the case of first integrals was treated in [IMV2]. Part of Lemma 4.6 is inspired in [GLR.] and in [DT2]. The example on retarded differential equations was taken, as a very special case of the literature in this subject, from [M-P.N].

Appendix A

Equivariant Matrices

The purpose of this appendix is to prove Theorems 5.2 and 5.3 of Chapter 1. Several versions of these results are well known in the literature. However, most of them either do not give such a precise description or are based on much more sophisticated tools. The proof of the first result is inspired on the proof of Frobenius Theorem, as given in Pontrjagin's book [P].

Theorem A.1 (Cfr. Theorem 5.2 of Chapter 1). *Let V be a finite dimensional irreducible orthogonal representation, then exactly one of the following statements is satisfied.*

(a) *Any equivariant linear map A is of the form $A = \mu I$, i.e., V is an absolutely irreducible representation.*

(b) *There is only one equivariant map B, such that $B^2 = -I$, $B^T + B = 0$. Then, any equivariant linear map A has the form $A = \mu I + \nu B$. In this case, V has a complex structure for which $A = (\mu + i\nu)I$.*

(c) *There are precisely B_1, B_2, B_3 with the above properties. Then, $B_iB_j = -B_jB_i$ and $B_3 = B_1B_2$. Moreover, V has a quaternionic structure and any equivariant linear map can be written as $A = \mu I + \nu_1 B_1 + \nu_2 B_2 + \nu_3 B_3 = qI$, where $q = \mu + \nu_1 i_1 + \nu_2 i_2 + \nu_3 i_3$ is in $\mathbb{H}$.*

Proof. Let $\mathcal{C}$ be the set of equivariant matrices from V into itself. Define $\mathcal{D} = \{A \in \mathcal{C} : A = kI\}$ and $\mathcal{F} = \{A \in \mathcal{C} : A^2 = -k^2 I, A + A^T = 0\}$. It is clear that $\mathcal{C}$ and $\mathcal{D}$ are linear subspaces and that the three sets are closed under transposition, since $\gamma^T = \gamma^{-1}$.

Step (a). *$\mathcal{F}$ is a linear subspace.* In fact, if $A \in \mathcal{F}$, then $\alpha A \in \mathcal{F}$. Also, if A_1 and A_2 are in $\mathcal{F}$, then, from Corollary 5.1 (c) in Chapter 1, we get $A_1 + A_2 = \mu I + \nu B$, for some B in $\mathcal{F}$. However, $A_1 A_2^T + A_2 A_1^T$ is equivariant and self-adjoint, hence, from Schur's lemma, it belongs to $\mathcal{D}$, that is $A_1A_2 + A_2A_1 = kI$. Similarly, $BA_1 + A_1B$ is self-adjoint, hence equal to αI. On the other hand, $kI = A_1A_2 + A_2A_1 = A_1(\mu I + \nu B - A_1) + (\mu I + \nu B - A_1)A_1 = 2\mu A_1 + \nu\alpha I + 2k_1^2 I$, where $A_1^2 = -k_1^2 I$. Hence, $2\mu A_1 = (k - \nu\alpha - 2k_1^2)I$. Thus, if $\mu \neq 0$, we have $A_1 \in \mathcal{D} \cap \mathcal{F} = \{0\}$, which is not possible. This implies that $\mu = 0$, $A_1 + A_2 = \nu B$ is in $\mathcal{F}$.

Step (b). *If not empty, $\mathcal{F}$ has dimension* 1 *or* 3. Note first that if $\mathcal{F} = \phi$ then, from Corollary 5.1 (c) in Chapter 1, any A in $\mathcal{C}$ is in fact in $\mathcal{D}$ and this gives *(a)* of the theorem. Furthermore, if $\mathcal{F}$ is one-dimensional, then any element A in $\mathcal{F}$ is of the form αB, with B in $\mathcal{F}$ and $B^2 = -I$.

Let B_1, B_2 in $\mathcal{F}$ be such that $B_1^2 = B_2^2 = -I$. Then, B_1B_2 is in $\mathcal{C}$ and as such $B_1B_2 = \mu I + \nu B$, for some B in $\mathcal{F}$, with $B^2 = -I$. Multiplying by B_1, one has $\mu B_1 + B_2 = -\nu B_1 B$, and, from Step (a), one has that $\nu B_1 B$ is in $\mathcal{F}$.

If $\nu = 0$, $B_2 = -\mu B_1$ and $B_2^2 = -\mu^2 I = -I$ gives $B_2 = \pm B_1$. If, on the other hand, $\nu \neq 0$, then $B_1 B$ is in $\mathcal{F}$. Set $B_1' = B_1$, $B_2' = B$ and $B_3' = B_1 B = B_1' B_2'$. Then, since B_3' belongs to $\mathcal{F}$, one has that $B_3'^T + B_3' = 0$ and $B_3' B_3'^T = B_1 B B^T B_1^T = I$, hence $B_3'^2 = -I$. Furthermore, $B_1' B_2' = B_3' = -B_3'^T = -B_2' B_1'$, $B_1' B_3' = -B_2' = -B_3' B_1'$ and $B_3' B_2' = -B_1' = -B_2' B_3'$. Thus, the B_i's, dropping the primes, satisfy the anticommutativity properties of the theorem.

Now, these B_1, B_2, B_3 are linearly independent in $\mathcal{F}$: In fact, if $\lambda_1 B_1 + \lambda_2 B_2 + \lambda_3 B_3 = 0$, then, multiplying by B_1 one has $\lambda_2 B_3 - \lambda_3 B_1 = \lambda_1 I$. But, from the fact that $\mathcal{F}$ is a linear subspace, one gets that the left-hand side is in $\mathcal{F}$ and so $\lambda_1 I$ would be in $\mathcal{F}$, which is impossible, unless $\lambda_1 = 0$. A multiplication by B_1 will give $\lambda_2 B_2 = \lambda_3 I$ and $\lambda_2 = \lambda_3 = 0$, hence, B_i, $i = 1, 2, 3$ are linearly independent.

Finally, suppose that there is a B in $\mathcal{F}$, with $B^2 = -I$, which is not a linear combination of B_1, B_2, B_3. Then, as above, $B_j B = \mu_j I + \nu_j \tilde{B}_j$ and, by taking transposes, $B B_j = \mu_j I - \nu_j \tilde{B}_j$. Let, for some $\alpha \neq 0$, $\hat{B} = \alpha(B + \mu_1 B_1 + \mu_2 B_2 + \mu_3 B_3)$. Then, $\hat{B}$ is in $\mathcal{F}$ and it is easy to see that $\hat{B}^2 = -\alpha^2(1 - \mu_1^2 - \mu_1^2 - \mu_3^2)I = -k^2 I$. Hence, either $\hat{B}^2 = 0$ and $\hat{B}$ has a nontrivial kernel, in which case, from Schur's lemma, $\hat{B} = 0$ and B is a linear combination of B_1, B_2, B_3, contrary to the hypothesis; or, one may choose α such that $k^2 = 1$ and $\hat{B}^2 = -I$. Now, $B_j \hat{B} = \alpha(\nu_j \tilde{B}_j \pm \mu_k B_l \pm \mu_l B_k) = -\hat{B} B_j$, for $k \neq l \neq j$. Thus, $B_j \hat{B}$ belongs to $\mathcal{F}$. Furthermore, $(B_1 \hat{B}) B_3 = -(\hat{B} B_1) B_3 = \hat{B} B_2$, while $B_1(\hat{B} B_3) = -B_1(B_3 \hat{B}) = B_2 \hat{B} = -\hat{B} B_2$. That is $\hat{B} B_2 = 0$, which is not possible, since both are isomorphisms. This proves that any B is a linear combination of B_1, B_2, B_3 and finishes the proof of step (b). Note that, because of the associativity of the product of matrices, there is no equivalent to Cayley numbers.

It remains to make explicit the structure of V.

Step (c). Let B be in $\mathcal{F}$ such that $B^2 = -I$, then there is a basis for V such that $B = \begin{pmatrix} 0 & -I \\ I & 0 \end{pmatrix}$ and V has a complex structure such that any A in $\mathcal{C}$ has the form $A = \lambda I$, with λ in $\mathbb{C}$. Note that, since $(\det B)^2 = (-1)^{\dim V}$, then $\dim V = n = 2m$. In fact, take e_1 a unit vector, then, since $B + B^T = 0$, Be_1 is orthogonal to e_1 and also a unit vector. Choose e_2 orthogonal to $\{e_1, Be_1\}$, then Be_2 is orthogonal to $\{e_1, Be_1, e_2\}$, and so on...On the basis $\{e_1, e_2, \ldots, e_m, Be_1, Be_2, \ldots, Be_m\}$, B has the above form. Defining $z_j = x_j + ix_{m+j}$ and $Z = X + iY$, then $V \cong \mathbb{C}^m$. If $\gamma = \begin{pmatrix} \gamma_1 & \gamma_3 \\ \gamma_2 & \gamma_4 \end{pmatrix}$, then $\gamma B = B\gamma$ implies that $\gamma_3 = -\gamma_2$ and $\gamma_4 = \gamma_1$. Thus,

$\gamma \begin{pmatrix} X \\ Y \end{pmatrix} \equiv \tilde{\gamma} Z = (\gamma_1 + i\gamma_2)(X + iY)$ and $B \begin{pmatrix} X \\ Y \end{pmatrix} \equiv \tilde{B} Z = iI(X + iY)$, where $\tilde{B} = iI$. Hence, if A is in $\mathcal{F}$, with $A = \mu I + \nu B = (\mu + i\nu)I = \lambda I$, with λ in $\mathbb{C}$. This proves Part *(b)* of the theorem, in case $\mathcal{F}$ has dimension 1.

Step (d). If $\mathcal{F}$ has dimension 3, then V has a quaternionic structure and Part *(c)* of the theorem holds.

In fact, take a unit vector e_1, then $(e_1, B_1e_1, B_2e_1, B_3e_1)$ are orthogonal. Next, take e_2 orthogonal to that set. It is easy to see that the vectors $(e_2, B_1e_2, B_2e_2, B_3e_2)$ are all orthogonal to the first set and among themselves, by using the relations of anticommutation of the B_j's. This implies that $\dim V = 4m$ and, on the basis $\{e_1, \dots, e_m, B_1e_1, \dots, B_1e_m, B_2e_1, \dots B_2e_m, B_3e_1, \dots, B_3e_m\}$, B_j has the form of the Pauli matrices:

$$B_1 = \begin{pmatrix} 0 & -I & 0 & 0 \\ I & 0 & 0 & 0 \\ 0 & 0 & 0 & -I \\ 0 & 0 & I & 0 \end{pmatrix}, \quad B_2 = \begin{pmatrix} 0 & 0 & -I & 0 \\ 0 & 0 & 0 & I \\ I & 0 & 0 & 0 \\ 0 & -I & 0 & 0 \end{pmatrix}, \quad B_3 = \begin{pmatrix} 0 & 0 & 0 & -I \\ 0 & 0 & -I & 0 \\ 0 & I & 0 & 0 \\ I & 0 & 0 & 0 \end{pmatrix}.$$

Then, if $X = (X_0, X_1, X_2, X_3)^T$ is written as $\hat{X} = X_0 + i_1X_1 + i_2X_2 + i_3X_3$, an element of $\mathbb{H}^m$, with $i_j^2 = -1$, $i_ji_k + i_ki_j = 0$, $i_1i_2 = i_3$, one has that B_jX is written as $i_j\hat{X}$. Furthermore, if γ is written as a $(4m \times 4m)$-matrix (γ_{kl}), $k, l = 0, \dots, 3$, then the relations $\gamma B_j = B_j\gamma$ imply that

$$\gamma = \begin{pmatrix} \gamma_0 & -\gamma_1 & -\gamma_2 & -\gamma_3 \\ \gamma_1 & \gamma_0 & \gamma_3 & -\gamma_2 \\ \gamma_2 & -\gamma_3 & \gamma_0 & \gamma_1 \\ \gamma_3 & \gamma_2 & -\gamma_1 & \gamma_0 \end{pmatrix}$$

can be written as $\hat{\gamma} = \gamma_0 + i_1\gamma_1 + i_2\gamma_2 + i_3\gamma_3$, acting on $\hat{X}$ on the *right*: $\gamma X = \hat{X}\hat{\gamma} = (X_0 + i_1X_1 + i_2X_2 + i_3X_3)(\gamma_0 + i_1\gamma_1 + i_2\gamma_2 + i_3\gamma_3)$.

Then, any A in $\mathcal{C}$ may be written as $A = qI$, with $q = \mu + \nu_1 i_1 + \nu_2 I_2 + \nu_3 i_3$ in $\mathbb{H}$ and I is the identity on $\mathbb{H}^m$. Thus, $AX = q\hat{X}$ and $A(\gamma X) = q\gamma X = q\hat{X}\hat{\gamma}$, while $\gamma AX = (AX)\hat{\gamma} = q\hat{X}\hat{\gamma}$. □

One may give the general form of an equivariant linear map between finite dimensional representations.

Theorem A.2 (Cfr. Theorem 5.3 of Chapter 1). *Let V be decomposed as*

$$\bigoplus_{i=1}^{i=I} (V_i^{\mathbb{R}})^{n_i} \bigoplus_{j=1}^{j=J} (V_j^{\mathbb{C}})^{n_j} \bigoplus_{l=1}^{l=L} (V_l^{\mathbb{H}})^{n_l},$$

where $V_i^{\mathbb{R}}$ are the absolutely irreducible representations of real dimension m_i repeated n_i times, $V_j^{\mathbb{C}}$ are complex irreducible representations of complex dimension m_j repeated n_j times, while $V_l^{\mathbb{H}}$ are quaternionic representations of dimension (over $\mathbb{H}$) m_l

repeated n_l times. Then, there are bases of V such that any equivariant matrix has a block diagonal form

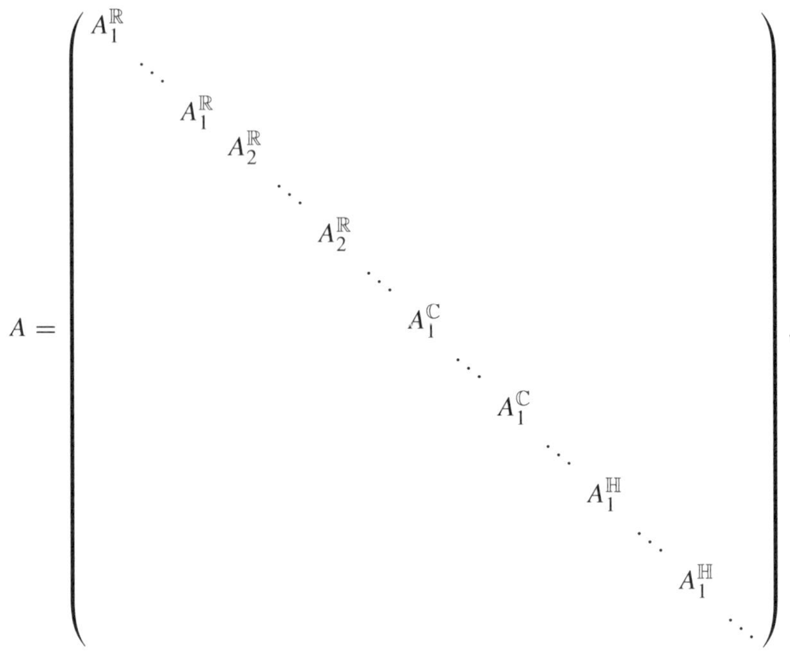

where $A_i^{\mathbb{R}}$ are real $n_i \times n_i$ matrices repeated m_i times, $A_j^{\mathbb{C}}$ are complex $n_j \times n_j$ matrices, repeated m_j times and $A_l^{\mathbb{H}}$ are $n_l \times n_l$ quaternionic matrices repeated m_l times. On the new basis, the equivariance of A and the action have the following form: γ is block diagonal on each subspace corresponding to the repetition of the same matrix, i.e., if $B_{n\times n}$ is repeated m times, on W corresponding to the same representation, then $\gamma = (\gamma_{ij} I)_{1\le i,j\le m}$, with γ_{ij} in $\mathbb{K} = \mathbb{R}, \mathbb{C}$ or $\mathbb{H}$, and I the identity on $\mathbb{K}^n$ where the product, for the quaternionic case, is on the right.

Proof. From the considerations of § 5 in Chapter 1, it is enough to consider A on equivalent subrepresentations of V. Take then $V = V_1 \oplus \cdots \oplus V_k$, where V_j are irreducible but with equivalent actions. Then, if $\tilde{\gamma} A = A\gamma$, A maps V into V and is similar to a matrix $\tilde{A}$ from $(V_1)^k$ into itself, such that $\gamma\tilde{A} = \tilde{A}\gamma$ and $\tilde{\gamma}$ acts orthogonally on V_1. Note that the similarity depends only on the actions, not on A. Hence, assume that there are bases in V and a norm such that γ is in $O(V)$ and has a diagonal form $\operatorname{diag}(\gamma, \dots, \gamma)$, since V_i are Γ-invariant.

Let $m = \dim_K V_i$, where $K = \mathbb{R}, \mathbb{C}$ or $\mathbb{H}$. Then, $\gamma|_{V_i}$ can be written as above, when considering the real matrix, or as $\hat{\gamma}$ for the K-structure: $A_{ij} : V_i \to V_j$ is $A_{ij} = \lambda_{ij} I$, with λ_{ij} in K and I the identity in K^m, $i, j = 1, \dots k$, on the basis of Theorem A.1.

Take a new basis for V by ordering the bases for V_1, $\{e_{11}, e_{12}, \dots, e_{1m}\}$, of V_2, $\{e_{21}, e_{22}, \dots, e_{2m}\}$, $\dots$ and of V_k, $\{e_{k1}, e_{k2}, \dots, e_{km}\}$, in the following way:

$\{e_{11}, e_{21}, \ldots, e_{k1}, e_{12}, e_{22}, \ldots, e_{2m}, \ldots, e_{km}\}$. It is easy to see that, on this new basis, A has the form

$$A = \begin{pmatrix} \Lambda & & 0 \\ & \Lambda & \\ 0 & & \Lambda \end{pmatrix},$$

where $\Lambda = (\lambda_{ij})_{1\le i,j\le k}$ is repeated m times on the diagonal. On the other hand, if $\gamma : V_l \to V_l$ has the form $(\gamma_{ij})_{1\le i,j\le m}$, then on the new basis $\gamma\big|_V = (\gamma_{ij} I)_{1\le i,j\le m}$, where I is the identity in K^k. The relation $\gamma A = A\gamma$ is maintained in the new basis: in fact, if K is $\mathbb{R}$ or $\mathbb{C}$, then γ_{ij} is a scalar which commutes with Λ. If $K = \mathbb{H}$, then the action is on the right and γq has to be interpreted as $\hat{q}\hat{\gamma}$ (one may also go back to the 4×4 real matrices, where γ_{ij} is as above and commutes with q). □

Appendix B

Periodic Solutions of Linear Systems

In this appendix we shall collect the results of Floquet theory needed in the book. Most of these results are well known, however the presentation given here will be slightly different.

Consider, in $\mathbb{R}^N$, the system

$$LX = \frac{d}{dt}X - A(t)X$$

where $A(t)$ is a continuous matrix, which is $(2\pi/p)$-periodic. The operator L is a continuous operator from $H^1(S^1)^N$ into $L^2(S^1)^N$. In terms of Fourier series, if $X(t) = \sum X_n e^{int}$ and $A(t) = \sum A_m e^{imt}$, then $A(t)X = \sum C_k e^{ikt}$, where

$$C_k = \sum A_l X_{k-l}$$

corresponds to a convolution.

Let $\Phi(t)$ be the fundamental matrix associated to L, i.e.,

$$\Phi' = A\Phi, \quad \Phi(0) = I,$$

($\Phi(t) = e^{At}$ if A is constant).

Then, $LX = Y$ if and only if

$$X(t) = \Phi(t)X(0) + \Phi(t)\int_0^t \Phi^{-1}(s)Y(s)\,ds.$$

Recall that the columns of $\Phi(t)$ are linearly independent solutions of $LX = 0$ and that $\det\Phi(t) = \exp(\int_0^t tr A(s)\,ds)$.

Hence, as an operator from H^1 into L^2, the equation $LX = Y$ will have a 2π-periodic solution if and only if

$$(I - \Phi(2\pi))X(0) = \Phi(2\pi)\int_0^{2\pi} \Phi^{-1}(s)Y(s)\,ds.$$

Lemma B.1. *One has the following isomorphism from* $\ker(L + \mu I)$ *in* H^1 *onto* $\ker(\Phi(2\pi) - e^{2\pi\mu I})$, *given by* $\ker\left(\frac{d}{dt} - A + \mu I\right) = \{X(t) = e^{-\mu t}\Phi(t)W$, *with* $W \in \ker(\Phi(2\pi) - e^{2\pi\mu}I)\}$.

Proof. By direct differentiation, it is easy to see that $e^{-\mu t}\Phi(t)$ is the fundamental matrix for $L+\mu I$. Hence $X(t) = e^{-\mu t}\Phi(t)X(0)$ belongs to $\ker(L+\mu I)$ if and only if $X(0)$ belongs to $\ker(e^{-2\pi\mu}\Phi(2\pi) - I)$. □

In particular, the multiplicity of $-\mu$ as eigenvalue of L is equal to the multiplicity of $e^{2\pi\mu}$ as eigenvalue of $\Phi(2\pi)$. The number $e^{2\pi\mu}$ is the *Floquet multiplier* of L, or of the *Poincaré return map* $\Phi(2\pi)$.

Remark B.1. Since $A(t)$ is $(2\pi/p)$-periodic, one has

$$\Phi'(t+\frac{2\pi}{p}) = A(t+\frac{2\pi}{p})\Phi(t+\frac{2\pi}{p}) = A(t)\Phi(t+\frac{2\pi}{p}),$$

hence $\Phi(t+2\pi/p)$ is a fundamental matrix and, as such, one has

$$\Phi(t+\frac{2\pi}{p}) = \Phi(t)\Phi(\frac{2\pi}{p}).$$

In particular, if p' divides p, one has

$$\Phi\left(\frac{2\pi}{p'}\right) = \Phi\left(\frac{2\pi}{p}\right)^{p/p'}.$$

Thus, the multiplicity of 0, as eigenvalue of L, is the sum of the multiplicities of the eigenvalues of the Poincaré map of first return $\Phi(\frac{2\pi}{p})$, which are p'th roots of unity. On the other hand, the elements of $\ker\big(\Phi(\frac{2\pi}{p}) - I\big)$ give $\big(\frac{2\pi}{p}\big)$-periodic solutions of $LX = 0$.

The L^2-adjoint of L is the operator

$$L^* = -\left(\frac{d}{dt} + A^T\right)$$

which has the fundamental matrix $\Psi(t) = \Phi^{-1}(t)^T$: since $\Phi^{-1}\Phi = 1$, one has $(\Phi^{-1})' = -\Phi^{-1}\Phi'\Phi^{-1} = -\Phi^{-1}A$.

Then, $LX = Y$ has a solution if and only if $\Phi(2\pi)\int_0^{2\pi}\Phi^{-1}(s)Y(s)ds$ belongs to $\mathrm{Range}(I - \Phi(2\pi))$, that is, if it is orthogonal, in $\mathbb{R}^N$, to all Z's in $\ker(I - \Phi(2\pi)^T) = \ker(\Phi^{-1}(2\pi)^T - I)$. Performing the scalar product in $\mathbb{R}^N$, one arrives at

$$\int_0^{2\pi} Y(s)^T\Phi^{-1}(s)^T Z ds = 0,$$

that is, $Y(t)$ is L^2-orthogonal to $Z(t) = \Phi^{-1}(s)^T Z$, any element of $\ker(L^*)$. This argument gives another proof of the fact that L is a Fredholm operator of index 0.

Assume now that $A(t)$ is smooth enough, then

Lemma B.2. $\ker(L+\mu I)^\alpha$ *in* H^1 *and* $\ker(\Phi(2\pi)-e^{2\pi\mu}I)^\alpha$ *in* $\mathbb{R}^N$ *are isomorphic. In fact,* $\ker\left(\frac{d}{dt}-A+\mu I\right)^\alpha = \{X(t)=e^{-\mu t}\Phi(t)\sum_0^{\alpha-1}W_k\frac{t^k}{k!}$, *with* W_k *uniquely determined by* W_0, $W_k \in \ker(\Phi(2\pi)-e^{2\pi\mu}I)^{\alpha-k}\}$.

Proof. Let $X(t)$ be in $\ker\left(\frac{d}{dt}-A+\mu I\right)^\alpha$ and define

$$Y(t)=e^{\mu t}\Phi^{-1}(t)X(t),$$

then, $(L+\mu I)X = e^{-\mu t}\Phi(t)Y'(t)$ and $(L+\mu I)^k X = e^{-\mu t}\Phi(t)Y^{(k)}(t)$. Thus, if $(L+\mu I)^\alpha X = 0$, one has $Y^{(\alpha)}(t)=0$ and

$$Y(t)=\sum_0^{\alpha-1} W_l\frac{t^l}{l!}.$$

One needs that $e^{-\mu t}\Phi(t)Y^{(k)}(t)$ belong to the space of 2π-periodic functions for $k=0,\dots,\alpha-1$. If B^{-1} is the matrix $e^{-2\pi\mu}\Phi(2\pi)$, this requirement amounts to solving the system

$$B^{-1}\left(\sum_{l=k}^{\alpha-1} W_l\frac{(2\pi)^{l-k}}{(l-k)!}\right) = Y^{(k)}(0) = W_k.$$

Hence, one has the linear relations

$$\begin{aligned}(B-I)W_{\alpha-1} &= 0\\ (B-I)W_{\alpha-2} &= 2\pi W_{\alpha-1}\\ (B-I)W_0 &= 2\pi W_1 + \frac{(2\pi)^2}{2!}W_2+\cdots+\frac{(2\pi)^{\alpha-1}}{(\alpha-1)!}W_{\alpha-1}.\end{aligned}$$

From here, one has that $W_{\alpha-k}$ belongs to $\ker(B-I)^k$ and that, for $k=1,\dots,\alpha$,

$$(B-I)^{k-1}W_{\alpha-k}=(2\pi)^{k-1}W_{\alpha-1}.$$

One may view the last $(\alpha-1)$-equations as a linear system for $W_1,\dots,W_{\alpha-1}$ in terms of $(B-I)W_0$. If this term is 0, then, from $(B-I)^{\alpha-1}W_0$, one has, if $\alpha>1$, that $W_{\alpha-1}=0$, $(B-I)^{k-2}W_{\alpha-k}=(2\pi)^{k-2}W_{\alpha-2}$, that is, the same system with α replaced by $\alpha-1$. But then, $(B-I)^{\alpha-2}W_0$ leads, if $\alpha>2$, to $W_{\alpha-2}=0$, and so on. Thus, if $(B-I)W_0=0$, one obtains that $W_1=\cdots=W_{\alpha-1}=0$. Thus implies that the system for $W_1,\dots,W_{\alpha-1}$ is invertible and these vectors are uniquely determined by W_0, in fact by $(B-I)W_0$. □

If A is constant, then $e^{-\mu t}\Phi(t)=e^{(A-\mu)t}$ which will have 1 as eigenvalue if and only if A has $\mu\pm in$ as eigenvalue. If A is taken in Jordan form, then on a Jordan block, $A - \mu I = inI + J$, where J is the upper diagonal. Then, on a block of size α, one has

$$e^{(A-\mu)t} = e^{int}\left(I + tJ + \cdots + \frac{t^{\alpha-1}}{(\alpha-1)!}J^{\alpha-1}\right).$$

Furthermore, $X(t) = e^{(A-\mu)t}\sum_0^{\alpha-1} W_k \frac{t^k}{k!}$, can be expressed as

$$\begin{aligned} X(t) = e^{int}\Big[&W_0 + t(W_1 + JW_0) + \frac{t^2}{2!}(W_2 + 2JW_1 + W_0) + \cdots \\ &+ \frac{t^k}{k!}\Big(W_k + kJW_{k-1} + \cdots + \tbinom{k}{l}J^{k-l}W_l + \cdots + J^k W_0\Big) \\ &+ \frac{t^{\alpha-1}}{(\alpha-1)!}(W_{\alpha-1} + \cdots + J^{\alpha-1}W_0)\Big]. \end{aligned}$$

The requirement of periodicity determines $W_1, \ldots, W_{\alpha-1}$ in terms of W_0. It is not difficult, but tedious, to check that these are the same as the other set of conditions. Then, one has

$$X(t) = \sum X_n e^{int}, \quad \text{with } (inI - A + \mu I)^{\alpha_n} X_n = 0,\ \alpha = \max \alpha_n,$$

a result which, of course, follows directly by looking at Fourier series.

If A is non-constant, then by complexifying A, one has that $\Phi(t) = P(t)e^{Rt}$, where $P(t)$ is 2π-periodic and R has, as eigenvalues, the Floquet exponents.

Bibliography

We first give a list of the references in the text or closer to it in spirit. It follows a list of further readings on the subject.

References

[AY] Alexander, J. C. and J. A. Yorke, 1978. Global bifurcation of periodic orbits. Amer. J. Math. **100**, 263–292.

[B] Bartsch, T., 1993. *Topological Methods for variational problems with symmetries.* Lecture Notes in Math. **1560**, Springer-Verlag.

[Br] Bredon, G. E., 1980. *Introduction to compact transformation groups.* Academic Press, New York.

[BD] Bröcker, T. and T. tom Dieck, 1985. *Representations of compact Lie groups.* Grad. Texts in Math. **98**, Springer-Verlag, New York.

[CM-P] Chow, S. N. and J. Mallet-Paret, 1978. The Fuller index and global Hopf bifurcation. J. Differential Equations **29**, 66–85.

[Da] Dancer, E. N., 1985. A new degree for S^1-invariant gradient mappings and applications. Ann. Inst. H. Poincaré Anal. Non Linéaire **2**, 329–370.

[DT1] Dancer, E. N. and J. F. Toland, 1990. Degree theory for orbits of prescribed period of flows with a first integral. Proc. London Math. Soc. **60**, 549–580.

[DT2] Dancer, E. N. and J. F. Toland, 1991. Equilibrium states in the degree theory of periodic orbits with a first integral. Proc. London Math. Soc. **61**, 564–594.

[DGJW] Dylawerski, G., Geba, K., Jodel, J. and W. Marzantowicz, 1991. An S^1-equivariant degree and the Fuller index. Ann. Pol. Math. **52**, 243–280.

[Fi] Fiedler, B., 1988. *Global Bifurcation of periodic solutions with symmetry.* Lecture Notes in Math. **1309**, Springer-Verlag, Berlin.

[F] Fuller, F. B., 1967. An index of fixed point type for periodic orbits. Amer. J. Math. **89**, 133–148.

[G] Geba, K., 1997. Degree for gradient equivariant maps and equivariant Conley index. *Topological Nonlinear Analysis* II (M. Matzeu and A. Vignoli, Eds.). Progr. Nonlinear Differential Equations Appl. **27**, Birkhäuser, Boston, 247–272.

[GKW] Geba, K., W. Krawcewicz and J. Wu, 1994. An equivariant degree with applications to symmetric bifurcation problems: Part I: construction of the degree. Bull. London Math. Soc. **69**, 377–398.

[GLR] Gokhberg I., P. Lancaster and L. Rodman, 1983. *Matrices and indefinite scalar products.* Birkhäuser, Basel.

[GS] Golubitsky, M. and D.G. Schaeffer, 1986. *Singularities and groups in bifurcation theory I.* Appl. Math. Sci. **51**, Springer-Verlag.

[Gr] Greenberg, G., 1967. *Lectures on algebraic topology.* Benjamin, New York.

[I0] Ize, J., 1985. Obstruction theory and multiparameter Hopf bifurcation. Trans. Amer. Math. Soc. **289**, 757–792.

[I1] Ize, J., 1988. Necessary and sufficient conditions for multiparameter bifurcation. Rocky Mountain J. Math. **18**, 305–337.

[I] Ize, J., 1994. Topological bifurcation. *Topological Nonlinear Analysis* (M. Matzeu and A. Vignoli, Eds.), Progr. Nonlinear Differential Equations Appl. **15**, Birkhäuser, Boston, 341–463.

[I2] Ize, J., 2000. Two mechanical systems and equivariant degree theory. *Recent trends in nonlinear analysis* (J. Appell, Ed.). Progr. Nonlinear Differential Equations Appl. **40**. Birkhäuser, Boston, 177–190.

[IMPV] Ize, J., I. Massabó, J. Pejsachowicz and A. Vignoli, 1985, Structure and dimension of global branches of solutions to multiparameter nonlinear equations. Trans. Amer. Math. Soc. **291**, 383–435.

[IMV0] Ize, J., I. Massabó and A. Vignoli, 1986. Global results on continuation and bifurcation for equivariant maps. NATO Adv. Sci. Inst. Ser. C Math. Phys. Sci. **173**, 75–111.

[IMV1] Ize, J., I. Massabó and A. Vignoli, 1989. Degree theory for equivariant maps I. Trans. Amer. Math. Soc. **315**, 433–510.

[IMV2] Ize, J., I. Massabó and A. Vignoli, 1992. *Degree theory for equivariant maps: The general S^1-action.* Mem. Amer. Math. Soc. **481**.

[IV1] Ize, J., and A. Vignoli, 1993. Equivaraint degree for abelian action. Part I: Equivariant homotopy groups. Topol. Methods Nonlinear Anal. **2**, 367–413.

[IV2] Ize, J., and A. Vignoli, 1996. Equivariant degree for abelian actions. Part II: Index Computations. Topol. Methods Nonlinear Anal. **7**, 369–430.

[IV3] Ize, J., and A. Vignoli, 1999. Equivariant degree for abelian actions. Part III: Orthogonal maps. Topol. Methods Nonlinear Anal. **13**, 105–146.

[Jo] Johnson, D. L., 1980. *Topics in the theory of group presentations*. London Math. Soc. Lect. Notes Ser. **42**. Cambridge University Press.

[K] Kato, T., 1966. *Perturbation theory for linear operators*. Springer-Verlag, Berlin.

[KZ] Krasnosel'skii, M. A. and P. P. Zabreiko, 1984. *Geometrical Methods in Nonlinear Analysis*. Grundlehren Math. Wiss. **263**, Springer-Verlag.

[KW] Krawcewicz, W. and J. Wu, 1997. *Theory of degrees with applications to bifurcations and differential equations*. Canadian Mathematical Society Series of Monographs and Advanced Texts. John Wiley & Sons, Inc., New York.

[KB] Kushkuley, A. and Z. Balanov, 1996. *Geometric methods in degree theory for equivariant maps*. Lecture Notes in Math. **1632**. Springer-Verlag, Berlin.

[M-PN] Mallet-Paret, J. and R. Nussbaum, 1992. Boundary layer phenomena for differential-delay equations with state dependent time delays. Arch. Ration. Mech. Anal. **120**, 99–146.

[N] Namboodiri, V., 1983. Equivariant vector fields on spheres. Trans. Amer. Math. Soc. **278**, 431–460.

[P] Pontrjagin, L., 1946. *Topological groups*. Princeton University Press, Princeton.

[R] Rybicki, S., 1994. S^1-degree for orthogonal maps and its applications to bifurcation theory. Nonlinear Anal. **23**, 83–102.

[Z] Zabrejko, P. P., 1996. Rotation of vector fields: definition, basic properties and calculations. *Topological Nonlinear Analysis* II. (M. Matzeu and A. Vignoli, Eds.). Progr. Nonlinear Differential Equations Appl. **27**, Birkhäuser, Boston, 449–601.

Further Reading

Adams, J. F., 1982. Prerequisites (on equivariant stable homotopy theory) for Carlsson's lecture. Lecture Notes in Math. **1091**, 483–532.

Agarwal, R. P., M. Meehan and D. O'Regan., 2001. *Fixed point theory and applications*. Cambridge Tracts in Math. **141**. Cambridge University Press.

Akhmerov, R. R., M. I. Kamenskii, A. S. Potapov and A. E. Rodkina, 1992. *Measures of Noncompactness and Condensing Operators.* Operator Theory: Advances and Applications, Vol. **55**, Birkhäuser, Basel.

Alexander, J. C., 1978. Bifurcation of zeros of parametrized functions. J. Funct. Anal. **29**, 37–53.

Alexander, J. C., 1980. Calculating bifurcation invariants as elements of the homotopy of the general linear group. II. J. Pure Appl. Algebra **17**, 117–125.

Alexander, J. C., 1981. A primer on connectivity. Lecture Notes in Math. **886**, 445–483. Springer-Verlag, Berlin.

Alexander, J. C., and S. S. Antman, 1981. Global and local behavior of bifurcating multidimensional continua of solutions for multiparameter nonlinear eigenvalue problems. Arch. Ration. Mech. Anal. **76**, 339–354.

Alexander, J. C., and S. S. Antman, 1983. Global behavior of solutions of nonlinear equations depending on infinite dimensional parameter. Indiana Univ. Math. J. **32**, 39–62.

Alexander, J. C., and P. M. Fitzpatrick, 1979. The homotopy of certain spaces of nonlinear operators and its relation to global bifurcation of the fixed points of parametrized condensing operators. J. Funct. Anal. **34**, 87–106.

Alexander, J. C., I. Massabó and J. Pejsachowicz, 1982. On the connectivity properties of the solution set of infinitely parametrized families of vector fields. Boll. Un. Mat. Ital. A (6) **1**, 309–312.

Alexander, J. C., and J. A. Yorke, 1976. The implicit function theorem and global methods of cohomology. J. Funct. Anal. **21**, 330–339.

Alexander, J. C., and J. A. Yorke, 1977. Parametrized functions, bifurcation and vector fields on spheres. *Problems of the asymptotic theory of nonlinear oscillations.* Naukova Dumka, **275**, 15–17.

Alexander, J. C., and J. A. Yorke, 1978. Calculating bifurcation invariants as elements of the general linear group I. J. Pure Appl. Algebra, **13**, 1–8.

Alligood, K. T., J. Mallet-Paret and J. A. Yorke, 1981. Families of periodic orbits: local continuability does not imply global continuability. J. Diff. Geometry, **16**, 483–492.

Alligood, K. T., and J. A. Yorke, 1986. Hopf bifurcation: the appearance of virtual periods in cases of resonance. J. Differential Geom. **64**, 375–394.

Amann, H., 1976. Fixed point equations and Nonlinear eigenvalue problems in ordered Banach spaces. SIAM Reviews, **18**, 620–709.

Amann, H., and E. Zehnder, 1980. Periodic solutions of asymptotically linear Hamiltonian systems, Manuscripta Math. **32**, 149–189.

Amann, H., 1990. *Ordinary differential equations: an introduction to nonlinear analysis.* De Gruyter Stud. Math. **13**, Walter de Gruyter, Berlin

Ambrosetti, A., 1994. Variational methods and nonlinear problems: classical results and recent advances. *Topological Nonlinear Analysis.* M. Matzeu and A. Vignoli Eds. Birkhäuser. Progr. Nonlinear Differential Equations Appl. **15**. 1–36.

Antmann, S. S., 1995. *Nonlinear problems in elasticity.* Springer-Verlag, New York.

Appell, J., 1981. Implicit function, nonlinear integral equations and the measure of noncompactness of superposition operator. J. Math. Anal. Appl. **83**, 251–261.

Appell, J. (ed.), 2000. *Recent trends in nonlinear analysis.* Progr. Nonlinear Differential Equations Appl. **40**. Birkhäuser, Basel.

Appell, J. and P. Zabrejko, 1983. On a theorem of M.A. Krasnosiel'skii. Nonlinear Anal. **7**, 695–706.

Appell, J., and P. Zabrejko, 1990. *Nonlinear superposition operators.* Cambridge Tracts in Math. **95**, Cambridge University Press.

Arnold, V. I., 1972. Lectures on bifurcation and versal systems. Russ. Math. Surveys. **27**, 54–113.

Arnold, V. I., 1978. *Mathematical Methods of Classical Mechanics.* Grad. Texts in Math. **60**. Springer-Verlag, New York.

Arnold, V. I., 1983. *Geometrical methods in the theory of differential equations.* Springer-Verlag, New York.

Balanov, Z., 1997. Equivariant Hopf theorem. Nonlinear Anal. **30**, 3463–3474.

Balanov, Z., and W. Krawcewicz, 1999. Remarks on the equivariant degree theory. Topol. Methods Nonlinear Anal. **13**, 91–103.

Balanov, Z., Krawcewicz, W. and A. Kushkuley, 1998. Brouwer degree, equivariant maps and tensor powers. Abstr. Appl. Anal. **3** (3–4), 401–409.

Balanov, Z., W. Krawcewicz and H. Steinlein, 2001. Reduced $SO(3) \times S^1$-equivariant degree with applications to symmetric bifurcation problems: the case of one free parameter. Nonlinear Anal. **47**, 1617–1628.

Balanov, Z., and A. Kushkuley, 1995. On the problem of equivariant homotopic classification. Arch. Math. **65**, 546–552.

Bartsch, T., 1988. A global index for bifurcation of fixed points. J. Reine Angew. Math. **391**, 181–197.

Bartsch, T., 1988. The role of the J-homomorphism in multiparameter bifurcation theory. Bull. Sci. Math. **112**, 177–184.

Bartsch, T., 1990. A simple proof of the degree formula for $\mathbb{Z}/p$-equivariant maps. Comment. Math. Helv. **65**, 85–95.

Bartsch, T., 1991. The global structure of the zero set of a family of semilinear Fredholm maps. Nonlinear Anal. **17**, 313–332.

Bartsch, T., 1992. The Conley index over a space, Math. Z, **209**, 167–177.

Bartsch, T., and M. Clapp, 1990. Bifurcation theory for symmetric potential operators and the equivariant cup-length. Math. Z. **204**, 341–356.

Benci, V., 1981. A geometrical index for the group S^1 and some applications to the study of periodic solutions of ordinary equations. Comm. Pure Appl. Math. **34**, 393–432.

Benci, V., 1994. Introduction to Morse theory: a new approach. *Topological Nonlinear Analysis.* M. Matzeu and A. Vignoli Eds. Birkhäuser. Progr. Nonlinear Differential Equations Appl. **15**. 37–177.

Benjamin, T. B., 1976. Applications of Leray–Schauder degree theory to problems in hydrodynamic stability. Math. Proc. Cambridge Philos. Soc. **79**, 373–392.

Berger, M. S., 1973. Applications of global analysis to specific nonlinear eigenvalue problems. Rocky Mountain J. Math. **3**, 319–354.

Berger, M. S., 1977. *Nonlinearity and Functional Analysis.* Academic Press.

Berger, M. S., and M. Berger, 1968. *Perspectives in nonlinearity. An introduction to nonlinear analysis.* Benjamin Inc.

Berger, M. S., and E. Podolak, 1974. On nonlinear Fredholm operator equations. Bull Amer. Math. Soc. **80**, 861–864.

Berger, M. S. and D. Westreich, 1973. A convergent iteration scheme for bifurcation theory on Banach spaces. J. Math. Anal. Appl. **43**, 136–144.

Böhme, R., 1972. Die Lösung der Verzweigungsgleichungen für nichtlineare Eigenwertprobleme. Math. Z. **127**, 105–126.

Borisovich, Y. G., 1979. Topology and nonlinear functional analysis. Russian Math. Surveys **37**, 14–23.

Borisovich, Y. G., 1992. Topological characteristics of infinite dimensional mappings and the solvability of nonlinear boundary value problems. Proc. Steklov Inst. Math. **3**, 43–50.

Borisovich, Y. G., V. G. Zvyagin and Y. I. Sapronov, 1977. Non linear Fredholm maps and the Leray Schauder theory. Russian Math. Surveys **32**, 1–54.

Brezis, H., 1996. Degree theory: old and new. *Topological Nonlinear Analysis II* (M. Matzeu and A. Vignoli, eds.). Progr. Nonlinear Differential Equations Appl. **27**, Birkäuser, 87–108.

Broer, H. W., Lunter, G. A. and G. Vegter, 1998. Equivariant singularity theory with distinguished parameters: two case studies of resonant Hamiltonian systems. Physica D **112**, 64–80.

Brouwer, L. E. J., 1912. Über Abbildungen von Mannigfaltigkeiten. Math. Ann. **71**, 97–115.

Browder, F. E., 1970. Nonlinear eigenvalue problems and group invariance. *Functional Analysis and related fields*. Springer-Verlag, New York, 1–58.

Browder, F. E., 1976. *Nonlinear operators in Banach Spaces*. Proc. Sympos. Pure Math. **18**, vol. 2, Amer. Math. Soc., Providence, RI.

Browder, F. E. and W. V. Petryshyn, 1969. Approximation methods and generalized topological degree for nonlinear mappings in Banach spaces. J. Funct. Anal. **3** 217–245.

Browder, F. E. and R. D. Nussbaum, 1968. The Topological degree for non compact nonlinear mappings in Banach spaces. Bull. Amer. Math. Soc. **74**, 671–676.

Brown, R. F., 1967. *The Lefschetz Fixed Point Theorem*. Foresman and Co., London.

Brown, R. F., 1993. *A topological introduction to nonlinear analysis*. Birkhäuser, Boston.

Buchner, M., Marsden, J. and S. Schecter, 1983. Applications of the blowing up construction and algebraic geometry to bifurcation problems. J. Differential Geom. **48**, 404–433.

Cantrell, R. S., 1984. Multiparameter bifurcation problems and topological degree, J. Differential Equations **52**, 39–51.

Cerami, G., 1986, Symmetry breaking for a class of semilinear elliptic problems. Nonlinear Anal. **10**, 1–14.

Cesari, L., 1976, Functional Analysis, Nonlinear Differential Equations and the alternative method. Lecture Notes in Pure Appl. Math. **Vol. 19**, Marcel Decker, New York.

Chang, K. C., 1986. Applications of homology theory to some problems in Differential Equations. Proc. Sympos. Pure Math. **45**, 253–262.

Chang, K. C., 1993. *Infinite dimensional Morse theory*. Progr. Nonlinear Differential Equations Appl. **6**, Birkhäuser, Boston.

Chenciner, A., and G. Iooss, 1979. Bifurcations de tores invariants. Arch. Ration. Mech. Anal. **69**, 109–198.

Chiapinelli, R., and C. A. Stuart, 1978. Bifurcation when the linearized problem has no eigenvalues. J. Differential Equations **30**, 296–307.

Chossat, P., and G. Iooss, 1994. *The Couette–Taylor problem.* Appl. Math. Sci. **102**. Springer-Verlag, New York.

Chossat, P., R. Lauterbach and I. Melbourne, 1990. Steady-state bifurcation with $O(3)$-symmetry. Arch. Rational Mech. Anal., **113**, 313–376.

Chow, S. N., and J. K. Hale, 1982. *Methods of Bifurcation theory.* Grundlehren Math. Wiss. **251**, Springer-Verlag, New York, Berlin.

Chow, S. N., J. Mallet-Paret and J. A. Yorke, 1978. Global Hopf bifurcation from a multiple eigenvalue. Nonlinear Anal. **2**, 753–763.

Cicogna, G., 1984. Bifurcation from topology and symmetry arguments. Boll. Un. Mat. Ital. A (6) **3**, 131–138.

Ciocci, M. A., and A. Vanderbauwhede, 1998. Bifurcation of periodic orbits for symplectic mappings. J. Differ. Equations Appl. **3**, 485–500.

Clapp, M., and W. Marzantowicz, 2000. Essential equivariant maps and Borsuk–Ulam theorems. J. London Math. Soc. **61**, 950–960.

Coddington, E. A., and N. Levinson, 1955. *Theory of Ordinary Differential Equations.* McGraw-Hill, New York.

Conley, C., 1978. *Isolated invariant sets and the Morse index.* CBMS Reg. Conf. Ser. Math. **38**, Amer. Math. Soc., Providence, RI.

Crandall, M. G., and P. H. Rabinowitz, 1971. Bifurcation from simple eigenvalues J. Funct. Anal. **8**, 321–340.

Crandall, M. G., and P. H. Rabinowitz, 1977. The Hopf bifurcation theorem in infinite dimensions. Arch. Rat. Mech. Anal. **67**, 53–72.

Cronin, J., 1964. *Fixed points and topological degree in Nonlinear Analysis.* Math Surveys **11**, Amer. Math. Soc. Providence.

Cronin, J., 1994. *Differential equations. Introduction and qualitative theory.* Pure Appl. Math. **180**, Marcel Dekker, Inc., New York

Damon, J., 1994. Applications of singularity theory to the solutions of nonlinear equations. In *Topological Nonlinear Analysis* (M. Matzeu and A. Vignoli, eds.). Progr. Nonlinear Differential Equations Appl. **15**, Birkhäuser, 178–302.

Dancer, E. N., 1971. Bifurcation theory in real Banach space. Proc. London Math. Soc. **23**, 699–774.

Dancer, E. N., 1973. Bifurcation theory for analytic operators. Proc. London Math. Soc. **26**, 359–384.

Dancer, E. N., 1980. On the existence of bifurcating solution in the presence of symmetries. Proc. Royal Soc. Edinburg A **85**, 321–336.

Dancer, E. N., 1982. Symmetries, degree, homotopy indices and asymptotically homogeneous problems. Nonlinear Anal. **6**, 667–686.

Dancer, E. N., 1983. On the indices of fixed points of mappings in cones and applications, J. of Math. Anal. and Appl. **91**, 131–151.

Dancer, E. N., 1984. Perturbation of zeros in the presence of symmetries. J. Austral. Math. Soc. **36**, 106–125.

Dancer, E. N., 1994. Fixed point index calculations and applications. *Topological Nonlinear Analysis.* M. Matzeu and A. Vignoli Eds. Birkhäuser. Progr. Nonlinear Differential Equations Appl. **15**, 303–340.

Deimling, K., 1974. *Nichtlineare Gleichungen und Abbildungsgrade.* Hochschultext. Berlin-Heidelberg-New York: Springer-Verlag.

Deimling, K., 1985. *Nonlinear Functional Analysis*, Springer-Verlag, Berlin.

Dellnitz, M., I. Melbourne and J. E. Marsden, 1992. Generic bifurcation of Hamiltonian vector fields with symmetry. Nonlinearity **15**, 979–996.

Ding, W. Y., 1985. Generalizations of the Borsuk Theorem. J. Math. Anal. Appl. **110**, 553–567.

Dold, A., 1972. *Lectures on algebraic topology.* Springer-Verlag, Berlin.

Dold, A., 1983. Simple proofs of Borsuk–Ulam results. Contemp. Math. **19**, 65–69.

Dold, A., 1983. *Fixpunkttheorie.* Kurseinheiten 1–7. Hagen: Fernuniversität-Gesamthochschule, Fachbereich Mathematik und Informatik.

Dugundji, J., and A. Granas, 1982. *Fixed point theory I.* Warzawa: PWN-Polish Scientific.

Eells, J., 1970. Fredholm structures. *Nonlinear functional analysis* **18**, 62–85. Amer. Math. Soc., Providence, RI.

Eisenack, G., and Ch. Fenske, 1978. *Fixpunkttheorie.* Bibliographisches Institut, Mannheim, Wien, Zürich.

Ekeland, I., 1990. *Convexity methods in Hamiltonian mechanics.* Ergeb. Math. Grenzgeb. **19**, Springer-Verlag, Berlin.

Elworthy, K. D., and A. J. Tromba, 1970. Degree theory on Banach manifolds. "Nonlinear Functional Analysis". Proc. Symp. Pure Math. Vol. **18/1**, AMS, 86–94.

Erbe, L., K. Geba, W. Krawcewicz and J. Wu, 1992. S^1-degree and global Hopf bifurcation theory of functional differential equations. J. Differential Geom. **98**, 277–298.

Fadell, E. R., and P. H. Rabinowitz, 1978. Generalized cohomological index theories for Lie group actions with application to bifurcation questions for Hamiltonian systems. Invent. Math. **45**, 139–174.

Fadell, E., S. Husseini. and P. Rabinowitz, 1982. Borsuk–Ulam theorems for arbitrary S^1-action and applications. Trans. Amer. Math. Soc. **274**, 345–360.

Fadell, E., and G. Fournier (eds.)., 1981. *Fixed point theory.* Lecture Notes in Math. **886**. Springer-Verlag, Berlin, Heidelberg, New York.

Fadell, E. and S. Husseini, 1987. Relative cohomological index theories. Adv. in Math. **64**, 1–31.

Fenske, C., 1989. An index for periodic orbits of functional differential equations. Math. Ann. **285**, 381–392.

Field, M., 1996, *Symmetry breaking for compact Lie groups.* Mem. Amer. Math. Soc. **574**, Amer. Math. Soc., Providence, RI.

Field, M., 1997. Symmetry breaking for equivariant maps. *Algebraic groups and Lie groups* (G. I. Lehrer, ed.). Cambridge University Press. 219–253.

Field, M., 1997. Symmetry breaking for equivariant maps. Austral. Math. Soc. Lect. Ser. **9**, 219-253.

Field, M. J., and R. W. Richardson, 1992. Symmetry breaking and branching problems in equivariant bifurcation theory, I. Arch. Mat. Mech. Anal. **118**, 297–348.

Fiedler, B., 1985. An index for global Hopf bifurcation in parabolic systems, J. Reine Angew. Math, **359**, 1–36.

Fiedler, B., 1986. Global Hopf bifurcation of two-parameters flows, Arch. Rat. Mech. Anal. **94**, 59–81.

Fiedler, B., and S. Heinze, 1996. Homotopy invariants of time reversible periodic orbits. I. Theory. J. Diff. Equations **126**, 184–203.

Fitzpatrick, P. M., 1970. A generalized degree for uniform limits of A-proper mappings. J. Math. Anal. Appl. **35**, 536–552.

Fitzpatrick, P. M., 1972. A-proper mappings and their uniform limits. Bull. Amer. Math. Soc. **78**, 806–809.

Fitzpatrick, P. M., 1974. On the structure of the set of solutions of equations involving A-proper mappings. Trans. Amer. Math. Soc. **189**, 107–131.

Fitzpatrick, P. M., 1988. Homotopy, linearization and bifurcation. Nonlinear Anal. **12**, 171–184.

Fitzpatrick, P. M., I. Massabó and J. Pejsachowicz, 1983. Complementing maps, continuation and global bifurcation. Bull. Amer. Math. Soc. **9**, 79–81.

Fitzpatrick, P. M., I. Massabó and J. Pejsachowicz, 1983. Global several parameters bifurcation and continuation theorems, a unified approach via complementing maps. Math. Ann. **263**, 61–73.

Fitzpatrick, P. M., I. Massabó and J. Pejsachowicz, 1986. On the covering dimension of the set of solutions of some nonlinear equations, Trans. Amer. Math. Soc. **296**, 777–798.

Fitzpatrick, P. M., and J. Pejsachowicz, 1988. The fundamental group of the space of linear Fredholm operators and the global analysis of semilinear equations. Contemp. Math. **72**, 47–87.

Fitzpatrick, P. M., and J. Pejsachowicz, 1990. A local bifurcation theorem for C^1-Fredholm maps. Proc. Amer. Math. Soc. **109**, 995–1002.

Fitzpatrick, P. M., and J. Pejsachowicz, 1991. Parity and generalized multiplicity. Trans. Amer. Math. Soc. **326**, 281–305.

Fitzpatrick, P. M., and J. Pejsachowicz, 1991. Nonorientability of the index bundle and several parameter bifurcation. J. Funct. Anal. **98**, 42–58.

Fitzpatrick, P. M., and J. Pejsachowicz, 1993. *The Leray–Schauder theory and fully non-linear elliptic boundary value problems.* Mem. Amer. Math. Soc. 101, **483**, Amer. Math. Soc., Providence, RI.

Fitzpatrick, P. M., J. Pejsachowicz and P. J. Rabier, 1990. Topological degree for nonlinear Fredholm operators. C. R. Acad. Sci. Paris **311**, 711–716.

Fitzpatrick, P. M., Pejsachowicz, J. and Rabier, P .J., 1993. The degree of proper C^2 Fredholm maps: covariant theory. Topol. Methods Nonlinear Anal. **3**, 325–367.

Fitzpatrick, P. M., Pejsachowicz, J. and L. Recht, 1999. Spectral flow and bifurcation of critical points of strongly indefinite functionals I. General theory. J. Funct. Anal. **162**, 52–95.

Fonseca, I., and W. Gangbo, 1995. *Degree theory in analysis and applications.* Oxford Lecture Series in Mathematics and its Applications **2**. Clarendon Press.

Friedman, A., 1969, *Partial differential equations.* Holt, Rinehart and Winston, New York.

Fucik, S., J. Necas and V. Soucek, 1973. *Spectral Analysis of Nonlinear Operators.* Lecture Notes in Math. **343**. Springer-Verlag, Berlin.

Furi, M., M. Martelli and A. Vignoli, 1980. On the Solvability of nonlinear operator equations in normed spaces, Ann. Mat. Pura Appl. **124**, 321–343.

Gaines, R. E., and J. L. Mawhin, 1977. *Coincidence degree and nonlinear differential equations.* Lecture Notes in Math. **568**, Springer-Verlag.

Gavalas, G. R., 1968. *Nonlinear differential equations of chemically reacting systems.* Springer Tracts in Nat. Phil. **17**, Springer-Verlag.

Geba, K., and A. Granas, 1973. Infinite dimensional cohomology theories. J. Math. Pures Appl. **52**, 147–270.

Geba, K., and W. Marzantowicz, 1993. Global bifurcation of periodic orbits. Topol. Methods Nonlinear Anal. **1**, 67–93.

Geba, K., I. Massabó and A. Vignoli, 1986. Generalized topological degree and bifurcation. Nonlinear Functional Analysis and its Applications. Reidel, 55–73.

Geba, K., I. Massabó and A. Vignoli, 1990. On the Euler characteristic of equivariant gradient vector fields. Boll. Un. Mat. Ital. A (7) **4**, 243–251.

Golubitsky, M., D. G. Schaeffer and I. N. Stewart, 1988. *Singularities and groups in bifurcation theory II*. Springer-Verlag.

Granas, A., 1962. *The theory of compact vector fields and some of its applications to topology of functional spaces.* Rozprawy Mat. **30**, Warzawa.

Granas, A., 1972. The Leray–Schauder index and the fixed point theory for arbitrary ANRs. Bull. Soc. Math. France **100**, 209–228.

Guckenheimer, J., and P. Holmes, 1983. *Nonlinear oscillations, dynamical systems and bifurcation of vector fields*, Springer-Verlag.

Hale, J. K., 1977. *Theory of Functional Differential Equations.* Springer-Verlag, New York.

Hale, J. K., and J. C. F. De Oliviera, 1980. Hopf bifurcation for functional equations. J. Math. Anal. Appl. **74**, 41–59.

Hassard, B. J., Kazarinoff, N. D. and Y.-H. Wan, 1981. *Theory and applications of Hopf bifurcation.* Cambridge University Press.

Hauschild, H., 1977. Äquivariante Homotopie I. Arch. Math. **29**, 158-167.

Hauschild, H., 1977. Zerspaltung äquivarianter Homotopiemengen. Math. Ann. **230**, 279–292.

Heinz, G., 1986. Lösungsverzweigung bei analytischen Gleichungen mit Fredholm-operator vom Index Null. Math. Nachr. **128**, 243–254.

Hetzer, G., and V. Stallbohm, 1978. Global behaviour of bifurcation branches and the essential spectrum. Math. Nachr. **86**, 347-360.

Hoyle, S. C., 1980. Local solutions manifolds for nonlinear equations, Nonlinear Anal. **4**, 285–295.

Hoyle, S. C., 1980. Hopf bifurcation for ordinary differential equations with a zero eigenvalue. J. Math. Anal. Appl. **74**, 212–232.

Hu, S. T., 1959. *Homotopy theory.* Academic Press, New York.

Husemoller, D., 1974. *Fiber bundles.* Springer-Verlag, Berlin

Hyers, D., G. Isac and T. M. Rassias, 1997. *Topics in nonlinear analysis and applications.* World Scientific.

Iooss, G., 1979. *Bifurcation of maps and applications*, North Holland.

Iooss, G., and D. D. Joseph, 1980. *Elementary stability and Bifurcation Theory.* Springer-Verlag.

Ize, J., 1976. *Bifurcation theory for Fredholm operators.* Mem. Amer. Math. Soc. **174**.

Ize, J., 1979, Periodic solutions for nonlinear parabolic equations, Comm. Partial Differential Equations **12**, 1299–1387.

Ize, J., 1982, Introduction to bifurcation theory. *Differential equations.* Lecture Notes in Math. **957**, Springer-Verlag, 145–202.

Ize, J., 1997. Connected sets in multiparameter bifurcation. Nonlinear Anal.. **30**, 3763–3774.

Jaworowski, J., 1976. Extensions of G-maps and Euclidean G-retracts. Math. Z. **146**, 143–148.

Jaworowski, J., 1981. An equivariant extension theorem and G-retracts with a finite structure. Manuscripta Math. **35**, 323–329.

Keller, J. B., and S. Antman (eds.), (1969). *Bifurcation theory and nonlinear eigenvalue problems.* Benjamin.

Kielhöfer, H., 1979. Hopf bifurcation at multiple eigenvalues. Arch. Rat. Mech. Anal. **69**, 53–83.

Kielhöfer, H., 1985. Multiple eigenvalue bifurcation for potential operators, J. Reine Angew. Math. **358**, 104–124.

Kielhöfer, H., 1986. Interaction of periodic and stationary bifurcation from multiple eigenvalues. Math Z. **192**, 159–166.

Kielhöfer, H., 1988. A bifurcation theorem for potential operators. J. Funct. Anal. **77**, 1–8.

Kielhöfer, H., 1992. Hopf bifurcation from a differentiable viewpoint. J. Differential Geom. **97**, 189–232.

Kirillov, A. R., 1976. *Elements of the theory of representations.* Springer-Verlag, New York.

Komiya, K., 1988. Fixed point indices of equivariant maps and Moebius inversion. Invent. Math. **91**, 129–135.

Kosniowski C., 1974. Equivariant cohomology and stable cohomotopy. Math. Ann. **210**, 83–104.

Krasnosel'skii, M. A., 1964. *Positive solutions of operator equations.* Noordhoff.

Krasnosel'skii, M. A., 1964. *Topological methods in the theory of nonlinear integral equations.* Pergamon Press, MacMillan, New York.

Krasnosel'skii, M. A., 1995. *Asymptotics of nonlinearities and operator equations.* Oper. Theory Adv. Appl. **76**, Birkhäuser, Basel.

Krawcewicz, W., and P. Vivi, 2000. Equivariant degree and normal bifurcations. B.N. Prasad birth centenary commemoration volume, II. Indian J. Math. **42** (1), 55–67.

Krawcewicz, W., J. Wu and H. Xia, 1993. Global Hopf bifurcation theory for condensing fields and neutral equations with applications to lossless transmission problems. Canadian Appl. Math. Quart. **1**, 167–220.

Krawcewicz, W., and H. Xia, 1996. Analytic definition of an equivariant degree. Russian Math. **40**, 34–49.

Krawcewicz, W., and J. Wu, 1999. Theory and applications of Hopf bifurcations in symmetric functional-differential equations. Nonlinear Anal. **35** (7), 845–870.

Kryszewski, W., and A. Szulkin, 1997. An infinite dimensional Morse theory with applications. Trans. Amer. Math. Soc. **349**, 3181–3234.

Küpper, T. ,and D. Riemer, 1979. Necessary and sufficient conditions for bifurcation from the continuous spectrum. Nonlinear Anal. **3**, 555–561.

Laloux, B., and J. Mawhin, 1977. Multiplicity, Leray–Schauder formula and bifurcation. J. Differential Geom. **24**, 309–322.

Lari-Lavassani, A., W. F. Langford and K. Huseyin, 1994. Symmetry breaking bifurcations on multidimensional fixed point subspaces. Dynam. Stability Systems **9**, 345–373.

Lee, C. N., and A. Wasserman, 1975. *On the groups JO* (G). Mem. Amer. Mat. Soc. **159**, Amer. Mat. Soc., Providence, Rhode Island.

Le V. K., and K. Schmitt, 1997. *Global bifurcation in variational inequalities.* Appl. Math. Sci. **123**, Springer-Verlag.

Leray, J., and J. Schauder, 1934. Topologie et equations fonctionnelles. Ann. Sci. École Norm. Sup. **51**, 45–78.

Lewis L. G., J. P. May, M. Steinberger and J. E. McClure, 1986. *Equivariant stable homotopy theory.* Lecture Notes in Math. **1213**, Springer-Verlag.

Liu, J. Q., and Z. Q. Wang, 1993. Remarks on subhamonics with minimal periods of Hamiltonian systems. Nonlinear Anal. **20**, 803–821.

Ljusternik, L., and L. Schnirelmann, 1934. *Methodes Topologiques dans les problèmes variationels*. Herman, Paris.

Lloyd, N. G., 1978. *Degree theory*. Cambridge Tracts in Math. **73**, Cambridge University Press.

Ma, T., 1972. *Topological degrees of set-valued compact fields in locally convex spaces*. Rozprawy Mat. **42**, Warszawa.

Maciejewski, A., and S. Rybicki, 1998. Global bifurcations of periodic solutions of Hénon-Heiles system via degree for S^1-equivariant orthogonal maps. Rev. Math. Phys. **10** (8) 1125–1146.

Magnus, R. J., 1976. A generalization of multiplicity and the problem of bifurcation. Proc. London Math. Soc. **32**, 251–278.

Magnus, R. J., 1976, On the local structure of the zero set of a Banach space valued mapping. J. Funct. Anal. **22**, 58–72.

Mallet-Paret, J., R. Nussbaum and P. Paraskevopoulos, 1994. Periodic solutions for functional-differential equations with multiple state-dependent time lags. Topol. Methods Nonlinear Anal., **3**, 101–162.

Mallet-Paret, J., and J. A. Yorke, 1982. Snakes: oriented families of periodic orbits, their sources, sinks and continuation. J. Differential Equations **43**, 419–450.

Marsden, J. E., and M. McCracken, 1976. *The Hopf bifurcation and its applications*. Springer-Verlag.

Marzantowicz, W., 1981. On the nonlinear elliptic equations with symmetries. J. Math. Anal. Appl. **81**, 156–181.

Marzantowicz, W., 1989. A G-Lusternik-Schnirelman category of a space with an action of a compact Lie group. Topology **28**, 403–412.

Marzantowicz, W., 1994. Borsuk–Ulam theorem for any compact Lie group. J. London Math. Soc. **49**, 195–208.

Massabó, I., and J. Pejsachowicz, 1984. On the connectivity properties of the solutions set of parametrized families of compact vector fields, J. Funct. Anal. **59**, 151–166.

Matsuoka, T., 1983. Equivariant function spaces and bifurcation points. J. Math. Soc. Japan **35**, 43–52.

Matzeu, M., and A. Vignoli (eds.), 1994. *Topological Nonlinear Analysis*. Progr. Nonlinear Differential Equations Appl. **15**. Birkhäuser.

Matzeu, M., and A. Vignoli (eds.), 1996. *Topological Nonlinear Analysis II*. Progr. Nonlinear Differential Equations Appl. **27**, Birkhäuser.

Mawhin, J., 1971. *Equations non linéaires dans les espaces de Banach.* (Nonlinear equations in Banach spaces). Rapport No. 39, Seminaires de Mathématique Appliquée et Mécanique. Louvain, Belgique: Vander editeur; Université Catholique de Louvain.

Mawhin, J., 1976. *Nonlinear boundary value problems for ordinary differential equations: from Schauder theorem to stable homotopy* - Seminaires de Mathématique Appliquée et Mécanique. Rapport No.86. Louvain-La-Neuve: Université Catholique de Louvain, Institut de Mathématique Pure et Appliquée.

Mawhin, J., 1977. *Topological degree methods in nonlinear boundary value problems.* CBMS Reg. Conf. Ser. Math. **40**, Amer. Math. Soc., Providence, RI.

Mawhin, J., and M. Willem, 1989. *Critical Point Theory and Hamiltonian Systems.* Springer-Verlag, New York.

Mawhin, J., 1999. Leray–Schauder degree: a half century of extensions and applications. Topol. Methods Nonlinear Anal. **14**, 195–228.

Milojevic, P. S., 1986. On the index and the covering dimension of the solution set of semilinear equations, Proc. Symp. Pure Math. **45**, 2, Amer. Math. Soc., Providence, RI, 183–205.

Morse, M., 1934. *The Calculus of Variations in the large.* Amer. Math. Soc.

Montaldi, J., M. Roberts and I. Stewart, 1990. Existence of nonlinear normal modes of symmetric Hamiltonian systems. Nonlinearity **3**, 695–730.

Nirenberg, L., 1971. An application of generalized degree to a class of nonlinear problems, 3rd. Colloq. Anal. Funct., Liege, Centre Belge de Recherches Math. 57–73.

Nirenberg. L., 1974. *Topics in nonlinear functional analysis.* Lecture Notes Courant Institute, New York University.

Nirenberg. L., 1981. Variational and topological methods in nonlinear problems. Bull. Amer. Math. Soc. **4**, 267–302.

Nirenberg, L., 1981. Comments on Nonlinear Problems. Le Matematiche **36**, 109–119.

Nishimura, T., T. Fukuda and K. Aoki, 1989. An algebraic formula for the topological types of one parameter bifurcation diagrams. Arch. Rat. Mech. **108**, 247–266.

Nitkura, Y., 1982. Existence and bifurcation of solutions for Fredholm operators with nonlinear perturbations. Nagoya Math. J. **86**, 249–271.

Nussbaum, R. D., 1972. Degree theory for local condensing maps. J. Math. Anal. Appl. **37**, 741–766.

Nussbaum, R. D., 1975. A global bifurcation theorem with applications to functional differential equations. J. Funct. Anal. **19**, 319–338.

Nussbaum, R. D., 1977. Some generalizations of the Borsuk–Ulam theorem. Proc. London Math. Soc. **35**, 136–158.

Nussbaum, R. D., 1978. A Hopf bifurcation theorem for retarded functional differential equations. Trans. Amer. Math. Soc. **238**, 139–163.

Nussbaum, R. D., 1978. *Differential-delay equations with two time lags.* Mem. Amer. Math. Soc. **205**, Amer. Math. Soc., Providence, RI.

Nussbaum, R. D., 1985. *The fixed point index and some applications.* - Séminaire de Mathématiques Supérieures. Séminaire Scientifique OTAN **94**. Université de Montréal. Les Presses de l'Université de Montréal.

Pascali, D., 1974. *Nichtlineare Operatoren.* (Operatori neliniari.). Editura Academiei Republicii Socialiste Romania, Bucuresti.

Parusinski, A., 1990. Gradient homotopies of gradient vector fields. Studia Math. **46**, 73–80.

Peitgen, H. O., and H. O. Walter, 1979. (eds.): *Functional differential equations and approximation of fixed points.* Lecture Notes in Math. **730**, Springer-Verlag.

Pejsachowicz, J., 1988. K-theoretic methods in bifurcation theory. Contemp. Math. **72**, 193–205.

Pejsachowicz, J., and P. Rabier, 1998. Degree theory for C^1 Fredholm mappings of index 0. J. Anal. Math. **76**. 289–319.

Peschke, G., 1994. Degree of certain equivariant maps into a representation sphere. Topology Appl. **59**, 137–156.

Petryshyn, W. V., 1970. Nonlinear equations involving noncompact operators. "Nonlinear Functional Analysis" Proc. Symp. Pure Math. **18/1**, Amer. Math. Soc., Providence, RI, 206–223.

Petryshyn, W. V., 1975. On the approximation solvability of equations involving A-proper and pseudo-A-proper mappings. Bull. Amer. Math. Soc. **81**, 223–312.

Petryshyn, W. V., 1978. Bifurcation and asymptotic bifurcation for equations involving A-proper mappings with applications to differential equations. J. Differential Equations **28**, 124–154.

Petryshyn, W. V., 1992. *Approximation-solvability of nonlinear functional and differential equation*, M. Dekker.

Petryshyn, W. V., 1995. *Generalized topological degree and semilinear equations.* Cambridge Tracts in Math. **117**. Cambridge University Press.

Petryshyn, W. V., and P. M. Fitzpatrick, 1974. A degree theory, fixed point theorems and mapping theorems for multivalued noncompact mappings. Trans. Amer. Math. Soc. **194**, 1–25.

Pimbley, G., 1969. *Eigenfunction branches of nonlinear operators and their bifurcation.* Lectures Notes in Math. **104**, Springer-Verlag.

Poincaré, H., 1885. Les figures equilibrium, Acta Math. **7**, 259–302.

Prieto, C., and H. Ulrich, 1991. Equivariant fixed point index and fixed point transfer in non-zero dimension. Trans. Amer. Math. Soc. **328**, 731–745.

Rabier, P. J., 1989. Generalized Jordan chains and two bifurcation theorems of Krasnoselskii. Nonlinear Anal. **13**, 903–934.

Rabier, P. J., 1991. Topological degree and the theorem of Borsuk for general covariant mappings with applications. Nonlinear Anal. **16**, 393–420.

Rabinowitz, P. H., 1971. Some global results for nonlinear eigenvalue problems. J. Funct- Anal- **7**, 487–513.

Rabinowitz, P. H., 1973. Some aspects of nonlinear eigenvalue problems. Rocky Mountain J. Math. **3**, 161–202.

Rabinowitz, P. H., 1975. *Theorie du Degree Topologique et Applications* (Lectures Notes).

Rabinowitz, P. H., 1977. A bifurcation theorem for potential operators. J. Funct. Anal. **25**, 412–424.

Rabinowitz, P. H., 1986. *Minimax methods in critical point theory with application to differential equations.* CBMS Reg. Conf. Ser. Math. **65**, Amer. Math. Soc., Providence, R.I.

Rabinowitz, P. H., 1994. Critical point theory. *Topological Nonlinear Analysis.* M. Matzeu and A. Vignoli Eds. Birkhäuser. Progr. Nonlinear Differential Equations Appl. **15**. 464–513.

Rados, G., 1885. *Zur Theorie der Congruenzen höheren Grades.* Kronecker Published.

Rothe, E. H., 1986. *Introduction to Various Aspects of Degree Theory in Banach Spaces.* Mathematical Surveys and Monographs, Vol. **23**, Amer. Math. Soc., Providence, RI.

Rubinstein, R. L., 1976. On the equivariant homotopy of spheres. Dissertationes Math. **134**, 1–48.

Ruf, B., 1996. Singularity theory and bifurcation phenomena. *Topological Nonlinear Analysis II.* M. Matzeu and A. Vignoli Eds. Birkhäuser. Progr. Nonlinear Differential Equations Appl. **27**, 315–396.

Rybicki, S., 1997. Applications of degree for S^1-equivariant gradient maps to variational problems with S^1-symmetries. Topol. Methods Nonlinear Anal. **9**, 383–417.

Rybicki, S., 1998. On periodic solutions of autonomous Hamiltonian systems via degree for S^1-equivariant gradient maps, Nonlinear Anal. **34**, 537–569.

Rybicki, S., 1998. On bifurcation from infinity for S^1-equivariant potential operators. Nonlinear Anal. **31**, 343–361.

Rybicki, S., 2001. Degree for S^1-equivariant strongly indefinite functionals. Nonlinear Anal. **43**, 1001–1017.

Sadovskii, B. N., 1972. Limit-compact and condensing operators. Russian Math. Surveys **27**, 85–155.

Sattinger, D. H., 1971. Stability of bifurcating solutions by Leray–Schauder degree, Arch. Rational Mech. Anal. **43**, 154–166.

Sattinger, D. H., 1973. *Topics in Stability and Bifurcation Theory.* Lecture Notes in Math. **309**. Springer-Verlag.

Sattinger, D. H., 1978. Group representation theory, bifurcation theory and Pattern formation, J. Funct. Anal. **28**, 58–101.

Sattinger, D. H., 1979. *Group theoretic methods in bifurcation theory.* Lecture Notes in Math. **762**, Springer-Verlag.

Sattinger, D. H., 1983. *Branching in the presence of symmetry.* John Wiley.

Sburlan, S., 1983. *The topological degree. Lectures on nonlinear equations.* (Gradul topologic Lectii asupra equatiilor neliniare). Editura Academiei Republicii Socialiste România, Bucuresti.

Schmidt, E., 1908. Zur Theorie der linearen und nichtlinearen Integralgleichungen, III. Math. Ann. **65**, 370–399.

Schmitt, K., 1982. *A study of eigenvalue and bifurcation problems for nonlinear elliptic partial differential equations via topological continuation methods.* Inst. Math. Pure et Appl. Louvain-la-Neuve.

Schmitt, K., and Z. Q. Wang, 1991. On bifurcation from infinity for potential operators. Differential Integral Equations **4**, 933–943.

Schwartz, J. T., 1969. *Nonlinear Functional Analysis.* Gordon and Breach, New York.

Smart, D. F., 1980. *Fixed point theorems.* Cambridge Tracts in Math. **66**, Cambridge University Press.

Smoller, J., and A. G. Wasserman, 1990. Bifurcation and symmetry-breaking. Invent. Math. **100**, 63–95.

Spanier, E., 1966. *Algebraic topology.* McGraw-Hill, New York.

Stakgold, I., 1971. Branching solutions of nonlinear equations, SIAM Rev. **13**, 289–332.

Steinlein, H., 1985. Borsuk's antipodal theorem and its generalizations and applications: a survey. Topol. et Anal. Non Lineaire. Press de l'Université de Montreal, 166–235.

Struwe, M., 1990. *Variational Methods.* Springer-Verlag, Berlin.

Stuart, C. A., 1973. Some bifurcation theory for k-set contractions. Proc. London Math. Soc. **27**, 531–550.

Stuart, C. A., 1983. Bifurcation from the essential spectrum. Lecture Notes in Math. **1017**, Springer-Verlag, 575–596.

Stuart, C. A., 1987. Bifurcation from the continuous spectrum in $L^p(\mathbb{R})$, Inter Sem. Numer. Math. **79**, Birkhäuser, 307–318.

Stuart, C. A., 1996. Bifurcation from the essential spectrum. *Topological Nonlinear Analysis II.* M. Matzeu and A. Vignoli Eds. Birkhäuser. Progr. Nonlinear Differential Equations Appl. **27**, 397–444.

Szulkin, A., 1997. Index theories for indefinite functionals and applications. Pitman Res. Notes Ser. **365**, 89–121.

Szulkin, A., and W. Zou, 2001. Infinite dimensional cohomology groups and periodic of assymptotically linear Hamiltonian systems. J. Differential Equations **174**, 369–391.

Takens, F., 1972. Some remarks on the Böhme-Berger bifurcation theorem. Math. Z. **125**, 359–364.

Temme, N. M.(ed.), 1976. *Nonlinear analysis.* Vol. 1, 2. Amsterdam: Mathematisch Centrum.

Toland, J., 1976. Global bifurcation for k-set contractions without multiplicity assumptions. Quart. J. Math. **27**, 199–216.

tom Dieck, T., 1979. *Transformation groups and representation theory.* Lecture Notes in Math. **766**, Springer-Verlag.

tom Dieck, T., 1987. *Transformation Groups.* De Gruyter Stud. Math. **8**, Walter de Gruyter, Berlin.

Tornehave, J., 1982. Equivariant maps of spheres with conjugate orthogonal actions. Algebraic Topology, Proc. Conf. London Ont. 1981, Canadian Math. Soc. Conf. Proc. **2**, 275–301.

Ulrich, H., 1988. *Fixed point theory of parametrized equivariant maps.* Lecture Notes in Math. **1343**. Springer-Verlag.

Vainberg, M. M., 1973. *Variational method and method of monotone operators in the theory of nonlinear equations.* John Wiley.

Vainberg, M. M., and V. A. Trenogin, 1974. *Theory of branching of solutions of nonlinear equations.* Noordhoff, Leyden.

Vanderbauwhede, A., 1982. *Local bifurcation and symmetry.* Pitman Res. Notes in Math. **75**.

Vignoli, A., 1992. L'Analisi Nonlineare nella teoria della biforcazione. Enciclopedia delle Scienze Fiziche dell' Inst. dell'Enciclopedia Italiana, Vol. 1, 134-144.

Wang, Z. Q., 1989. Symmetries and the calculation of degree. Chinese Ann. Math. Ser. B **10**, 520–536.

Webb, J. R. L. and S. C. Welsh, 1986. Topological degree and global bifurcation. Proc. Symp. Pure Math. **45/II**, Amer. Math. Soc. Providence, RI, 527–531.

Welsh, S. C., 1988. Global results concerning bifurcation for Fredholm maps of index zero with a transversality condition. Nonlinear Anal. **12**, 1137–1148.

Werner, B., 1990. Eigenvalue problems with the symmetry of a group and bifurcation. NATO Adv. Sci. Inst. Ser. C Math. Phys. Sci. **313**, 71–88.

Westreich, D., 1972. Banach space bifurcation theory, Trans. Amer. Math. Soc. **171**, 135–156.

Westreich, D., 1977. Bifurcation at double characteristic values. J. London Math. Soc. (2) **15**, 345–350.

Whitehead, G. W., 1942. On the homotopy groups of spheres and rotation groups. Ann. of Math. **43**, 634–640.

Whitehead, G. W., 1978. *Elements of homotopy theory.* Grad. Texts in Math. **61**, Springer-Verlag.

Whyburn, G. T., 1964. *Topological analysis.* 2nd. ed. Princeton Math. Ser. 23, Princeton University Press.

Wu, J., 1996. *Theory and applications of partial differential functional equations.* Appl. Math. Sci. **119**, Springer-Verlag.

Yakubovich, V. A., and V. M. Ztarzhinskii, 1975. *Linear differential equations with periodic coefficients.* Vol. 2. John Wiley.

Zarantonello, E. H., 1971. *Contributions to nonlinear functional analysis.* Academic Press.

Zeidler, E., 1984. *Nonlinear Functional Analysis and its Applications*, Vol. III. Springer-Verlag.

Zeidler, E., 1985. *Nonlinear Functional Analysis and its Applications*, Vol. IV. Springer-Verlag.

Zeidler, E., 1993. *Nonlinear functional analysis and its applications.* Volume I: *Fixed-point theorems.* Translated from the German by Peter R. Wadsack. 2. corr. printing. Springer-Verlag, New York.

Zhang, Z. Q., 1998. The equivariant Sard-Smale theorem. (Chinese) J. Lanzhou Univ. Nat. Sci. **34** (2), 1–6.

Zhang, Z. Q., 1999. The equivariant Borsuk–Ulam theorem. (Chinese) J. Lanzhou Univ. Nat. Sci. **35** (2), 7–11.

Zhang, Z. Q., A zero-orbit index formula for 1-set-contractive equivariant fields. (Chinese. English), J. Lanzhou Univ. Nat. Sci. **36** (5), 5–9.

Zhang, Z. Q., and H. Zhao, 2000. Calculation of the degree of equivariant set-contractive fields. (Chinese) J. Lanzhou Univ. Nat. Sci. **36** (3), 1–4.

Zvyagin, V. G., 1999. On a degree theory for equivariant $\Phi_0 C^I BH$-mappings. (Russian) Dokl. Akad. Nauk **364** (2), 155–157.

Index